Nanometal Oxides in Horticulture and Agronomy

Nanomaterial-Plant Interactions

Nanometal Oxides in Horticulture and Agronomy

Edited by

Li Xinghui
College of Horticulture, Nanjing Agricultural University, Nanjing, Jiangsu Province, P.R. China

Rajiv Periakaruppan
College of Horticulture, Nanjing Agricultural University, Nanjing, Jiangsu Province, P.R. China; Department of Biotechnology, PSG College of Arts & Science, Coimbatore, Tamil Nadu, India

Remya Mohanraj
Houston Community College, Houston, TX, United States

Sugapriya Dhanasekaran
Department of Molecular Analytics, Institute of Bioinformatics, SSE - Saveetha Institute of Medical and Technical Sciences, Chennai, Tamil Nadu, India

Series Editors

Vijay Pratap Singh

Durgesh Kumar Tripathi

Jose R. Peralta-Videa

ELSEVIER

ACADEMIC PRESS
An imprint of Elsevier

Academic Press is an imprint of Elsevier
125 London Wall, London EC2Y 5AS, United Kingdom
525 B Street, Suite 1650, San Diego, CA 92101, United States
50 Hampshire Street, 5th Floor, Cambridge, MA 02139, United States
The Boulevard, Langford Lane, Kidlington, Oxford OX5 1GB, United Kingdom

ISBN: 978-0-323-91809-1

For Information on all Academic Press publications
visit our website at https://www.elsevier.com/books-and-journals

Publisher: Nikki P. Levy
Acquisitions Editor: Nancy J. Maragioglio
Editorial Project Manager: Andrea R. Dulberger
Production Project Manager: Fahmida Sultana
Cover Designer: Mark Rogers

Typeset by MPS Limited, Chennai, India

Contents

List of contributors

Salwan Ali Abed Department of Environmental Science, College of Science, University of Al-Qadisiyah, Al Diwaniyah, Iraq

P. Akila Vaishnavi Department of Microbiology, Nadar Saraswathi College of Arts and Science, Theni, Tamil Nadu, India

Tawfiq Mohammad Al-Antary Department of Plant Protection, School of Agriculture, The University of Jordan, Amman, Jordan

Syed Azmal Ali Cell Biology and Proteomics Lab, Animal Biotechnology Center, ICAR-NDRI, Karnal, Haryana, India

Vidhya Arumugam Department of Biotechnology, A.V.C. College (Autonomous), Mayiladuthurai, Tamil Nadu, India; Department of Biotechnology, Bon Secours College for Women, Thanjavur, Tamil Nadu, India

P.R Chandhni National Dairy Research Institute Karnal, Karnal, Haryana, India

Rajkuberan Chandrasekaran Department of Biotechnology, Karpagam Academy of Higher Education, Coimbatore, Tamil Nadu, India

Jeyapragash Danaraj Centre for Ocean Research (DST-FIST Sponsored Centre), MoES—Earth Science and Technology Cell (Marine Biotechnological Studies), Col. Dr. Jeppiaar Research Park, Sathyabama Institute of Science and Technology, Chennai, Tamil Nadu, India

Sugapriya Dhanasekaran Department of Molecular Analytics, Institute of Bioinformatics, SSE - Saveetha Institute of Medical and Technical Sciences, Chennai, Tamil Nadu, India

P. Divya Department of Chemistry, KGiSL Institute of Technology, Saravanampatti, Coimbatore, Tamil Nadu, India

P. Geethamani Department of Chemistry, KGiSL Institute of Technology, Saravanampatti, Coimbatore, Tamil Nadu, India

Alaa Yousef Ghidan Department of Nanotechnologies and Microbiology, Research and Development Center, The Higher Council for Science and Technology, Amman, Jordan

M. Helan Soundra Rani Department of Biotechnology, Sri Ramakrishna College of Arts & Science, Coimbatore, Tamil Nadu, India

Daraksha Iram National Dairy Research Institute Karnal, Karnal, Haryana, India

M. Jayanthi Department of Biotechnology, Vels Institute of Science, Technology and Advanced Studies (VISTAS), Chennai, Tamil Nadu, India

Priya Kalyan National Dairy Research Institute Karnal, Karnal, Haryana, India

Gurjeet Kaur Centre for Healthy Brain Ageing, School of Psychiatry, University of New South Wales, Sydney, NSW, Australia; Mark Wainwright Analytical Centre, Bioanalytical Mass Spectrometry Facility, University of New South Wales, Sydney, NSW, Australia

Harpreet Kaur National Dairy Research Institute Karnal, Karnal, Haryana, India

Subasankari Kumaresan Department of Biochemistry, Dhanalakshmi Srinivasan College of Arts and Science for Women, Perambalur, Tamil Nadu, India

Yosuva Mariasingarayan Centre for Advanced Study in Marine Biology, Annamalai University, Parangipettai, Tamil Nadu, India

Iryna Matsukevich Institute of General and Inorganic Chemistry, National Academy of Sciences of Belarus, Minsk, Belarus

Deepti Mittal Nanosafety Lab, Division of Biochemistry, ICAR-NDRI, Karnal, Haryana, India

Remya Mohanraj Houston Community College, Houston, TX, United States

Anil Kumar Moola Department of Biotechnology, Aditya Degree and PG College, Kakinada, Andhra Pradesh, India; Department of Entomology, College of Agriculture, Food and Environment, Agriculture Science Center North, University of Kentucky, Lexington, KY, United States

Kapil Singh Narayan New Bolton Center, School of Veterinary Medicine, Department of Clinical Studies, University of Pennsylvania, Kennett Square, PA, United States

Vignesh Kumar Palanisamy Department of Chemistry, KGiSL Institute of Technology, Saravanampatti, Coimbatore, Tamil Nadu, India

Rajiv Periakaruppan College of Horticulture, Nanjing Agricultural University, Nanjing, Jiangsu Province, P.R. China; Department of Biotechnology, PSG College of Arts & Science, Coimbatore, Tamil Nadu, India

G. Priyanka Department of Biotechnology, Sri Ramakrishna College of Arts & Science, Coimbatore, Tamil Nadu, India

Sikha Rana National Dairy Research Institute Karnal, Karnal, Haryana, India

Elena Romanovskaia Department of Materials Science and Engineering, University of Virginia, Charlottesville, VA, United States

Valentin Romanovski Center of Functional Nano-Ceramics, National University of Science and Technology "MISIS," Moscow, Russia; Department of Materials Science and Engineering, University of Virginia, Charlottesville, VA, United States

Hasna Abdul Salam Department of Biotechnology, Karpagam Academy of Higher Education, Coimbatore, Tamil Nadu, India

Manish Singh Sansi National Dairy Research Institute Karnal, Karnal, Haryana, India

Selvam Sathish Department of Biotechnology, Bharathidasan University, Tiruchirappalli, Tamil Nadu, India

Parul Singh National Dairy Research Institute Karnal, Karnal, Haryana, India

M. Suganthi Department of Biotechnology, Vels Institute of Science, Technology and Advanced Studies (VISTAS), Chennai, Tamil Nadu, India

M. Thenmozhi Department of Biotechnology, Vels Institute of Science, Technology and Advanced Studies (VISTAS), Chennai, Tamil Nadu, India

N.K. Udaya Prakash Department of Biotechnology, Vels Institute of Science, Technology and Advanced Studies (VISTAS), Chennai, Tamil Nadu, India

P. Vanathi Department of Biotechnology, Sri Ramakrishna College of Arts & Science, Coimbatore, Tamil Nadu, India

Venkatachalam Vasudevan Department of Biotechnology, A.V.C. College (Autonomous), Mayiladuthurai, Tamil Nadu, India; Department of Biotechnology, Bharathidasan University, Tiruchirappalli, Tamil Nadu, India

Subha Priya Venkateswaran Department of Biotechnology, Sri Krishna Arts and Science College, Kuniamuthur, Coimbatore, Tamil Nadu, India

D. Vidhya TRM Biotech Private Limited, Namakkal, Tamil Nadu, India

Li Xinghui College of Horticulture, Nanjing Agricultural University, Nanjing, Jiangsu Province, P.R. China

Karmveer Yadav Nanosafety Lab, Division of Biochemistry, ICAR-NDRI, Karnal, Haryana, India

Sameena Zanab Chaudhary Charan Singh University Meerut, Meerut, Uttar Pradesh, India

Preface

The need to feed the world well demands more pronounced ecofriendly solutions with innovative technological approaches to achieve sustainable development of agriculture. Nanometal oxides have many unique properties such as nontoxicity, biocompatibility, biodegradability, absorption ability, and bioactivity, which enables them to improve the plant defense and enhance the crop productivity without harming the environment. Metal oxide nanomaterials also serve as an efficient delivery system for controlled and sustained release of agrochemicals and genetic materials in plants. In addition, the nanometal oxides play a vital role to control the water and soil pollution. Moreover, these influence the soil nutritional quality and improve the microbe's growth. Great progress has been made in the field of nanometal oxides in horticulture and agronomy.

Nanometal Oxides in Horticulture and Agronomy, a book containing 17 chapters, introduces the main research achievements in the field. It is an important repository for researchers, academicians, and scientists to enhance and boost their knowledge for further research in the field of metal oxide nanomaterials, agriculture, and environment.

The monograph was written by academic scholars from universities and research institutions in China, India, United States, Belarus, Russian Federation, and Jordan. On behalf of all the editors, we thank all participants in this book for their excellent work here. It is the result of international cooperation, especially the Belt and Road cooperation.

Li Xinghui
on behalf of the editors

Acknowledgments

This monograph research was supported by Ministry of Science and Technology of the People's Republic of China for the National Key Research and Development Program (2022YFD1601800), Ministry of Education of the People's Republic of China for the national first-class undergraduate specialty construction project (202214), Ministry of Science and Technology of the People's Republic of China for the national project plan for foreign experts of China (G20200010081), Ministry of Agriculture and Rural Affairs of the People's Republic of China for the project of international institute of tea industry innovation for "the Belt and Road" (2021) and the China Agriculture Research System of MOF and MARA (CAR-19), Jiangsu Provincial People's Government for the priority academic program development of Jiangsu higher education institutions and Yunnan Province Science and Technology Department for Expert Workstation Program (202105AF150045).

Chapter 1

Metal oxide nanomaterials: an introductory overview

Rajiv Periakaruppan[1,2], Li Xinghui[1], Remya Mohanraj[3] and Sugapriya Dhanasekaran[4]
[1]College of Horticulture, Nanjing Agricultural University, Nanjing, Jiangsu Province, P.R. China, [2]Department of Biotechnology, PSG College of Arts & Science, Coimbatore, Tamil Nadu, India, [3]Houston Community College, Houston, TX, United States, [4]Department of Molecular Analytics, Institute of Bioinformatics, SSE - Saveetha Institute of Medical and Technical Sciences, Chennai, Tamil Nadu, India

Chapter outline

1.1 Introduction

In recent years, nanotechnology has created applications in various fields such as medicine, agriculture, and the environment. Nanoparticles exhibit distinctive properties and are different in their dimensions, structures, shapes, compositions, uniformity states, and agglomeration (Sengul and Asmatulu, 2020). There are different types of nanomaterials, namely metal/metal-oxide-based nanomaterials, carbon-based composite-based nanoparticles, composite-based nanomaterials, and organic-based nanomaterials. Metal oxide nanomaterials have important advantages such as a rapid and simple preparation process, an easy method to get the desired size offering porosity, high stability and shape, and easy incorporation of the hydrophobic and hydrophilic system for various applications (Sanchez-Moreno et al., 2018). Metal and metal oxide nanomaterials possess more physiochemical properties (optical, surface, electrical and thermal properties) than their native bulk materials. Metal/metal-oxide-based nanomaterials are produced through the addition of oxidizing/precipitating and reducing agents in their synthesis (Sanchez-Dominguez et al., 2009). Metal and metal oxide nanoparticles have distinctive properties with regard to their

Nanometal Oxides in Horticulture and Agronomy. DOI: https://doi.org/10.1016/B978-0-323-91809-1.00010-X

sorption behaviors, chemical reduction, and magnetic properties, among others. Metal oxide nanomaterials are vital materials for industrial applications and are employed in various processes and products as well as cosmetics, sunscreens, paints, solar panels, catalysts, fuel additives, glass polish, and medicinal products (Larsen et al., 2016).

There are two types of approaches to the synthesis of nanomaterials: (1) a bottomup approach that organizes atomic/molecular components in hierarchical nano-complexes, and (2) a topdown approach that controls the formation of nanomaterials and adds and eliminates the thin layers of bulk material (Chavali and Nikolova, 2019). The physical and chemical methods for the synthesis of nanomaterials follow the bottomup approach. The physical and chemical methods are pulsed laser deposition, molecular beam epitaxy, ion implantation (II), chemical vapor deposition, atomic layer deposition, pulse and physical vapour deposition, spray pyrolysis, sonochemical electrochemical process, hydrothermal and photochemical processes, sol-gel, microemulsion routes, and microwaves and are typically expensive, labor-intensive apart from being harmful to the environment (Khashan et al., 2016; Musić et al., 2003; Wilson, 2018; Girigoswami, 2018).

In recent times, researchers have begun focusing on green synthesis owing to its environment-friendly approach. In this method, the extracts of plants, algae, fungi, and bacteria are used as reducing, capping, and stabilizing agents for the synthesis of nanomaterials. The extracts possess various bioactive compounds and metabolites. Biological-based nanomaterials have attracted considerable interest as an up-and-coming technology to reduce the toxicity of nanoparticles commonly associated with conventional chemical synthesis approaches (Jeevanandam et al., 2016; Marquis et al., 2016). Different methods are used for the synthesis of nanoparticles that can change and control the physicochemical properties of nanoparticles.

At present, nanotechnology is one of the chief technologies offering scope for constantly growing applications in the agricultural sector and it could become the most useful technology in the future. Metal oxide nanomaterials are applied as nano-fungicides, nano-fertilizers, nano-pesticides, and plant-growth promoters. Nanotechnology deals with the fabrication and usage of metal oxide with nanoscale dimensions for better yield in agriculture. For that rationale, this chapter was aimed at exploring new techniques for synthesis benefitting the features of the metal oxide nanoparticles in agriculture.

1.2 Application of metal oxide nanomaterials in agriculture

Metal oxide nanomaterials are considered a new effective input to emergent agricultural production via implementing nutrient efficiency and progress of plant protection practices. Nano-agriculture involves the usage of nanoparticles in agriculture. Metal oxide nanomaterials may have a solution to overcome various problems like disease detection and monitoring of plant growth in horticulture and agronomy (Elfeky et al., 2013). The application of nano nutrients, nanosensors, and nano-herbicides/fungicides/pesticides is already well-developed in the agricultural sector through nanotechnology approaches. The lower concentration of metal oxide nanomaterials exhibited better antimicrobial activity as compared to bulk metal oxide. Hence it could prevent and control diseases and the growth of phytopathogens in plants. Nanofertilizers such as Nitrogen, Phosphorus, Potassium, Iron, Magnesium, Zinc, Copper, Molybdenum, and carbon nanotubes show their effectiveness in their superior release and targeted delivery. Nano-pesticides such as Silver, Copper, Silica dioxide, zinc oxide, and nano-formulations have better broad-spectrum pest-defence efficiency in comparison with conventional pesticides (Chhipa, 2017).

Nano-fertilizer has become most popular in agriculture, and its ability to increase crop yield, enhance soil fertility, reduce pollution and make a favorable environment for plants and microbes (Ahmed et al., 2012) has endeared itself to the researchers. Nano-biosensors can be successfully utilized for sensing soil pH, insects, moisture, pests, toxins and mycotoxins, nutritional constituents, etc. The controlled and proper employment of nano-biosensor can maintain sustainable agriculture by improving crop production (Rai et al., 2012; Arora, 2018). Nanomaterial-based smart delivery systems and nanosensors regulate the controlled release of agrochemicals and site-targeted delivery of various macromolecules needed for enhancing disease resistance, efficient nutrient utilization, and improved defence in an eco-friendly manner (Panpatte et al., 2016).

Various types of metal oxide nanomaterials act as nano-fertilizers and give tolerance capacity against salinity in plants. Bio-synthesized metal oxide nanomaterials boost antioxidant mechanisms and detoxification, and develop a tolerance for abiotic stress-induced oxidative damage in plants (Alabdallah and Hasan, 2021). Foliar application of zinc oxide nanomaterials could be a promising practice for enhancing drought stress tolerance in eggplant cultivated in salt-alkaline soil (Semida et al., 2021). The cerium oxide nanoparticles can improve photosynthesis in salt-stressed plants (Rossi et al., 2017).

A tissue culture investigation was carried out to assess the impact of silicon dioxide nanoparticles on the growth, photosynthesis, and protective enzymes of a bamboo species (*Arundinaria pygmaea*) under three varied heavy metal stresses, namely copper (Cu), manganese (Mn), and cadmium (Cd) by Emamverdian et al. (2020). They reported that photosynthetic parameters of bamboo species (*Arundinaria pygmaea*) under Cu and Mn toxicity were promoted by silicon dioxide nanoparticles.

1.3 Conclusion

The metal oxide nanomaterials are one of the new materials that are used in almost all fields, particularly agriculture. Development and application of metal-oxide-based agricultural products in nanoforms are one of the preferred choices for extensive use to increase agricultural yield worldwide. Nanomaterials are gradually used as agrochemical components (fertilizers or pesticides) via foliar spraying. Metal-oxide-based nano-fertilizers and nanpesticides show better efficiency in nutrient release and delivery and broad-spectrum pest protection.

References

Ahmed, S., Niger, F., Kabir, M.H., Chakrabarti, G., Nur, H.P., Imamul Huq, S.M., 2012. Development of slow-release nano fertilizer. In: Proceedings of the International Workshop on Nanotechnology, Dhaka, Bangladesh.

Alabdallah, N.M., Hasan, M.M., 2021. Plant-based green synthesis of silver nanoparticles and its effective role in abiotic stress tolerance in crop plants. Saudi J. Biol. Sci. 28 (10), 5631–5639.

Arora, K., 2018. Advances in nano based biosensors for food and agriculture. Nanotechnol. Food Security Water Treat. 1–52.

Chavali, M.S., Nikolova, M.P., 2019. Metal oxide nanoparticles and their applications in nanotechnology. SN Appl. Sci. 1 (6), 1–30.

Chhipa, H., 2017. Nanofertilizers and nanopesticides for agriculture. Environ. Chem. Lett. 15 (1), 15–22.

Elfeky, S.A., Mohammed, M.A., Khater, M.S., Osman, Y.A., Elsherbini, E., 2013. Effect of magnetite nano-fertilizer on growth and yield of *Ocimum basilicum* L. Int. J. Indig. Med. Plants 46 (3), 1286–11293.

Emamverdian, A., Ding, Y., Mokhberdoran, F., Ahmad, Z., Xie, Y., 2020. Determination of heavy metal tolerance threshold in a bamboo species (*Arundinaria pygmaea*) as treated with silicon dioxide nanoparticles. Glob. Ecol. Conserv. 24, e01306.

Girigoswami, K., 2018. Toxicity of metal oxide nanoparticles. Cellular and Molecular Toxicology of Nanoparticles. Springer, Cham, pp. 99–122.

Jeevanandam, J., Chan, Y.S., Danquah, M.K., 2016. Biosynthesis of metal and metal oxide nanoparticles. ChemBioEng Rev. 3 (2), 55–67.

Khashan, K.S., Sulaiman, G.M., Abdulameer, F.A., 2016. Synthesis and antibacterial activity of CuO nanoparticles suspension induced by laser ablation in liquid. Arab. J. Sci. Eng. 41 (1), 301–310.

Larsen, S.T., Jackson, P., Poulsen, S.S., Levin, M., Jensen, K.A., Wallin, H., et al., 2016. Airway irritation, inflammation, and toxicity in mice following inhalation of metal oxide nanoparticles. Nanotoxicology 10 (9), 1254–1262.

Marquis, G., Ramasamy, B., Banwarilal, S., Munusamy, A.P., 2016. Evaluation of antibacterial activity of plant mediated CaO nanoparticles using *Cissus quadrangularis* extract. J. Photochem. Photobiol. B: Biol. 155, 28–33.

Musić, S., Dragčević, Đ., Maljković, M., Popović, S., 2003. Influence of chemical synthesis on the crystallization and properties of zinc oxide. Mater. Chem. Phys. 77 (2), 521–530.

Panpatte, D.G., Jhala, Y.K., Shelat, H.N., Vyas, R.V., 2016. Nanoparticles: the next generation technology for sustainable agriculture. Microbial Inoculants in Sustainable Agricultural Productivity. Springer, New Delhi, pp. 289–300.

Rai, V., Acharya, S., Dey, N., 2012. Implications of nanobiosensors in agriculture. J. Biomater. Nanobiotechnol. Available from: https://doi.org/10.4236/jbnb.2012.322039.

Rossi, L., Zhang, W., Ma, X., 2017. Cerium oxide nanoparticles alter the salt stress tolerance of Brassica napus L. by modifying the formation of root apoplastic barriers. Environ. Pollut. 229, 132–138.

Sanchez-Dominguez, M., Boutonnet, M., Solans, C., 2009. A novel approach to metal and metal oxide nanoparticle synthesis: the oil-in-water microemulsion reaction method. J. Nanopart. Res. 11, 1823. Available from: https://doi.org/10.1007/s11051-009-9660-8.

Sanchez-Moreno, P., Ortega-Vinuesa, J.L., Peula-Garcia, J.M., Marchal, J.A., Boulaiz, H., 2018. Smart drug-delivery systems for cancer nanotherapy. Curr. Drug. Targets 19 (4), 339–359.

Semida, W.M., Abdelkhalik, A., Mohamed, G., El-Mageed, A., Taia, A., El-Mageed, A., et al., 2021. Foliar application of zinc oxide nanoparticles promotes drought stress tolerance in eggplant (*Solanum melongena* L.). Plants 10 (2), 421.

Sengul, A.B., Asmatulu, E., 2020. Toxicity of metal and metal oxide nanoparticles: a review. Environ. Chem. Lett. 18 (5), 1659–1683.

Wilson, N., 2018. Nanoparticles: environmental problems or problem solvers? Bioscience 68 (4), 241–246.

Chapter 2

Why metal oxide nanoparticles are superior to other nanomaterials for agricultural application?

Valentin Romanovski[1,2] and Rajiv Periakaruppan[3,4]
[1]*Center of Functional Nano-Ceramics, National University of Science and Technology "MISIS," Moscow, Russia,* [2]*Department of Materials Science and Engineering, University of Virginia, Charlottesville, VA, United States,* [3]*College of Horticulture, Nanjing Agricultural University, Nanjing, Jiangsu Province, P.R. China,* [4]*Department of Biotechnology, PSG College of Arts & Science, Coimbatore, Tamil Nadu, India*

Chapter outline

2.1 Introduction

The production and use of nanosized particles is a relevant area of study in the recent decades. Of course, this is due to innovations, as well as the fact that nanosized materials or materials modified by nanosized particles significantly improve the characteristics of the initial materials. The use of metals or metal oxides significantly increases the expected effect compared to bigger particle size. In addition, new approaches to synthesis appear, and these, in turn, give rise to the production of new materials with new properties (Khort et al., 2020a; Lei et al., 2015). In the course of sustainability and circular economy wastes can be considered for recovery (Romanovskaia et al., 2021) and

Nanometal Oxides in Horticulture and Agronomy. DOI: https://doi.org/10.1016/B978-0-323-91809-1.00014-7

production of valuable materials (Akinwande et al., 2022a,b; Kamarou et al., 2020, 2021a,b,c; Hurynovich et al., 2021; Romanovski et al., 2021a). It is of interest to use some wastes as a source for nanometal oxides: water treatment wastes (Propolsky et al., 2020; Ramanouski and Andreeva, 2012; Romanovski, 2020, 2021; Romanovski et al., 2021b; Romanovskii and Khort, 2017; Romanovskii and Martsul, 2009, 2012), galvanic sludge (Martsul et al., 2012a,b, 2013), spent platting electrolytes (Zalyhina et al., 2021a,b), and other metal-containing wastes. Among most green methods of synthesis, solution combustion synthesis (Khort et al., 2020b; Matsukevich et al., 2021; Romanovski et al., 2020; Romanovskiy et al., 2018; Sdobnyakov et al., 2020) and "green" synthesis based on plant materials (Archana et al., 2022; Alghuthaymi et al., 2021; Periakaruppan et al., 2021) can be mentioned.

Agriculture includes a system of various elements interconnected with each other, forming a single whole. The goal is to increase yields. The elements of this system are the use of nanofertilizers, pesticides, herbicides, watering and feeding plants, in some cases water and wastewater treatment, the use of sensors for operational control, etc.

Here we try to summarize the data about positive and negative effects of different nanometal oxides used in agricultural systems and lack of materials, and will analyze different causes and aspects of positive and negative effects. In this chapter, we have not tried to present the results of numerous experiments on the use of various nanometal oxides for seed germination or plant growth. This is the subject of many review articles.

2.2 Nanoparticles in the agricultural system

The number of works devoted to the use of nanomaterials in agriculture is constantly increasing. This is due to various factors: (1) the growth of the world's population contributes to the interest in increasing agricultural production; (2) development of new promising nanomaterials; (3) reduction in the cost of obtaining nanomaterials.

Most of the works focus on the synthesis and evaluation of the effect of the obtained particles on plants, the study of the interaction mechanisms of nanoparticles with plants, and the migration of particles inside plants. Studying the science of toxicology, we can say that all substances are toxic, depending on the dose. It is known that metals are important trace elements for humans, animals, microorganisms, and plants. At the same time, when using them, it is necessary to separate doses and effects with their deficiency, optimal amount, and excess amount. Based on the information in published articles, it should be concluded that it is time to structure the existing knowledge, analyze it comprehensively, and move from laboratory research to field research.

The directions of the use of nanoparticles in agriculture include a wide range of applications (Fig. 2.1). Firstly, it should be noted that the system of agriculture includes, in addition to the main elements, a number of auxiliary

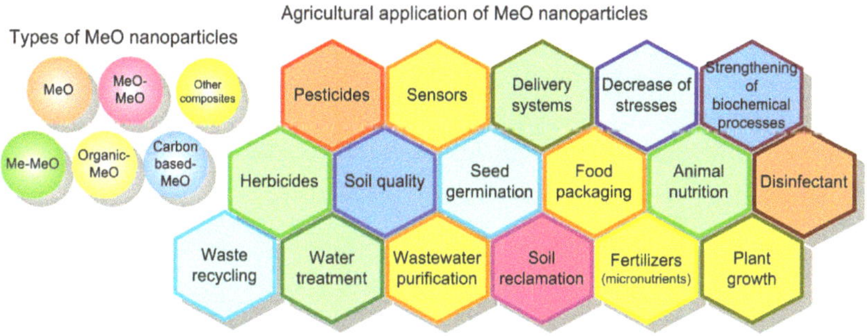

FIGURE 2.1 Types of nanometal oxide nanoparticles and their agricultural application.

ones; for example, water treatment, the use of sensors for monitoring the state of the environment, and the condition of plants, etc.

When choosing one or another material for use in agriculture, it is first necessary to decide what goal is to be achieved. Next, you need to have information about which substances can increase the efficiency or quality of the desired process. When using nanosized metal oxides, an initial comprehensive analysis of possible effects should be carried out; for example, the fact that substances often have complex effects should be taken into account, and positively solving one problem can exacerbate the second. To improve the process of using the considered particles, they can be used as composite materials, carriers, etc.

2.2.1 Nanoparticles for inactivation of pathogenic microorganisms, nanopesticides, nanoherbicides, and nanofungicides

In addition to microorganisms that have a positive effect on soil condition and plant growth, there are many microorganisms that have negative effects. This leads to a decrease in yield, product quality, and finally could lead to the death of plants. The use of nanoparticles may open up a new area of their application for these purposes. There are known inactivating properties of many metal oxide particles. Their multifunctionality can be an advantage in using them.

Metal oxide nanoparticles can also be considered as replacement for highly toxic organic pesticides and herbicides. The use of nanometal oxides will make it possible to solve various environmental issues associated with the use of existing highly toxic compounds.

2.2.2 Soil improvers and nanofertilizers

Nanofertilizers based on metal oxides will provide plants with the necessary nutrients, ensure targeted use, reduce their leaching from soils, and reduce the frequency of fertilization into the soil and the load on soils. The use of metal oxide nanoparticles also makes it possible to increase the efficiency of their absorption by plants. Here, it is important to take into account not only the dose, but also the possibility of using additional options for use, for example, the mobilization of particles on the surface of biochars, etc.

2.2.3 Nanosensors

Metal oxide nanoparticles can be used as nanosensors both in the main and auxiliary agricultural processes. In addition to the main use of metal oxide nanoparticles as sensors for monitoring the quality of nutrients in water and soil, the number of pollutants, and environmental physical parameters, they can also be used as biosensors. Nanoparticles of metal oxides are introduced into plants, which allow remote monitoring of the state of plants.

2.3 Main aspects of the use of nanometal oxides and interactions with plants

In agriculture (without considering a number of auxiliary processes), nanometal oxides can be used as microelements. Trace elements are called chemical elements necessary for the normal life of plants and animals, and are used by them in micro quantities compared to the main components of nutrition. However, the biological role of trace elements is great. Trace elements are necessary for the construction of enzyme systems (biocatalysts). Among the trace elements, iron, manganese, zinc, molybdenum, cobalt, etc. are of the greatest importance. Accordingly, their deficiency causes a decrease in the speed and processes responsible for the development of the body (Fig. 2.2). Finally, the plants cannot realize their potential and produce a low and not always high-quality yield. Trace(micro-) elements cannot be replaced by other substances. For example, plants can only use microelements mostly in a water-soluble form. Insoluble forms should be transferred into a soluble form, for example, when interacting with humic acids. These processes go very slowly, and with abundant watering of the soil, a significant part of the mobile forms of microelements is washed out. However, the use of nanoparticles has the advantage of bypassing this step.

The main role of trace elements to improve the quality and quantity of a crop is as follows:

- With the necessary amount of microelements, plants have the ability to synthesize a full range of enzymes that will allow higher use of energy, water, and nutrition (N, P, K), and give a higher yield,
- Trace elements and enzymes based on them help to enhance the regenerative activity of tissues and prevent diseases,

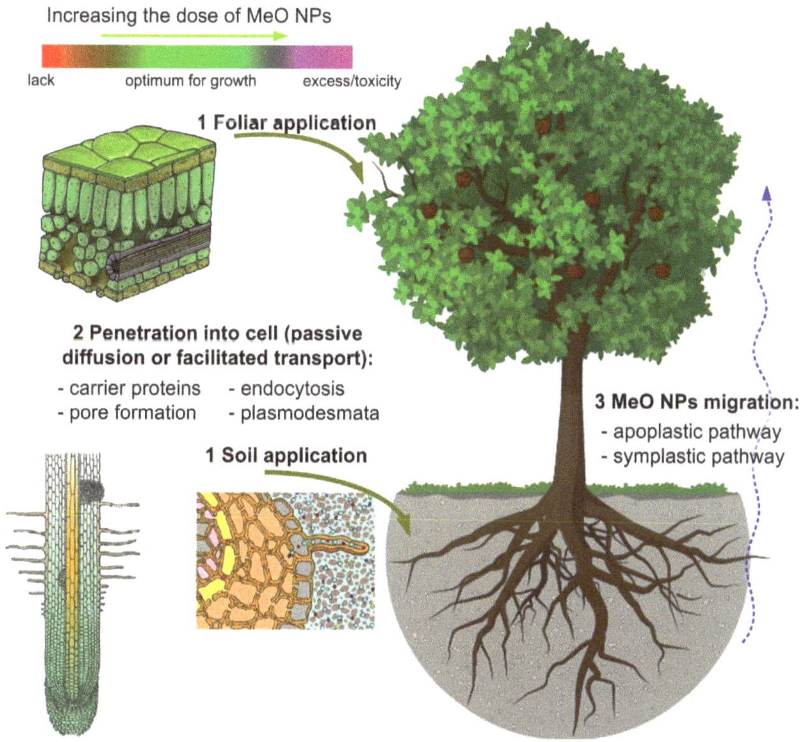

FIGURE 2.2 Application, penetration, and migration of MeO nanoparticles.

- Under trace elements deficiency, a state of physiological depression and a general susceptibility of plants to parasitic diseases is created due to the decreased immunity of plants.

It should also be noted that:

- The simultaneous intake of macro- and microelements, especially phosphorus and zinc, nitrate, nitrogen, and molybdenum is optimal.
- During the entire growing season, plants need essential microelements; some microelements are not reutilized, that is, they are not reused in plants. They do not move from old organs to younger ones.
- Trace elements can be applied both at root top dressing and by spraying. During root nutrition of plants, an apronetic concentration gradient is observed. The concentration of these substances in the plant decreases from bottom to top.
- Prophylactic doses of biologically active microelements do not really affect the total content of microelements in the soil but have a positive effect on the condition of plants.

For a long time, people have been using various soil-improving additives, mainly in the form of fertilizers, to increase the growth and development of plants. At the same time, as noted above, sometimes nutrients become inaccessible to absorption by plants due to the adverse effects of various factors. Those who work with inorganic fertilizers know that often a significant amount of the heavy metals in the composition of inorganic fertilizers comes to the soil. At the same time, when choosing the direction of using municipal wastewater treatment sludge as fertilizers, people are often faced with stringent standards for the content of heavy metals in them, which, however, is much lower than that applied with traditional mineral fertilizers. In recent years, various additives based on biomass substrate and containing nanoparticles have been increasingly developed (Prasad et al., 2017). It is noted (Shang et al., 2019) that they can become a link between nutrients and microstructure of bio cells. It is known that nanoparticles are able to stimulate plant metabolism (Giraldo et al., 2014; DeRosa et al., 2010; Nair et al., 2010). The use of nanotechnology in agriculture is of interest primarily to improve the growth and productivity of crops, as well as to minimize the use of traditional chemical fertilizers (Lee et al., 2008; Siddiqui and Al-Whaibi, 2014; Josko and Oleszczuk, 2013).

One of the processes where nanomaterials can be used is seed germination. The most important aspect here is to prevent the development of unwanted microorganisms. Accordingly, at this stage, nanostructured materials having a bactericidal effect can be used. Many studies are devoted specifically to the production and use of silver nanoparticles (Davies, 2009). However, the suspensions used are stable only in low concentrations, less than 0.1 g/L. During seed germination, positive effects were noted by using nanosized oxides of silicon and titanium (Manjaiah et al., 2019; Lu et al., 2002).

Positive effects are also noted (increased germination, improved growth, photosynthesis, etc.) for plant growth when using silver nanoparticles (Shelar and Chavan, 2015; Sharma et al., 2012). The effect of silver nanoparticles (AgNPs) in a concentration range of 20–60 mg/L was studied in Sadak (2019). AgNPs improved the plant growth parameters such as shoot length, number of leaves, and dry shoot weight, and also increased some biochemical aspects such as photosynthesis. The positive effect of AgNPs on the development of the root system has been shown in Jasim et al., 2017. This, in turn, led to an increase in productivity and a number of other positive effects. In addition to silver, the influence of other substances was also studied, for example, zinc oxide increases the growth of biomass, develops the root system, accelerates photosynthesis (Rossi et al., 2019; Du et al., 2019; Tirani et al., 2019) and increases the nutrient content (Raliya and Tarafdar, 2013). It is widely known that not only silver, but also zinc and copper oxides (Malandrakis et al., 2019; Giannousi et al., 2013) and even magnesium oxide (Imada et al., 2016) have bactericidal effect. Some nanosized substances are also noted as giving positive effects on the plants' growth: iron sulfide (Srivastava et al., 2014a,b; Das et al., 2016), titanium oxide (Rafique et al., 2014; Rizwan et al., 2019), silicon oxide (Rizwan et al., 2019), hematite (Askary et al., 2017), and magnetite (Konate et al., 2017). The use of a number of these materials increases the resistance of plants to stressful conditions (Torabian et al., 2017; Tripathi et al., 2017).

However, there are possible toxic effects of nanosized particles of metals and metal oxides on plants: iron and iron oxides (Martinez-Fernandez et al., 2016), zinc oxide (Priester et al., 2017; Zafar et al., 2017), and copper oxide (Singh and Kumar, 2016), which manifest in reduced uptake and translocation of nutrients and root hydraulic conductivity, decreased root elongation, reduced chlorophyll content, inhibited growth, imbalanced nutrients in shoots, etc. Also, metals and metal oxides nanosized particles can have a toxic effect on soil microorganisms: iron and iron oxides (Shah et al., 2014; Jiling et al., 2016), zinc oxide (You et al., 2017), and copper and copper oxide (Kumar et al., 2012; Xu et al., 2015), which could affect negatively on soil microbial community.

To ensure safety, it is necessary to know the migration paths of nanoparticles in plants, so that, along with a positive effect on increasing productivity, it is necessary to ensure safety when consuming plants and fruits for food by animals or humans.

Other ways of using nanometal particles in agriculture, such as nanosensors, materials for water treatment and wastewater treatment, are bone processes, so the choice of materials for them will be discussed in detail in the relevant chapters.

2.4 Conclusions and further perspectives

Some nanoparticles have a positive effect, while others have a negative effect in agriculture, depending on a number of reasons:

- First, it is necessary to determine the scope of metal oxide nanoparticles (fertilizer, herbicide, pesticide, nanosensor, material for water treatment and wastewater treatment, etc.), since some are good for inactivating microorganisms, others help to increase the level of chlorophyll, and others are good semiconductors and promising for use as sensors or substances for water treatment or wastewater treatment, and so on;
- Second, what is the optimal dose of the material for the highest efficiency of the selected process (e.g., for resistance to certain stressful conditions), since at a higher concentration, nanometallic particles can have a detrimental effect on the plant, or lead to soil and groundwater pollution;
- The functional effect of the material in the selected area; for example, we need to decide which function we need to stimulate and in which plant.

Prospects for the use of nanometal oxides in agriculture are as follows:

- Development of the most environmentally friendly methods of synthesis;
- Increasing the efficiency of delivery of nanometal oxides to the target place of their use (depending on the positive effects created when using one or another nanometal oxide);
- The use of materials with a balanced complex effect;
- Search for new less toxic nanometallic oxides to replace more toxic ones;
- Study of mechanisms and effects in the complex use of nanometal oxides in agriculture;
- Transition from conducting experiments in Petri dishes to full-scale ones;
- Generalization of materials on understanding the processes of migration of nanometallic oxides in the systems under study;
- Studying the issues of accumulation, toxicology, and avoidance of the negative effects of the use of nanometal oxides in agriculture;
- As a consequence of the previous paragraph, this is the determination of the optimal doses of use and methods for applying nanometal oxides in agriculture.

References

Akinwande, A.A., Balogun, O.A., Romanovski, V., Ademati, A.O., Adetula, Y.V., 2022a. Recycling of synthetic waste wig fiber in the production of cement-adobe for building envelope: physio-hydric properties. Environ. Sci. Pollut. Res. Available from: https://doi.org/10.1007/s11356-022-18649-6.

Akinwande, A.A., Balogun, O.A., Romanovski, V., Danso, H., Kamarou, M., Ademati, A.O., 2022b. Mechanical performance and Taguchi optimization of kenaf fiber/cement-paperboard composite for interior application. Environ. Sci. Pollut. Res. Available from: https://doi.org/10.1007/s11356-022-19449-8.

Alghuthaymi, M.A., Rajkuberan, C., Santhiya, T., Krejcar, O., Kuča, K., Periakaruppan, R., et al., 2021. Green synthesis of gold nanoparticles using *Polianthestuberosa L.* Flor. Extract. Plants. 10 (11), 2370.

Archana, P., Janarthanan, B., Bhuvana, S., Rajiv, P., Sharmila, S., 2022. Concert of zinc oxide nanoparticles synthesized using cucumismelo by green synthesis and the antibacterial activity on pathogenic bacteria. Inorg. Chem. Commun. 109255.

Askary, M., Mehdi Talebi, S., Amino, F., Bangan, A.D.B., 2017. Effects of iron nanoparticles on *Menthapiperita* L. under salinity stress. Biologija 64, 65–75.

Das, C.K., Srivastava, G., Dubey, A., Roy, M., Jain, S., Sethy, N.K., et al., 2016. Nano-iron pyrite seed dressing: a sustainable intervention to reduce fertilizer consumption in vegetable (beetroot, carrot), spice (fenugreek), fodder (alfalfa), and oilseed (mustard, sesamum) crops. Nanotechnol. Environ. Eng. 1, 2.

Davies, J.C., 2009. Nanotechnology oversight: an agenda for the new administration. Project on Emerging Technologies. Woodrow Wilson International Center for Scholars, Washington, DC.

DeRosa, M.C., Monreal, C., Schnitzer, M., Walsh, R., Sultan, Y., 2010. Nanotechnology in fertilizers. Nat. Nanotechnol. 5, 91.

Du, W., Yang, J., Peng, Q., Liang, X., Mao, H., 2019. Comparison study of zinc nanoparticles and zinc sulphate on wheat growth: from toxicity and zinc biofortification. Chemosphere. 227, 109–116.

Giannousi, K., Avramidis, I., Dendrinou-Samara, C., 2013. Synthesis, characterization and evaluation of copper based nanoparticles as agrochemicals against Phytophthorainfestans. RCS Adv. 3, 21743–21752.

Giraldo, J.P., Landry, M.P., Faltermeier, S.M., McNicholas, T.P., Iverson, N.M., Boghossian, A. A., et al., 2014. Plant nanobionics approach to augment photosynthesis and biochemical sensing. Nat. Mater. 13 (4), 400–408.

Hurynovich, A., Kwietniewski, M., Romanovski, V., 2021. Evaluation of the possibility of utilization of sewage sludge from wastewater treatment plant—case study. Desalination Water Treat. 227, 16–25.

Imada, K., Sakai, S., Kajihara, H., Tanaka, S., Ito, S., 2016. Magnesium oxide nanoparticles induce systemic resistance in tomato against bacterial wilt disease. Plant Pathol. 65, 551–560.

Jasim, B., Thomas, R., Mathew, J., Radhakrishnan, E.K., 2017. Plant growth and diosgenin enhancement effect of silver nanoparticles in fenugreek (*Trigonellafoenum-graecum* L.). Saudi Pharm. J. 25, 443–447.

Jiling, C., Youzhi, F., Xiangui, L., Junhua, W., 2016. Arbuscularmycorrhizal fungi alleviate the negative effects of iron oxide nanoparticles on bacterial community in rhizospheric soils. Front. Environ. Sci. 4, 10.

Josko, I., Oleszczuk, P., 2013. Influence of soil type and environmental conditions on ZnO, TiO2 and Ni nanoparticles phytotoxicity. Chemosphere. 92, 91–99.

Kamarou, M., Kuzmenkov, M., Korob, N., Kwapinski, W., Romanovski, V., 2020. Structurally controlled synthesis of calcium sulphatedihydrate from industrial wastes of spent sulphuric acid and limestone. Environ. Technol. Innov. 17, 100582.

Kamarou, M., Korob, N., Hil, A., Moskovskikh, D., Romanovski, V., 2021a. Low-energy technology for producing anhydrite in the $CaCO_3$—H_2SO_4—H_2O system derived from industrial wastes. J. Chem. Technol. Biotechnol. 96 (7), 2065–2071.

Kamarou, M., Korob, N., Kwapinski, W., Romanovski, V., 2021b. High-quality gypsum binders based on synthetic calcium sulfate dihydrate produced from industrial wastes. J. Ind. Eng. Chem. 100, 324–332.

Kamarou, M., Korob, N., Romanovski, V., 2021c. Structurally controlled synthesis of synthetic gypsum derived from industrial wastes: sustainable approach. J. Chem. Technol. Biotechnol. 96 (11), 3134–3141.

Khort, A., Romanovski, V., Lapitskaya, V., Kuznetsova, T., Yusupov, K., Moskovskikh, D., et al., 2020a. Graphene@metalnanocomposites by solution combustion synthesis. Inorg. Chem. 59 (9), 6550–6565.

Khort, A., Romanovski, V., Leybo, D., Moskovskikh, D., 2020b. CO oxidation and organic dyes degradation over graphene-Cu and graphene-CuNi catalysts obtained by solution combustion synthesis. Sci. Rep. 10, 16104.

Konate, A., He, X., Zhang, Z., Ma, Y., Zhang, P., Alugongo, G.M., et al., 2017. Magnetic (Fe_3O_4) nanoparticles reduce heavy metals uptake and mitigate their toxicity in wheat seedling. Sustainability 9, 1e16.

Kumar, N., Shah, V., Walker, V.K., 2012. Influence of a nanoparticle mixture on an arctic soil community. Environ. Toxicol. Chem. 31, 131–135.

Lee, W.M., An, Y.J., Yoon, H., Kweon, H.S., 2008. Toxicity and bioavailability of copper nanoparticles to the terrestrial plants mungbean (Phaseolusradiatus) and wheat (Triticumaestivum): plant agar test for water-insoluble nanoparticles. Environ. Toxicol. Chem. 27 (9), 1915–1921.

Lei, J.C., Zhang, X., Zhou, Z., 2015. Recent advances in MXene: preparation, properties, and applications. Front. Phys. 10 (3), 276–286.

Lu, C., Zhang, C., Wen, J., Wu, G., Tao, M.X., 2002. Research of the effect of nanometer materials on germination and growth enhancement of *Glycine max* and its mechanism. Soybean Sci. 21, 168–171.

Malandrakis, A.A., Kavroulakis, N., Chrysikopoulos, C.V., 2019. Use of copper, silver and zinc nanoparticles against foliar and soil-borne plant pathogens. Sci. Total. Environ. 670, 292–299.

Manjaiah, K.M., Mukhopadhyay, R., Paul, R., Datta, S.C., Kumararaja, P., Sarkar, B., 2019. Clay minerals and zeolites for environmentally sustainable agriculture. In: Mariano, M., Binoy, S., Alessio, L. (Eds.), Modified Clay and Zeolite Nanocomposite Materials. Elsevier, Amsterdam, The Netherlands, pp. 309–329.

Martinez-Fernandez, D., Barroso, D., Komarek, M., 2016. Root water transport of *Helianthus annuus L.* under iron oxide nanoparticle exposure. Environ. Sci. Pollut. Res. 23, 1732–1741.

Martsul, V.N., Zalygina, O.S., Shibeka, L.A., Likhacheva, A.V., Ramanouski, V.I., 2012a. Some applications of galvanic manufacture waste. Proceedings of BSTU. Chemistry and Technology of Inorganic Substances 3, pp. 66–70.

Martsul, V.N., Likhacheva, A.V., Shibeka, L.A., Zalygina, O.S., Ramanouski, V.I., Khodin, V. V., 2012b. Inventory of galvanic sludge and deposits of the treatment facilities formed at the enterprises of the Republic of Belarus. Proceedings of BSTU. Chemistry and Technology of Inorganic Substances, 3, pp. 71–77.

Martsul, V.N., Zalygina, O.S., Likhacheva, A.V., Romanovski, V.I., 2013. Treatment of electroplating shop wastewater at enterprises of the Republic of Belarus. Proceedings of BSTU. Chemistry and Technology of Inorganic Substances, 3, pp. 57–62.

Matsukevich, I., Kulak, A., Palkhouskaya, V., Romanovski, V., Jo, J.H., Aniskevich, Y., et al., 2021. Comparison of different synthesis methods for $Li_2MTi_3O_8$ (M—Co, Cu, Zn) as electrode materials. J. Chem. Technol. Biotechnol. 97 (4), 1021–1026. Available from: https://doi.org/10.1002/jctb.6992.

Nair, R., Varghese, S.H., Nair, B.G., Maekawa, T., Yoshida, Y., Kumar, D.S., 2010. Nanoparticulate material delivery to plants. Plant. Sci. 179, 154–163.

Periakaruppan, R., Chen, X., Thangaraj, K., Jeyaraj, A., Nguyen, H.H., Yu, Y., et al., 2021. Utilization of tea resources with the production of superparamagnetic biogenic iron oxide nanoparticles and an assessment of their antioxidant activities. J. Clean. Prod. 278, 123962.

Prasad, R., Bhattacharyya, A., Nguyen, Q.D., 2017. Nanotechnology in sustainable agriculture: recent developments, challenges, and perspectives. Front. Microbiol. 8, 1014.

Priester, J.H., Shelly, C.M., Katherine, E., Yuan, G., Ying, W., Roger, M.N., et al., 2017. Damage assessment for soybean cultivated in soil with either CeO2 or ZnO manufactured nanomaterials. Sci. Total. Environ. 579, 1756–1768.

Propolsky, D., Romanovskaia, E., Kwapinski, W., Romanovski, V., 2020. Modified activated carbon for deironing of underground water. Environ. Res. 182, 108996.

Rafique, R., Arshad, M., Khokhar, M.F., Qazi, I.A., Hamza, A., Virk, N., 2014. Growth response of wheat to titania nanoparticles application. NJES 7, 42–46.

Raliya, R., Tarafdar, J.C., 2013. ZnO nanoparticle biosynthesis and its effect on phosphorous-mobilizing enzyme secretion and gum contents in Clusterbean (*Cyamopsistetragonoloba* L.). Agric. Res. 2, 48–57.

Ramanouski, V.I., Andreeva N.A., 2012. Purification of washing waters of iron removal stations. Proceedings of BSTU. Chemistry and Technology of Inorganic Substances, 3, pp. 62–65.

Rizwan, M., Ali, S., ur Rehman, M.Z., Malik, S., Adrees, M., Qayyum, M.F., et al., 2019. Effect of foliar applications of silicon and titanium dioxide nanoparticles on growth, oxidative stress, and cadmium accumulation by rice (*Oryza sativa*). Acta Physiol. Plant. 41, 35.

Romanovskaia, E., Romanovski, V., Kwapinski, W., Kurilo, I., 2021. Selective recovery of vanadium pentoxide from spent catalysts of sulfuric acid production: sustainable approach. Hydrometallurgy. 200, 105568.

Romanovski, V., 2020. New approach for inert filtering media modification by using precipitates of deironing filters for underground water treatment. Environ. Sci. Pollut. Res. 27, 31706–31714.

Romanovski, V., 2021. Agricultural waste based-nanomaterials: green technology for water purifications. In: Abd-Elsalam, K., Zahid, M. (Eds.), Aquananotechnology. 1st Edition. Applications of Nanomaterials for Water Purification. Elsevier, pp. 577–595.

Romanovski, V., Kolosov, A.Y., Khort, A.A., Myasnichenko, V.S., Podbolotov, K.B., Savina, K. G., et al., 2020. Features of Cu-Ni nanoparticle synthesis: experiment and computer simulation. Phys. Chem. Asp. Study Clust. Nanostruct. Nanomater. 12, 293–309.

Romanovski, V., Klyndyuk, A., Kamarou, M., 2021a. Green approach for low-energy direct synthesis of anhydrite from industrial wastes of lime mud and spent sulfuric acid. J. Environ. Chem. Eng. 9 (6), 106711.

Romanovski, V., Romanovskaia, E., Moskovskikh, D., Kuskov, K., Likhavitski, V., Mehmet, F. A., et al., 2021b. Recycling of iron-rich sediment for surface modification of filters for underground water deironing. J. Environ. Chem. Eng. 9 (4), 105712.

Romanovskii, V.I., Khort, A.A., 2017. Modified anthracites for deironing of underground water. J. Water Chem. Technol. 39, 299–304.

Romanovskii, V.I., Martsul, V.N., 2009. Distribution of heteroatoms of synthetic ion exchangers in pyrolysis products. Russ. J. Appl. Chem. 82, 836–839.

Romanovskii, V.I., Martsul, V.N., 2012. Functional group distribution over the surface and in the bulk of particles of spent ion exchangers in the course of mechanochemical destruction. Russ. J. Appl. Chem. 85, 371–376.

Romanovskiy, V.I., Khort, A.A., Podbolotov, K.B., Sdobnyakov, N.Y., Myasnichenko, V.S., Sokolov, D.N., 2018. One-step synthesis of polymetallic nanoparticles in air invironment. Izv.Vyssh. Uchebn. Zaved. Khim. Khim. Tekhnol. 61, 43–48.

Rossi, L., Fedenia, L.N., Sharifan, H., Ma, X., Lombardini, L., 2019. Effects of foliar application of zinc sulfate and zinc nanoparticles in coffee (*Coffeaarabica* L.) plants. Plant. Physiol. Biochem. 135, 160–166.

Sadak, M.S., 2019. Impact of silver nanoparticles on plant growth, some biochemical aspects, and yield of fenugreek plant (Trigonella foenum-graecum). Bull. Natl Res. Cent. 43 (1), 1–6.

Sdobnyakov, N., Khort, A., Myasnichenko, V., Podbolotov, K., Romanovskaia, E., Kolosov, A., et al., 2020. Solution combustion synthesis and Monte Carlo simulation of the formation of CuNi integrated nanoparticles. Comput. Mater. Sci. 184, 109936.

Shah, V., Collins, D., Walker, V.K., Shah, S., 2014. The impact of engineered cobalt, iron, nickel and silver nanoparticles on soil bacterial diversity under field conditions. Environ. Res. Lett. 9 (2), 024001.

Shang, Y., Hasan, M., Ahammed, G.J., Li, M., Yin, H., Zhou, J., 2019. Applications of nanotechnology in plant growth and crop protection: a review. Molecules 24 (14), 2558.

Sharma, P., Bhatt, D., Zaidi, M.G., Saradhi, P.P., Khanna, P.K., Arora, S., 2012. Silver nanoparticle-mediated enhancement in growth and antioxidant status of *Brassica juncea*. Appl. Biochem. Biotechnol. 167, 2225–2233.

Shelar, G.B., Chavan, A.M., 2015. Myco-synthesis of silver nanoparticles from Trichodermaharzianum and its impact on germination status of oil seed. Biolife 3, 109–113.

Siddiqui, M.H., Al-Whaibi, M.H., 2014. Role of nano-SiO2 in germination of tomato (Lycopersicumesculentum seeds. Mill.). Saudi J. Bio. Sci 21, 13–17.

Singh, D., Kumar, A., 2016. Impact of irrigation using water containing CuO and ZnO nanoparticles on *Spinach oleracea* grown in soil media. Bull. Environ. Contam. Toxicol. 97, 548–553.

Srivastava, G., Das, A., Kusurkar, T.S., Roy, M., Airan, S., Sharma, R.K., et al., 2014a. Iron pyrite, a potential photovoltaic material, increases plant biomass upon seed pretreatment. Mater. Express 4, 23–31.

Srivastava, G., Das, C.K., Das, A., Singh, S.K., Roy, M., Kim, H., et al., 2014b. Seed treatment with iron pyrite (FeS2) nanoparticles increases the production of spinach. RSC Adv. 4, 58495–58504.

Tirani, M.M., Haghjou, M.M., Ismaili, A., 2019. Hydroponic growntobacco plants respond to zinc oxide nanoparticles and bulk exposures by morphological, physiological and anatomical adjustments. Funct. Plant. Biol. 46, 360–375.

Torabian, S., Zahedi, M., Khoshgoftar, A.H., 2017. Effects of foliar spray of nano-particles of FeSO4 on the growth and ion content of sunflower under saline condition. J. Plant. Nutr. 40, 615–623.

Tripathi, D.K., Singh, S., Singh, V.P., Prasad, S.M., Dubey, N.K., Chauhan, D.K., 2017. Silicon nanoparticles more effectively alleviated UV-B stress than silicon in wheat (Triticumaestivum) seedlings. Plant. Physiol. Biochem. 110, 70–81.

Xu, C., Peng, C., Sun, L., Zhang, S., Huang, H., Chen, Y., et al., 2015. Distinctive effects of TiO2 and CuO nanoparticles on soil microbes and their community structures in flooded paddy soil. Soil. Biol. Biochem. 86, 24–33.

You, T., Liu, D., Chen, J., Yang, Z., Dou, R., Gao, X., et al., 2017. Effects of metal oxide nanoparticles on soil enzyme activities and bacterial communities in two different soil types. J. Soils Sediment. 18 (1), 211–221.

Zafar, H., Ali, A., Zia, M., 2017. CuO nanoparticles inhibited root growth from *Brassica nigra* seedlings but induced root from stem and leaf explants. Appl. Biochem. Biotechnol. 181, 365–378.

Zalyhina, V., Cheprasova, V., Belyaeva, V., Romanovski, V., 2021a. Pigments from spent Zn, Ni, Cu, and Cd electrolytes from electroplating industry. Environ. Sci. Pollut. Res. 28, 32660–32668.

Zalyhina, V., Cheprasova, V., Romanovski, V., 2021b. Pigments from spent chloride-ammonium zinc plating electrolytes. J. Chem. Technol. Biotechnol. 96 (10), 2767–2774.

Chapter 3

Effect of metal oxide nanoparticles on biochemical pathways in plants

Daraksha Iram[1], Manish Singh Sansi[1], Parul Singh[1], P.R Chandhni[1], Sameena Zanab[2], Sikha Rana[1], Priya Kalyan[1], Kapil Singh Narayan[3], Harpreet Kaur[1], Deepti Mittal[4], Karmveer Yadav[4], Gurjeet Kaur[5,6] and Syed Azmal Ali[7]

[1]*National Dairy Research Institute Karnal, Karnal, Haryana, India,* [2]*Chaudhary Charan Singh University Meerut, Meerut, Uttar Pradesh, India,* [3]*New Bolton Center, School of Veterinary Medicine, Department of Clinical Studies, University of Pennsylvania, Kennett Square, PA, United States,* [4]*Nanosafety Lab, Division of Biochemistry, ICAR-NDRI, Karnal, Haryana, India,* [5]*Centre for Healthy Brain Ageing, School of Psychiatry, University of New South Wales, Sydney, NSW, Australia,* [6]*Mark Wainwright Analytical Centre, Bioanalytical Mass Spectrometry Facility, University of New South Wales, Sydney, NSW, Australia,* [7]*Cell Biology and Proteomics Lab, Animal Biotechnology Center, ICAR-NDRI, Karnal, Haryana, India*

Chapter outline

Nanometal Oxides in Horticulture and Agronomy. DOI: https://doi.org/10.1016/B978-0-323-91809-1.00018-4

3.1 Introduction

Globally, nanoscience and nanotechnology research has surged in popularity during the past decade. At the nanometer length scale, nanoparticles (NPs) may now be used as structural blocks to create complex and new phases of condensed matter. Nanoscale materials have garnered considerable interest due to their structural and behavioral differences with atoms and molecules and bulk materials (Rosei, 2004). As a result, nanotechnology (which encompasses metal NPs) is a significant source of innovation with significant economic implications such as advanced nanobiosensors, clustered regularly interspaced short palindromic repeats (CRISPR)-based gene transfer technologies, and others (Ali et al., 2021, Singh and Ali, 2021). While nanotechnology applications and processes may benefit human health and the environment, concerns have emerged and risen worldwide about the potential harm they may cause. NPs are defined by Auffan et al. (2009) as ingredients with a diameter of less than 100 nm in any one dimension. NPs naturally occur in rocks, clays, and waste products of bacteria. Although it has been employed as a metal colorant since antiquity, the systematic design and manufacture of NPs for various applications began recently. It is possible to create NPs with unique physical and chemical characteristics. Materials, sizes, and forms vary widely among engineered NPs; they also contain synthetic surface molecules that distinguish them from naturally occurring NPs (Radad et al., 2012). In particular, metal oxide nanoparticles (MONPs) exhibit unique characteristics due to their limited size that influence any material's basic structural properties, unlike bulk material (Ayyub, 1998). To produce metal and MONPs, respectively, reducing or oxidizing/precipitating agents are preferred. Reactivity of NPs with biomolecules is determined by several variables, including their physiochemical characteristics and production process (Teske and Detweiler, 2015; Wang et al., 2016a,b).

Additionally, solute conditions, including hydrogen ion concentration, agglomeration, and aggregation, affect the reactivity of NPs with biomolecules. Moreover, biomolecules itself is influenced by the NPs through the exchange of energies within thermodynamic, kinetic, and physicochemical limits (Hotze et al., 2010). Since NPs are likely to maintain the bulk material's essential properties, it is crucial to address the influence of bulk material while examining the interaction of NPs with biological systems. A metalloid, such as silicon, has been found to be beneficial to plants, despite heavy metals being harmful to them (Yadav, 2010; Tubana et al., 2016; Helaly et al., 2017). With the rising usage of NPs in commercial goods, many firms are generating proprietary NPs to market their services and products but are ignorant of the purposeful or unintentional release of these NPs into the ecosystem. Only a few of the many NPs are widely used and constitute a hazard to the environment. Silver nanoparticles (Ag NPs) and zinc oxide nanoparticles (ZnO NPs) are projected to be released substantially and become ubiquitous across Europe, depending on per capita use (Dumont et al., 2015). Numerous mathematical models are being developed that underpin experimental research to enhance our basic knowledge of environmental destiny and its interaction with organisms (Baalousha et al., 2016). Plants, the primary contributor to the environment, must be considered when examining the influence of NPs. Plants are continually exposed to the environment, including NPs produced by humans. This is significant because when NPs are used excessively and their quantity increases in the environment, plants and animals will become sources of NPs for humans.

NPs are influenced by their environment, so their aggregation state, oxidation status, secondary phase precipitation, and so on may vary depending on the environment (Levard et al., 2012). As a consequence of the composition of NPs, different plant responses to the same NP may be seen, such as changes in reactivity, penetration, and translocation within the plant, among other things. It was found that encapsulating NPs changes plant responses compared to exposure to bare NPs (Barrios et al., 2016).

Several research studies published in the last several years have confirmed the trophic transmission of specific NPs via a food chain on land or in water (Unrine et al., 2012; Tangaa et al., 2016). However, the impacts of nonpoint source pollution on plants have not been well researched, even though plants are producers and play an important function in the environment (Zuverza-Mena et al., 2017). Research of NPs is complicated by a lack of suitable detection instruments that can detect real-time concentration in the environment (Navratilova et al., 2015; Mahdi et al., 2017; Hadioui et al., 2015; Navratilova et al., 2015; Mahdi et al., 2017). Depending on their physiochemical features, NPs may have both a good and detrimental influence on plants, according to a study conducted on various plants (Fig. 3.1) (Tripathi et al., 2015, 2017; Wang et al., 2016a,b). Several investigations have discovered that NPs have a deleterious influence on plants, lowering a plant's ability to grow and pigment production (Tripathi et al., 2017). However, elegantly designed NPs are being used as growth stimulators, nanopesticides, nanofertilizers, soil-improving agents, and sensors (Mittal et al., 2020), as well as monitoring several agricultural factors in the field to improve agricultural crop production (Fraceto et al., 2016; Wang et al., 2016a,b). In modern days, the majority of the study on the environmental effects of industrial NPs on plants has been carried out as a result of the increasing attention on the subject (Table 3.1). The following parts will address the presence of various NPs in the surroundings, their effect on plants, a brief overview, and the primary mechanism of their action.

3.1.1 Effect of nanoparticles on plants

Several researches published in the past several years have established the trophic transmission of different NPs across a food chain on the ground or in

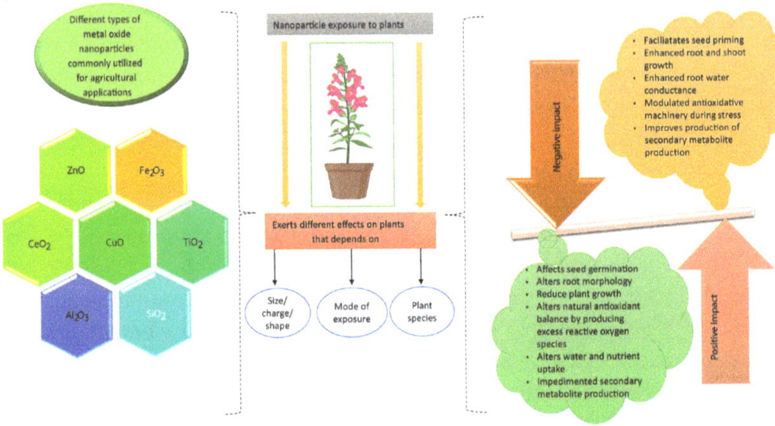

FIGURE 3.1 Possible effects of metal and metal oxide nanoparticles on overall plant development. NPs' size, composition, concentration, physical and chemical features, application technique, and plant species all influence these consequences.

the water (Judy et al., 2012; Hawthorne et al., 2014; De la Torre-Roche et al., 2015). Although plants are producers and have a detrimental environmental effect, their interactions with NPs have not been extensively researched (Rico et al., 2011; Feng et al., 2013). Research of NPs is complicated by a lack of suitable detection instruments for NPs in the environment (Mahdi et al., 2017). Inductively coupled plasma mass spectrometry (ICP-MS) is a prominent example of dependable approaches available when it comes to identifying NPs, (Hadioui et al., 2015; Navratilova et al., 2015). It was found that NPs may have a helpful or harmful consequence on plants based on their physiology (Nhan et al., 2015; Wang et al., 2016a,b). Numerous researches have shown that NPs have a detrimental consequence on plants, resulting in reduced growth and pigment production (Landa et al., 2016). However, intelligently designed NPs enhance agricultural crop output as growth stimulants, nanopesticides, nanofertilizers, soil enhancers, and sensors for observing and recording various agricultural data (Fraceto et al., 2016; Wang et al., 2016a, b). Most studies on the effects of industrial NPs on plants have conceded in recent years, owing to an increase in interest in the subject matter. After that, we'll go through the many types of NPs that may be found in the environment, their impact on plants, and a quick description and explanation of the basic method by which NPs can influence a plant (Fig. 3.2).

3.1.2 Vascular plant response to metal nanoparticles

Metal NPs harmful effects on plants may be due to chemical toxicity and stress or stimulated by the particle's surface, size, and form (Brunnert et al., 2006). Because of their increased specific surface area and reactivity,

TABLE 3.1 List of metal oxide nanoparticles that have been researched on various plant species at various concentrations and with varying degrees of influence.

S. No	Nanoparticle	Size (nm)	Concentration	Plant	Impact	References
1	CuO	30	0.025, 0.25, 0.5, 1, 5 mg/L	*Elodea densa*	Catalase and superoxide dismutase activities increase by 1.5–2 times. Stimulated photosynthesis up to 0.25 mg/L level, whereas suppressing it above 1 mg/L concentration.	Nekrasova et al. (2011)
2.	CeO_2	8	500 mg/kg	*Oryza sativa*	Under NP influence, rice grain contains less Fe, S, prolamin, glutelin, lauric acid, valeric acid, and starch in comparison to control. NP could compromise the quality of rice grain	Rico et al. (2013c)
3.	ZnO	25	400, 1000, 2000 mg/L	*Arachis hypogaea*	Zn as a micronutrient can be delivered to plant through NP. Up to 1000 mg/L, the NP promoted seed germination and growth vigor, whereas 2000 mg/L was considered toxic for the plant.	Prasad et al. (2012)
4.	Fe_3O_4	10	5, 10, 15, 20 mg/L	*Triticum aestivum*	NP exposure did not alter germination, plant growth, and chlorophyll content. Plant exposed to NP showed a favorable response to prevent oxidative damage	Iannone et al. (2016)
5.	TiO_2	0.25–4	2500 mg/dm	*Spinacia oleracea*	TiO_2 NPs have been shown to increase spinach growth and nitrogen absorption	Yang et al. (2006)
6.	CeO_2	7	0, 500, 1000, 2000, 4000 mg/L	Soybean	Genotoxicity recorded at 2000 and 4000 mg/L concentration; a new band in the roots' RAPD profile was observed	López-Moreno et al. (2010)

(Continued)

TABLE 3.1 (Continued)

S. No	Nanoparticle	Size (nm)	Concentration	Plant	Impact	References
7.	Al_2O_3	13	2000 mg/L	Maize, cucumber, carrots, cabbage	Reduced root growth	Yang et al. (2006)
8.	SiO_2	10–20	50 and 100 mg/L	*Strawberry*	Strawberry exposure in nutrient sol improved vegetative growth, increased proline chlorophyll, improved epicuticular wax layer, leaf relative water content, and improved canopyTemperature	Avestan et al. (2019)
9.	NiO	23–34	25, 50, 100, 250, 500, 1000, 2000 mg/L	*Solanum lycopersicum*	NiO induces apoptosis in tomato root cells, increase in ROS, antioxidants, and mitochondrial membrane potential, and triggers the release of caspase-3 proteases from mitochondria.	Faisal et al. (2013)
10.	TiO_2 and ZnO	30	0, 5, 10, 20, 40, and 80 mg/kg	*Hordeum vulgare L.*	The antioxidant enzyme activities are diversely affected, but no changes in seed germination and root elongation have been observed; ZnO was more toxic than TiO2.	Dogaroglu and Köleli (2017)
11.	CuO	10–50	0, 20, 50, 100, 200, 500 mg/L	Mung bean	Reduced biomass and root length at all concentrations; reduced chlorophyll content above 100 mg/L; no changes in carotenoid content; increased H_2O_2 and lipid peroxidation; increased ROS production with an increase in concentration; modulations in gene expression	Nair et al. (2014)

12.	CdO	7–60	$2.03 \pm 0.45 \times 105$ particles cm^{-3}	*Hordeum vulgare*	No change in total chlorophyll concentration, with a minor change in Fv/Fm with (3) treatment. Increase in total amino acids in all three cases with the maximum in (3) treatment	Vecerova et al. (2016)
13.	*n*CeO_2	25–50	0–2000 mg Ce/L	*Holcus lanatus and dicots Lychnis-flos-cuculi and Diplotaxis tenuifolia*	*n*CeO_2 had a stimulating effect in the early stages of the development of the plants	Lizzi et al. (2020)
15.	*n*CeO_2	8 ± 1	62.5, 125, 250, and 500 mg/L	*Oryza sativa*	nCeO_2 modulate the antioxidant defense system by decreasing H_2O_2 content in root and shoot of rice seedlings and enhances the enzyme activities concentration-dependent manner	Rico et al. (2013c)
16.	Cr_2O_3	15–30	0–100 mg/L	*Triticum aestivum*	Growth chamber inhibition of seed germination, biomass, and shoot and root length	Vajpayee et al. (2011)
17.	ZnO NP	18 ± 8	100–500 mg	*Capsicum chinense* L. var. *Chichen Itza*	Germination is increased with concentration; radicule length ↑ at 300 mg/L	García-López et al. (2018)
18.	NiO	23, 34	25, 50, 100, 250, 500, 1000, 2000 mg/L	*Solanum lycopersicum*	NiO induces apoptosis in tomato root cells, increases in ROS, antioxidants, and mitochondrial membrane potential, and triggers the release of caspase-3 proteases from mitochondria	Faisal et al. (2013)
19.	ZnO and TiO_2	35 to 50	0, 50, 100, and 150 mg/L	*Solanum melongena L.*	Accelerated germination through using these nanoparticles may be an open chance for breaking down the seeds dormancy of plants	Younes et al. (2020)
20.	CuO	40	10, 50, 100, 150, 200 mg/L	*Lemna minor*	Enhanced activity of peroxidase, catalase, superoxide dismutase; increased lipid peroxidation and inhibition of plant growth	Song et al. (2016)

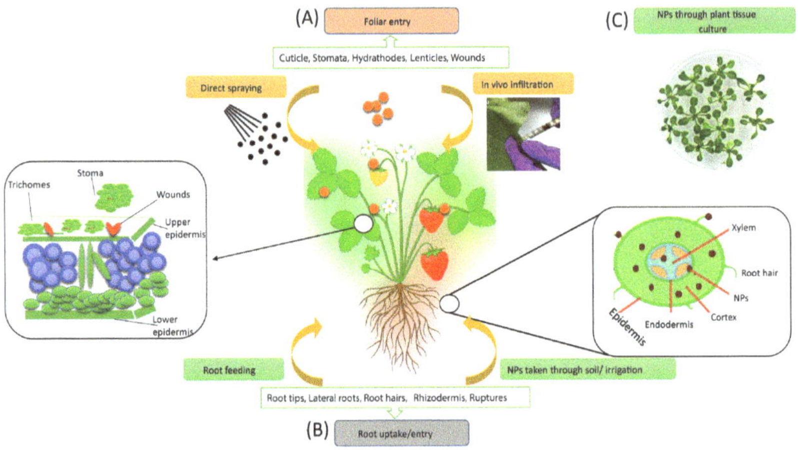

FIGURE 3.2 Diagrams depicting the various processes of NP administration and translocation in plants are shown here. (A) The nanoparticle may be given to the foliar or root systems of plants, depending on their needs. (B and C) They then move along symplastic or apoplastic paths to reach different organs and tissues once they have penetrated the outer layers of the skin.

NPs exhibit markedly different characteristics from bulk materials of the same composition, potentially resulting in enhanced bioavailability and toxicity (Zhao et al., 2011). Furthermore, toxicity of metal NPs involves the production of hydroxyl radicals as a result of visible light producing extracellular reactive oxygen species (ROS) that have the potential to damage cell membranes, leading to changes in membrane permeability and an increased possibility of NP entrance into the cell (Kim et al., 2007). However, others have shown that metal and MONPs promote the development of higher plants (Mittal et al., 2020).

3.1.2.1 Metal oxide nanoparticles have a negative impact on plants

Beadle and coworkers observed that NPs induced formation of new large holes as a route for the internalization of bigger NPs through membranes. According to the researchers, Fe_3O_4, TiO_2, and carbon NPs had a detrimental impact on the germination rate of cucumber plant seeds, root elongation, and germination index of cucumber plants (Mushtaq, 2011). In their study, Lee et al. investigated the effects of four MONPs (Al_2O_3, SiO_2, iron oxide, and zinc oxide) in three concentrations (400, 2000, and 4000 mg/dm^3) on seed germination, root elongation, and leaf number in *Arabidopsis thaliana*. They discovered that phytotoxicity of NPs was reduced in a dose-dependent manner. Chlorophyll levels were significantly decreased in plants treated with 1.0 mg/dm^3 CuO NPs, while plants treated with 0.2 mg/dm^3 soluble Cu had no impact on this assimilation pigment (Shi et al., 2011). The observed results signify that the Cu concentration of fronds exposed to CuO NPs was fourfold higher than fronds treated with a soluble Cu equivalent dosage. Cupric ions produced by Cu NPs had no impact, suggesting that Cu NPs caused evident toxicity. The Cu NP bioaccumulation increased with increasing concentrations of Cu NPs, and particle aggregation in cells was seen using transmission electron microscopy (TEM) and energy-dispersive X-ray spectroscopy (EDS) (Lee et al., 2008). Researchers found that CuO NPs harm DNA in agricultural and grassland plants (Atha et al., 2014). After ZnO NPs treatments, Zn^{2+} accumulation was much more significant in *Cucumis sativus* than following other NPs treatments, indicating that this plant has a robust defense mechanism against ZnO NPs. Increased metal absorption by plants is most likely due to root exudates' capacity to modify the characteristics and behavior of ZnO NPs in the solution. Increased antioxidant enzyme activity was observed in plant root tissues treated with CuO and ZnO NPs. Phytotoxicity of cobalt oxide nanoparticles (Co_3O_4 NPs) has been shown owing to their extensive adsorption into the root system, while the toxicity of zinc oxide nanoparticles (ZnO NPs) has been demonstrated because of their substantial deposition in both cellular and chromosomal compartments. Co_3O_4 NPs may adsorb into onion roots and block water channels, while ZnO NPs may pierce onion roots in large quantities, affecting cellular metabolism and cell division stages. Because acetic acid significantly enhanced the dissolution of La_2O_3 NPs in vitro, the researchers hypothesized that acids produced from plant roots are responsible for NP degradation at the root surface, a critical factor in phytotoxicity of La_2O_3 NPs.

3.1.2.2 Metal oxide nanoparticles benefit plants

Seeds immersed in a solution containing high-strength TiO_2-NPsNPs (0.25−4) had a beneficial consequence on germination and seedling growth, with 2500 mg/dm^3 TiO_2 exerted the highest effect. TiO_2 NPs have the ability to increase spinach growth and nitrogen absorption (Yang et al., 2006). The addition of 0.25% TiO_2 NPs (rutile) to spinach chloroplasts increased the oxygen evolution rate and improved chloroplast coupling. TiO_2 NPs (rutile) were shown to boost the activities of Mg^{2+}-ATPase and chloroplast coupling factor ICF1-ATPase on the thylakoid membranes, and they were also found to protect chloroplasts from ageing when exposed to long-term light (Hong et al., 2005a,b). Significant improvements in *Lemna minor* biomass accumulation were seen after the application of alumina nanoparticles (Al_2O_3 NPs) with an average size of 20 nm and increased root length and the number of fronds per colony, as well as better photosynthetic efficiency. This rise in biomass accumulation was linked with improved photosynthetic light reaction efficiency (Juhel et al., 2011). Aluminum (Al) NPs also helped grow radish and enhance rape root elongation (Ling and Xing, 2007). Another study performed by Lee et al. (2010) confirmed the positive effect of Al_2O_3 NPs (applied at 400, 2000, and 4000 mg/dm^3) on *A. thaliana* due to increased root elongation that the action of inert Al_2O_3 NPs could contribute to improving gas transfer and preventing water loss. Masarovicova and Králová, 2013 performed field tests to determine the impact of iron oxide (Fe_2O_3) NPs sprayed on soybean agronomic characteristics. They found that Fe_2O_3 NPs at a dosage of 0.75 g/dm^3 enhanced leaf and pod dry weight. The highest grain production was obtained with a 0.5 g/dm^3 Fe_2O_3 NPs 48% increase over the control. The discovery of Fe_2O_3 NPs facilitated the transmission of photosynthesis and Fe to peanut leaves (Liu et al., 2005). After 15 days, metal NPs such as silica (Si), palladium (Pd), gold (Au), and Cu NPs significantly affected lettuce plant development, as shown by a rise in the shoot/root ratio (relative to the control) (Shah and Belozerova, 2009).

3.1.3 Techniques for characterization of metal nanoparticles

To monitor the formation and characterization of metal NPs, various experimental methods are used.

3.1.3.1 Uultraviolet–visible

It is possible to identify and characterize NMs using ultraviolet–visible (UV–VIS) spectroscopy, which is considered a good tool because of its optical characteristics, which are sensitive to the size, shape, concentration, aggregation state, and refractive index at the surface of the NPs (Banasiuk et al., 2020). Accumulation is expressed as a rise in the intensity of the red and near-infrared portions of the electromagnetic spectrum due to this process. The reduction of metal ions (Ag^+, Au^+, and Pd^{2+}) to metal NPs following exposure to plant leaf extracts was followed by a change in color and, as a result, a shift in the UV–VIS spectrum, as shown in the study by Petla et al. (2011).

3.1.3.2 Fourier transform infrared spectroscopy

To identify or classify organic substances, a spectrum of molecular vibrations is generated and employed. This spectrum contains information on the chemical bonds and molecular structure of a material. In certain cases, the substance may be recognized by comparing its Fourier transform infrared spectroscopy spectra with those of already cataloged materials, as in the identification of the biomolecules responsible for metal ion reduction and capping of bioreduced metal NPs synthesized using plant extract (Petla et al., 2011; Edalati et al., 2016).

3.1.3.3 Transmission electron microscopy and scanning transmission electron microscopy

TEM and scanning transmission electron microscopy (STEM) are used to image the crystallographic structure on an atomic-scale level (Vijayakumar et al., 2011). Techniques similar to photography with an electron beam are used to detect the crystallographic phase, and crystallographic orientation can also be used to photograph a sample (by diffraction mode experiments). Scanning tunneling microscopes (STMs) allow researchers to examine surface details at the atomic level. Acceptable resolution is believed to be achieved using an STM with 0.1 nm and 0.01 nm lateral resolution depth resolution, respectively (Kaviya et al., 2012; Lin et al., 2010).

3.1.3.4 Atomic force microscopy

Atomic force microscopy (AFM) is a method for surface topography imaging in which images of near-atomic resolution are created, allowing measuring surface roughness at the angstrom scale. In addition to presenting a surface image, AFM can not only show a surface image but also provide quantitative measurements of feature sizes, such as step heights and other dimensions, (Raghunandan et al., 2009; Gopinath et al., 2012).

3.1.3.5 X-ray diffraction

X-ray diffraction (XRD) is a flexible technique used for identifying the crystallographic constituent of elements' arrangement and structure in regular and artificial materials. When a combination of altered segments is present, the final diffractogram is produced by joining different data layers (Stoycheva and Murdjeva, 2006; Aromal et al., 2012).

3.1.3.6 Energy-dispersive X-ray spectroscopy or EDX

EDS is a systematic technique used for measuring the composition or characterization of material at an elemental level under investigation. An X-ray source interacts with a sample, and the results of this analysis are used to create the image. A significant part of its ability to characterize elements is based on the basic assumption of having a unique composition of individual elements, resulting in different peaks in its X-ray spectra (Bankar et al., 2010; Noruzi et al., 2012).

3.1.3.7 X-ray absorption near edge structure and extended X-ray absorption fine structure

They have been effectively used to study the structures and behaviors of nanoscale materials with great success. It is possible to analyze the alteration of an element's absorption coefficient at a certain core level in a chemical environment by using the X-ray absorption near edge structure (XANES) technique. In the fields of chemical bonding, electronic structure, and surface chemistry, it has been employed extensively in the research process. However, extended X-ray absorption fine structure (EXAFS) has been thoroughly used to investigate the atomic confirmation for arrangement and randomness. It is an excellent choice for categorizing the arrangements and identifying the local structures of amorphous materials (Corrias et al., 2000; Novgorodov et al., 1998).

3.1.4 Mechanism of nanoparticles uptake in plants

The absorption and dispersion of oxide NPs in plants have long been a source of dispute and difficulty for scientists. The exact method of NP translocation is uncertain, whether nanocrystals are generated within the plant and then moved via the membrane or whether they are made outside the plant and then transmitted through the membrane (Husen and Siddiqi, 2014). The absorption, storage, and alteration of oxide NPs in plant cellular systems hint at a probable NP penetration mechanism. Understanding how nutrients are absorbed and translocated into plants may create a particular micronutrient and NPs transport route (Zhang et al., 2012). Usually, the plant roots absorb water from the soil before passing through the xylem components. The xylem structures are essential in regulating the flow and velocity of water during transit. As a result, different xylem structures may indicate unique mineral and NP absorption rates (Ma et al., 2010), and the transit of MONPs is believed to occur via the epidermis, cortex, and plant's stele.

3.1.5 Ways of uptake and translocation of nanoparticles by plants

The plants were exposed to NPs in two distinct ways: root and foliar application. Because the leaves of several plant species are coated in a waxy cuticle to prevent water loss and unregulated interchange of critical solutes, the cuticle is regarded as the first natural barrier to the entry of NPs into cells (Pollard et al., 2008). According to previous studies, nonpolar solutes are absorbed and translocated across the cuticle through diffusion and permeation; polar solutes, on the other hand, are absorbed and translocated via polar aqueous pores with sizes ranging from 0.6 to 4.8 nm (Popp et al., 2005; Eichert et al., 2008). However, others suggested that NPs less than 4.8 nm may pass straight through the cuticle through the cuticular channel, while NPs bigger than 5 nm need foliar absorption and accumulation. On the other hand, the method by which these NPs are absorbed remains unknown.

Additionally, leaf cuticles vary according to life stage, environment, and pest and disease damage; thus, in some circumstances, the initial barrier of cuticles against NPs may be bypassed. Moreover, many studies have shown that the cuticular pathway absorbs and stores hydrophilic substances through stomatal pores. Stomata have a diameter of about 250 nm, and owing to their unique physiological role, the stomatal aperture serves as a carrier for NPs penetration (Burkhardt and Eiden, 1994). Therefore only the stomatal route confirms the foliar transport of external NPs/interior tissues to the leaf surface (Eichert et al., 2008). When NPs are present in high quantities, they cause damage to the cellular membrane and wall of plants, enabling them to enter and disturb various plant functions (Mirzajani et al., 2013). NPs can migrate inside a plant's tissue via underground or aboveground components, such as connections between the root system and lesions. To be taken up and delivered, NPs must overcome a range of chemical and physiological obstacles. When NPs interact with plants, the cell wall is the initial obstacle they must overcome.

Plant cellular membranes are made of cellulose and allow for tiny particles while limiting the passage of bigger ones. As a result, smaller NPs can more readily penetrate through this barrier than bigger NPs. The plant cell wall has a size exclusion between the measure of 5 and 20 nm; therefore the penetration potential by the similarly sized NPs increases (Dietz and Herth, 2011). Specific NPs have been discovered to cause the formation of bigger craters on the wall of the cell, enabling NPs within desired size range to pass through (Navarro et al., 2008; Kurepa et al., 2010). NPs may be transported from the cell wall to other plant tissues through endocytosis and symplastic transport (Etxeberria et al., 2006). According to the study, size, magnitude, and zeta potential of NPs all play a role in controlling their transit through the plant and interactions between them. NPs may affect various factors influencing plant growth and development, including delivering nutrients, regulating gene expression, and interfering with a range of oxidative processes, resulting in oxidative bursts (Liu and Lal, 2015; Nair and Chung, 2014). The predominance of NPs in the excess generates ROS and disrupts the oxidative process, while other interacting pathways are yet to be decoded and need significant further study to understand.

Under the influence of NP-based stress conditions, the rate of carbon fixation is decreased, while higher photoinhibition possibly directs the photosystem toward the excessive generation of ROS (Foyer and Noctor, 2005). It has a good track record that once ROS are generated as a consequence of NPs contact, they interact with cells, resulting in biomolecular modification, as well as impairment (Van Breusegem and Dat, 2006). Numerous studies found that plants challenged by NPs have been demonstrated to have greater lipid peroxidation and DNA damage (Atha et al., 2012; Belava et al., 2017; Saha and Dutta Gupta, 2017), indicating that the plant–NPs interaction results in lipid peroxidation. Furthermore, increased generation of ROS may result in apoptosis or necrosis, both of which end in plant cell death (Rastogi and Pospíšil, 2012) (Fig. 3.3).

Despite their destructive character, it is well established that ROS play a role in various cellular functions, including environmental stress tolerance (Sharma et al., 2012). If the antioxidants generated cannot regulate the ROS, it oxidizes the cells' macromolecules, causing them to die (Sharma et al., 2012), eventually killing the plant. According to recent research (O'Brien and Benková, 2013), phytohormones also affect plant stress response signaling. Numerous studies on *Cucurbita pepo*, *Citrullus lanatus*, *Allium porrum*, *A. thaliana*, and *Lactuca sativa* have previously been published that support

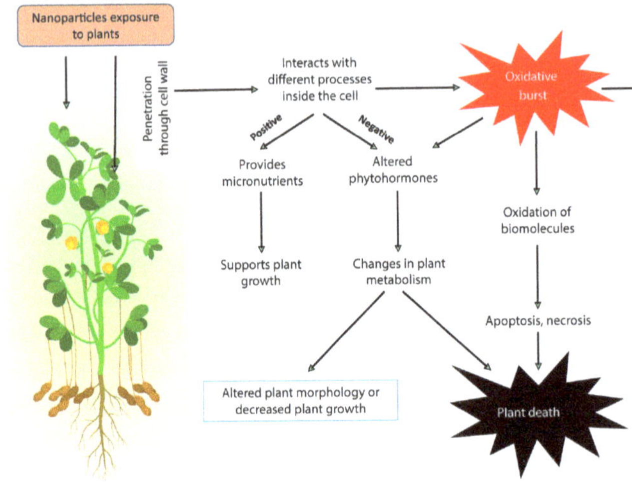

FIGURE 3.3 A general mechanism for nanoparticle-plant interactions has been described.

the foliar route of NP absorption, including their aggregation and accumulation in stomata and subsequent transport to deeper plant tissues (Corredor et al., 2009; Kurepa et al., 2010; Zhao et al., 2016). Even though foliar absorption of NPs has been shown in many studies, there is no evidence to suggest that the phloem translocation route for NPs in plants is involved. In general, the results indicate that NPs may be transported uphill through the xylem and downhill via the phloem, but the mechanisms behind these translocations via the xylem and phloem remain unclear. The NPs are gathered and carried by the roots, which may be released into the rhizosphere and phyllosphere (Lv et al., 2019).

The translocation, stability, as well as accumulation of NPs are strongly influenced by the shape and functioning of plant cells (Mittal et al., 2020). The physiochemical properties like charge, particle size, and hydrophobicity of the NPs all contributed to the absorption translocation and adhering to the cell surface (Kaphle et al., 2018). Capillary pressures, osmotic pressure, and direct transit via root epidermal cells contribute to the permeation of NPs (3–5 nm in diameter) into plant roots. The root epidermal cells contain minute pores and are only partially permeable, making it hard for bigger NPs to get through. After piercing the cellular membrane and walls, NPs are carried via extracellular gaps until they reach the central vascular cylinder through the xylem's unidirectional upward movement. NPs translocate between cells through plasmodesmata and sometimes cluster in the cytoplasm, but they are transported to shoots once they reach the xylem. NPs may sometimes penetrate the leaves through the cuticles, a significant barrier to NPs bigger than 5 nm. NPs bigger than 10 nm enter the stomata and are always transported through the plant's symplastic and apoplastic channels. The size, concentration, and type of NPs are all important determinants in the absorption and translocation of these particles.

3.1.6 A general mechanism of nanoparticles–plant interface pathways

Plant metabolism may be affected by NPs in various ways, including by supplying micronutrients, influencing gene expression, or interfering with various processes, resulting in ROS bursts in the plant's metabolism (Fig. 3.3) (Hossain et al., 2015). As described earlier, if NPs are present in excess, they generate ROS and disrupt the oxidative process, while other kinds of interacting pathways are yet to be decoded and need significant further study to understand. Consequently, the oxidative component of the equation is shown in Fig. 3.3 and discussed below in more detail. The NPs can potentially disrupt the mitochondria and chloroplast electron transport chain (etc), resulting in an oxidative burst and an increase in ROS levels (Jiang et al., 2014; Pakrashi et al., 2014). Henceforth, the wide range of stress factors decreases the rate of carbon fixation, resulting in increased photoinhibition, which may cause the photosystem to overproduce superoxide anion radicals and H_2O_2, potentially causing the photosystem to malfunction (Foyer and Noctor, 2005). It is well established that once ROS are generated as a consequence of NPs contact, they interact with complete cellular machinery and cause damage (Van Breusegem and Dat, 2006). Numerous studies have demonstrated a rise in plant oxidative damage and DNA fragmentation exposed to NPs, suggesting that the plant–NP interaction causes lipid peroxidation (Cvjetko et al., 2017). Increased ROS generation may result in cell death in plants (Rastogi and Pospíšil, 2012). Despite their destructive character, it is well established that ROS play a role in various cellular functions, including environmental stress tolerance (Sharma et al., 2012). Because ROS performs various tasks, the cellular system has evolved a very effective defense system for accurately counterbalance its effect levels.

The antioxidant process produces both enzymatic and nonenzymatic proteins (Yadav et al., 2021). Plants tend to exert an antioxidant response during stress (Rastogi and Pospíšil, 2010; Sharma et al., 2012) and generate a comparable reaction when exposed to NPs, thus (Faisal et al., 2013; Costa and Sharma, 2016) indicating that NPs mediated induction of the antioxidant system. ROS are inextricably linked to hormone signaling and synergize one another's activity (Kwak et al., 2006). In mitochondria and chloroplasts, excessive expansion of reactive species is detrimental to the functioning of these organelles because of the aerobic environment in which the electron transport chain is found (Foyer and Shigeoka, 2011). Numerous studies have shown that NPs affect the quantity and activity of photosynthetic pigments in plants (Perreault et al., 2014; Tripathi et al., 2017). NPs at large doses may negatively impact photosynthesis, causing plant development to be slowed or mortality to occur. Additionally, NPs contribute significantly to plant defense against abiotic stress by inhibiting antioxidative enzymatic activity and scavenging ROS (Rico et al., 2013a,b,c,d).

3.1.7 The overall effect of metal oxide nanoparticles on plants

Plants use antioxidative proteins and various other molecules to protect it from the cytotoxic effects of reactive species (Das and Roychoudhury, 2014). These components mitigate oxidative damage when exposed to MONPs. NMs cytotoxic reactions are influenced by their concentration, particle size, and shape. For example, metal ions and MONPs, such as NPs from ZnO and CeO_2 are bioaccumulated by hydroponically grown soybean plants, influencing the soil and related microbes and the biomass (Ali et al., 2021). As some NPs, such as CeO_2 NPs, inhibit nitrogen fixation and plant development in soybeans, nitrogen-fixing bacteria suffer the most.

3.1.7.1 Iron oxide nanoparticles (Fe_3O_4/Fe_2O_3)

Iron (Fe) is vital for all life, including plants, because of its involvement in several physiological activities, including chlorophyll production and photosynthesis. Inadequate Fe absorption manifests itself in a variety of ways. Growth retardation, interveinal chlorosis, and reduced fitness are all indications. Additionally, sufficient iron levels in food crops are essential for avoiding Fe deficiency that leads to anemia, one of the most prevalent nutritional disorder worldwide. Although soil contains a significant amount of Fe, a large portion is bonded to soil particles. Fe is primarily available as the insoluble Fe^{3+} cation in high-pH and aerobic soils. As a consequence, these soils are prone to Fe^{2+} deficiency. On the other hand, an excess of Fe is toxic to cells. Consequently, plants must overcome an often limited supply of soil Fe by enhancing its mobility while limiting its absorption when it is plentiful. Fe_2O_3 NPs help crops that need Fe as a critical ingredient for growth and development. NMs have the potential to be utilized as agricultural fertilizers due to their unique physiochemical characteristics. Fertilizers using NPs are primarily utilized in hydroponics rather than in the field. According to research, either an excess or a shortage of Fe is detrimental to plants. Such nutritional deficiency may result in oxidative stress, which activates scavenging systems to neutralize ROS. Oxidative stress may not occur in the presence of Fe_2O_3 NPs or EDTA-Fe at the recommended exposure doses owing to antioxidant enzyme activity and MDA content. Plants treated with Fe_2O_3 NPs may produce more ROS, a signaling molecule that stimulates plant development. Plant hormones, which are generated in minute quantities by plants, act as signaling molecules that control numerous growth and development processes. In peanut plants, the phytohormone abscisic acid

(ABA) inhibits growth, causes senescence, and lowers metabolic activity. Coincidentally, following Fe_2O_3 NPs treatment, the ABA concentration of peanut plants decreased. Numerous marine and terrestrial animals and plants are cytotoxic to superparamagnetic iron oxide NPs (SPION), although they generate ROS (Mushtaq, 2011). SPION and other noxious NPs have the potential to bioaccumulate in animals, posing a threat to biodiversity (Siddiqi and Husen, 2017). Fe_3O_4 NPs inhibited cucumber seedling development and root elongation over a broad dose range (500, 2500, and 5000 g/mL) (Zhu et al., 2008a,b). When given at 30, 100, and 500 mg/L concentrations, the 25 nm NPs exhibited minimal absorption and increased root elongation. It inhibits the activity of aquaporins and results in oxidative stress. After exposure to Fe_3O_4 NPs, tomato plants grew quicker and longer, but their green biomass was reduced. Although Fe_3O_4 was not extensively transferred to stems and leaves of tomato, a high quantity of Fe was found in the fruit and roots of the plants. According to several research studies, the bioavailability of Fe_2O_3 NPs in *Bacillus thuringiensis* (Bt) has been observed for genetic engineered cotton, which may represent a danger to the ecosystem (Van Nhan et al., 2016). The quantity of Fe in the number of branches and roots grew as the amount of Fe_2O_3 NPs increased.

3.1.7.2 Cerium oxide (CeO_2) nanoparticles

While catalase and ascorbate activity were detected in maize (*Zea mays* L.) treated with cerium oxide (CeO_2) NPs, no lipid peroxidation or other physiological problems were identified (Zhao et al., 2012a). Rice exposed with CeO_2 NPs (0–500 mg/L) exhibited no increase in lipid peroxidation, while increased ion leakage was observed (Rico et al., 2013a,b,c,d). Corn plants cultivated in the presence of CeO_2 NPs produced H_2O_2 in their parenchyma cells, vascular tissue, and epidermal cells for 3 weeks, according to the results of the research. The activity of the antioxidant enzymes catalase and ascorbate peroxidase was also improved in the shoots. Heat shock protein 70 (HSP70) expression was found to be enhanced in the plants roots exposed to 400 and 800 mg CeO_2/kg, suggesting that the plants were responding to a systemic stress response. Increased enzyme and HSP70 activity result fromthe induced reaction with CeO_2 NPs.

3.1.7.3 Copper oxide (CuO) nanoparticles

The use of CuO NPs on *Elodea densa* (waterweed) has been found to favor the plant and promote photosynthesis when used in low doses. Plants have been proven to produce more ROS when exposed to NP (Shaw et al., 2014). Plants exposed to NPs generate much more oxidative molecules, indicating that the plants have activated the protective mechanism (Shaw et al., 2014). After being exposed to 0.2 mg/L of CuO NPs, a genetic level investigation on *A. thaliana* found that it did not affect the expression of genes involved in oxidative stress responses, sulfur absorption, glutathione production, or proline biosynthesis. Gene expression, on the other hand, is significantly increased at increasing expression (Nair and Chung, 2014). An accumulation of oxidatively changed, mutagenic DNA lesions in various plants was identified, demonstrating DNA damage induced by CuO NPs exposure. Additionally, electron transport, thylakoid counts per grana, photosynthetic rate, photosynthetic pigments, transpiration rate, and stomatal conductance were all inhibited by CuO NPs (Perreault et al., 2014). Additionally, when different types of plants were studied simultaneously, phytohormones were altered in response to CuO NPs (Van Nhan et al., 2016; Lee et al., 2008; Adhikari et al., 2012; Atha et al., 2012; Van Nhan et al., 2016; Wang et al., 2016a,b). The toxic impact on plants was higher with simultaneous treatment with UV light and CuO NPs (Regier et al., 2015). Due to the oxidative nature of CuO NPs, they are further dangerous than Cu. CuO NPs, on the other hand, are advantageous to *E. densa* and promote photosynthesis at low concentrations (Nekrasova et al., 2011). CuO NPs reduced shoot and root development in *Hordeum vulgare* seedlings in specific dose concentration treatment over time. The study indicates that Cu and CuO NPs are hazardous to plants at concentrations more than 0.2 mg/L, and impacts plant growth, physiology, and biochemistry.

3.1.7.4 Impact of titanium dioxide nanoparticles (TiO_2 NPs)

Despite the fact that inadequate data is available to examine the effect of titanium dioxide NPs (TiO_2 NPs) on plants. It seems that TiO_2 NPs may have both beneficial and harmful effects on plants (Fig. 3.2). Most studies using TiO_2 NPs at the foliar level have reported a beneficial impact due to their photocatalytic properties (Gao et al., 2008; Jaberzadeh et al., 2013; Raliya et al., 2015). For example, fresh and dried weights of spinach were enhanced by more than 60% when exposed to TiO_2 NPs (Gao et al., 2008). Photosynthesis increased the amount and activity of Rubisco activase. At the same time, foliar treatment of *Solanum lycopersicum* with TiO_2 NPs increased plant growth, fruit output, and chlorophyll content (Raliya et al., 2015). Further, they enhanced agronomic characteristics of *Triticum aestivum*, reducing water stress.

Additionally, roots were shown to contain the smaller TiO_2 NPs. Once TiO_2 NPs reach a diameter of 140 nm, they are believed to stop collecting in roots. A plants' response to TiO_2 NPs in hydroponic solution may differ from its response in contaminated soil. Physical impacts on the root-water transport system limit leaf development and transpiration in colloidal NP solutions (Asli and Neumann, 2009). TiO_2 NP is toxic to plants at higher concentrations, even in soil systems (Rafique et al., 2014). The phytotoxic reaction was equivalent to that of Ag NP or CuO NP, with a reduction in plant growth and mitotic index and an increase in ROS, antioxidant properties, and genotoxicity. According to the findings, the effect of TiO_2 NP on various plants is dosage-dependent, treatment method-dependent, and size-dependent. Additionally, the research found that the availability of TiO_2 NP to plants is route-dependent (foliar, soil, or water solution). Although it was shown that the photocatalytic activity of titanium is linked with the favorable effect of TiO_2 NP, the mechanism behind this connection is unclear. TiO_2 exhibits solid photocatalytic properties, which explains why it is frequently used in antiseptic and antibacterial formulations for decomposing organic pollutants and pathogens. Reduction of TiO_2 salt into TiO_2 NPs is one of the most processes.

TiO_2 NPs are also used as a nutrient to promote plant growth and development and enhance photosynthetic particles. As a consequence, crop growth and production are enhanced. Increased photosynthetic rate and chlorophyll production due to TiO_2 NPs application increased crop yield and yield-enhancing enzyme activity. Nanotechnology may be able to assist in resolving these problems. TiO_2 has lately been extensively utilized as an ecologically safe and clean photocatalyst due to its optical characteristics, chemical stability, and nontoxic nature. The bulk of TiO_2 NPs research has been conducted at the foliar level and has shown that it is beneficial to plants. During photosynthesis, the volume and activity of Rubisco activase increase. Foliar treatment of *Solanum lycopersicum* with TiO_2 NPs increased plant growth, fruit output, and chlorophyll content (Raliya et al., 2015), resulting in an increase of more than 60% in plant fresh and dry weight in the case of small-sizedNPs, that is 5 nm. When untreated and treated samples were examined, the concentration of Rubisco activase was enhanced by 42%, while its activities more than doubled. TiO_2NPs have been demonstrated to be hazardous to plants in high concentrations, even in soil systems. Ag NPs or CuO NPs have comparable effects on plant development, resulting in a reduction in the growth rate of plants, mitotic index, ROS, antioxidant activity, and genotoxicity.

3.1.8 Mechanism pathways

Numerous variables affect the absorption, transportation, and use of NPs by soil and plants. In soil, NPs may interact with bacteria and other components, either promoting or impeding their uptake. Numerous tissues and other barrier layers must be passed through before reaching the vascular tissues. These NPs adapt to either apoplastic or symplastic routes for upward and downward migration of the particles and a drastic change in the direction of the pathways. Endocytosis, pore formation, plasmodesmata, and carrier proteins may all be utilized to complete the internalization of the NPs into the cell. Because cuticles on the leaf surface impede gas exchange, roots mainly engage in mineral and other component absorption. Because plant species and different type of NP affects the method and effectiveness of absorption but with possibility of differential mechanism. The application technique significantly impacts how effectively the plant absorbs the NPs. As the type and chemical composition of the particle changes, additional surface modification through the coating and other methods enables modification of the NPs, resulting in a shift in crop uptake patterns.

Certain organic materials in the soil, for example, humic acid, may improve particle stability and increase the bioavailability of NPs, although other salts can have an antagonistic effect on these processes. Additionally, the presence of biological agents, such as bacteria, fungus, and other creatures capable of symbiotic interactions with the crop increases plants absorption of NPs. The way NPs move throughout the crop and interacts with plant components is important because it affects the regions where the final effect is perceived. Because NPs that travel only via the xylem will move from the root to the shoot and leaves, it should be applied over the roots. If, on the other hand, the NPs prefer the phloem route, it is more probable that it will accumulate in fruits and grains, requiring treatment through the foliage. Even though the routes are distinct, the cells will interact. After ingesting the NPs, the plant may choose an apoplastic or symplastic path. The extracellular space and xylem vessels are where the apoplastic process takes place.It is subsequently transported through xylem vessels to the central cylinder vessel and the aerial portion of the plant. The apoplastic route is focused on non-photosynthetic tissues and organs. After breaking past the cuticle barrier, the particle must either pass via hydrophilic or lipophilic aqueous pores or through cuticle wax in the event of foliar application.

3.1.9 Effects of nanoparticles and MONPs on plant morphology

Some of the NPs. including TiO_2, CeO_2, Fe_3O_4, and ZnO, are among the most studied. It has been observed that humic acids and pH effects interact with the activity of Fe_3O_4 NPs, raising the pH results in a greater degree of aggregation. TiO_2 has been found to stimulate nitrate reductase production in soybeans (Glycine max), thus increasing the plant's antioxidant system and water absorption. Seeds treated with TiO_2 increased dry weight by 73% and chlorophyll content by 45% (Mingfang et al., 2013). Due to its capacity to absorb water and use fertilizer, the combination of nano-sized TiO_2 and glycine max increases nitrate reductase, thus stimulating the antioxidant system and promoting germination and development. TiO_2 also enhances spinach growth during the anatase phase by boosting nitrate metabolism and nitrogen absorption (Stephen et al., 2012). It aids in the absorption of water and seed germination. TiO_2 is a nontoxic material. Particle aggregation and subsequent sedimentation are the primary causes of this phenomenon (Kurepa et al., 2010).

According to Zhu et al. (2008a,b), chlorophyll synthesis increased at lower Fe_2O_3 concentrations. Coating nanomaterials with tetramethyl ammonium increased chlorophyll production and plant growth, as demonstrated in maize (Zhu et al., 2008a,b). Zinc is a micronutrient that plants, animals, and humans require. Since this micronutrient is present in several products, including paints, cosmetics, and solar cells, it is straightforward for it to enter the environment, where it may cause damage to plants (Gupta et al., 2016). Plant root development was inhibited when seeds were soaked and incubated with a Zn/ZnO solution. According to Ahmed and coworkers, Zn^{2+} is more dangerous than ZnO NPs to *Alium cepa* roots (Ahmed et al., 2017). Cu/Cu_2O obstructs water channels, allowing for the entry of more radicals into the roots. Additionally, it disrupts cell division at different stages (Ghori et al., 2019).Toxicity and bioavailability of Cu NPs were investigated in mung bean and wheat plants. Plant agar was used as the growing media to ensure that all NPs were exposed equally. As a result, plant growth and length were adversely correlated with exposure concentrations to NPs. Higher concentrations of Cu NPs were available. The growth inhibition of a mung bean (*Phaseolus radiata*) seedling exposed to varying doses of Cu NPs was more sensitive than wheat (Lee et al., 2008). NPs, whether natural or synthetic, threaten ecosystems when their concentrations exceed a specific threshold. It is easily transmitted to humans via the food chain. Consequently, further assessment and analysis are necessary before these NPs may be utilized to help plants in the field. Only by increasing our knowledge and understanding will we be able to establish a good balance.

3.1.10 Impact of nanoparticles on the development of seedlings and germination

The earliest and most sensitive stage in the life cycle of higher plants is seed germination. There has been an increase in interest in research on how NPs may either favorably or adversely affect seed germination (Rossi et al., 2019). The most critical phases of a plant's life cycle are seed germination, radicle emergence, and main root expansion. ZnO is the most widely used MONPs because of its many properties. It could be utilized in a vast array of fields, from cosmetics to pharmaceuticals. Numerous research studies have shown that an adequate ZnO concentration in crops and plants enhances seed germination and seedling development. Treatment of ZnO NPs increased the germination efficiency of cucumber, mung bean, and chili pepper in a Petri dish (Awan et al., 2021). All onion, oat, and brassica species were shown to have a concentration range-dependent effect. ZnO NPs were shown to promote primary root extension at low doses but were found to be completely inhibitory at higher concentrations (Nandhini et al., 2019; George et al., 2019; Sun et al., 2020). Smaller ZnO NPs can enter roots and biotransform into ionic forms. These ionic forms may contribute to the imbalance in ROS/RNS. As a consequence, root meristem cells have decreased vitality. The impact of TiO_2 on sprouting has been demonstrated in *Alyssum homolocarpum*, *Nigella sativa*, and *Salvia mirzayanii* (Hatami et al., 2014). Additionally, tests with onions, white mustard (Hatami et al., 2014), and rice (Hatami et al., 2014) were conducted (Hatami et al., 2014). Lower doses (10–100 mg/L application) were shown to improve germination rates, while higher doses proved toxic. According to Cartiglione and group, TiO_2 treatment reduced root development and resulted in chromosomal aberrations in the apical meristem of the root tip of *Vicia narbonensis* seedlings (Castiglione et al., 2011). When plants are fed more significant amounts of Fe, they become dangerous owing to the Fenton reactions' excessive ROS synthesis (Wu et al., 2014). SiO_2 and TiO_2 application to soybean seedlings increased nitrate reductase activity, resulting in increased germination (Coman et al., 2019). Although higher doses of NPs have been proven toxic and harmful, more negligible levels always act as seed priming agents. Plants react to NPs differently depending on their shape and size (Rosa et al., 2013; Mittal et al., 2020).

3.1.11 Future perspective

The field application of NPs requires extensive research. While metal and MONPs in excess are harmful to plants, they may be helpful when present in trace quantities. As a result, increasing the quantity of NPs in the ecosystem may ultimately negatively influence farming. NPs of metal and MO have a significant effect on plants. However, more study is required to elucidate the chemical process through which plant NPs interact.

While metal and MONPs may positively influence agriculture, they should not be overlooked. Although most studies indicate that metal and MONPs cause morphological alterations in plants, there is a notable lack of standardization in NP phytotoxicity testing. Consequently, further study on the impacts of metal and MONPs on plant physiology and molecular biology is necessary. Furthermore, the nanoscience enthusiasm should be channeled toward educating the public about safe NP disposal. Additionally, research on the removal of NPs from agricultural soil and waste is required.

3.2 Conclusion

In today's society, NPs are extensively used and have become an essential component of human existence. However, the environment cannot be ignored owing to the needs of contemporary living. According to research, although metal and MONPs in excess are harmful to plants, they may be helpful when present in trace quantities. As a result, increasing the quantity of NPs may ultimately negatively influence agriculture. This chapter assembled strong data from the existing literature to demonstrate metal and MONPs' influence on plants. However, further study is necessary to understand the molecular process of plant NP interaction thoroughly. A limited studies indicate that metal and MONPs have positive effects in agriculture, but the processes are primarily understood, and the study is still in its infancy.

Consequently, much more study is necessary before NPs may be utilized in the field. The bulk of studies indicates that metal and MONPs alter plant morphology. Additionally, the research demonstrates a marked lack of consistency in the phytotoxic test for NPs. As a result, more analysis is necessary to ascertain the influence of MONPs on plant cellular systems. Nanoscience attracts significant research funding, some of which should be used to educate the public about proper NP disposal. Additionally, research on NPs removal from agricultural waste is necessary.

References

Adhikari, T., Kundu, S., Biswas, A.K., Tarafdar, J.C., Rao, A.S., 2012. Effect of copper oxide nano particle on seed germination of selected crops. J. Agric. Sci. Technol. A 2, 815–823.

Ahmed, B., Dwivedi, S., Abdin, M.Z., Azam, A., Al-Shaeri, M., Khan, M.S., et al., 2017. Mitochondrial and chromosomal damage induced by oxidative stress in Zn2+ ions, ZnO-bulk and ZnO-NPs treated *Allium cepa* roots. Sci. Rep. 7, 1–14.

Ali, S.A., Mittal, D., Kaur, G., 2021. In-situ monitoring of xenobiotics using genetically engineered whole-cell-based microbial biosensors: recent advances and outlook. World J. Microbiol. Biotechnol. 37 (5), 1–24.

Aromal, S.A., Vidhu, V.K., Philip, D., 2012. Green synthesis of well-dispersed gold Nanoparticles using Macrotyloma uniflorum. Spectrochim. Acta Part A: Mol. Biomol. Spectro 85, 99–104.

Asli, S., Neumann, P.M., 2009. Colloidal suspensions of clay or titanium dioxide Nanoparticles can inhibit leaf growth and transpiration via physical effects on root water transport. Plant Cell Env. 32, 577–584.

Atha, D.H., Wang, H., Petersen, E.J., Cleveland, D., Holbrook, R.D., Jaruga, P., et al., 2014. Correction to copper oxide nanoparticle mediated DNA damage in terrestrial plant models. Environ. Sci. Technol. 48, 12473.

Atha, D.H., Wang, H., Petersen, E.J., Cleveland, D., Holbrook, R.D., Jaruga, P., et al., 2012. Copper oxide nanoparticle mediated DNA damage in terrestrial plant models. Environ. Sci. Technol. 46, 1819–1827.

Auffan, M., Rose, J., Bottero, J.Y., Lowry, G.V., Jolivet, J.P., Wiesner, M.R., 2009. Towards a definition of inorganic Nanoparticles from an environmental, health and safety perspective. Nat. Nanotechnol. 4, 634–641.

Avestan, S., Ghasemnezhad, M., Esfahani, M., Byrt, C.S., 2019. Application of nano-silicon dioxide improves salt stress tolerance in strawberry plants. Agronomy 9, 246.

Awan, S., Shahzadi, K., Javad, S., Tariq, A., Ahmad, A., Ilyas, S., 2021. A preliminary study of influence of zinc oxide nanoparticles on growth parameters of *Brassica oleracea* var italic. J. Saudi Soci. Agricul. Sci. 20, 18–24.

Ayyub, P., 1998. Structure-property relations in oxide nanoparticles. Frontiers in Materials Modelling and Design. Springer, Berlin, Heidelberg, pp. 228–235.

Baalousha, M., Cornelis, G., Kuhlbusch, T.A.J., Lynch, I., Nickel, C., Peijnenburg, W.J.G.M., et al., 2016. Modeling nanomaterial fate and uptake in the environment: current knowledge and future trends. Environ. Sci: Nano 3, 323–345.

Banasiuk, R., Krychowiak, M., Swigon, D., Tomaszewicz, W., Michalak, A., Chylewska, A., et al., 2020. Carnivorous plants used for green synthesis of silver nanoparticles with broad-spectrum antimicrobial activity. Arab. J. Chem. 13, 1415–1428.

Bankar, A., Joshi, B., Kumar, A.R., Zinjarde, S., 2010. Banana peel extract mediated novel route for the synthesis of silver nanoparticles. Colloids Surf. A: Physicochem. Eng. Asp. 368, 58–63.

Barrios, A.C., Rico, C.M., Trujillo-Reyes, J., Medina-Velo, I.A., Peralta-Videa, J.R., Gardea-Torresdey, J.L., 2016. Effects of uncoated and citric acid coated cerium oxide nanoparticles, bulk cerium oxide, cerium acetate, and citric acid on tomato plants. Sci. Total. Environ. 563, 956–964.

Belava, V.N., Panyuta, O.O., Yakovleva, G.M., Pysmenna, Y.M., Volkogon, M.V., 2017. The effect of silver and copper nanoparticles on the wheat-*Pseudocercosporella herpotrichoides* pathosystem. Nanoscale Res. Lett. 12, 250.

Brunnert, I., Wick, P., Manserp, S., Grass, R.N., Limbach, L.K., Bruinink, A., et al., 2006. In vitro cytotoxicity of oxide Nanoparticles: comparison to asbestos, silica, and the effect of particle solubility. Env. Sci. Technol. 40, 4374–4381.

Burkhardt, J., Eiden, R., 1994. Thin water films on coniferous needles. Atmos. Environ. 28, 2001–2011.

Castiglione, M.R., Giorgetti, L., Geri, C., Cremonini, R., 2011. The effects of nano-TiO2 on seed germination, development and mitosis of root tip cells of *Vicia narbonensis* L. and *Zea mays* L. J. Nanop. Res. 13, 2443–2449.

Coman, V., Oprea, I., Leopold, L.F., Vodnar, D.C., Coman, C., 2019. Soybean interaction with engineered nanomaterials: a literature review of recent data. Nanomaterials. 9, 1248.

Corredor, E., Testillano, P.S., Coronado, M.J., Gonzalez-Melendi, P., Fernandez-Pacheco, R., et al., 2009. Nanoparticle penetration and transport in living pumpkin plants: in situ subcellular identification. BMC Plant. Biol. 9, 45–53.

Corrias, A., Ennas, G., Mountjoy, G., Paschina, G., 2000. An x-ray absorption spectroscopy study of the Fe K edge in nano-sized maghemite and in Fe2O3–SiO2 nanocomposites. Phys. Chem. 2, 1045–1050.

Costa, M.V.J.D., Sharma, P.K., 2016. Effect of copper oxide nanoparticles on growth, morphology, photosynthesis, and antioxidant response in *Oryza sativa*. Photosynthetica 54, 110–119.

Cvjetko, P., Milošić, A., Domijan, A.M., Vinković Vrček, I., Tolić, S., Peharec Štefani, P., et al., 2017. Toxicity of silver ions and differently coated silver nanoparticles in *Allium cepa* roots. Ecotoxicol. Environ. Saf. 137, 18–28.

Das, K., Roychoudhury, A., 2014. Reactive oxygen species (ROS) and response of antioxidants as ROS-scavengers during environmental stress in plants. Front. Environ. Sci. 2, 53.

De la Torre Roche, R., Servin, A., Hawthorne, J., Xing, B., Newman, L.A., Ma, X., et al., 2015. Terrestrial trophic transfer of bulk and nanoparticle La2O3 does not depend on particle size. Environ. Sci. Technol. 49, 11866–11874.

Dietz, K.-J., Herth, S., 2011. Plant nanotoxicology. Trends Plant Sci. 16, 582–589.

Dogaroglu, Z.G., Köleli, N., 2017. TiO2 and ZnO nanoparticles toxicity in barley (*Hordeum vulgare* L.). Clean. Soil. Air Water 45, 1700096.

Dumont, E., Johnson, A.C., Keller, V.D., Williams, R.J., 2015. Nanosilver and nano zinc-oxide in surface waters—exposure estimation for Europe at high spatial and temporal resolution. Environ. Pollut. 196, 341–349.

Edalati, K., Shakiba, A., Vahdati-Khaki, J., Zebarjad, S.M., 2016. Low-temperature hydrothermal synthesis of ZnO nanorods: Effects of zinc salt concentration, various solvents and alkaline mineralizers. Mater. Res. Bull. 74, 374–379.

Eichert, T., Kurtz, A., Steiner, U., Goldbach, H.E., 2008. Size exclusion limits and lateral heterogeneity of the stomatal foliar uptake pathway for aqueous solutes and water-suspended nanoparticles. Physiol. Plant. 134, 151–160.

Etxeberria, E., Gonzalez, P., Pozueta-Romero, J., Romero, J.P., 2006. Fluid phase endocytic uptake of artificial nano-spheres and fluorescent quantum dots by sycamore cultured cells: evidence for the distribution of solutes to different intracellular compartments. Plant. Signal. Behav. 1, 196–200.

Faisal, M., Saquib, Q., Alatar, A.A., Al-Khedhairy, A.A., Hegazy, A.K., Musarrat, J., 2013. Phytotoxic hazards of NiO-nanoparticles in tomato: a study on mechanism of cell death. J. Hazard. Mater. 250–251, 318–332. Available from: https://doi.org/10.1016/j.jhazmat.2013.01.063.

Feng, Y., Cui, X., He, S., Dong, G., Chen, M., Wang, J., et al., 2013. The role of metal nanoparticles in influencing arbuscular mycorrhizal fungi effects on plant growth. Environ. Sci. Technol. 47, 9496–9504.

Foyer, C.H., Noctor, G., 2005. Redox homeostasis and antioxidant signaling: a metabolic interface between stress perception and physiological responses. Plant Cell 17, 1866–1875.

Foyer, C.H., Shigeoka, S., 2011. Understanding oxidative stress and antioxidant functions to enhance photosynthesis. Plant Physiol. 155, 93–100.

Fraceto, L.F., Grillo, R., de Medeiros, G.A., Scognamiglio, V., Rea, G., Bartolucci, C., 2016. Nanotechnology in agriculture: which innovation potential does it have? Front. Environ. Sci. 4, 20.

Gao, F., Liu, C., Qu, C., Zheng, L., Yang, F., Su, M., et al., 2008. Was improvement of spinach growth by nano-TiO(2) treatment related to the changes of Rubisco activase? Biometals 21, 211–217.

García-López, I.J., Lira-Saldivar, R.H., Zavala-García, F., Olivares-Sáenz, E., Niño-Medina, G., Angélica Ruiz-Torres, N., et al., 2018. Effects of zinc oxide nanoparticles on growth and antioxidant enzymes of Capsicum chinense. Toxicol. Environ. Chem. 100, 560–572.

George, D., Maheswari, P.U., Begum, K.M.S., 2019. Synergic formulation of onion peel quercetin loaded chitosan-cellulose hydrogel with green zinc oxide nanoparticles towards controlled release, biocompatibility, antimicrobial and anticancer activity. Int. J. Biol. Macromol. 132, 784–794.

Ghori, N.H., Ghori, T., Hayat, M.Q., Imadi, S.R., Gul, A., Altay, V., et al., 2019. Heavy metal stress and responses in plants. Int. J. Environ. Sci. Technol. 16, 1807–1828.

Gopinath, V., Mubarak, Ali, D., Priyadarshini, S., Priyadharsshini, N.M., Thajuddin, N., et al., 2012. Colloids Surf. B Biointerfaces 1, 69–74.

Gupta, N., Ram, H., Kumar, B., 2016. Mechanism of zinc absorption in plants: uptake, transport, translocation and accumulation. Rev. Environ. Sci. Biotechnol. 15, 89–109.

Hadioui, M., Merdzan, V., Wilkinson, K.J., 2015. Detection and characterization of ZnO Nanoparticles in surface and waste waters using single particle ICPMS. Environ. Sci. Technol. 49, 6141–6148.

Hatami, M., Ghorbanpour, M., Salehiarjomand, H., 2014. Nano-anatase TiO_2 modulates the germination behavior and seedling vigority of some commercially important medicinal and aromatic plants. J. Biol. Environ. Sci. 8 (22), 52–53.

Hawthorne, J., De la Torre Roche, R., Xing, B., Newman, L.A., Ma, X., Majumdar, S., et al., 2014. Particle-size dependent accumulation and trophic transfer of cerium oxide through a terrestrial food chain. Environ. Sci. Technol. 48, 13102–13109.

Helaly, M.N., El-Hoseiny, H., El-Sheery, N.I., Rastogi, A., Kalaji, H.M., 2017. Regulation and physiological role of silicon in alleviating drought stress of mango. Plant Physiol. Biochem. 118, 31–44.

Hong, F., Yang, F., Liu, C., Gao, Q., Wan, Z., Gu, F., et al., 2005a. Influences of nano-TiO2 on the chloroplast aging of spinach under light. Biol. Trace Elem. Res. 104, 249–260.

Hong, F., Zhou, J., Liu, C., Yang, F., Wu, C., Zheng, L., et al., 2005b. Effect of nano-TiO2 on photochemical reaction of chloroplasts of spinach. Biol. Trace Elem. Res. 105, 269–279.

Hossain, Z., Mustafa, G., Komatsu, S., 2015. Plant responses to nanoparticle stress. Int. J. Mol. Sci. 16, 26644–26653.

Hotze, E.M., Phenrat, T., Lowry, G.V., 2010. Nanoparticle aggregation: challenges to understanding transport and reactivity in the environment. J. Environ. Qual. 39, 1909–1924.

Husen, A., Siddiqi, K.S., 2014. Carbon and fullerene nanomaterials in plant system. J. Nanobiotechnol 12 (1), 1–10.

Iannone, M.F., Groppa, M.D., de Sousa, M.E., van Raap, M.B.F., Benavides, M.P., 2016. Impact of magnetite iron oxide nanoparticles on wheat (*Triticum aestivum* L.) development: evaluation of oxidative damage. Environ. Exp. Bot. 131, 77–88. Available from: https://doi.org/10.1016/j.envexpbot.2016.07.004.

Jaberzadeh, A., Moaveni, P., Moghadam, H.R.T., Zahedi, H., 2013. Influence of bulk and nanoparticles titanium foliar application on some agronomic traits, seed gluten and starch contents of wheat subjected to water deficit stress. Not.Bot. Horti Agrobo 41, 201–207.

Jiang, H.S., Qiu, X.N., Li, G.B., Li, W., Yin, L.Y., 2014. Silver nanoparticles induced accumulation of reactive oxygen species and alteration of antioxidant systems in the aquatic plant *Spirodela polyrhiza*. Environ. Toxicol. Chem. 33, 1398–1405.

Judy, J.D., Unrine, J.M., Rao, W., Bertsch, P.M., 2012. Bioaccumulation of gold nanomaterials by *Manduca sexta* through dietary uptake of surface contaminated plant tissue. Environ. Sci. Technol. 46, 12672–12678.

Juhel, G., Batisse, E., Hugues, Q., Daly, D., van Pelt, F.N., O'Halloran, J., et al., 2011. Alumina nanoparticles enhance growth of lemna minor. Aquat. Toxicol. 105, 328–336.

Kaphle, A., Navya, P.N., Umapathi, A., Daima, H.K., 2018. Nanomaterials for agriculture, food and environment: applications, toxicity and regulation. Environ. Chem. Lett. 16, 43–58.

Kaviya, S., Santhanalakshmi, J., Viswanathan, B., 2012. Biosynthesis of silver nano-flakes by *Crossandra infundibuliformis* leaf extract. Mat. Lett. 67, 64–66.

Kim, J.S., Kuk, E., Yu, K.N., Kim, J.H., Park, S.J., Lee, H.J., et al., 2007. Nanomed.: Nanotechnol. Biol. Med. 3, 95–101.

Kurepa, J., Paunesku, T., Vogt, S., Arora, H., Rabatic, B.M., Lu, J., et al., 2010. Uptake and distribution of ultrasmall anatase TiO2 Alizarin red S nanoconjugates in *Arabidopsis thaliana*. Nano Lett. 10, 2296–2302.

Kwak, J.M., Nguyen, V., Schroeder, J.I., 2006. The role of reactive oxygen species in hormonal responses. Plant Physiol. 141, 323–329.

Landa, P., Cyrusova, T., Jerabkova, J., Drabek, O., Vanek, T., Podlipna, R., 2016. Effect of metal oxides on plant germination: phytotoxicity of Nanoparticles, bulk materials, and metal ions. Water Air Soil. 227, 448.

Lee, C.W., Mahendra, S., Zodrow, K., Li, D., Tsai, Y.C., Braam, J., et al., 2010. Developmental phytotoxicity of metal oxide Nanoparticles to *Arabidopsis thaliana*. Environ. Toxicol. Chem.: An. Int. J. 29, 669–675.

Lee, W.Á., An, Y.J., Yoon, H., Kweon, H.S., 2008. Toxicity and bioavailability of copper nanoparticles to the terrestrial plants mung bean (*Phaseolus radiatus*) and wheat (*Triticum aestivum*): plant agar test for water-insoluble nanoparticles woo-mi. Env. Toxicol. Chem. 27, 1915–1921.

Levard, C., Hotze, E.M., Lowry, G.V., Brown Jr., G.E., 2012. Environmental transformations of silver nanoparticles: impact on stability and toxicity. Environ. Sci. Technol. 46, 6900–6914.

Lin, L., Wang, W., Huang, J., Li, Q., Sun, D., Yang, X., et al., 2010. Nature factory of silver nanowires: plant-mediated synthesis using broth of Cassia fistula leaf. Chem. Eng. J. 162 (2), 852–858.

Ling, D.H., Xing, B.S., 2007. Phytotoxicity of nanoparticles: inhibition of seed germination and root elongation. Environ. Pollut. 150, 243–250.

Liu, R., Lal, R., 2015. Potentials of engineered nanoparticles as fertilizers for increasing agronomic productions. Sci. Total. Environ. 514, 131–139.

Liu, X.M., Zhang, F.D., Zhang, S.Q., He, X.S., Fang, R., Feng, Z., et al., 2005. Plant. Nutr. Fert. Sci. 11, 14–18.

Lizzi, D., Mattiello, A., Piani, B., Fellet, G., Adamiano, A., Marchiol, L., 2020. Germination and early development of three spontaneous plant species exposed to nanoceria (nCeO2) with different concentrations and particle sizes. Nanomaterials 10 (12), 2534.

López-Moreno, M.L., de la Rosa, G., Hernández-Viezcas, J.A., Castillo-Michel, H., Botez, C.E., Peralta-Videa, J.R., et al., 2010. Evidence of the differential biotransformation and genotoxicity of ZnO and CeO_2 nanoparticles on soybean (*Glycine max*) plants. Environ. Sci. Technol. doi:10.1021/es903891g.

Lv, J., Christie, P., Zhang, S., 2019. Uptake, translocation, and transformation of metal-based nanoparticles in plants: recent advances and methodological challenges. Environ. Sci. Nano. 6, 41–59.

Ma, X., Geisler-Lee, J., Deng, Y., Kolmakov, A., 2010. Interactions between engineered nanoparticles (enanoparticles) and plants: phytotoxicity, uptake and accumulation. Sci. Total. Environ. 408, 3053–3061.

Mahdi, K.N.M., Petersc, R.J.B., Klumpp, E., Bohme, S., Ploeg, M.V.D., Ritsema, C., et al., 2017. Silver nanoparticles in soil: aqueous extraction combined with single-particle ICP-MS for detection and characterization. Environ. Nanotechnol. Monit. Manage 7, 24–33.

Masarovicova, E., Králová, K., 2013. Metal nanoparticles and plants. Ecol. Chem. Engin. 20, 9.

Mingfang, Q., Yufeng, L., Tianlai, L., 2013. Nano-TiO2 improve the photosynthesis of tomato leaves under mild heat stress, biological trace element research. Biolo. Trace Element. Res. 1 (323–328), 2013.

Mirzajani, F., Askari, H., Hamzelou, S., Farzaneh, M., Ghassempour, A., 2013. Effect of silver nanoparticles on *Oryza sativa* L. and its rhizosphere bacteria. Ecotoxicol. Environ. Saf. 88, 48–54.

Mittal, D., Kaur, G., Singh, P., Yadav, K., Ali, S.A., 2020. Nanoparticle-based sustainable agriculture and food science: recent advances and future outlook. Front. Nanotechnol. 2, 10.

Mushtaq, Y.K., 2011. Effect of nanoscale Fe3O4, TiO_2 and carbon particles on cucumber seed germination. J. Environ. Sci. Health, Part. A 46, 1732–1735.

Nair, P.M.G., Kim, S.H., Chung, I.M., 2014. Copper oxide nanoparticle toxicity in mung bean (*Vigna radiata* L.) seedlings: physiological and molecular level responses of in vitro grown plants. Acta Physiol. Plant. 36, 2947–2958.

Nair, P.M., Chung, I.M., 2014. Impact of copper oxide nanoparticles exposure on *Arabidopsis thaliana* growth, root system development, root lignificaion, and molecular level changes. Environ. Sci. Pollut. Res. Int. 21, 12709–12022.

Nandhini, M., Rajini, S.B., Udayashankar, A.C., Niranjana, S.R., Lund, O.S., Shetty, H.S., et al., 2019. Biofabricated zinc oxide nanoparticles as an eco-friendly alternative for growth promotion and management of downy mildew of pearl millet. Crop. Prot. 121, 103–112.

Navarro, E., Piccapietra, F., Wagner, B., Marconi, F., Kaegi, R., Odzak, N., et al., 2008. Toxicity of silver nanoparticles to *Chlamydomonas reinhardtii*. Environ. Sci. Technol. 42, 8959–8964.

Navratilova, J., Praetorius, A., Gondikas, A., Fabienke, W., von-der Kammer, F., Hofmann, T., 2015. Detection of engineered copper Nanoparticles in soil using single particle ICP-MS. Int. J. Environ. Res. Public. Health 12, 15756–15768.

Nekrasova, G.F., Ushakova, O.S., Ermakov, A.E., Uimin, M.A., 2011. Effects of copper (II) ions and copper oxide nanoparticles on *Elodea densa* planch. Russ. J. Ecol. 42, 458–463.

Nhan, L.V., Ma, C., Rui, Y., Liu, S., Li, X., Xing, B., et al., 2015. Phytotoxic mechanism of nanoparticles: destruction of chloroplasts and vascular bundles and alteration of nutrient absorption. Sci. Rep. 5, 11618.

Noruzi, M., Zare, D., Davoodi, D., 2012. A rapid biosynthesis route for the preparation of gold nanoparticles by aqueous extract of cypress leaves at room temperature. Spectrochim. Acta Part. A: Mol. Biomol. Spectrosc. 94, 84–88.

Novgorodov, B.N., Kochubey, D.I., Vargaftik, M.N., 1998. EXAFS study of ultradisperse noble metal Nanoparticles structure phase conversion. Nucl. Instrum. Methods Phys. Rese. Sec. A: Accel., Spectrom., Detect. Assoc. Equip. 405 (2–3), 351–354.

O'Brien, J.A., Benková, E., 2013. Cytokinin cross-talking during biotic and abiotic stress responses. Front. Plant. Sci. 4, 451.

Pakrashi, S., Jain, N., Dalai, S., Jayakumar, J., Chandrasekaran, P.T., Raichur, A.M., et al., 2014. In vivo genotoxicity assessment of titanium dioxide nanoparticles by *Allium cepa* root tip assay at high exposure concentrations. PLoS One 9, e87789.

Perreault, F., Samadani, M., Dewez, D., 2014. Effect of soluble copper released from copper oxide nanoparticles solubilisation on growth and photosynthetic processes of *Lemnagibba.* L. Nanotoxicology 8, 374–382.

Petla, R.K., Vivekanandhan, S., Misra, M., Mohanty, A.K., Satyanarayana, N., 2011. Soybean (*Glycine max*) leaf extract based green synthesis of palladium nanoparticles. J. Biomater. Nanobiotechnol. doi: 10.4236/jbnb.2012.31003.

Pollard, M., Beisson, F., Li, Y., Ohlrogge, J.B., 2008. Building lipid barriers: biosynthesis of cutin and suberin. Trends Plant Sci. 13, 236–246.

Popp, C., Burghardt, M., Friedmann, A., Riederer, M., 2005. Characterization of hydrophilic and lipophilic pathways of *Hedera helix* L. cuticular membranes: permeation of water and uncharged organic compounds. J. Exp. Bot. 56, 2797–2806.

Prasad, T.N.V.K.V., Sudhakar, P., Sreenivasulu, Y., Latha, P., Munaswamy, V., Raja Reddy, K., et al., 2012. Effect of nanoscale zinc oxide particles on the germination, growth and yield of peanut. J. Plant. Nutr. 35, 906–927. Available from: https://doi.org/10.1080/01904167.2012.663443.

Radad, K., Al-Shraim, M., Moldzio, R., Rausch, W.D., 2012. Recent advances in benefits and hazards of engineered Nanoparticles. Environ. Toxicol. Pharmacol. 34, 661–672.

Rafique, R., Arshad, M., Khokhar, M.F., Qazi, I.A., Hamza, A., Virk, N., 2014. Growth response of wheat to titania nanoparticles application. NUST J. Engin. Sci. 7, 42–46. Available online at. Available from: http://journals.nust.edu.pk/index.php/njes/article/view/133.

Raghunandan, D., Basavaraja, S., Mahesh, B., Balaji, S., Manjunath, S.Y., Venkataraman, A., 2009. Biosynthesis of stable polyshaped gold nanoparticles from microwave-exposed aqueous extracellular anti-malignant guava (*Psidium guajava*) leaf extract. Nanobiotechnology 5, 34–41.

Raliya, R., Nair, R., Chavalmane, S., Wang, W.N., Biswas, P., 2015. Mechanistic evaluation of translocation and physiological impact of titanium dioxide and zinc oxide nanoparticles on the tomato (*Solanum lycopersicum* L.) plant. Metallomics 7, 1584–1594.

Rastogi, A., Pospíšil, P., 2010. Effect of exogenous hydrogen peroxide on biophoton emission from radish root cells. Plant Physiol. Biochem. 48, 117–123.

Rastogi, A., Pospíšil, P., 2012. Production of hydrogen peroxide and hydroxyl radical in potato tuber during the necrotrophic phase of hemibiotrophic pathogen *Phytophthora infestans* infection. J. Photochem. Photobiol. B Biol. 117, 202–206.

Regier, N., Cosio, C., von-Moos, N., Slaveykova, V.I., 2015. Effects of copper-oxide nanoparticles, dissolved copper and ultraviolet radiation on copper bioaccumulation, photosynthesis and oxidative stress in the aquatic macrophyte *Elodea nuttallii*. Chemosphere 128, 56–61.

Rico, C.M., Majumdar, S., Duarte-Gardea, M., Peralta-Videa, J.R., Gardea- Torresdey, J.L., 2011. Interaction of nanoparticles with edible plants and their possible implications in the food chain. J. Agric. Food Chem. 59, 3485–3498.

Rico, C.M., Hong, J., Morales, M.I., Zhao, L., Barrios, A.C., Zhang, J.Y., et al., 2013a. Effect of cerium oxide nanoparticles on rice: a study involving the antioxidant defense system and in vivo fluorescence imaging. Environ. Sci. Technol. 47, 5635–5642.

Rico, C.M., Morales, M.I., Barrios, A.C., McCreary, R., Hong, J., Lee, W.Y., et al., 2013b. Effect of cerium oxide nanoparticles on the quality of rice (*Oryza sativa* L.) grains. J. Agric. Food Chem. 61, 11278–11285. Available from: https://doi.org/10.1021/jf404046v.

Rico, C.M., Hong, J., Morales, M.I., Zhao, L., Barrios, A.C., Zhang, J.Y., et al., 2013c. Effect of cerium oxide nanoparticles on rice: a study involving the antioxidant defense system and in vivo fluorescence imaging. Environ. Sci. & Technol. 47 (11), 5635–5642.

Rico, C.M., Morales, M.I., McCreary, R., Castillo-Michel, H., Barrios, A.C., Hong, J., et al., 2013d. Cerium oxide Nanoparticles modify the antioxidative stress enzyme activities and macromolecule composition in rice seedlings. Environ. Sci. Technol. 47, 14110–14118.

Rosei, F., 2004. Nanostructured surfaces: challenges and frontiers in nanotechnology. J. Physics: Condens. Matter 16, S1373.

Rossi, L., Fedenia, L.N., Sharifan, H., Ma, X., Lombardini, L., 2019. Effects of foliar application of zinc sulfate and zinc nanoparticles in coffee (*Coffea arabica* L.) plants. Plant Physiol. Biochem. 135, 160–166.

Saha, N., Dutta Gupta, S., 2017. Low-dose toxicity of biogenic silver nanoparticles fabricated by *Swertia chirata* on root tips and flower buds of *Allium cepa*. J. Hazard. Mater. 330, 18–28.

Shah, V., Belozerova, I., 2009. Influence of metal Nanoparticles on the soil microbial community and germination of lettuce seeds. Water. Air Soil. Pollu. 197, 143–148.

Sharma, P., Jha, A.B., Dubey, R.S., Pessarakli, M., 2012. Reactive oxygen species, oxidative damage, and antioxidative defense mechanism in plants under stressful conditions. J. Bot. 2012, 217037.

Shaw, A.K., Ghosh, S., Kalaji, H.M., Bosa, K., Brestic, M., Zivcak, M., et al., 2014. Nano-CuO stress induced modulation of antioxidative defense and photosynthetic performance of Syrian barley (*Hordeumvulgare* L.). Environ. Experim. Botany. 102, 37–47.

Shi, J., Abid, A.D., Kennedy, I.M., Hristova, K.R., Silk, W.K., 2011. To duckweeds (*Landoltia punctata*), nanoparticulate copper oxide is more inhibitory than the soluble copper in the bulk solution. Environ. Pollu 159, 1277–1282.

Siddiqi, K.S., Husen, A., 2017. Plant response to engineered metal oxide nanoparticles. Nanoscale. Res. Lett. 12, 1–18.

Singh, P., Ali, S.A., 2021. Impact of CRISPR-Cas9-based genome engineering in farm animals. Veterinary Sci. 8 (7), 122.

Song, G., Hou, W., Gao, Y., Wang, Y., Lin, L., Zhang, Z., et al., 2016. Effects of CuO nanoparticles on Lemna minor. Bot. Stud. 57, 3. Available from: https://doi.org/10.1186/s40529-016-0118-x.

Stephen, G.W., Li, H., Jennifer, H., Da-Ren, C., In-Chul, K., Yinjie, J.T., 2012. Phytotoxicity of metal oxide nanoparticles is related to both dissolved metals ions and adsorption of particles on seed surfaces. J. Pet. Environ. Biotechnol. 4.

Stoycheva, M.V., Murdjeva, M.A., 2006. Antimicrobial therapy of salmonelloses—current state and perspectives. Folia Med 48, 5–10.

Sun, L., Wang, Y., Wang, R., Wang, R., Zhang, P., Ju, Q., et al., 2020. Physiological, transcriptomic, and metabolomic analyses reveal zinc oxide nanoparticles modulate plant growth in tomato. Environ. Sci. Nano. 7, 3587–3604.

Tangaa, S.R., Selck, H., Winther-Nielsen, N., Khan, F.R., 2016. Trophic transfer of metal-based nanoparticles in aquatic environments: a review and recommendations for future research focus. Environ. Sci. Nano. 3, 966–981.

Teske, S.S., Detweiler, C.S., 2015. The biomechanisms of metal and metaloxide nanoparticlesnanoparticles' interactions with cells. Int. J. Environ. Res. Public. Health 12, 1112–1134.

Tripathi, D.K., Singh, S., Singh, S., Srivastava, P.K., Singh, V.P., Singh, S., et al., 2017. Nitric oxide alleviates silver nanoparticles (AgNanoparticles)-induced phytotoxicity in Pisumsativumseedlings. Plant Physiol. Biochem. 110, 167–177.

Tripathi, D.K., Singh, V.P., Prasad, S.M., Chauhan, D.K., Dubey, N.K., 2015. Silicon nanoparticles (SiNp) alleviate chromium (VI) phytotoxicity in *Pisum sativum* (L.) seedlings. Plant Physiol. Biochem. 96, 189–198.

Tubana, B.S., Babu, T., Datnoff, L.E., 2016. A review of silicon in soils and plants and its role in US agriculture: history and future perspectives. Soil. Sci. 181, 393–411.

Unrine, J.M., Shoults-Wilson, W.A., Zhurbich, O., Bertsch, P.M., Tsyusko, O.V., 2012. Trophic transfer of Au nanoparticles from soil along a simulated terrestrial food chain. Environ. Sci. Technol. 46, 9753–9760.

Vajpayee, P., Khatoon, I., Patel, C.B., et al., 2011. Adverse effects of chromium oxide nanoparticles on seed germination and growth in *Triticum aestivum* L. J. Biomed. Nanotechnol. 7 (1), 205–206.

Van Breusegem, F., Dat, J.F., 2006. Reactive oxygen species in plant cell death. Plant Physiol. 141, 384–390.

Van Nhan, L., Ma, C., Rui, Y., Cao, W., Deng, Y., Liu, L., et al., 2016. The effects of Fe2O3 nanoparticles on physiology and insecticide activity in non-transgenic and Bt-transgenic cotton. Frontiers. Plant. Sci. 6, 1263.

Večeřová, K., Večeřa, Z., Docekal, B., Oravec, M., Pompeiano, A., Tříska, J., et al., 2016. Changes of primary and secondary metabolites in barley plants exposed to CdO nanoparticles. Environ. Pollut. 218, 207–218. Available from: https://doi.org/10.1016/j.envpol.2016.05.013.

Vijayakumar, R., Devi, V., Adavallan, K., Saranya, D., 2011. Green synthesis and characterization of gold nanoparticles using extract of anti-tumor potent *Crocus sativus*. Phys. E: Low-Dimen. Sys. Nanost 44, 665–671.

Wang, P., Lombi, E., Zhao, F.J., Kopittke, P.M., 2016a. Nanotechnology: a new opportunity in plant sciences. Trends Plant Sci. 21, 699–712.

Wang, Z., Xu, L., Zhao, J., Wang, X., White, J.C., Xing, B., 2016b. CuO nanoparticle interaction with *Arabidopsis thaliana*: toxicity, parent progeny transfer, and gene expression. Environ. Sci. Technol. 50, 6008–6016.

Wu, H., Yin, J.J., Wamer, W.G., Zeng, M., Lo, Y.M., 2014. Reactive oxygen species-related activities of nano-iron metal and nano-iron oxides. J. Food. Drug. Anal. 22, 86–94.

Yadav, K., Ali, S.A., Mohanty, A.K., Muthusamy, E., Subaharan, K., Kaul, G., 2021. MSN, MWCNT and ZnO nanoparticle-induced CHO-K1 cell polarisation is linked to cytoskeleton ablation. J. Nanobiotechnol. 19 (1), 1–24.

Yadav, S.K., 2010. Heavy metals toxicity in plants: an overview on the role of glutathione and phytochelatins in heavy metal stress tolerance of plants. S. Afr. J. Bot. 76, 167–179.

Yang, F., Hong, F., You, W., Liu, C., Gao, F., Wu, C., et al., 2006. Influence of nano-anatase TiO2 on the nitrogen metabolism of growing spinach. Biol. Trace. Elem. Res. 110, 179–190.

Younes, N.A., Hassan, H.S., Elkady, M.F., Hamed, A.M., Dawood, M.F., 2020. Impact of synthesized metal oxide nanomaterials on seedlings production of three Solanaceae crops. Heliyon 6 (1), e03188.

Zhang, P., Ma, Y.H., Zhang, Z.Y., He, X., Guo, Z., Tai, R.Z., et al., 2012. Env. Sci. Technol. 46, 1834–1841.

Zhao, L., Ortiz, C., Adeleye, A.S., Hu, Q., Zhou, H., Huang, Y., et al., 2016. Metabolomics to detect response of lettuce (Lactuca sativa) to Cu(OH)2 nanopesticides: Oxidative stress response and detoxification mechanisms. Environ. Sci. Technol. 50, 9697–9707.

Zhao, L., Peralta-Videa, J.R., Ren, M., Varela-Ramirez, A., Li, C., Hernandez-Viezcas, J.A., et al., 2012a. Transport of Zn in a sandy loam soil treated with ZnO Nanoparticles and uptake by corn plants: electron microprobe and confocal microscopy studies. Chem. Engin. J. 184, 1–8.

Zhao, F., Zhao, Y., Liu, Y., Chang, X., Chen, C., Zhao, Y., 2011. Cellular uptake, intracellular trafficking, and cytotoxicity of nanomaterials. Small 7, 1322–1337.

Zhu, H., Han, J., Xiao, J.Q., Jin, Y., 2008a. Uptake, translocation, and accumulation of manufactured iron oxide nanoparticles by pumpkin plants. J. Env. Monit. 10, 713–717.

Zhu, H., Han, J., Xiao, J.Q., Jin, Y., 2008b. Uptake, translocation, and accumulation of manufactured iron oxide nanoparticles by pumpkin plants. J. Environ. Monit. 6, 713–717.

Zuverza-Mena, N., Martínez-Fernández, D., Du, W., Hernandez-Viezcas, J.A., Bonilla-Bird, N., López-Moreno, M.L., et al., 2017. Exposure of engineered nanomaterials to plants: insights into the physiological and biochemical responses-a review. Plant Physiol. Biochem. 110, 236–264.

Chapter 4

Nano metal oxide-based delivery system for weed control

Rajiv Periakaruppan[1,2], P. Vanathi[3], M. Helan Soundra Rani[3] and Salwan Ali Abed[4]

[1]*College of Horticulture, Nanjing Agricultural University, Nanjing, Jiangsu Province, P.R. China,* [2]*Department of Biotechnology, PSG College of Arts & Science, Coimbatore, Tamil Nadu, India,* [3]*Department of Biotechnology, Sri Ramakrishna College of Arts & Science, Coimbatore, Tamil Nadu, India,* [4]*Department of Environmental Science, College of Science, University of Al-Qadisiyah, Al Diwaniyah, Iraq*

Chapter outline

4.1 Introduction

Weeds are harmful nonnative species that produce a destructive exterior from their unique habitat (Heap and Duke, 2018), which brings about 10%–100% depletion in crop yields. Systematic control methods, as well as mechanical control such as tillage, hand pick off, biotic control, and chemical administration, have been used in agricultural environments (Catton et al., 2016). Each of the conservative approaches of weed control has its advantages and disadvantages, which make some techniques prominent with farmers and others comparatively less; chemical weed control has been the most preferred

Nanometal Oxides in Horticulture and Agronomy. DOI: https://doi.org/10.1016/B978-0-323-91809-1.00017-2

system of weed control over several decades. Herbicides play an important role in ensuring the safety and productivity of food at minimal cost due to decrease in labor charges, as well as land preparation cost (Melander, 1995). The chemical constituents of herbicides prevent the development of plants or eradicate them completely. New-fangled herbicides are artificial substances that change the endogenous hormone content in plants, which can impede the progression of impartial crops. Even though weeds are extinguished by herbicides, overuse of herbicides inhibits plant growth, along with being harmful for human beings. Besides, it has the maximum chance of exterminating the useful bacteria, particularly N_2-fixing bacteria, Rhizobium-associated species also lead to a decline in *Aspergillus* species responsible for plant development. Therefore, detoxification of that herbicide or finding a more capable method for the wide-range extinction of synthetic-based herbicides for sustainable progress in agricultural has become imperative (Owen, 2016).

In recent years, the application of nanoparticles (NPs) in ensuring agriculture development has been classified as follows: in the form of nanofertilizers for enhancing the agricultural efficiency and in the form of nanoherbicides for eliminating the weeds obstructing the evolution of crops. Nanotechnology has the capability for the active transportation of biological or synthetic herbicides prepared through nanomaterials-based agrochemical formulations or nanosized particles (Owen, 2016). Phytonanotechnology is an innovative technology used to improve the agronomic application that is protection of crop yield. Moreover, it offers simple, environmentally friendly, stable and cost-effective methods with biocompatibility and scalability through the use of polar solvents such as water as a reducing agent. Diverse forms of green-based NPs are used for better exclusion of weeds. We can exploit the special characteristics of these methods like biodegradability, permeability, solubility, crystallinity, and thermal stability in order to have a greater affinity to the target by affording a higher surface area. The widely available herbicides not only destroy sectors of weeds above the ground, but also control and prevent the growth of the underground sections such as rhizomes and tubers, which facilitate subsequent generation of weeds (Dashora and Sharma, 2018). Plant-based herbicides encapsulated in nanosized materials aids in effective spraying by reducing the losses of spraying materials, which are mixed with the particles of soil that harm weeds or weed seeds. Generally, herbicides are spread by a foliar spray method, which does not destroy completely perennial weeds such as *Cynodon dactylon*, *Cyperus* sp., and *Solanum elaeagnifolium*, but nano-based herbicides disrupt the structure and function of the chloroplast, along with impeding lipid biosynthesis, interfering with cell-division by disrupting the mitotic system, or hindering the growth of plants (Wakabayashi and Böger, 2004). Thus, the application of encapsulated herbicide compounds through root absorption is a simple and more effective method compared to foliar absorption because roots do not possess cuticles that are present in leaves.

Specific particles of herbicides encapsulated within nanocarriers target receptors in the weed roots, thus penetrating the weed roots and disrupting the glycolysis process occurring in roots, leading to the death of specific weeds (Hess, 2018). Furthermore, herbicides synthesized using nanocarriers are beneficial in decreasing the conservational influence triggered through chemical herbicides. Herbicide studies help decrease the nondirectional deadliness of the herbicides as compared to insecticide and fungicide nanocarrier studies. Progression of nanoherbicides similarly encompasses a wide assortment of nanomaterials (Han et al., 2010). For instance, clay films (mont morillonite) layered by means of pH reliant on polymer, core hollow shell (manganese carbonate) (Kanimozhi and Chinnamuthu, 2012), tubular alloy site and play kaolinite at a nanoscale level (Tan et al., 2015), amino-activated iron oxide NPs with magnetic power (Viirlaid et al., 2009), and alteration of rice husks into nanosized particles have been used (Chidambaram, 2016). Besides, nanoherbicides diminish toxicity with 80% viability against fibroblast cells of mouse, as compared to conventional herbicides of 64% viability. Alteration of rice husk into nanosized particles, when combined with 2,4-dichlorophenoxy acetic acid, has greater weed control action than the acid alone when applied on targeted plants (*Brassica* sp.) with less side-effects like the decrease in discharge in soil once the herbicide is attached to the rice husks in nanosized particles (Chidambaram, 2016).

Nanomaterials are relatively new to the atmosphere and mostly occur in the form of minerals and metabolites from bacteria. It has been utilized since the prehistoric period as a stain for metals, but the methodical manufacturing of nanomaterials for several applications has started happening over the past few years (Maurer-Jones et al., 2013). NPs have distinct properties with different types of constituents and are manufactured in many dimensions and methods with a collection of artificial particles (Radad et al., 2012). Metal-oxide nanomaterials show various physiological and chemical properties that are more diverse compared to naturally occurring substances in numerous aspects, mainly exterior, thermal, visual, and electrical. Metal-oxide NPs are developed through inclusion of reducing and oxidizing agents in their synthesis process (Sanchez-Dominguez et al., 2009). Metal-oxide NPs of TiO_2, silver, ZnO, CeO_2, CuO, Al, Ni, and Fe are most frequently used NPs in industries and hence are frequently studied for their effects on diverse florae. Reliant on their chemical alignment, four main modules of NPs are as follows: carbon-coated nanomaterials like nanotubes and nanofibers of carbon, metal oxide-based nanomaterials (Ag, Zn, Fe, CeO, Cu, etc.), biological-based nanomaterials (liposomes, micelles, etc.), and nanocomposite-based nanomaterials (Pandey and Prajapati, 2018).

4.2 Nanomaterials used as herbicides to control weeds

Nanoherbicides are formulated using various NPs. Metal-based NPs improve the efficiency of herbicide activity and reduce the toxicity caused by conventional herbicides. Weed control by nanomaterials reduces the efficacy of

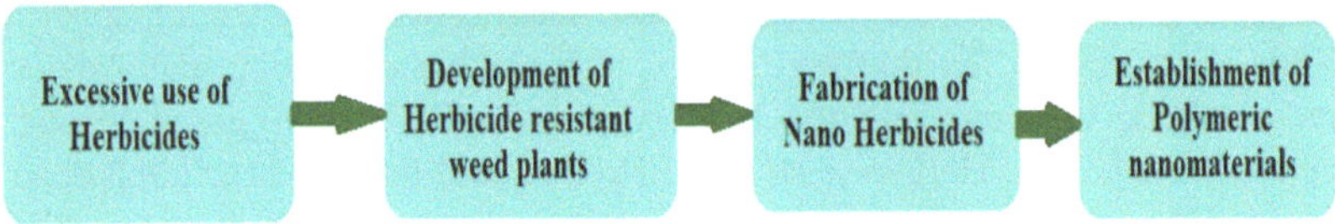

FIGURE 4.1 Evolutionary order of nanoherbicide development (Abigail and Chidambaram, 2017).

herbicide resistance potential and maintains their activity by prolonged release over long periods. Herbicides encapsulating molecules with nanomaterials act on specific receptors present in the roots of weed and suppress the growth of the weed (Seehra, 2017). Nanoherbicides developed using NPs enter through the root, translocate, and inhibit the glycolysis pathway that occurs in the root system. This will generate starvation of plant and reduces the growth of the plant and the plant will ultimately die (Manjunatha et al., 2016). An NP serves as a great carrier and creates nano-formulations together with herbicides. These NPs will overcome the chief disadvantage of conservative herbicide industry such as developing plants as herbicide-resistant plants (Fig. 4.1 and Table 4.2). These nanomaterials for herbicide development are mostly made of polymeric compounds that are biodegradable and have nontoxic metabolites (Seehra, 2017).

4.3 Engineered nanoparticles

It is identified that nearly all florae comprising fragrance, latex, phenols, flavonoids, and proteins are responsible for producing metal-oxide nanomaterials from the metallic briny. Though NPs are produced by physical or chemical methods, NPs synthesized from biological materials prevent the ecosystem from contamination. The outline and dimensions of NPs need to be precise, and the chosen form of nanomaterials are perhaps synthesized under controlled temperature and concentration of the medium. Biofabricated NPs can be divided into the metal and metal-oxide NPs.

4.4 Metal and metal-oxide nanoparticles

Remedy for ecological pollution due to the industrial waste and agricultural elements and pesticides and herbicide residues alike is probably metal-oxide NPs. The toxic metals namely titanium, cadmium, mercury, and lead produce toxic substances, which have adverse consequences for both plants and animals. In contrast, several optimistic properties of fabricated metal-oxide NPs are practically demonstrated. Plants and their extracts are friendly to numerous metal-oxide NPs manufactured organically, which is environmentally friendly, and their effect depends on the magnitude and form of the nanomaterials. For illustration, Zn is identified as an important component for both plants and animals; a suspension of 200 mg Zn/L exhibited plant toxicity in defined vegetation florae (Lin and Xing, 2007), even though such absorption is occasionally achieved in florae; the situation is different when a concentration of 1–4 mg Zn/L is maintained, which did not exhibit phytotoxicity, which causes hindrance to development of plants. This result can be effectively utilized in bulk production of budding bonsai and ornamental plants (Table 4.1).

4.5 Weed response to metal-oxide nanoparticles

4.5.1 ZnO

Zinc oxide (ZnO) NPs exhibited toxicity in a concentration range of 400–4000 mg/L in *Arabidopsis thaliana* (Lee et al., 2010). Seed germination and growth of plants were nearly inhibited in the presence of ZnO NPs. It may be due to the suspension of toxic metal ions (Murashov, 2006; Franklin et al., 2007). ZnO NPs were analyzed and observed in several parts of plant, it was and found that the NPs were captured and transferred to different parts of the plant, which get dispersed in epidermis, cortex, and vascular bundles at higher NP concentration. The absorption and accumulation of ZnO NPs in the root harms the root structure when more than optimal dosage was applied. ZnO NPs affect plant propagation and have adverse consequences on the elongation of root (Lin and Xing, 2008). ZnO NPs decrease photosynthetic efficacy and antioxidant activity; however, these NPs facilitate the production of reactive oxygen species (ROS) in a wheat plant (Tripathi et al., 2017). In *Allium cepa*, a higher concentration of ZnO NPs leads to DNA fragmentation (Ghosh et al., 2016).

4.5.2 TiO_2

With regard to the phytonanotechnology, *Azadirachta indica*, *Coriandrum* sp., *Ocimum sanctum, Nelumbolucifera*, and several other plants are compatible with the green chemistry methodologies (Begum et al., 2009; Philip, 2009; Badri Narayanan and Sakthivel, 2008). Phytochemical constituents such as terpenoids, flavones, ketones, aldehydes, and amides are responsible for the NP synthesis. In persistence with the efforts for synthesizing TiO_2-based NPs, it was proposed that TiO_2 hinders cell growth and N_2 fixative action and restricts the propagation and development in *Anabaena variabilis*, a synergetic nitrogen-fixing bacterium that enters through root hairs (epidermal cells) of host plants, where it is proliferated and reproduced, stimulating the development of root nodules (Rhizobium), widenings of plant and microbes cells in adjacent associations, where in nodes, the microbes convert free nitrogen into ammonia, which is further utilized by the host plant in their growth. TiO_2 NPs at concentrations of 110, 250, and 500 mg/L resulted

TABLE 4.1 Diverse consequences of nanoparticles on plants.

S. no.	Nanoparticles	Effects	Plants	References
1	TiO_2	The cell development, nitrogen fixation in roots, and germination activity are inhibited.	*Triticum aestivum* L., *Anabaena variabilis*	Cherchi and Gu (2010); Feizi et al. (2012)
2	Ag	Disturbs the mitosis metaphase, cell wall fragmentation, and breakdowns. It also reduces the shoot and root measurement and seed propagation activity.	*Allium cepa, T. aestivum, Lolium perenne, Cucurbita pepo*	El-Temsah and Joner 2012; Kumari et al. (2009); Dimkpa et al. (2013)
3	Zn	Inhibits the root development.	*Brassica napus, Cucumis sativus, Zea mays*	Lin and Xing (2007)
4	Al	Reduces root and seed germination.	*L. perenne, Z. mays*	
5	Al_2O_3	Decreases root length.	*Z. mays*	
6	Al_2O_3	Inhibits root development.	*Brassica oleracea, Daucus carota, C. sativus, Z. mays*	Yang and Watts (2005)
7	CeO_2	Inhibits shoot development.	*Lycopersiconesculentum, Z. mays*	Lee et al. (2010)
8	Cu	Decreases seedling development, reduces biomass, and inhibits root elongation and growth.	*Phaseolus radiatus, C. pepo*	Stampoulis et al. (2009), Lee et al. (2008)

in biomass and antioxidant defense decrease in *Xanthomonas perforans* (Paret et al., 2013), whereas an increase in the concentration of glucose-6-phosphate, glucose-1-phosphate, and isocitric acid was found, with decrease in concentration of sucrose and isomaltose. *A. thaliana* in interaction with TiO_2 at a concentration of 100–1000 mg/L exhibited an enrichment in root development, but with increase in concentration of TiO_2, a reduction in whole biomass and chlorophyll constituents was observed. TiO_2 NPs when applied in higher concentration cause increase in lipid peroxidation activity, affect the antioxidant response, and alter the genes, causing alterations in vitamin E (tocopherol) content (Szymańska et al., 2016). TiO_2 NPs at a concentration of 50 and 500 mg/kg results in amino acid agglomeration in *Triticum aestivum* (Wang et al., 2019). Besides, TiO_2 NPs form defense against bacteria and viruses (Sadeghi et al., 2017). It was also found that TiO_2 NPs inhibit the development of marine plants (Kim et al., 2011). In a research work involving the application of diverse metal-oxide NPs such as ZnO, Fe_3O_4, SiO_2, and Al_2O_3 on *A. thaliana*, these NPs were found to inhibit seed propagation, leaf amount, and root elongation when plants or seeds were exposed to NP concentrations ranging from 400 to 4000 mg/L (Lee et al., 2010).

4.5.3 Al_2O_3

Aluminum oxide (Al_2O_3) NPs are used in various industries, specifically in cosmetic care products. It is also used in various applications like dental and bone implants, military industry, and commercial industries as thermites, propellants, and coatings, etc. Aluminum has been revealed to be capable of eliminating pollutants from water (Salam et al., 2011; Xie et al., 2015). Aluminum NPs are preferably used as insecticidal compared to other NPs like ZnO, CuO, CeO, Fe, etc. (Kah and Hofmann, 2014). Nano-Al_2O_3 constrains root development in *Zea mays* and *Cucumis sativus*. It is believed that certain elements contained in Al_2O_3 NPs have adverse consequences on plants or even in humans (Murashov, 2006). Al_2O_3 NPs affect crop growth and yield. Plant toxicity of Al_2O_3 NPs was tested against five plant species (Yang and Watts, 2005). Al_2O_3 loaded with phenanthrene was found to cause less toxicity in root than pure Al_2O_3. It is also believed that Al_2O_3 NPs might lead to lethal effects on seed germination and root progression. Nevertheless, submicron alumina particles overloaded with phenanthrene do not display any significant consequences on seed germination and root development. The NPs accrued and further enhanced by the presence of phenanthreneNP reduce the phytotoxicity of the particles (Lavalley and Benaissa, 1985). Insignificant lethal effects have been noted with the absorption of Al_2O_3 at a concentration of 2000 mg/L, facilitating root elongation in *Z. mays* (Yang et al., 2015). Concentration in the range of 1.25–5 μM causes DNA damages in *A. cepa* (De Oliveira et al., 2016) and short root development in

C. sativus, *Brassica oleracea*, *Z. mays*, and *Daucus carota*. The expression of microRNA in tobacco plant increases under different stress conditions and also when exposed to a higher concentration of Al_2O_3 (Burklew et al., 2012).

4.5.4 CuO

The application of CuO NPs (CuO NPs) has improved significantly, mainly in the technical and engineering fields. However, when the plants are exposed to CuO NPs in the concentration range of 1.0–10.0 g/L, it results in inhibition of plant growth, change in plant's morphology, effects on ultrastructure of plants, and variations in photosynthetic pigments in the *Landoltiapunctate*. CuO NPs at a concentration of 10 mg/L cause root retardation and damage to *A. thaliana* (Tang et al., 2016). In hydroponic experiments of seeds in the plant *A. thaliana*, CuO NPs in a diameter range of 20–40 nm at a concentration of 20 and 50 mg/L showed its existence from root to seeds (Wang et al., 2016a,b). The experiment of CuO NPs on *Phaseolus radiates* and *T. aestivum* established that the damage incurred is due mostly to NPs rather than the liberation of cupric ions. CuO NPs decreased the development rate for both the plants (Lee et al., 2008). In *T. aestivum*, root elongation, hair proliferation, and the zones of division were inhibited with the exposure to CuO NPs at a concentration of 3, 10, 30, and 300 mg/kg with a diameter of <50 nm under treatment for seed experiments. The biochemical reactions occuring in plants due to exposure of plant to higher concentrations of NPs include the proffering of ROS in greater amounts, which disturbs the uptake, movement, and distribution of water in plants, causing oxidative stress and difficulties in photosynthesis and other various chemical reactions essential for the development of plants (Valavanidis et al., 2009). Several genetic variations were detected in exposure of *Nicotiana tabacum* and *Allium cepainon* to various NPs like CuO, TiO_2, Cd, Fe, etc. (Ghosh et al., 2010; Zhu et al., 2013). It is concluded that CuO NPs cause more damage to DNA compared to other NPs.

4.5.5 AgNPs

Silver NPs (AgNPs) are synthesized by the reduction of Ag^+ ions using various biological agents, and they are available in three forms such as Ag^+, $AgNO_3$, and Ag_2O. Ag NPs are not present in the form of Ag but anions in balance with cations, $AgNO_3 \rightarrow Ag^+ + NO^{-3}$. *A. thaliana* treated with silver NPs exhibited alterations in the plasma membrane, and also inhibiting the root elongation and leaf expansion (Sosan et al., 2016).When the Ag NPs are introduced at higher concentrations, the physiological characteristics like transpiration (Hawthorne et al., 2012), chlorophyll quantity (Mirzajani et al., 2013; Qian et al., 2013), and chlorophyll fluorescence (Tripathi et al., 2017) were affected. When Ag NPs at a concentration of 500 mg/L is applied to

Raphanus sativus, substantial changes in diverse biomolecules, proteins, lipids, pectin, and cellulose were detected (Zuverza-Mena et al., 2016). Plant growth hormones include auxin and cytokinin affected by means of Ag NPs (Yin et al., 2012; Vinković et al., 2017).

4.5.6 FeO

On exposure to FeO NPs at higher concentrations (3.2 mg/kg), biomass is reduced in both directly above ground and underground; FeO NPs at a concentration of 3.2 mg/kgin Clover induced decreased growth of the crops (Feng et al., 2013). The chemically synthesized nanomaterials have proven to be more toxic and hindered the plant growth even at lower concentrations (Hasan et al., 2020). The green algae *Parachlorella kessle* was grown well in biologically synthesized NPs compared to the chemically synthesized NPs.

4.5.7 CeO_2

Cerium dioxide NPs (CeO_2 NPs) mostly utilized in self-propelled manufacturing and semiconductor industries inhibit the cellular metabolism by the production of more ROS. CeO_2 NPs when coated with citric acid perform better on tomato plants compared to bare CeO_2 NPs (Barrios et al., 2016). The hydroponic experiment done on *Gossypium* sp. demonstrated that CeO_2 NPs extinguishes the vascular bundles in *Gossypium* sp., with a decrease in indole-3-acetic acid and abscisic acid. The research with CeO and ZnO NPs on *Glycine max* exhibited an upsurge in the class and yield of crops. ZnO NPs are absorbed by the plant and dispersed consistently over the plant tissues. All synthesized NPs may never be similarly active for all vegetative crops. When *G. max* was treated with CeO_2 NPs, leaf amounts were reduced with decrease in NP concentration. Even though, a smallest quantity of NPs exhibited retarded growth of the plants (Priestera et al., 2012). However, the stunted plants demonstrated growth when CeO_2 NPs were applied, but there was no increase in crop yield. Reduction in zinc, magnesium, Fe, and phosphorus levels in xylem sap were detected at a CeO_2 NP concentration of 100–500 mg/L with a diameter of 10 $\pm$ 3.2 nm by hydroponic treatments in transgenic cotton (Bt-29317) as well as conventional cotton (Jihe321) (Nhan et al., 2015). Furthermore, it shows considerable photocatalytic activity responsible for the deprivation of diverse carbon-based pollutants (Nagajyothi et al., 2020). CeO_2 inhibits the shoot development in *Z. mays* (Lee et al., 2010). *Cucumis sativus* exposed to CeO_2 NPs shows variations in the enzymatic activity of plants (Hong et al., 2016). It damages the chloroplast, vascular bundle destruction, and causes variation in uptake of nutrients (Prasad et al., 2015). On exposure of CeO_2 NPs to soybean plants at a concentration of 1 g/kg results in the accumulations of copper and phosphorus, whereas it

TABLE 4.2 Application of nanoparticles as herbicides in commercial crops.

S. no.	Nanoparticles	Nano herbicides utilized against weeds	References
1.	Ag nanoparticles encapsulated with chitosan and paraquate	*Eichhornia crassipes*	Namasivayam et al. (2014)
2.	Ag, Zn, Mn, Cu, Fe	Onion	Konotop et al. (2014)
3.	Cu	*Cucurbita pepo, Elodea densa*	Musante, White (2012); Nekrasova et al. (2011)
4.	CuO	*Radish, Lolium perenne* and *Rye grasses*	Atha et al. (2012)
5.	CuO	*Buckwheat, Cucumis, sativus*	Kim et al. (2012)

decreases the content of calcium in pods (Table 4.2) (Peralta-Videa et al., 2014).

Out of 53 heavy metals, more than 17 heavy metals are significant for living organisms, which include iron, copper, molybdenum, and manganese (Miransari, 2016). The remaining heavy metals lethal to the atmosphere include mercury (Hg), cadmium (Cd), arsenic (As), and chromium (Cr), which disturb the biochemical constraints in florae (Kumar et al., 2016). The occurrence of alloys in plants are significant for manufacturing protected products for humanoid feasting (Majeed et al., 2019). Heavy metal ions have the capability to exchange positively charged ions in binding sites, which deactivates the enzymes and fabricates ROS, leading to amino acid oxidation, protein deprivation, and DNA damage. Antioxidants include phenolic compounds, carotenoids, and glutathione, which bind the heavy metal ions, and reduce or impede the ROS (Maleki et al., 2017).

4.6 Smart delivery system of nanoparticles

Multispecies method with a single herbicide or frequent usage of herbicides in a similar mode of action in the cropped area resulted in poor control of weed plants and development of herbicide-resistant weed plants. Hence, target-specific herbicides enveloped with NPs designed at a particular receptor in weed plant roots and translocated to other parts of the plant will inhibit glycolysis and other root system pathways responsible for food reservation

(Fig. 4.2). This will prevent the weed plant from receiving sufficient food and thus inhibiting the growth (Chinnamuthu and Kokiladevi, 2007).

4.7 Mode of action of nanoherbicides

Although several researchers have investigated the impact of nanoherbicides on target organisms, there is very little research conducted on how the NPs

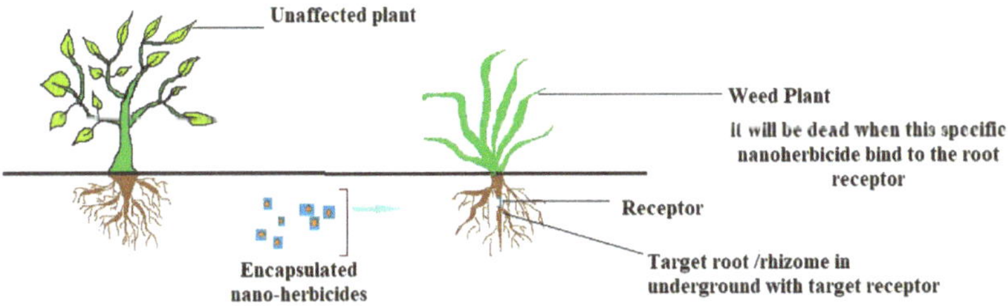

FIGURE 4.2 Smart delivery of nanoparticles that target the specific receptors of weed plants (Joshi et al., 2019).

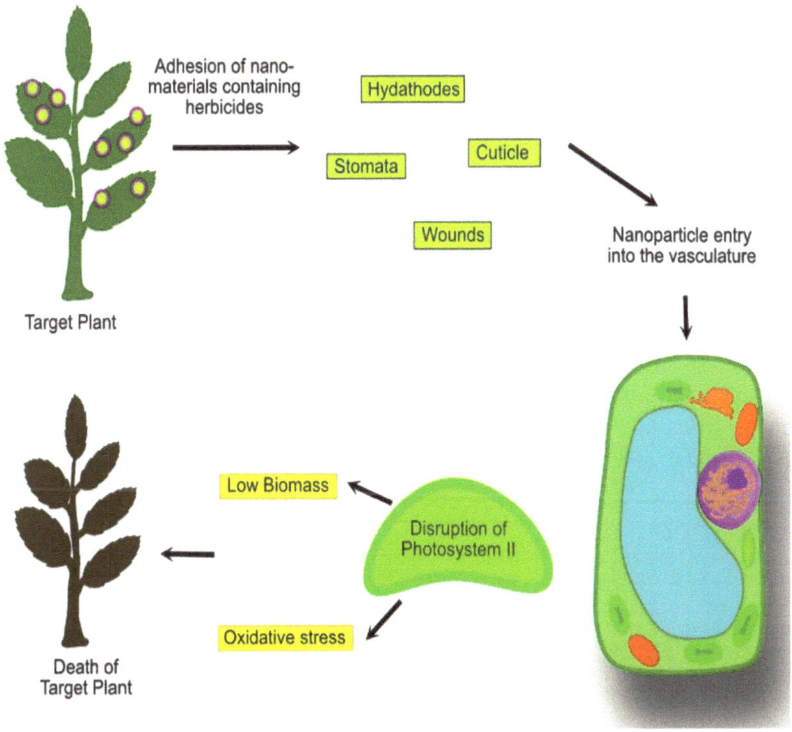

FIGURE 4.3 A schematic diagram showing the possible mechanism of action of nanoherbicides (Bombo et al., 2019; Avellan et al., 2021).

interact with the plants. The majority of herbicides are applied to plants by foliar spray besides soil application (Zimdahl, 2018). Scientists resorted to foliar spraying to develop the most effective herbicide composition employing nanocarrier systems. The foliar uptake of the NPs could be either through stomata, hydathodes, cuticle, trichomes, etc. (Avellan et al., 2021).

One of the mechanisms studied by Bombo et al. (2019) revealed that the nano-encapsulated atrazine enters the plant tissues through the stomata in the hydathodes to the mesophyll cells. Atrazine causes oxidative stress by inhibiting electron flow in photosystem II (Mou et al., 2011). Based on these studies, nano-atrazine also caused reduced activity of the photosystem II and produced oxidative stress (Bombo et al., 2019; Oliveira et al., 2015). The release of atrazine in an encapsulated form following their entry into the plant cells and subsequently their release into the vasculature and then into the cells resulted in decreased chloroplast activity (Bombo et al., 2019). The research work by Bombo et al. (2019) is a pioneer study that showed the interaction between plants and polymeric nanocapsules (Fig. 4.3).

4.8 Conclusion

Nanotechnology application in weed control would enhance the potency of plants, increase solubility, decrease the chemical products that are used as herbicides for controlling the weeds, and reduce the toxicity of the products. In this chapter, we have emphasized the present advances in the usage of NPs as herbicides without significant environmental impairment. The objective behing the use of nanomaterials in agriculture is to reduce the usage of chemical herbicides and increase the yield of other vegetative crops through weed management. Nanotechnology in agriculture has specific applications such as nanoherbicides to enhance the products and herbicide utilization level for increasing the productivity of the plants without affecting the various biotic and abiotic stresses. Nanomaterial application in the form of agrochemicals have attracted a lot of attention recently. The use of nanotechnology in agriculture fields and agribusiness will have its specific advantages in the economic, social, and ecological systems.

References

Abigail, E.A., Chidambaram, R., 2017. Nanotechnology in herbicide resistance. Nanostructured Materials—Fabrication to Applications. Intech Open, London, pp. 207—212.

Atha, D.H., Wang, H., Petersen, E.J., Cleveland, D., Holbrook, R.D., Jaruga, P., et al., 2012. Copper oxide nanoparticle mediated DNA damage in terrestrial plant models. Environ. Sci. Technol. 46 (3), 1819—1827.

Avellan, A., Yun, J., Morais, B.P., Clement, E.T., Rodrigues, S.M., Lowry, G.V., 2021. Critical review: role of inorganic nanoparticle properties on their foliar uptake and in planta translocation. Environ. Sci. Technol. 55 (20), 13417—13431.

Barrios, A.C., Rico, C.M., Trujillo-Reyes, J., Medina-Velo, I.A., Peralta-Videa, J.R., Gardea-Torresdey, J.L., 2016. Effects of uncoated and citric acid coated cerium oxide nanoparticles, bulk cerium oxide, cerium acetate, and citric acid on tomato plants. Sci. Total. Environ. 563, 956–964.

Begum, N.A., Mondal, S., Basu, S., Laskar, R.A., Mandal, D., 2009. Biogenic synthesis of Au and Ag nanoparticles using aqueous solutions of Black Tea leaf extracts. Colloids Surf. B: Biointerfaces 71 (1), 113–118.

Bombo, A.B., Pereira, A.E.S., Lusa, M.G., de Medeiros Oliveira, E., de Oliveira, J.L., Campos, E.V.R., et al., 2019. A mechanistic view of interactions of a nano-herbicide with target organism. J. Agric. Food Chem. 67 (16), 4453–4462.

Burklew, C.E., Ashlock, J., Winfrey, W.B., Zhang, B., 2012. Effects of aluminium oxide nanoparticles on the growth, development, and microRNA expression of tobacco (*Nicotiana tabacum*). PLoS One 7 (5), e34783.

Catton, H.A., Lalonde, R.G., Buckley, Y.M., De Clerck-Floate, R.A., 2016. Biocontrol insect impacts population growth of its target plant species but not an incidentally used nontarget. Ecosphere 7 (5), e01280.

Cherchi, C., Gu, A.Z., 2010. Impact of titanium dioxide nanomaterials on nitrogen fixation rate and intracellular nitrogen storage in *Anabaena variabilis*. Environ. Sci. Technol. 44 (21), 8302–8307.

Chidambaram, R., 2016. Application of rice husk nano sorbents containing 2, 4-dichlorophenoxyacetic acid herbicide to control weeds and reduce leaching from soil. J. Taiwan. Inst. Chem. Eng. 63, 318–326.

Chinnamuthu, C.R., Kokiladevi, E., 2007. Weed management through nanoherbicides. Appl. Nanotechnol. Agri. 10, 978–971.

Dashora, A., Sharma, K., 2018. Green synthesis of nanoparticles and their applications. Adv. Sci., Eng. Med. 10 (6), 523–541.

De Oliveira, C.R., Fraceto, L.F., Rizzi, G.M., Salla, R.F., Abdalla, F.C., Costa, M.J., et al., 2016. Hepatic effects of the clomazone herbicide in both its free form and associated with chitosan-alginate nanoparticles in bullfrog tadpoles. Chemosphere 149, 304–313.

Dimkpa, C.O., McLean, J.E., Martineau, N., Britt, D.W., Haverkamp, R., Anderson, A.J., 2013. Silver nanoparticles disrupt wheat (*Triticum aestivum* L.) growth in a sand matrix. Environ. Sci. Technol. 47 (2), 1082–1090.

El-Temsah, Y.S., Joner, E.J., 2012. Impact of Fe and Ag nanoparticles on seed germination and differences in bioavailability during exposure in aqueous suspension and soil. Environ. Toxicol. 27 (1), 42–49.

Feizi, H., Rezvani Moghaddam, P., Shahtahmassebi, N., Fotovat, A., 2012. Impact of bulk and nanosized titanium dioxide (TiO_2) on wheat seed germination and seedling growth. Biol. Trace Elem. Res. 146 (1), 101–106.

Feng, Y., Cui, X., He, S., Dong, G., Chen, M., Wang, J., et al., 2013. The role of metal nanoparticles in influencing arbuscular mycorrhizal fungi effects on plant growth. Environ. Sci. Technol. 47 (16), 9496–9504.

Franklin, N.M., Rogers, N.J., Apte, S.C., Batley, G.E., Gadd, G.E., Casey, P.S., 2007. Comparative toxicity of nanoparticulate ZnO, bulk ZnO, and ZnCl2 to a freshwater microalga (*Pseudo kirchneriella subcapitata*): the importance of particle solubility. Environ. Sci. Technol. 41 (24), 8484–8490.

Ghosh, M., Bandyopadhyay, M., Mukherjee, A., 2010. Genotoxicity of titanium dioxide (TiO2) nanoparticles at two trophic levels: plant and human lymphocytes. Chemosphere 81 (10), 1253–1262.

Ghosh, M., Jana, A., Sinha, S., Jothiramajayam, M., Nag, A., Chakraborty, A., et al., 2016. Effects of ZnO nanoparticles in plants: cytotoxicity, genotoxicity, deregulation of antioxidant defenses, and cell-cycle arrest. Mutat. Res./Genet. Toxicol. Environ. Mutagen. 807, 25–32.

Han, Y.S., Lee, S.Y., Yang, J.H., Hwang, H.S., Park, I., 2010. Paraquat release control using intercalated montmorillonite compounds. J. Phys. Chem. Solids 71 (4), 460–463.

Hasan, M., Rafique, S., Zafar, A., Loomba, S., Khan, R., Hassan, S.G., et al., 2020. Physiological and anti-oxidative response of biologically and chemically synthesized iron oxide: *Zea mays* a case study. Heliyon 6 (8), e04595.

Hawthorne, J., Musante, C., Sinha, S.K., White, J.C., 2012. Accumulation and phytotoxicity of engineered nanoparticles to *Cucurbita pepo*. Int. J. Phytoremediat. 14 (4), 429–442.

Heap, I., Duke, S.O., 2018. Overview of glyphosate-resistant weeds worldwide. Pest. Manag. Sci. 74 (5), 1040–1049.

Hess, F.D., 2018. Herbicide effects on plant structure, physiology, and biochemistry. Pesticide Interactions in Crop Production. CRC Press, pp. 13–34.

Hong, J., Wang, L., Sun, Y., Zhao, L., Niu, G., Tan, W., et al., 2016. Foliar applied nanoscale and microscale CeO2 and CuO alter cucumber (*Cucumis sativus*) fruit quality. Sci. Total. Environ. 563, 904–911.

Joshi, H., Choudhary, P., Mundra, S.L., 2019. Future prospects of nanotechnology in agriculture. Int. J. Chem. Stud. 7 (2), 957–963.

Kah, M., Hofmann, T., 2014. Nanopesticide research: current trends and future priorities. Environ. Int. 63, 224–235.

Kanimozhi, V., Chinnamuthu, C.R., 2012. Engineering core/hallow shell nanomaterials to load herbicide active ingredient for controlled release. Res. J. Nanosci. Nanotechnol. 2 (2), 58–69.

Kim, E., Kim, S.H., Kim, H.C., Lee, S.G., Lee, S.J., Jeong, S.W., 2011. Growth inhibition of aquatic plant caused by silver and titanium oxide nanoparticles. Toxicol. Environ. Health Sci. 3 (1), 1–6.

Kim, S., Lee, S., Lee, I., 2012. Alteration of phytotoxicity and oxidant stress potential by metal oxide nanoparticles in *Cucumis sativus*. Water, Air, Soil. Pollut. 223 (5), 2799–2806.

Konotop, Y.O., Kovalenko, M.S., Ulynets, V.Z., Meleshko, A.O., Batsmanova, L.M., Taran, N. Y., 2014. Phytotoxicity of colloidal solutions of metal-containing nanoparticles. Cytol. Genet. 48 (2), 99–102.

Kumar, R., Mishra, R.K., Mishra, V., Qidwai, A., Pandey, A., Shukla, S.K., et al., 2016. Detoxification and tolerance of heavy metals in plants. Plant Metal Interaction. Elsevier, pp. 335–359.

Kumari, M., Mukherjee, A., Chandrasekaran, N., 2009. Genotoxicity of silver nanoparticles in *Allium cepa*. Sci. Total. Environ. 407 (19), 5243–5246.

Lavalley, J.C., Benaissa, M., 1985. Infrared study of surface modes on alumina, Studies in Surface Science and Catalysis, Vol. 21. Elsevier, pp. 251–261.

Lee, C.W., Mahendra, S., Zodrow, K., Li, D., Tsai, Y.C., Braam, J., et al., 2010. Erratum: Developmental phytotoxicity of metal oxide nanoparticles to *Arabidopsis thaliana*. Environ. Toxicol. Chem. 29 (6), 1399–1399.

Lee, W.M., An, Y.J., Yoon, H., Kweon, H.S., 2008. Toxicity and bioavailability of copper nanoparticles to the terrestrial plants mung bean (*Phaseolus radiatus*) and wheat (*Triticum aestivum*): plant agar test for water-insoluble nanoparticles. Environ. Toxicol. Chem.: An. Int. J. 27 (9), 1915–1921.

Lin, D., Xing, B., 2007. Phytotoxicity of nanoparticles: inhibition of seed germination and root growth. Environ. Pollut. 150 (2), 243–250.

Lin, D., Xing, B., 2008. Root uptake and phytotoxicity of ZnO nanoparticles. Environ. Sci. Technol. 42 (15), 5580–5585.

Majeed, A., Muhammad, Z., Siyar, S., 2019. Assessment of heavy metal induced stress responses in pea (*Pisum sativum* L.). Acta Ecol. Sin. 39 (4), 284–288.

Maleki, M., Ghorbanpour, M., Kariman, K., 2017. Physiological and antioxidative responses of medicinal plants exposed to heavy metals stress. Plant. Gene 11, 247–254.

Manjunatha, S.B., Biradar, D.P., Aladakatti, Y.R., 2016. Nanotechnology and its applications in agriculture: a review. J. Farm. Sci. 29 (1), 1–13.

Maurer-Jones, M.A., Gunsolus, I.L., Murphy, C.J., Haynes, C.L., 2013. Toxicity of engineered nanoparticles in the environment. Anal. Chem. 85 (6), 3036–3049.

Melander, B., 1995. Impact of drilling date on *Aperaspica-venti L.* and *Alopecurus myosuroides* Huds, in winter cereals. Weed Res. 35 (3), 157–166.

Miransari, M. (Ed.), 2016. Academic Press, Cambridge, MA, pp. 197–216.

Mirzajani, F., Askari, H., Hamzelou, S., Farzaneh, M., Ghassempour, A., 2013. Effect of silver nanoparticles on *Oryza sativa* L. and its rhizosphere bacteria. Ecotoxicol. Environ. Saf. 88, 48–54.

Mou, R.X., Chen, M.X., Cao, Z.Y., Zhu, Z.W., 2011. Simultaneous determination of triazine herbicides in rice by high-performance liquid chromatography coupled with high resolution and high mass accuracy hybrid linear ion trap-orbitrap mass spectrometry. Anal. Chim. Acta 706 (1), 149–156.

Murashov, V., 2006. Comments on Particle surface characteristics may play an important role in phytotoxicity of alumina nanoparticles, Yang, L., Watts, DJ, Toxicol. Lett., 2005, 158, 122–132. Toxicol. Lett. 2 (164), 185–187.

Musante, C., White, J.C., 2012. Toxicity of silver and copper to *Cucurbita pepo*: differential effects of nano and bulk-size particles. Environ. Toxicol. 27 (9), 510–517.

Nagajyothi, P.C., Prabhakar Vattikuti, S.V., Devarayapalli, K.C., Yoo, K., Shim, J., Sreekanth, T.V.M., 2020. Green synthesis: photocatalytic degradation of textile dyes using metal and metal oxide nanoparticles-latest trends and advancements. Crit. Rev. Environ. Sci. Technol. 50 (24), 2617–2723.

Namasivayam, K.R.S., Aruna, A., Gokila, 2014. Evaluation of silver nanoparticles-chitosan encapsulated synthetic herbicide paraquate (AgNp-CS-PQ) preparation for the controlled release and improved herbicidal activity against *Eichhornia crassipes*. Res. J. Biotechnol. 9 (9), 19–27.

Narayanan, K.B., Sakthivel, N., 2008. Coriander leaf mediated biosynthesis of gold nanoparticles. Mater. Lett. 62 (30), 4588–4590.

Nekrasova, G.F., Ushakova, O.S., Ermakov, A.E., Uimin, M.A., Byzov, I.V., 2011. Effects of copper (II) ions and copper oxide nanoparticles on *Elodea densa* Planch. Russ. J. Ecol. 42 (6), 458–463.

Nhan, L.V., Ma, C., Rui, Y., Liu, S., Li, X., Xing, B., et al., 2015. Phytotoxic mechanism of nanoparticles: destruction of chloroplasts and vascular bundles and alteration of nutrient absorption. Sci. Rep. 5 (1), 1–13.

Oliveira, H.C., Stolf-Moreira, R., Martinez, C.B., Sousa, G.F., Grillo, R., De Jesus, M.B., et al., 2015. Evaluation of the side effects of poly (epsilon-caprolactone) nanocapsules containing atrazine toward maize plants. Front. Chem. 3, 61.

Owen, M.D., 2016. Diverse approaches to herbicide-resistant weed management. Weed Sci. 64 (S1), 570–584.

Pandey, R.K., Prajapati, V.K., 2018. Molecular and immunological toxic effects of nanoparticles. Int. J. Biol. Macromol. 107, 1278–1293.

Paret, M.L., Vallad, G.E., Averett, D.R., Jones, J.B., Olson, S.M., 2013. Photocatalysis: effect of light-activated nanoscale formulations of TiO2 on *Xanthomonas perforans* and control of bacterial spot of tomato. Phytopathology 103 (3), 228–236.

Peralta-Videa, J.R., Hernandez-Viezcas, J.A., Zhao, L., Diaz, B.C., Ge, Y., Priester, J.H., et al., 2014. Cerium dioxide and zinc oxide nanoparticles alter the nutritional value of soil cultivated soybean plants. Plant. Physiol. Biochem. 80, 128–135.

Philip, D., 2009. Biosynthesis of Au, Ag and Au–Ag nanoparticles using edible mushroom extract. Spectrochim. Acta Part. A: Mol. Biomol. Spectrosc. 73 (2), 374–381.

Prasad, R., Kumar, M., Varma, A., 2015. Role of PGPR in soil fertility and plant health. In: Plant-Growth-Promoting Rhizobacteria (PGPR) and Medicinal Plants, Springer, Cham, pp. 247–260.

Priestera, J.H., Gea, Y., Mielkea, R.E., Horsta, A.M., Moritzb, S.C., Espinosae, K., et al., 2012. Soybean susceptibility to manufactured nanomaterials with evidence for food quality and soil fertility interruption. Proc. Natl. Acad. Sci. USA 109, 14734–14735.

Qian, H., Peng, X., Han, X., Ren, J., Sun, L., Fu, Z., 2013. Comparison of the toxicity of silver nanoparticles and silver ions on the growth of terrestrial plant model *Arabidopsis thaliana*. J. Environ. Sci. 25 (9), 1947–1956.

Radad, K., Al-Shraim, M., Moldzio, R., Rausch, W.D., 2012. Recent advances in benefits and hazards of engineered nanoparticles. Environ. Toxicol. Pharmacol. 34 (3), 661–672.

Sadeghi, R., Rodriguez, R.J., Yao, Y., Kokini, J.L., 2017. Advances in nanotechnology as they pertain to food and agriculture: benefits and risks. Annu. Rev. Food Sci. Technol. 8, 467–492.

Salam, M.A., Makki, M.S., Abdelaal, M.Y., 2011. Preparation and characterization of multi-walled carbon nanotubes/chitosan nanocomposite and its application for the removal of heavy metals from aqueous solution. J. Alloy. Compd. 509 (5), 2582–2587.

Sanchez-Dominguez, M., Boutonnet, M., Solans, C., 2009. A novel approach to metal and metal oxide nanoparticle synthesis: the oil-in-water microemulsion reaction method. J. Nanopart. Res. 11 (7), 1823–1829.

Seehra, M. (Ed.), 2017. Nanostructured Materials: Fabrication to Applications. IntechOpen, London.

Sosan, A., Svistunenko, D., Straltsova, D., Tsiurkina, K., Smolich, I., Lawson, T., et al., 2016. Engineered silver nanoparticles are sensed at the plasma membrane and dramatically modify the physiology of *Arabidopsis thaliana* plants. Plant. J. 85 (2), 245–257.

Stampoulis, D., Sinha, S.K., White, J.C., 2009. Assay-dependent phytotoxicity of nanoparticles to plants. Environ. Sci. Technol. 43 (24), 9473–9479.

Szymańska, R., Kołodziej, K., Ślesak, I., Zimak-Piekarczyk, P., Orzechowska, A., Gabruk, M., et al., 2016. Titanium dioxide nanoparticles (100–1000 mg/l) can affect vitamin E response in *Arabidopsis thaliana*. Environ. Pollut. 213, 957–965.

Tan, D., Yuan, P., Annabi-Bergaya, F., Dong, F., Liu, D., He, H., 2015. A comparative study of tubular halloysite and platy kaolinite as carriers for the loading and release of the herbicide amitrole. Appl. Clay Sci. 114, 190–196.

Tang, Y., He, R., Zhao, J., Nie, G., Xu, L., Xing, B., 2016. Oxidative stress-induced toxicity of CuO nanoparticles and related toxico genomic responses in *Arabidopsis thaliana*. Environ. Pollut. 212, 605–614.

Tripathi, A., Liu, S., Singh, P.K., Kumar, N., Pandey, A.C., Tripathi, D.K., et al., 2017. Differential phytotoxic responses of silver nitrate ($AgNO_3$) and silver nanoparticle (AgNps) in *Cucumis sativus* L. Plant. Gene 11, 255–264.

Valavanidis, A., Vlachogianni, T., Fiotakis, C., 2009. 8-hydroxy-2′-deoxyguanosine (8-OHdG): a critical biomarker of oxidative stress and carcinogenesis. J. Environ. Sci. Health Part. C. 27 (2), 120–139.

Viirlaid, E., Riiberg, R., Mäeorg, U., Rinken, T., 2009. Glyphosate attachment on amino activated carriers for sample stabilization and concentration. Agron. Res. 13, 1152–1159.

Vinković, T., Novák, O., Strnad, M., Goessler, W., Jurašin, D.D., Parađiković, N., et al., 2017. Cytokinin response in pepper plants (*Capsicum annuum* L.) exposed to silver nanoparticles. Environ. Res. 156, 10–18.

Wakabayashi, K.O., Böger, P., 2004. Phytotoxic sites of action for molecular design of modern herbicides (Part 1): the photosynthetic electron transport system. Weed Biol. Manag. 4 (1), 8–18.

Wang, X., Yang, X., Chen, S., Li, Q., Wang, W., Hou, C., et al., 2016a. Zinc oxide nanoparticles affect biomass accumulation and photosynthesis in Arabidopsis. Front. Plant. Sci. 6, 1243.

Wang, Z., Xu, L., Zhao, J., Wang, X., White, J.C., Xing, B., 2016b. CuO nanoparticle interaction with *Arabidopsis thaliana*: toxicity, parent-progeny transfer, and gene expression. Environ. Sci. Technol. 50 (11), 6008–6016.

Wang, Y., Jiang, F., Ma, C., Rui, Y., Tsang, D.C., Xing, B., 2019. Effect of metal oxide nanoparticles on amino acids in wheat grains (*Triticum aestivum*) in a life cycle study. J. Environ. Manag. 241, 319–327.

Xie, Y., Huang, Q., Liu, M., Wang, K., Wan, Q., Deng, F., et al., 2015. Mussel inspired functionalization of carbon nanotubes for heavy metal ion removal. RSC Adv. 5 (84), 68430–68438.

Yang, L., Watts, D.J., 2005. Particle surface characteristics may play an important role in phytotoxicity of alumina nanoparticles. Toxicol. Lett. 158 (2), 122–132.

Yang, Z., Chen, J., Dou, R., Gao, X., Mao, C., Wang, L., 2015. Assessment of the phytotoxicity of metal oxide nanoparticles on two crop plants, maize (*Zea mays* L.) and rice (*Oryza sativa* L.). Int. J. Environ. Res. Public. Health 12 (12), 15100–15109.

Yin, L., Colman, B.P., McGill, B.M., Wright, J.P., Bernhardt, E.S., 2012. Effects of silver nanoparticle exposure on germination and early growth of eleven wetland plants.

Zhu, X., Hondroulis, E., Liu, W., Li, C.Z., 2013. Biosensing approaches for rapid genotoxicity and cytotoxicity assays upon nanomaterial exposure. Small 9 (9-10), 1821–1830.

Zimdahl, R.L., 2018. Fundamentals of Weed Science. Academic Press.

Zuverza-Mena, N., Armendariz, R., Peralta-Videa, J.R., Gardea-Torresdey, J.L., 2016. Effects of silver nanoparticles on radish sprouts: root growth reduction and modifications in the nutritional value. Front. Plant. Sci. 7, 90.

Chapter 5

Nano metal oxide-based delivery system for insect and pest control

Rajkuberan Chandrasekaran[1] and Rajiv Periakaruppan[2,3]
[1]Department of Biotechnology, Karpagam Academy of Higher Education, Coimbatore, Tamil Nadu, India, [2]College of Horticulture, Nanjing Agricultural University, Nanjing, Jiangsu Province, P.R. China, [3]Department of Biotechnology, PSG College of Arts & Science, Coimbatore, Tamil Nadu, India

Chapter outline

5.1 Introduction

The growing human population necessitates the farmers to produce sufficient food to ensure food security. However, the challenges faced by the agriculturalist in an environment are climate change, soil pollution, insects, and pest damage (Popp et al., 2013). In particular, more than 10,000 insects damage the food, but 10% of the insects are considered as major pests (Dhaliwal et al., 2010). In the 19th century, the importance of crop pests and damaging patterns were assessed, and a module was developed to reduce crop damage. Later, new synthetic pesticides and insecticides were intensively utilized to combat the insects, but the toxicity associated with pesticides is needed to be a concern. In the 20th century, Bt cotton was successfully launched in India against insects. In such a prevailing situation, the production of foods in a sustainable approach is indeed one. To counteract various abiotic and biotic factors, the crop protection strategies should change their dimension to design and develop a new chemical + biological entity to protect the crop from insects with minimum environmental impact.

Nanometal Oxides in Horticulture and Agronomy. DOI: https://doi.org/10.1016/B978-0-323-91809-1.00009-3

The era of the green revolution is a golden period for researchers and farmers which led to an increase in yield, the development of resistant varieties, reduced green gas emissions, and an environmentally benign approach (Qaim, 2020). At this juncture, new hyphenated technology, so-called nanotechnology, dominated the scientific world with its intriguing physicochemical properties. This nanotechnology is applied in all physical, chemical, and biological sectors, having wide applications such in electric batteries, solar cells, electrochemical reactions, catalysts, antimicrobial, anticancer, antiplasmodial, antiinflammatory, biosensor, drug delivery, and dye degradation (Patil and Chandrasekaran, 2020).

During the last decade, nanotechnology became momentum in agriculture practices due to the awareness and adoption of new hyphenated technology in sustainable development. This nanotechnology lends credence to the farmers owing to the unique distinct physicochemical properties and vast diversity of applications. Nanotechnology in agriculture boost plant productivity and protect against pest, fertilizers, stimulants, gene transfer, biosensors, food storing, and packaging (Sekhon, 2014). This review specifically emphasizes the role of metal oxide in delivering pesticides and insecticides against crop pests and their deeds.

5.2 Synthesis of metal oxide nanoparticles

Metal oxide materials are noble materials with intriguing physicochemical properties. These properties of metal oxide nanoparticles (NPs) were strongly governed and influenced by oxide's crystal, structure, composition, native defects, doping, processing, and growth parameters (Grilli, 2020). Henceforth, these oxide NPs have received substantial attention from researchers to utilize NPs widely in passive optics, solar cells, gas sensors, catalysis, corrosion protection, environmental protection, and optoelectronic devices (Laurent et al., 2018).

Synthesis of metal oxide NPs is an important phenomenon because several unique properties developed by nano metal oxide particles are directly proportional to the size and the high density of sites at the corners or high surface area compared with bulk structure (Chavali and Nikolova, 2019); so the development of size based nano oxide particles is an impactful process. In the synthesis process, two distinct processes are employed topdown approach and the bottomup approach (Table 5.1).

The topdown approach is a process where the NPs are derived from the bulk phase of solid materials, that is, the process of slicing/scaling down the bulk materials into nano dimensions. The topdown approach is broadly categorized into two groups: mechanical and lithographic techniques. For instance, using mechanical milling, Lee et al. fabricated the 1 μm α-Fe_2O_3 powder (99.9%) into 15 nm α-Fe_2O_3 powder in 100 h (Lee et al., 2001). Likewise, in the mechanochemical method, various metal oxides have been synthesized, such asFe_2O_3 NPs (Seyedi et al., 2015), CeO_2 (Li et al., 2005), ZnO (Lu et al., 2008), and $CoFe_2O_4$ (Shi et al., 2000). Priyadarshana et al. (2015) synthesized magnetite NPs from natural iron oxide ore from Sri Lanka using oleic acid as a surface modifier in the wet

TABLE 5.1 Nanomaterial synthesis approaches.

Topdown approach	Bottomup approach
Mechanical milling	Chemical vapor deposition
Etching	Sol-gel process
Laser ablation	Laser pyrolysis
Sputtering	Chemical reduction
Electro-explosion	Green synthesis

grinding method. Co_3O_4 nano oxide was conceivably synthesized from core CO_3O_4 powder by the etching method (Zhou et al., 2016). Though the topdown approach produces uniform size and structures, the method is cost expensive, employing heat, electricity, pressure, and energy.

The bottomup approach is another method to produce oxide NPs where the synthetic process starts from the self-assembly of atoms and molecules (Arole and Munde, 2014). The approach comprises three domains to produce nano oxide NPs liquid-phase synthesis, vapor-phase synthesis, and green synthesis. In liquid-phase synthesis, a sonochemical method is employed to produce nano metal oxide. ZnONPs in the size range were produced from zinc acetate and sodium hydroxide (Hosseini and Khodadadipoor, 2018). In the coprecipitation method, alkaline precipitation of $FeCl_3$ and $FeCl_2$ produces superparamagnetic iron oxide NPs of spherical shape with a diameter of 8 nm (Massart and Cabuil, 1987). ε-Fe_2O_3 in the size range of 19.8–48 nm was synthesized by the solvothermal method (Mishra et al., 2014). In the synthesis of TiO_2NPs by hydrothermal method, titanium tetraisopropoxide was hydrolyzed using HNO_3 as a peptizing agent to produce TiO_2NPs (Kim et al., 2005). Another important aspect of the bottomup approach is green synthesis. This method uses biological resources like plants, bacteria, and fungus to produce nano oxide NPs. These synthesis methods are distinguished from other method in the bottomup approach due to the nonusage of hazardous chemicals and reagents, nontoxicity, avoiding high pressure and energy, cost inexpensive, and environmentally friendly. Over the past decades, numerous studies have been reported on the synthesis of NPs using leaves, fruits, stems, roots, flowers, latex, and tubers (Akhtar et al., 2013). Likewise, actinomycetes, fungi, and bacterium were also deployed to synthesize NPs. The NPssynthesized from bioresources showed multifaceted properties like a polydisperse, cubic, hexagonal, spheroid, triangle, and oval sizes ranging from 0 to 200 nm. The plant contains active constituents such as alkaloids, flavonoids, terpenoids, terpenes, sterol, proteins, and carbohydrates, which might act as a reducing and stabilizing agent in synthesizing metal oxide NPs. In contrast, microbes utilize intra/extracellular metabolites, enzymes, and cofactors to produce metal oxide NPs (Marslin et al., 2018).

As mentioned above, a wide variety of synthesis methods are available for the metal oxide, but the synthesis process is required depending on the vast technological applications. All the physical and chemical methods produce high crystal quality metal oxide with precise size and shape, but the metal oxide synthesized from biological resources is challenging. For instance, the cubic crystalline structure of ZnO can be achieved only through very high pressures. Therefore, it is needed for detailed analysis of metal oxide NP synthesis, physicochemical properties, and its applications in diversified ranges.

5.3 Applications of metal oxide nanoparticles

Globally, almost all researchers drive their research into nano-based concepts and applications, due to their unique optical, physical, and chemical properties. Henceforth, metal/metal oxide NPs have been employed in all the of physical, chemical, and biological sciences sectors. The metal oxide NPs are applied on the rationale of two classes: (1) high surface area materials and (2) materials takingadvantage of the size dependence of physical properties. For example, in supercapacitors, nanomaterials are employed based on large surface area per weight or film thickness for their function (Zhao et al., 2011). Here, the NPs are relevant to geometrical consideration, whereas they do not have unique size-dependent physical properties. In the other class, in tunable light wavelength, ZnO is applied based on the size dependence of luminescence wavelength (Raji and Gopchandran, 2017). Therefore, properties and size of the NPs are important factors for the applications of metal oxide NPs.

Developing nano-based gas sensors is one of the high-priority areas in gas sensor research. $BaTiO_3$−CuO-sputtered thin films were utilized for carbon dioxide sensing (Herrán et al., 2008), TiO_2-rGO was used for carbon monoxide sensing (Bandi et al., 2018), and Nb_2O_5 and CeO_2 semiconductors were utilized for O_2 sensors (Takeuchi, 1988). Likewise, variant facet metal oxide NPs are extensively used in methane gas, ozone gas, nitric oxide, and dioxide gas sensors (Chavali and Nikolova, 2019). The above metal oxide composite exhibited effective activity in sensing gas with advantages like a fast response, ease of handling, low cost, a wide range of target gases, and long life.

Another most promising application of MONPs is rechargeable batteries; in lithium-ion batteries, the role of metal oxide NPs has gauned utmost importance. Likewise, in solar cells, Cu_2O cells act as a photovoltaic material and are widely used (Anderson et al., 2014). Similarly, barium titanate ($BaTiO_3$), amorphous indium−zinc oxide (IZO), and indium tin oxide (ITO) were the notable metal oxides used in solar cells (Chavali and Nikolova, 2019). The investigation of nano metal oxide as catalysts to overcome environmental problems is intensively studied. For example, cobalt oxide nanocrystals are used as catalysts for chlor-alkali electrolysis system (Liang et al., 2012), iron oxide NPs in catalytic oxidation of phenolic and aniline compounds (Zhang et al., 2009), and tin oxide NPs for the reduction and photodegradation of organic compounds (Bhattacharjee et al., 2015). The abovementioned metal oxide NPs act as catalysts to accelerate the reaction and increase product yields due to the high surface area and active sites.

5.4 Biomedical applications of metal oxide nanoparticles

Another prominent application of metal oxide NPs is in biomedicine. Though different metal oxide NPs have existed, NPs ferric oxide (Fe_2O_3), ferrous oxide (Fe_3O_4) ZnO, TiO_2, and CuO are safe for applications in mammals (Beyth et al., 2015). Owing to the growing antibiotic resistance in gram-positive and negative bacterium, it is an urgent need to develop alternative antimicrobial products for the management of infectious diseases. At this juncture, nanotechnology played an immense task as bactericidal, fungicidal, and insecticidal agents. Azam et al. (2012) synthesized ZnO, CuO, and Fe_2O_3 nano oxide particles through the sol-gel combustion method and evaluated them against *Bacillus subtilis, Pseudomonas aeruginosa, Staphylococcus aureus*, and *Escherichia coli*. The synthesized metal oxide NPs performed superior antagonistic activity against the tested pathogens in the following order of efficacy ($ZnO > CuO > Fe_2O_3$). Similarly, the same ZnO, CuO, and Fe_2O_3 NPs were fabricated through biogenic sources from green tea leaves *Camellia sinensis* (Senthilkumar and Sivakumar, 2014), *Abutilon indicum* (Ijaz et al., 2017), *Lagenaria siceraria* (Kanagasubbulakshmi and Kadirvelu, 2017), and evaluated against the *E. coli* and *S.aureus*. The aforementioned NPs performed splendid activity against the tested pathogens. This is due to the capping of phytochemicals coated on the surface of oxide NPs. Likewise, metal oxide NPs also played a precise role in cancer therapy, immunotherapy, tissue regeneration, wound healing, diagnostics, dentistry, magnetic resonance image scanning, and biosensors (Nikolova and Chavali, 2020).

5.5 Environmental applications of metal oxides

Nanotechnology restores the quality of air, soil, and water contaminated by pollutants at low cost, high efficiency, and eco-friendly (Tsekhmistrenko et al., 2020). In particular, the nanotechnology's role in agriculture is huge, and its potential impact is significant in the agroecosystem. Though robust techniques and devices still exist in modern practices, the usage of the nano-based concept in agriculture is need for the hour.

5.6 Role of metal oxide nanoparticles in agriculture

China is one of the leading countries in using nano-based concepts in agriculture, with 28 patents developed between 2011 and 2015 (Yata et al., 2018). Globally, about 20%–40% of crops are damaged by pathogens and insects annually (Sharma et al., 2017). Using synthetic pesticides and insecticides intensively might cause insect resistance, soil impotence, failure in nutrient recycling, toxicity accumulation, and global warming (Tripathi et al., 2020). To curb the toxicity of synthetic chemicals, nano-based delivery of pesticides/insecticides/fertilizers/micronutrients/DNA/proteins was keenly performed in various plants and crops (Fig. 5.1). The success of nano-based concepts in plants depends upon the plants, nanomaterials, and other factors also.

CuO and CaO were synthesized through a synthetic route and assayed for their pesticide activities against cotton leafworm (*Spodoptera littoralis*) (Ayoub et al., 2018). The so-called metal oxide NPs exert a toxic effect with $LC_{50} = 232.75$ mg/L for CuO and $LC_{50} = 129.03$ mg/L for CaO. Moreover, the author demonstrated that CuO and CaO damage the mid-gut epithelium and cuticle layer.

Bulb mite *Rhizoglyphus robini* Claparede (Acari: Acaridae) is a major pest for tuber plants. By using the corm dip bioassay method, metal oxide NPs titanium dioxide (TiO_2) and iron oxide (Fe_2O_3), magnesium oxide (MgO), copper oxides (CuO), and zinc oxide (ZnO) were investigated against bulb mite (Korghond et al., 2021). From the assay, the CuO exhibited high contact toxicity than other metal oxide NPs. Further, the authors assessed the sublethal effects in adult females at LC_{25} concentration. The metal oxide NPs, particularly MgO and TiO_2, exert prolong egg incubation, larval period, and nymphal period compared with control. At the same time, the generation time (T), the finite rate of increase (λ),

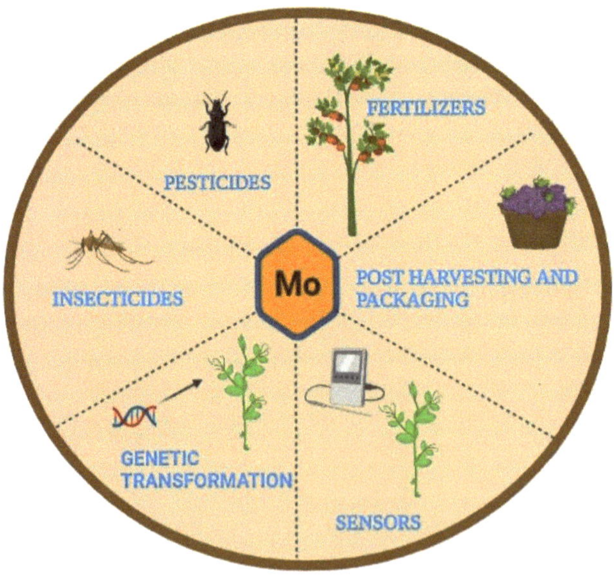

FIGURE 5.1 Metal oxide (Mo) nanoparticles and their application in agriculture.

the intrinsic rate of increase (r_m), and the net reproductive rate (R_0) were varied among the metal oxide NP treatments.

In another report, SiO_2, TiO_2, and ZnO NPs were synthesized, and different concentrations (10^{-1}, 10^{-3}, and 10^{-5}) of oxide particles were diluted with a standard concentration of eugenol (1 mg/mL). In the lure dispenser, the rate of discharge of eugenol in various concentrations of oxide NPs was investigated. At 10^{-5} TiO_2 concentration, the maximum slow and sustained release was achieved, and a high number of the fruit fly was trapped for up to 12 weeks in mango orchards (Dharanivasan et al., 2017). This study showed that nano oxide-based attractants are one of the easy ways to attract a mass number of insects in integrated pest management.

Sharma et al. (2017) synthesized copper selenide (Cu_2-xSe) NPs by the arrested precipitation method and decorated them with reduced graphene oxide (rGO). To the rGO-Cu_2-xSe composite pesticide, chlorpyrifos was loaded by the organic/water solvent evaporation method. The aforementioned pesticide nanocomposite was evaluated preliminarily in the HEK293 cell line to understand the toxicity of the composite. The composite triggers toxicity as it causes more mortality in the cell line. Further, the leaves were treated with different concentrations of pesticide nanocomposite and assayed on the third instars larvae of *Pieris rapae*. After 48 h, the pesticide nanocomposite showed 80% mortality than the chlorpyrifos emulsion. Further, the rGO-Cu_2-xSe composite did not exert toxicity to the larvae as it was healthy and has not shown any abnormal distribution after feeding leaves.

Pesticide pyridaben (Pyr), chlorpyrifos (Chl), and beta-cyfluthrin (Cyf) were conjugated with rGO through π-π interactions (Wang et al., 2019a). The three rGO-pesticides nanocomposite was assayed for acaricidal activity against *Tetranychus truncates* by leaf dip bioassay method. The rGO-Pyr exhibited toxicity and caused mortality up to 86%, while bare Pyr exhibited mortality up to 34%. Likewise, rGO-Chl 87%, bare Chl 42% and rGO-Cyf-87%, bare Cyf-55%. All three rGO-pesticides performed superior activity while compared with bare pesticides. Further, the LC_{50} of Chl 3.21 μg/mL rGO-Chl 2.07 μg/mL; Cyf 3.19 μg/mL rGO-Cyf 1.80 μg/mL; and Pyr 3.29 μg/mL rGO-Pyr 2.11. These results showed that pesticides with graphene oxide increase the pesticide performance in the assay.

Similarly, β-cyfluthrin (Cyf), Monosultap (Mon), and imidacloprid (Imi) pesticide were conjugated with rGO (Wang et al., 2019b) (Fig. 5.1). The so-called rGO-Cyf, rGO-Mon, rGO-Imi were exclusively characterized by transmission electron microscopy and atomic force microscopy. The nanocomposite was assayed for toxicity in larvae of the lepidopteran insect Asian corn borer (ACB). The assay portrayed that rGO-Cyf, rGO-Mon, and rGO-Imi display 2.1-, 1.51-, and 1.83-fold higher toxicity than bare insecticide in ACB (Fig. 5.2).

Photocatalyst metal oxide W/TiO_2 was synthesized by a simple hydrothermal method at a low synthetic temperature. Avermectins (AVMs) are biological agents produced by *Streptomyces avermitilis* widely used to control parasites and pests. AVM microcrystals were fabricated through the

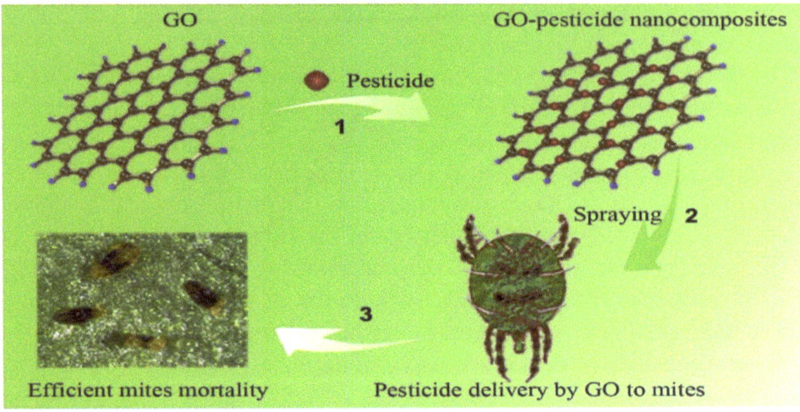

FIGURE 5.2 Synergistic interaction of pesticide pyridaben (Pyr), chlorpyrifos (Chl), and beta-cyfluthrin (Cyf) against two spider mites *Tetranychus truncatus and T. urticae. Figure need to be get copyright permission.*

crystallization procedure. To this AVM microparticles, polyelectrolyte multilayers of chitosan/alginate (CHI/ALG) were deposited through a layer-by-layer self-assembly technique. The so-called AVM−(CHI/ALG)$_5$ complex was doped with W/TiO_2 to form W/TiO2−AVM−(CHI/ALG)$_5$. Adult stage *Martianus dermestoides* was employed to study the toxicity of photodegradable insecticide. The toxicity was higher in the photodegradable insecticide with extended-release of AVM (Guan et al., 2011).

CuO, MnO, SiO, TiO_2, and ZnO were applied as foliar sprays to protect watermelon against *Fusarium* wilt (Peréz et al., 2019). Plants treated with CuONPs were significantly larger than the other metal oxide NP treatment. At the same time, the Cu in root was also increased in CuONPs treated plants. Likewise, CuONPs treated plants yield more fruits than other treated metal oxide NPs. Molecular studies showed an upregulation of pathogenicity-related and polyphenol oxidase proteins in the watermelon plants.

Thus, from the above-mentioned examples, it is clear that metal oxide-mediated pesticide delivery will be the new tool for pest management practices. The nanometal oxide will minimize the toxicity of pesticides and enhance the efficiency of the nano pesticide complex.

5.7 Future perspective and outlook

Scientists developing new modern materials and their subsequent release into the environment need a proper dynamics study. As known NPs have a high surface area and active sites, any pesticides/insecticides can easily be conjugated/doped with NPs. This facilitates the function and efficacy of the chemicals effectively. At the same time, nano-mediated pesticide delivery will improve the availability, controlled and sustained release, targeted drug delivery, protection from photolysis, hydrolysis, and degradation of soil.

Nano oxide particles will pose toxicity to the plants, which increases the production of reactive oxygen species (ROS), followed by apoptosis or necrosis, and finally, plant death (Jastrzębska et al., 2012). Therefore, researchers should drive to use nano oxide particles with the least concentration in agriculture practices. Before releasing it into the environment, the researchers should conduct preliminary studies on toxicity with plants, non-target crops/organisms, soil health, and other parameters. There is an urgent need to study the use of metal oxide in understanding plant physiology and molecular biology.

At present, a large number of pesticides/insecticide manufacturing companies exist worldwide. Industrial production of nano-derived pesticides should be encouraged, and infrastructure facilities should be adopted. However, a sustainable government should follow a green policy in implementing new techniques for the agriculture sectors to avoid negative effects. The government should focus on academic institutes, R & D Laboratories, and nongovernment organizations to develop such metal oxide-based pesticides to support subsidies.

Finally, the farmers handling this product should be aware of using what and so. They should be given proper guidance and counseling for handling the metal oxide NPs. Success is not an easy task. It is a stepping stone process; for the commercial success of metal oxide NPs in pesticide delivery, it needs a large array of studies in invitro, pilot, greenhouse, and farm.

5.8 Conclusion

Growing population and plant stress damage crop production severely. Conventional pesticides/fertilizers/insecticides were routinely practiced, but the negative effects worsened and threatened the environment alarmingly. Indeed, it is an urgent need to cop out the pesticides/insecticides to deliver in the form of smart technology. Thus, the chapter exclusively covered the metal oxide NP-mediated pesticide delivery comprehensively.

References

Akhtar, M.S., Panwar, J., Yun, Y.S., 2013. Biogenic synthesis of metallic nanoparticles by plant extracts. ACS Sustain. Chem. Eng. 1 (6), 591–602.

Anderson, A.Y., Bouhadana, Y., Barad, H.N., Kupfer, B., Rosh-Hodesh, E., Aviv, H., et al., 2014. Quantum efficiency and bandgap analysis for combinatorial photovoltaics: sorting activity of Cu–O compounds in all-oxide device libraries. ACS Comb. Sci. 16 (2), 53–65.

Arole, V.M., Munde, S.V., 2014. Fabrication of nanomaterials by top-down and bottom-up approaches-an overview. J. Mater. Sci. 1, 89–93.

Ayoub, H.A., Khairy, M., Elsaid, S., Rashwan, F.A., Abdel-Hafez, H.F., 2018. Pesticidal activity of nanostructured metal oxides for generation of alternative pesticide formulations. J. Agric. Food Chem. 66 (22), 5491–5498.

Azam, A., Ahmed, A.S., Oves, M., Khan, M.S., Habib, S.S., Memic, A., 2012. Antimicrobial activity of metal oxide nanoparticles against gram-positive and gram-negative bacteria: a comparative study. Int. J. Nanomed. 7, 6003–6009.

Bandi, S., Hastak, V., Peshwe, D.R., Srivastav, A.K., 2018. In-situ $$\hbox {TiO} _ {2} $$ TiO 2–rGO nanocomposites for CO gas sensing. Bull. Mater. Sci. 41 (5), 1–5.

Beyth, N., Houri-Haddad, Y., Domb, A., Khan, W., Hazan, R., 2015. Alternative antimicrobial approach: nano-antimicrobial materials. Evidence-based Complement. Alternat. Med. 2015.

Bhattacharjee, A., Ahmaruzzaman, M., Sinha, T., 2015. A novel approach for the synthesis of SnO2 nanoparticles and its application as a catalyst in the reduction and photodegradation of organic compounds. Spectrochim. Acta Part A: Mol. Biomol. Spectrosc. 136, 751–760.

Chavali, M.S., Nikolova, M.P., 2019. Metal oxide nanoparticles and their applications in nanotechnology. SN Appl. Sci. 1 (6), 1–30.

Dhaliwal, G.S., Jindal, V., Dhawan, A.K., 2010. Insect pest problems and crop losses: changing trends. Indian. J. Ecol. 37 (1), 1–7.

Dharanivasan, G., Sithanantham, S., Kannan, M., Chitra, S., Kathiravan, K., Janarthanan, S., 2017. Metal oxide nanoparticles assisted controlled release of synthetic insect attractant for effective and sustainable trapping of fruit flies. J. Clust. Sci. 28 (4), 2167–2183.

Grilli, M.L., 2020. Metal oxides. Metals 10 (6), 820.

Guan, H.N., Chi, D.F., Yu, J., Zhang, S.Y., 2011. Novel photodegradable insecticide W/TiO2/Avermectin nanocomposites obtained by polyelectrolytes assembly. Colloids Surf. B: Biointerfaces 83 (1), 148–154.

Herrán, J., Mandayo, G.G., Castaño, E., 2008. Solid state gas sensor for fast carbon dioxide detection. Sens. Actuators B: Chem. 129 (2), 705–709.

Hosseini, S.G., Khodadadipoor, Z., 2018. Sonochemical synthesis of ZnO, NiO and α-Fe_2O_3 nanoparticles and their catalytic activity for thermal decomposition of ammonium perchlorate.

Jastrzębska, A.M., Kurtycz, P., Olszyna, A.R., 2012. Recent advances in graphene family materials toxicity investigations. J. Nanopart. Res. 14 (12), 1–21.

Kanagasubbulakshmi, S., Kadirvelu, K., 2017. Green synthesis of iron oxide nanoparticles using Lagenaria siceraria and evaluation of its antimicrobial activity. Def. Life Sci. J. 2 (4), 422–427.

Kim, J.H., Noh, B.H., Lee, G.D., Hong, S.S., 2005. Hydrothermal synthesis of titanium dioxide using acidic peptizing agents and their photocatalytic activity. Korean J. Chem. Eng. 22 (3), 370–374.

Korghond, G.T., Sahebzadeh, N., Allahyari, H., Ramroodi, S., 2021. Acute toxicity and sublethal effects of metal oxide nanoparticles against the bulb mite. Syst. Appl. Acarol. 26 (4), 788–800.

Laurent, S., Boutry, S., Muller, R.N., 2018. Metal oxide particles and their prospects for applications. *Iron Oxide Nanoparticles for Biomedical Applications*. Elsevier, pp. 3–42.

Lee, J.S., Lee, C.S., Oh, S.T., Kim, J.G., 2001. Phase evolution of Fe2O3 nanoparticle during high energy ball milling. Scr. Mater. 44 (8–9), 2023–2026.

Li, Y.X., Chen, W.F., Zhou, X.Z., Gu, Z.Y., Chen, C.M., 2005. Synthesis of CeO2 nanoparticles by mechanochemical processing and the inhibiting action of NaCl on particle agglomeration. Mater. Lett. 59 (1), 48–52.

Liang, Y., Wang, H., Diao, P., Chang, W., Hong, G., Li, Y., et al., 2012. Oxygen reduction electrocatalyst based on strongly coupled cobalt oxide nanocrystals and carbon nanotubes. J. Am. Chem. Soc. 134 (38), 15849–15857.

Lu, J., Ng, K.M., Yang, S., 2008. Efficient, one-step mechanochemical process for the synthesis of ZnO nanoparticles. Ind. Eng. Chem. Res. 47 (4), 1095–1101.

Marslin, G., Siram, K., Maqbool, Q., Selvakesavan, R.K., Kruszka, D., Kachlicki, P., et al., 2018. Secondary metabolites in the green synthesis of metallic nanoparticles. Materials 11 (6), 940.

Massart, R., Cabuil, V., 1987. Effect of some parameters on the formation of colloidal magnetite in alkaline-medium-yield and particle-size control. J. Chim. Phys. Phys.-Chim. Biol. 84 (7–8), 967–973.

Mishra, D., Arora, R., Lahiri, S., Amritphale, S.S., Chandra, N., 2014. Synthesis and characterization of iron oxide nanoparticles by solvothermal method. Prot. Met. Phys. Chem. Surf. 50 (5), 628–631.

Nikolova, M.P., Chavali, M.S., 2020. Metal oxide nanoparticles as biomedical materials. Biomimetics 5 (2), 27.

Patil, S., Chandrasekaran, R., 2020. Biogenic nanoparticles: a comprehensive perspective in synthesis, characterization, application and its challenges. J. Genet. Eng. Biotechnol. 18 (1), 1–23.

Peréz, C.D., De La Torre Roche, R., Zuverza-Mena, N., Ma, C., Shen, Y., White, J.C., et al., 2019. Metalloid and metal oxide nanoparticles suppress sudden death syndrome of soybean. J. Agric. Food Chem. 68 (1), 77–87.

Popp, J., Pető, K., Nagy, J., 2013. Pesticide productivity and food security. A review. Agron. Sustain. Dev. 33 (1), 243–255.

Priyadarshana, G., Kottegoda, N., Senaratne, A., De Alwis, A., Karunaratne, V., 2015. Synthesis of magnetite nanoparticles by top-down approach from a high purity ore. J. Nanomaterials 2015.

Qaim, M., 2020. Role of new plant breeding technologies for food security and sustainable agricultural development. Appl. Econ. Perspect. Policy 42 (2), 129–150.

Raji, R., Gopchandran, K.G., 2017. ZnO nanostructures with tunable visible luminescence: effects of kinetics of chemical reduction and annealing. J. Sci.: Adv. Mater. Devices 2 (1), 51–58.

Sekhon, B.S., 2014. Nanotechnology in agri-food production: an overview. Nanotechnol. Sci. Appl. 7, 31.

Senthilkumar, S.R., Sivakumar, T., 2014. Green tea (Camellia sinensis) mediated synthesis of zinc oxide (ZnO) nanoparticles and studies on their antimicrobial activities. Int. J. Pharm. Pharm Sci. 6 (6), 461–465.

Seyedi, M., Haratian, S., Khaki, J.V., 2015. Mechanochemical synthesis of Fe2O3 nanoparticles. Procedia Mater. Sci. 11, 309–313.

Sharma, S., Singh, S., Ganguli, A.K., Shanmugam, V., 2017. Anti-drift nano-stickers made of graphene oxide for targeted pesticide delivery and crop pest control. Carbon 115, 781–790.

Shi, Y., Ding, J., Yin, H., 2000. CoFe2O4 nanoparticles prepared by the mechanochemical method. J. Alloy. Compd. 308 (1–2), 290–295.

Takeuchi, T., 1988. Oxygen sensors. Sens. Actuators 14 (2), 109–124.

Tripathi, S., Srivastava, P., Devi, R.S., Bhadouria, R., 2020. Influence of synthetic fertilizers and pesticides on soil health and soil microbiology. Agrochemicals Detection, Treat. Remediation. Butterworth-Heinemann, pp. 25–54.

Tsekhmistrenko, O.S., Bityutskyy, V.S., Tsekhmistrenko, S.I., Kharchishin, V.M., Melnichenko, O.M., Rozputnyy, O.I., et al., 2020. Nanotechnologies and environment: a review of pros and cons. Ukr. J. Ecol. 10 (3).

Wang, X., Xie, H., Wang, Z., He, K., 2019a. Graphene oxide as a pesticide delivery vector for enhancing acaricidal activity against spider mites. Colloids Surf. B: Biointerfaces 173, 632–638.

Wang, X., Xie, H., Wang, Z., He, K., Jing, D., 2019b. Graphene oxide as a multifunctional synergist of insecticides against lepidopteran insect. Environ. Sci.: Nano 6 (1), 75–84.

Yata, V.K., Tiwari, B.C., Ahmad, I., 2018. Nanoscience in food and agriculture: research, industries and patents. Environ. Chem. Lett. 16 (1), 79–84.

Zhang, S., Zhao, X., Niu, H., Shi, Y., Cai, Y., Jiang, G., 2009. Superparamagnetic Fe3O4 nanoparticles as catalysts for the catalytic oxidation of phenolic and aniline compounds. J. Hazard. Mater. 167 (1–3), 560–566.

Zhao, X., Sánchez, B.M., Dobson, P.J., Grant, P.S., 2011. The role of nanomaterials in redox-based supercapacitors for next generation energy storage devices. Nanoscale 3 (3), 839–855.

Zhou, Y., Dong, C.K., Han, L.L., Yang, J., Du, X.W., 2016. Top-down preparation of active cobalt oxide catalyst. ACS Catal. 6 (10), 6699–6703.

Chapter 6

Impact of metal oxide nanomaterials on seed germination, plant growth and development

Subha Priya Venkateswaran[1], Vignesh Kumar Palanisamy[2], P. Divya[2], P. Geethamani[2] and Rajiv Periakaruppan[3]

[1]Department of Biotechnology, Sri Krishna Arts and Science College, Kuniamuthur, Coimbatore, Tamil Nadu, India, [2]Department of Chemistry, KGiSL Institute of Technology, Saravanampatti, Coimbatore, Tamil Nadu, India, [3]Department of Biotechnology, PSG College of Arts & Science, Coimbatore, Tamil Nadu, India

Chapter outline

Nanometal Oxides in Horticulture and Agronomy. DOI: https://doi.org/10.1016/B978-0-323-91809-1.00008-1

6.1 Introduction

Nanomaterials can be classified as either natural or synthetic. Various environmental processes (physical, chemical, and biological) contribute to the natural formation of nanomaterials. Photo-oxidation, mineralization, physical fragmentation, weathering of minerals, gas-solid nucleation in the atmosphere, redox, and precipitation reactions all lead to the generation of natural nanomaterials (Lespes et al., 2020).

In contrast to natural nanomaterials, many kinds of nanomaterials are produced by human activity, either intentionally or unintentionally. Carbon-based nanoparticles (NPs), metal-based nanomaterials, dendrimers, and nanocomposites are the four basic categories of synthetic nanomaterials (Fig. 6.1). Carbon-based nanomaterials are composed of carbon, and they form different shapes such as ellipsoids, hollow tubes, cylinders, etc. Nanotubes are comprised of carbon that has been rolled into a tube shape from a flat plane. They offer industrial products with exciting properties like strength, structure, conduction of heat, and electricity, which are of great technological use in developing improved batteries. Metal-based nanomaterials are pure metals or their compounds in the size of submicron scale. They have a metal core made of inorganic metal/metal oxide surrounded by an organic or inorganic material or metal oxide shell (Khan, 2020). Dendrimers are a well-defined family of nanomaterials that are homogeneous, monodisperse macromolecules with symmetric branches/dendrons originating from a central core (Abbasi et al., 2014; Mittal et al., 2021).

Metallic nanoparticles (MNPs) have a metal core made of inorganic metal or metal oxide surrounded by an organic or inorganic material or metal

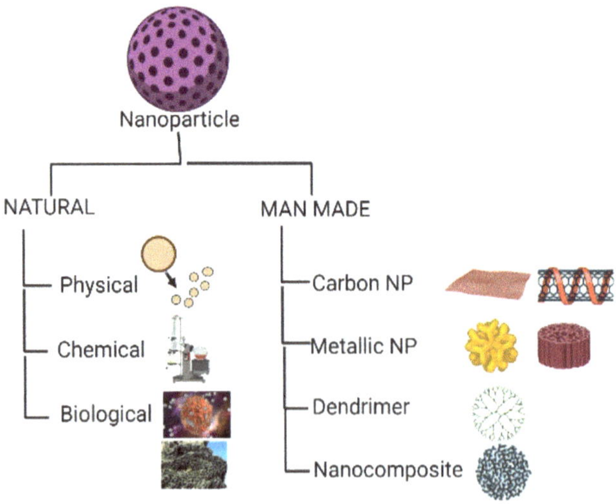

FIGURE 6.1 Types of nanomaterials.

oxide shell. Although globally there have been significant advancements in the manufacturing and use of MNPs, there is still a lack of knowledge regarding the toxicity-related impact of these on human health and the environment (Khan, 2020). NPs find various applications in medicine, agriculture, pollution control, electronics, and textiles. Sustainable and resilient agriculture is possible with the use of nanomaterials. Nanomaterials can boost seed germination, growth, and nutrition in crops (Cervantes-Avilés et al., 2021). Amidst the numerous benefits observed, research into the route of absorption, its impact on the chemical composition of NPs, their destiny, and alterations inside biological systems is still in its early stages.

6.2 Metal oxide nanomaterials

Metal oxide nanomaterials (MONMs) play an essential role in many scientific areas. MONMs exhibit unique optical and electrical properties compared to bulk materials. From a scientific and technological standpoint, these nanomaterials constitute a crucial class of materials. In terms of electronic structure, physical, chemical, and electromagnetic properties, nanomaterials based on metal oxides represent a diverse family of materials.

Apart from the numerous industrial applications, MONMs are also used in various household products. Some of the medical applications of MONMs include cancer treatment, tissue engineering, MRI, etc. They are also used in paints and as industrial catalysts. Additionally, they find significant applications in agriculture and the treatment of environmental pollutants (Prasanna et al., 2019).

Metal oxide nanoparticles (MONPs) are manufactured by adding precipitating or oxidizing agents. MONPs such as titanium dioxide, silver oxide, zinc oxide, cerium dioxide, copper oxide, and iron oxide are used in industries. These NPs, therefore, find their way into the environment through multiple routes, which may impact plant growth and development in several ways (Rastogi et al., 2017).

MONMs are effective antimicrobial agents capable of acting against several multidrug-resistant bacteria (Niño-Martínez et al., 2019).

6.3 Effect of nanomaterials on plants

Increased use of metal and MONPs has resulted in their entry into the environment as a pollutant. The fate of such materials in the environment is not entirely understood and is, therefore, an emerging research area (Kanwar et al., 2019).

Nowadays, nanomaterials are used in agriculture as pesticides, fertilizers, and other related products, called nano-enabled agricultural products. Understanding their dissolution, aggregation, and stability in root exudates and soil leachate will aid in the creation of considerably more effective nanoagrochemicals. Cervantes-Avilés et al. (2021) discovered that when NPs (nano-CeO_2, nano-Mn_3O_4, nano-$Cu(OH)_2$, nano-MoO_3) came into contact with root exudates and soil leachate, there was an alteration in the bioavailability and delivery efficiency.

6.4 Metal oxide nanoparticles and their impacts

6.4.1 Titanium dioxide (TiO_2) nanoparticles

Titanium dioxide is one of the highly exploited nanomaterials. It is used in paints, paper, toothpaste, cosmetics, and many other consumer products. Studies suggest that TiO_2 nanoparticles (TiO_2 NPs) when applied on crop plants, combat the effects caused by stress, and help in enhancing the growth and development of plants (Chaudhary and Singh, 2020). Studies on wheat plants have proved that TiO_2 NPs of moderate concentrations were not harmful to the plants. At the same time, these NPs enhanced the activities of antioxidant enzymes such as SOD, APX, POD, and CAT (Alharby et al., 2021). Gohari et al. (2020) have reported an effective response of TiO_2 NPs on Moldavian balm plants exposed to salinity stress. TiO_2 NPs lower H_2O_2 levels in these plants, thereby increasing the antioxidant enzyme activity. This stimulates secondary metabolite synthesis while simultaneously activating the plant's defense mechanism, resulting in stress protection and growth promotion. Silva et al. (2020) found that *Triticum aestivum* plants treated with TiO_2 NPs had developmental damage. The metabolic pathways of plants were also severely affected when subjected to TiO_2 NP treatments. Moreover, these NPs pose severe toxicological effects on bacteria that aid in promoting plant growth. Direct contact with TiO_2 NPs inhibits the growth of nitrogen-fixing bacteria, phosphate solubilizers, and biofilm-forming organisms (Chavan et al., 2020).

6.4.2 Silicon dioxide (SiO_2) nanoparticles

Silicon (Si) is a nonessential element for plants, comparable to cadmium (Cd), arsenic (As), lead (Pb), and others. However, because Si is favorable to plant development, it is also known as a quasi-essential element. Si can increase the antioxidant activities of plants, thereby increasing their stress tolerance. They can also decrease the uptake of soil pollutants by plant roots (Rajput et al., 2021b). Many researchers have employed silicon dioxide nanoparticles (SiO_2 NPs) to decrease stress in agricultural plants, which has resulted in increased growth and production. Tomato plants treated with SiO_2 NPs are more resistant to *Orobanche ramosa* (broomrape) infection through improved structural and metabolic responses, according to Madany et al. (2020). At a concentration of 200 mg/L SiO_2 NPs, an increase in proline and catalase activity with enhanced lipid peroxidation was noticed. Decreased levels of chlorophyll a, chlorophyll b, carotenoids, root and shoot fresh, and dry weight were also observed. However, lower SiO_2 concentrations in the range of 50 and 100 mg/L have shown positive and negative effects (Karimi and Mohsenzadeh, 2016).

SiO_2 nanofertilizers show a positive impact on the growth and yield of cucumber. They increase the N and P uptake and decrease the Na uptake by these plants (Yassen et al., 2017). Emamverdian et al. (2020) performed a tissue culture of bamboo (*Pleioblastus pygmaeus*) to study the impacts of SiO_2 NPs on plant growth under Pb pollution. It has been observed that plant growth is enhanced by SiO_2 NPs by increasing the antioxidant levels in plants.

6.4.3 Iron oxide (Fe_3O_4) nanoparticles

Iron oxide nanoparticles (Fe_3O_4 NPs) have unique mobility, reaction, and toxicity, all of which can be harmful to human health and the environment. The phytotoxicity of Fe_3O_4 NPs on *Lemna minor* was studied by Souza et al. (2019), which revealed an increase in lipid peroxidation and a decrease in chlorophyll content with increasing dosage of Fe_3O_4 NPs. Fe_3O_4 NPs, according to a study conducted by Kokina et al. (2020) in yellow medick (*Medicago falcata* L.), can be used to increase the resistance of plants to fungal infections, such as powdery mildew. Fe_3O_4 NPs pose toxicity on soil microbes affecting their carbon and nitrogen mineralization properties. This results in a decline in plant-available nitrogen in spinach (Kamran et al., 2020). Fe_3O_4 NPs at concentrations of 1, 2, and 4 mg/L generate modest genotoxicity and have a favorable effect on the growth and maturation of rocket (*Eruca sativa* Mill.) seedlings, suggesting that NPs may increase plants' capacity to withstand environmental challenges (Plaksenkova et al., 2019). These NPs enhance the callus formation and regeneration pattern of the obstinate *Cicer arietinum* L., which makes the Fe_3O_4 NPs a likely component to be used in the tissue culture of recalcitrant legume species (Irum et al., 2020).

6.4.4 Zinc oxide (ZnO) nanoparticles

Kokina et al. (2020) have demonstrated the impacts of zinc oxide nanoparticles (ZnO NPs) on barley seedlings. It has been found that ZnO NPs enhance seed germination, shoot and root elongation, whereas root cell viability and genomic template stability were decreased. H_2O_2 stress levels were also enhanced by ZnO NPs. A study conducted on tomato plants showed that root and shoot growth was inhibited by ZnO NPs. The levels of chlorophyll a and b and the photosynthetic efficiency of tomato plants were significantly reduced, whereas Zn^{2+} did not show such an effect. Therefore, it has been considered that the toxicity is attributed to the decreased chlorophyll levels, and damaged photochemical system but not because of Zn (Wang et al., 2018). Rajput et al. (2021a) studied the toxicity of ZnO NPs on physiological and anatomical indices of spring barley, which showed inhibition of growth due to damage in the chloroplasts and their structural organization. ZnO NPs are highly efficient in enhancing the development of *Brassica oleracea var italica* (broccoli) when compared to $ZnSO_4$. This makes the ZnO NPs a promising candidate for use as nanosupplements and nanofertilizers (Awan et al., 2021). ZnO NPs have also been used as a plant growth booster to enhance the productivity of *Portulaca oleracea L.* (purslane), which has been demonstrated as a better alternative to ZnO.

6.4.5 Gallium oxide (Ga_2O_3) nanoparticles

UV detection, flat panel displays, photocatalysis, gas sensors, UV filters, and optoelectronic devices employ gallium oxide nanoparticles (Ga_2O_3 NPs), which have specific characteristics in optics, photoelectric conversion, and electrical fields (Shi and Qiao, 2021). Plants can selectively absorb gallium from soil which is evident from the comparatively lower concentrations of Ga in soil than plants. In plants, Ga inhibits the uptake of iron, leading to Fe deficiency (Johnson and Barton, 2007). Jensen et al. (2018) have studied the mobility of Ga in the plant-soil system and its potential to enter the food chain. Although the phytotoxicity of gallium has been demonstrated by a few studies, the effects of gallium oxide and its NPs, in particular, have not yet been studied. Further investigations are needed to identify the possible impacts of Ga_2O_3 NPs on plant growth and development.

6.4.6 Nickel oxide (NiO) nanoparticles

Treatment of *Nigella arvensis* L. with 50 mg/L nickel oxide nanoparticles (NiO NPs) increased the plant biomass and antioxidant activities. Still, higher concentrations led to a decrease in biomass and an inhibition of antioxidant activities (Chahardoli et al., 2020). Extensive research in the roots of *Allium cepa* L. has revealed that NiO NP concentration played a significant role in inducing toxicity by disrupting the delicate balance of reactive oxygen species (ROS)-nitric oxide, which became lethal for the cell at higher doses of the NiO NP, resulting in the accumulation of stress markers and the necessary termination of cellular mechanisms (Manna et al., 2021). Torbati (2018) has shown that high concentrations of NiO NPs are more toxic than bulk NiO, which is evident from the reduction in photosynthetic pigments, enhanced production of phenol, flavonoid, and Malondialdehyde (MDA) content leading to high ROS formation in *L. minor* and *Spirodela polyrhiza*. NiO NPs have been found to alter the photosynthetic electron transport mechanism in the aquatic plant *Lemna gibba* (Oukarroum et al., 2015). Chung et al. (2019) have suggested that the application of NiO NPs promotes harmful effects and increases the levels of phytochemicals such as glucosinolates and phenolic compounds in Chinese cabbage seedlings.

6.4.7 Copper oxide (CuO) nanoparticles

Copper oxide nanoparticles (CuO NPs) possess unique properties that make them more important than any other MONPs (Akintelu et al., 2020).

Increasing the concentration of CuO NPs poses several adverse effects on crop plants. Decreased growth and productivity, inhibition of root elongation, decreased biomass, delayed development, damage of cell membrane, chlorophyll synthesis, etc. are some of the toxic effects observed (Arif et al., 2018). CuO NPs have been shown to present less toxicity to bell pepper, and, when compared to bulk CuO, have not demonstrated significant differences in stem elongation, dry biomass, leaf area, and chlorophyll content (Rawat et al., 2018). The treatment of the *Lens culinaris* plant with CuO NPs leads to enhanced plant defense enzymes, increased antioxidant levels, lipid peroxidation, and proline content (Sarkar et al., 2020). The results obtained by Pelegrino et al. (2020) demonstrate that CuO NPs of concentration around 20 μg/mL show beneficial effects on the growth and development of lettuce. CuO NPs can be potentially genotoxic to plants, as displayed by the studies of Ghosh et al. (2017) on *Lathyrus sativus L.* It has to be noted that the genotoxic effects posed by this NP are similar to that of the impact caused by X-ray and gamma rays. Hydrogels embedded with CuO NPs have been used to enhance the growth and nutrient supply (P, Mg, Zn, and Mn) in *Fusarium oxysporum* infected lettuce (Shang et al., 2021).

6.4.8 Silver oxide (Ag_2O) nanoparticles

Studies performed by Singh and Kumar (2015) showed that silver oxide nanoparticles (Ag_2O NPs) decrease the dry weight, root length, and shoot length of *Vigna radiata*. Low concentrations of Ag_2O NPs reduce the activity of antioxidant enzymes such as SOD, POD, GR, and CAT, thereby playing an essential role in the drought stress tolerance of *Sanguisorba officinalis* and *Agropyron cristatum* (Jahantab et al., 2021).

6.4.9 Cadmium oxide (CdO) nanoparticles

When drought-stressed *S. officinalis* and *A. cristatum* were exposed to cadmium oxide nanoparticles (CdO NPs) of concentration up to 60 mg/L, the NP had a positive response on the enzyme activities (CAT, POD, SOD, and GR). In contrast, at higher concentrations (90–120 mg/L), it poses phytotoxicity and decreases enzyme activities (Jahantab et al., 2021). Treatment of *V. radiata* seeds with CdO NPs reduced the germination rate and caused accumulation of Cd. This reduced the shoot and root length as well as the biomass of plants. Interestingly, it has also been observed that the addition of sawdust to the growth medium minimized these effects and promoted plant growth and development (Subpiramaniyam et al., 2021). The accumulation of CdO NPs is concentration and humidity dependent in *V. radiata* (Jung et al., 2020).

6.4.10 Zirconia (ZrO_2) nanoparticles

Zirconia, also called zirconium dioxide, occurs in its natural form in the mineral baddeleyite, and can also be synthetically produced from zircon. It is a dielectric material that can be used as an insulator in transistors. It has also found many applications in industries (Keiteb et al., 2016). Studies by Doskocz et al. (2018) showed that zirconia nanoparticles (ZrO_2 NPs) were toxic to *Desmodesmus quadricauda* but not any other algal species tested (*Raphidocelis subcapitata*, *Microcystis aeruginosa*, and *L. minor*). It has also been shown that the nanoform of these substances poses more threats to plants than the bulk forms. The phytotoxicity of ZrO_2 NPs was tested on maize. The results of seed germination and root elongation showed that ZrO_2 NPs exerted neither positive nor negative influence on the parameters studied (Karunakaran et al., 2016). Uptake and accumulation of ZrO_2 were noticed when wheat plants were exposed to ZrO_2 NPs. This treatment has also damaged the roots to a significant extent (Zhang et al., 2019). Another study performed by Derbalah et al. (2019) has found that ZrO_2 NPs induce resistance to root rot disease caused by the fungi *Rhizoctonia solani* in cucumbers. This is evident from the expression of regulatory and defense genes in NP-exposed plants.

6.4.11 Aluminum oxide (Al_2O_3) nanoparticles

Aluminum oxide nanoparticles (Al_2O_3 NPs) are used extensively in industries and cosmetics. Sharma and Uttam (2019) have proved that the exposure of *T. aestivum* L. to Al_2O_3 NPs leads to a significant increase in the spectra of various biochemical parameters of leaf such as cellulose, hemicellulose, lignin, pectin, lipids, amino acids, and proteins. This indicates the toxic effects of Al_2O_3 NPs on the cell wall and cell membrane constituents of wheat leaves. In tobacco BY-2 cell suspension cultures, these NPs lead to programmed cell death as evident from the loss of mitotic potential, increased caspase-like activity, and DNA fragmentation (Poborilova et al., 2013). Treatment of *N. arvensis* L. with 50 and 100 mg/L of Al_2O_3 NPs resulted in increased biomass which was reduced when the concentration was increased further. An increase in antioxidant levels was also noticed upon Al_2O_3 NP treatment (Chahardoli et al., 2020). Chowdhury et al. (2020) have pointed out that Al_2O_3 NPs repress the growth of roots in plants like carrot, soybean, cucumber, cabbage, and corn. The aquatic plants *D. quadricauda*, *R. subcapitata*, and *M. aeruginosa* showed toxic symptoms when exposed to Al_2O_3 NPs. However, other algal species and *L. minor* were not affected (Doskocz et al., 2018). Compared to bulk Al_2O_3, the Al_2O_3 NPs have been shown to exert higher genotoxicity and oxidative stress in *A. cepa* (De et al., 2016).

6.4.12 Cobalt oxide (Co_3O_4) nanoparticles

The findings of Jahani et al. (2020) revealed that low cobalt oxide nanoparticles (Co_3O_4 NPs) had a favorable influence on growth indices. Still, elevated levels produced severe oxidative damage and triggered defensive responses through the organization of phenolic compounds and the antioxidative system. Treatment of *A. cepa* with Co_3O_4 NPs caused extensive adsorption of the NPs in the root system, leading to severe inhibition of the root elongation (Ghodake et al., 2011). Another investigation conducted by Faisal et al. (2016) in *Solanum melongena* L. showed that these NPs reduce seed germination and root growth. They also cause swelling of mitochondria and abnormalities in the nitric oxide signaling pathway, leading to DNA damage and cell death. Generation of ROS and physical damage to cell membrane are the other symptoms of phytotoxicity in algae (Sharan and Nara, 2019). Co_3O_4 NPs improve the elongation of roots of radish seedlings. The NPs were found to adsorb to the surface of seeds and release metal ions near the seeds, thereby causing phytotoxicity. This was severe than that of free metal ions in terms of Co_3O_4 NPs (Wu et al., 2012).

6.4.13 Magnesium oxide (MgO) nanoparticles

Magnesium oxide nanoparticles (MgO NPs) are adsorbents used in effluent and heavy metal treatment procedures. Treatment of textile and tannery effluents with MgO NPs has reduced their toxicity, and this is evident from the increased germination rate of *Zea mays* and *Vicia faba* (Fouda et al., 2021).

6.5 Rare earth oxide nanoparticles

The constant quest to introduce new technological innovations for the betterment of life has resulted in the utilization of rare earth elements (REE) for nanomaterial synthesis. This has revealed the excellent properties and applications of various elements that were almost ignored in the past. REEs (a.k.a. rare earth metals or rare earth oxides) include scandium (Sc), yttrium (Y), lanthanum (La), cerium (Ce), praseodymium (Pr), neodymium (Nd), promethium (Pm), samarium (Sm), europium (Eu), gadolinium (Gd), terbium (Tb), dysprosium (Dy), holmium (Ho), erbium (Er), thulium (Tm), ytterbium (Yb), and lutetium (Lu). The NPs of all these elements have been synthesized and characterized. Sc_2O_3, Y_2O_3, La_2O_3, CeO, CeO_2, Pr_6O_{11}, Nd_2O_3, Sm_2O_3, Eu_2O_3, Gd_2O_3, Tb_4O_7, Dy_2O_3, Ho_2O_3, Er_2O_3, Tm_2O_3, Yb_2O_3, and Lu_2O_3 are the various rare earth oxide nanoparticles synthesized (Ahmad et al., 2018).

6.5.1 Lanthanum oxide (La_2O_3) nanoparticles

The use of lanthanum oxide nanoparticles (La_2O_3 NPs) has drastically increased, paving the way for their entry into natural habitats. Liu et al. (2018) investigated the phytotoxicity of La_2O_3 NPs on *Zea mays* L. and observed that 5 mg/L of La_2O_3 was toxic to the plants. Decreases in shoot and root biomass, root length, chlorophyll content, and photosynthetic rate were noticed. Another study in cucumber indicated the transition of La_2O_3 NPs to needle-like La_2PO_4 nanoclusters in the intercellular space and middle lamella of the roots. The phytotoxicity is attributed to the dissolution of La_2O_3 NPs at the nano-bio interface due to the organic acids that have been excreted by roots (Ma et al., 2011). La_2O_3 NPs have been shown to cause severe toxicity in gram-positive bacteria, whereas gram-negative organisms are least affected. These NPs disrupt the microbiota balance leading to severe phytotoxicity in plants (Zheng et al., 2021). When Xiao et al. (2021) analyzed the effects of La_2O_3 NPs in radish, they found that these inhibited the root growth, reduced photosynthetic assimilation and sucrose transport, altered the structure of cell wall, and disrupted cell adhesion leading to cracking of roots. In a similar study, La_2O_3 NPs inhibited root elongation, shoot elongation, root biomass, and shoot biomass of cucumber. The formation of ROS was induced, causing root cell death (Ma et al., 2015).

6.5.2 Cerium dioxide (CeO_2) nanoparticles

The production and utilization of cerium dioxide nanoparticles (CeO_2 NPs) have skyrocketed, causing concerns about their entrance into biological systems. Studies show that this NP aggravates stress in plants, thereby affecting their physiological, morphological, and biochemical attributes (Prakash et al., 2021). Cui et al. (2014) has assessed the toxicity of CeO_2 NPs on asparagus lettuce that has been cultured on a plant agar medium. This study shows that high quantities of CeO_2 NPs limit root development, induce membrane damage and lipid peroxidation, and alter superoxide dismutase levels. Soybean seedlings exposed to CeO_2 NPs did not show a significant difference in growth when compared to control plants. However, NP translocation from root to shoot and the seed was detected, suggesting that Ce might enter the food chain (Rodrigues et al., 2021). Ionomic analysis of *Phaseolus vulgaris* L. grown on a solid medium containing CeO_2 NPs has shown that these NPs altered the secondary metabolites, and also interfered with K, Mo, and Zn (Salehi et al., 2019a,b). Under greenhouse circumstances, pot experiments of *Holcus lanatus* L. and *Diplotaxis tenuifolia* L. DC demonstrated that CeO_2 NPs were translocated to leaf tissues, resulting in macronutrient and micronutrient imbalances (Lizzi et al., 2021).

6.5.3 Yttrium oxide (Y_2O_3) nanoparticles

Yttrium oxide nanoparticles (Y_2O_3 NPs) are widely used in medicine, agriculture, catalysis, magnetism, and military operations. The widespread use of these NPs has posed threats to the environment. According to Zhao et al. (2021), these NPs did not affect *Oryza sativa* L. seedlings at low concentrations. However, seed germination was delayed, chlorophyll content was decreased, and root activity was also impeded at larger doses. This study also suggests that the Y_2O_3 NPs, rather than the Y^{3+} ions, were responsible for these phytotoxicities. Maize seeds exposed to Y_2O_3 NPs inhibited seed vitality, shoot growth, and root growth (Gong et al., 2019) (Table 6.1).

TABLE 6.1 Impact of metal oxide nanomaterials on seed germination, plant growth and development.

MO NP	Size/dose	Plant	Impacts	References
ZnO	25 mg; 100 mL	*Vigna mungo* L.	Increased germination percentage, chlorophyll, reducing sugars, total sugars and protein; decreased root length, shoot length, number of leaves, length of leaves, width of leaves, fresh weight of plant, and dry weight of plant	Pavani et al. (2020)
ZnO	400 and 800 mg dm^{-3}	*Solanum lycopersicum* L	Inhibition of root and shoot growth, decreased chlorophylls a and b content, reduced photosynthetic efficiency, enhanced defense response by increasing activities of antioxidant enzymes	Wang et al. (2018)
CuO	6.5 ± 1.5 nm; 0.025 mg/mL	*Lens culinaris*	Roots show highest defense enzyme activity, antioxidants, and increased lipid peroxidation and proline content	Sarkar et al. (2020)
Fe_3O_4	80 nm; 500 mg/L	*Brassica juncea* L.	Can act as nanoadsorbents to reduce the toxicity of As in *Brassica juncea*	Praveen et al. (2018)
ZnO and CuO mixture	10–1000 mg/kg	*Raphanus sativus*	Reduced root length, shoot length, biomass; accumulation of Cu and Zn	Singh and Kumar (2018)
CeO_2	6.05 ± 0.49 nm	*Gossypium hirsutum* L.	Improved cotton leaf Na^+ exclusion, tolerance to salinity, chlorophyll content, and biomass	Liu et al. (2021)
CeO_2		Rice	Salinity tolerance, improved growth performance under high salt concentration	Zhou et al. (2021)
ZnO	9 nm	*Zea mays*	Accumulation in roots followed by phytotoxicity	Lv et al. (2021)
CeO_2	10.3 nm	*Arabidopsis thaliana*	Protects plants from abiotic stress by scavenging the reactive oxygen species	Wu et al. (2017)

(*Continued*)

TABLE 6.1 (Continued)

MO NP	Size/dose	Plant	Impacts	References
CuO	10 mg and 250 mg per plant	Lettuce and cabbage	Higher foliar uptake of Cu, decrease in biomass, photosynthesis, water content	Xiong et al. (2017)
Fe_3O_4	20–30 nm; 25 mg/L and 50 mg/L	Rice	NPs improved Fe uptake by arsenic (As) stressed plants, reduced As phytotoxicity and oxidative stress leading to improved photosynthesis and growth of rice plants.	Bidi et al. (2021)
ZnO	15–18 nm; 100–200 μM	*Triticum aestivum* L.	Affected the growth and photosynthetic efficiency of wheat but was overcome by adding nitric oxide	Tripathi et al. (2017)
CeO_2	50 mg/L	*Glycine max* (Linn.) Merrill.	CeO_2 NP increased Sb(III) accumulation in roots whereas Sb(III) and Sb(IV) decreased CeO_2 accumulation	Cao et al. (2020)
TiO_2	50 mg/L	*Glycine max* (Linn.) Merrill.	Sb increased the accumulation of TiO_2 NPs in leaves and roots	Cao et al. (2020)
CuO	50 mg/L	*Glycine max* (Linn.) Merrill.	Sb increased the accumulation of CuO NPs in leaves and roots	Cao et al. (2020)
ZnO	250 mg/L	Rice	Growth inhibition, reduced chlorophyl content	Chen et al. (2015)
TiO_2	50, 100 and 500 mg/L	*Oryza sativa* L.	Reduces the phytotoxicity of Cd but promoted migration of Cd to seeds	Zhang et al. (2020)
CdO	<50 nm	*Sanguisorba officinalis* and *Agropyron cristatum*	Positive effect on enzyme activity reduction at lower conentration but increased the enzyme activity at higher concetration	Jahantab et al. (2021)

ZnO	20 nm	*Sanguisorba officinalis* and *Agropyron cristatum*	Positive effect on enzyme activity reduction at lower conentration but increased the enzyme activity at higher concetration	Jahantab et al. (2021)
AgO	50 nm	*Sanguisorba officinalis* and *Agropyron cristatum*	Positive effect on enzyme activity reduction at lower conentration but increased the enzyme activity at higher concetration	Jahantab et al. (2021)
FeO	<50 nm	*Sanguisorba officinalis* and *Agropyron cristatum*	Positive effect on enzyme activity reduction at lower conentration but increased the enzyme activity at higher concetration	Jahantab et al. (2021)
CuO	100 mg	*Vigna radiata*	Higher accumulation of Cu in plant	Subpiramaniyam et al. (2021)
CuO	38 nm	*Arabidopsis thaliana*	Adverse effects on stomata, chlorophyll content, plant biomass, guard cells, and other cel organelles with an increase in ROS level	Azhar et al. (2020)
CuO	18.2 nm	*Coriandrum sativum* L.	Causes growth inhibition and cytotoxicity in plants	Pramanik et al. (2018)
CeO_2	500 mg/L	Cotton (conventional and transgenic)	NP attached to external surface of chloroplast ad ruptured it; destroyed vascular bundles, reduced Zn, Fe, Mg, and P levels in xylem sap and reduced IAA and ABA levels in roots	Nhan et al. (2015)
ZnO	20 μg/mL	*Allium cepa*	Severe accumulation in the cells causing chromosomal damage	Ghodake et al. (2011)

References

Abbasi, E., Aval, S.F., Akbarzadeh, A., Milani, M., Nasrabadi, H.T., Joo, S.W., et al., 2014. Dendrimers: synthesis, applications, and properties. Nanoscale Res. Lett. 9 (1), 1–10.

Ahmad, H.R., Zia-ur-Rehman, M., Sohail, M.I., ul Haq, M.A., Khalid, H., Ayub, M.A., et al., 2018. Effects of rare earth oxide nanoparticles on plants. Nanomaterials in Plants, Algae, and Microorganisms. Academic Press, pp. 239–275.

Akintelu, S.A., Folorunso, A.S., Folorunso, F.A., Oyebamiji, A.K., 2020. Green synthesis of copper oxide nanoparticles for biomedical application and environmental remediation. Heliyon 6 (7), e04508.

Alharby, H.F., Rizwan, M., Iftikhar, A., Hussaini, K.M., ur Rehman, M.Z., Bamagoos, A.A., et al., 2021. Effect of gibberellic acid and titanium dioxide nanoparticles on growth, antioxidant defense system and mineral nutrient uptake in wheat. Ecotoxicol. Environ. Saf. 221, 112436.

Arif, N., Yadav, V., Singh, S., Tripathi, D.K., Dubey, N.K., Chauhan, D.K., et al., 2018. Interaction of copper oxide nanoparticles with plants: uptake, accumulation, and toxicity. Nanomaterials in Plants, Algae, and Microorganisms. Academic Press, pp. 297–310.

Awan, S., Shahzadi, K., Javad, S., Tariq, A., Ahmad, A., Ilyas, S., 2021. A preliminary study of influence of zinc oxide nanoparticles on growth parameters of Brassica oleracea var italic. J. Saudi Soc. Agric. Sci 20 (1), 18–24.

Azhar, W., Khan, A.R., Muhammad, N., Liu, B., Song, G., Hussain, A., et al., 2020. Ethylene mediates CuO NP-induced ultrastructural changes and oxidative stress in *Arabidopsis thaliana* leaves. Environ. Sci. Nano 7 (3), 938–953.

Bidi, H., Fallah, H., Niknejad, Y., Tari, D.B., 2021. Iron oxide nanoparticles alleviate arsenic phytotoxicity in rice by improving iron uptake, oxidative stress tolerance and diminishing arsenic accumulation. Plant. Physiol. Biochem. 163, 348–357.

Cao, W., Gong, J., Zeng, G., Song, B., Zhang, P., Li, J., et al., 2020. Potential interactions between three common metal oxide nanoparticles and antimony (III/V) involving their uptake, distribution, and phytotoxicity to soybean. ACS Sustain. Chem. Eng. 8 (27), 10125–10141.

Cervantes-Avilés, P., Huang, X., Keller, A.A., 2021. Dissolution and aggregation of metal oxide nanoparticles in root exudates and soil leachate: implications for nanoagrochemical application. Environ. Sci. Technol.

Chahardoli, A., Karimi, N., Ma, X., Qalekhani, F., 2020. Effects of engineered aluminum and nickel oxide nanoparticles on the growth and antioxidant defense systems of *Nigella arvensis* L. Sci. Rep. 10 (1), 1–11.

Chaudhary, I.J., Singh, V., 2020. Titanium dioxide nanoparticles and its impact on growth, biomass and yield of agricultural crops under environmental stress: a review. Res. J. Nanosci. Nanotechnol. 10, 1–8.

Chavan, S., Sarangdhar, V., Nadanathangam, V., 2020. Toxicological effects of TiO2 nanoparticles on plant growth promoting soil bacteria. Emerg. Contam. 6, 87–92.

Chen, J., Liu, X., Wang, C., Yin, S.S., Li, X.L., Hu, W.J., et al., 2015. Nitric oxide ameliorates zinc oxide nanoparticles-induced phytotoxicity in rice seedlings. J. Hazard. Mater. 297, 173–182.

Chowdhury, M.A., Shuvho, M.B.A., Hossain, M.I., Ali, M.O., Kchaou, M., Rahman, A., et al., 2020. Multiphysical analysis of nanoparticles and their effects on plants. Biotechnol. Appl. Biochem.

Chung, I.M., Venkidasamy, B., Thiruvengadam, M., 2019. Nickel oxide nanoparticles cause substantial physiological, phytochemical, and molecular-level changes in Chinese cabbage seedlings. Plant. Physiol. Biochem. 139, 92–101.

Cui, D., Zhang, P., Ma, Y., He, X., Li, Y., Zhang, J., et al., 2014. Effect of cerium oxide nanoparticles on asparagus lettuce cultured in an agar medium. Environ. Sci.: Nano 1 (5), 459–465.

De, A., Chakrabarti, M., Ghosh, I., Mukherjee, A., 2016. Evaluation of genotoxicity and oxidative stress of aluminium oxide nanoparticles and its bulk form in *Allium cepa*. Nucl. 59 (3), 219–225.

Derbalah, A., Elsharkawy, M.M., Hamza, A., El-Shaer, A., 2019. Resistance induction in cucumber and direct antifungal activity of zirconium oxide nanoparticles against Rhizoctonia solani. Pestic. Biochem. Physiol. 157, 230–236.

Doskocz, N., Załęska Radziwiłł, M., Affek, K., Łebkowska, M., 2018. Effects of selected nanoparticles on aquatic plants. Desalination Water Treat. 117, 42–48.

Emamverdian, A., Ding, Y., Mokhberdoran, F., Xie, Y., Zheng, X., Wang, Y., 2020. Silicon dioxide nanoparticles improve plant growth by enhancing antioxidant enzyme capacity in bamboo (*Pleioblastus pygmaeus*) under lead toxicity. Trees 34 (2), 469–481.

Faisal, M., Saquib, Q., Alatar, A.A., Al-Khedhairy, A.A., Ahmed, M., Ansari, S.M., et al., 2016. Cobalt oxide nanoparticles aggravate DNA damage and cell death in eggplant via mitochondrial swelling and NO signaling pathway. Biol. Res. 49 (1), 1–13.

Fouda, A., Hassan, S.E.D., Saied, E., Hamza, M.F., 2021. Photocatalytic degradation of real textile and tannery effluent using biosynthesized magnesium oxide nanoparticles (MgO-NPs), heavy metal adsorption, phytotoxicity, and antimicrobial activity. J. Environ. Chem. Eng. 9 (4), 105346.

Ghodake, G., Seo, Y.D., Lee, D.S., 2011. Hazardous phytotoxic nature of cobalt and zinc oxide nanoparticles assessed using *Allium cepa*. J. Hazard. Mater. 186 (1), 952–955.

Ghosh, B., Datta, A.K., Pramanik, A., Kumbhakar, D.V., Das, D., Paul, R., et al., 2017. Mutagenic effectivity of cadmium sulphide and copper oxide nanoparticles on some physiological and cytological attributes of *Lathyrus sativus L.* Cytologia 82 (3), 267–271.

Gohari, G., Mohammadi, A., Akbari, A., Panahirad, S., Dadpour, M.R., Fotopoulos, V., et al., 2020. Titanium dioxide nanoparticles (TiO_2 NPs) promote growth and ameliorate salinity stress effects on essential oil profile and biochemical attributes of *Dracocephalum moldavica*. Sci. Rep. 10 (1), 1–14.

Gong, C., Wang, L., Li, X., Wang, H., Jiang, Y., Wang, W., 2019. Responses of seed germination and shoot metabolic profiles of maize (*Zea mays* L.) to Y_2O_3 nanoparticle stress. RSC Adv. 9 (47), 27720–27731.

Irum, S., Jabeen, N., Ahmad, K.S., Shafique, S., Khan, T.F., Gul, H., et al., 2020. Biogenic iron oxide nanoparticles enhance callogenesis and regeneration pattern of recalcitrant *Cicer arietinum L.* PLoS One 15 (12), e0242829.

Jahani, M., Khavari-Nejad, R.A., Mahmoodzadeh, H., Saadatmand, S., 2020. Effects of cobalt oxide nanoparticles (Co3O4 NPs) on ion leakage, total phenol, antioxidant enzymes activities and cobalt accumulation in *Brassica napus L.* Not. Bot. Horti Agrobot. Cluj-Napoca 48 (3), 1260–1275.

Jahantab, E., Farzadmehr, J., Abadi, N., Yazdanshenas, H., 2021. Effect of metal oxide nanoparticles on plant enzymatic activities under drought. Res. Sq. Available from: https://doi.org/10.21203/rs.3.rs-664294/v1.

Jensen, H., Gaw, S., Lehto, N.J., Hassall, L., Robinson, B.H., 2018. The mobility and plant uptake of gallium and indium, two emerging contaminants associated with electronic waste and other sources. Chemosphere 209, 675–684.

Johnson, G.V., Barton, L.L., 2007. Inhibition of iron deficiency stress response in cucumber by rare earth elements. Plant. Physiol. Biochem. 45 (5), 302–308.

Jung, E.S., Sivakumar, S., Hong, S.C., Yi, P.I., Jang, S.H., Suh, J.M., 2020. Influence of relative humidity on germination and metal accumulation in *Vigna radiata* exposed to metal-based nanoparticles. Sustainability 12 (4), 1347.

Kamran, M., Ali, H., Saeed, M.F., Bakhat, H.F., Hassan, Z., Tahir, M., et al., 2020. Unraveling the toxic effects of iron oxide nanoparticles on nitrogen cycling through manure-soil-plant continuum. Ecotoxicol. Environ. Saf. 205, 111099.

Kanwar, M.K., Sun, S., Chu, X., Zhou, J., 2019. Impacts of metal and metal oxide nanoparticles on plant growth and productivity. Nanomaterials and Plant Potential. Springer, Cham, pp. 379–392.

Karimi, J., Mohsenzadeh, S., 2016. Effects of silicon oxide nanoparticles on growth and physiology of wheat seedlings. Russ. J. Plant Physiol. 63 (1), 119–123.

Karunakaran, G., Suriyaprabha, R., Rajendran, V., Kannan, N., 2016. Influence of ZrO_2, SiO_2, Al_2O_3 and TiO_2 nanoparticles on maize seed germination under different growth conditions. IET Nanobiotechnol. 10 (4), 171–177.

Keiteb, A.S., Saion, E., Zakaria, A., Soltani, N., 2016. Structural and optical properties of zirconia nanoparticles by thermal treatment synthesis. J. Nanomater 2016.

Khan, S.A., 2020. Metal nanoparticles toxicity: role of physicochemical aspects. Metal Nanoparticles for Drug Delivery and Diagnostic Applications. Elsevier, pp. 1–11.

Kokina, I., Plaksenkova, I., Jermaļonoka, M., Petrova, A., 2020. Impact of iron oxide nanoparticles on yellow medick (*Medicago falcata L.*) plants. J. Plant. Interact. 15 (1), 1–7.

Lespes, G., Faucher, S., Slaveykova, V.I., 2020. Natural nanoparticles, anthropogenic nanoparticles, where is the frontier? Front. Environ. Sci. 8, 71.

Liu, J., Li, G., Chen, L., Gu, J., Wu, H., Li, Z., 2021. Cerium oxide nanoparticles improve cotton salt tolerance by enabling better ability to maintain cytosolic K + /Na + ratio. J. Nanobiotechnol. 19 (1), 1–16.

Liu, Y., Xu, L., Dai, Y., 2018, February. Phytotoxic effects of lanthanum oxide nanoparticles on maize (*Zea mays* L.). In IOP Conference Series: Earth and Environmental Science (Vol. 113, No. 1, p. 012020). IOP Publishing.

Lizzi, D., Mattiello, A., Adamiano, A., Fellet, G., Gava, E., Marchiol, L., 2021. Influence of cerium oxide nanoparticles on two terrestrial wild plant species. Plants 10 (2), 335.

Lv, Z., Sun, H., Du, W., Li, R., Mao, H., Kopittke, P.M., 2021. Interaction of different-sized ZnO nanoparticles with maize (*Zea mays*): accumulation, biotransformation and phytotoxicity. Sci. Total. Environ. 796, 148927.

Ma, Y., He, X., Zhang, P., Zhang, Z., Guo, Z., Tai, R., et al., 2011. Phytotoxicity and biotransformation of La2O3 nanoparticles in a terrestrial plant cucumber (*Cucumis sativus*). Nanotoxicology 5 (4), 743–753.

Ma, Y., Zhang, P., Zhang, Z., He, X., Li, Y., Zhang, J., et al., 2015. Origin of the different phytotoxicity and biotransformation of cerium and lanthanum oxide nanoparticles in cucumber. Nanotoxicology 9 (2), 262–270.

Madany, M.M., Saleh, A.M., Habeeb, T.H., Hozzein, W.N., AbdElgawad, H., 2020. Silicon dioxide nanoparticles alleviate the threats of broomrape infection in tomato by inducing cell wall fortification and modulating ROS homeostasis. Environ. Sci.: Nano 7 (5), 1415–1430.

Manna, I., Sahoo, S., Bandyopadhyay, M., 2021. Effect of engineered nickel oxide nanoparticle on reactive oxygen species–nitric oxide interplay in the roots of *Allium cepa* L. Front. Plant. Sci. 12.

Mittal, P., Saharan, A., Verma, R., Altalbawy, F., Alfaidi, M.A., Batiha, G.E.S., et al., 2021. Dendrimers: a new race of pharmaceutical nanocarriers. BioMed. Res. Int. 2021.

Nhan, L.V., Ma, C., Rui, Y., Liu, S., Li, X., Xing, B., et al., 2015. Phytotoxic mechanism of nanoparticles: destruction of chloroplasts and vascular bundles and alteration of nutrient absorption. Sci. Rep. 5, 11618.

Niño-Martínez, N., Salas Orozco, M.F., Martínez-Castañón, G.A., Torres Méndez, F., Ruiz, F., 2019. Molecular mechanisms of bacterial resistance to metal and metal oxide nanoparticles. Int. J. Mol. Sci. 20 (11), 2808.

Oukarroum, A., Barhoumi, L., Samadani, M., Dewez, D., 2015. Toxic effects of nickel oxide bulk and nanoparticles on the aquatic plant *Lemna gibba L.* BioMed. Res. Int. 2015.

Pavani, K.V., Beulah, M., Sai Poojitha, G.U., 2020. The effect of zinc oxide nanoparticles (ZnO NPs) on *Vigna mungo L.* seedling growth and antioxidant activity. Nanosci. Nanotechnol.-Asia 10 (2), 117–122.

Pelegrino, M.T., Kohatsu, M.Y., Seabra, A.B., Monteiro, L.R., Gomes, D.G., Oliveira, H.C., et al., 2020. Effects of copper oxide nanoparticles on growth of lettuce (*Lactuca sativa L.*) seedlings and possible implications of nitric oxide in their antioxidative defense. Environ. Monit. Assess. 192 (4), 1–14.

Plaksenkova, I., Jermaļonoka, M., Bankovska, L., Gavarāne, I., Gerbreders, V., Sledevskis, E., et al., 2019. Effects of Fe3O4 nanoparticle stress on the growth and development of rocket *Eruca sativa*. J. Nanomater., 2019.

Poborilova, Z., Opatrilova, R., Babula, P., 2013. Toxicity of aluminium oxide nanoparticles demonstrated using a BY-2 plant cell suspension culture model. Environ. Exp. Bot. 91, 1–11.

Prakash, V., Peralta-Videa, J., Tripathi, D.K., Ma, X., Sharma, S., 2021. Recent insights into the impact, fate and transport of cerium oxide nanoparticles in the plant-soil continuum. Ecotoxicol. Environ. Saf. 221, 112403.

Pramanik, A., Datta, A.K., Das, D., Kumbhakar, D.V., Ghosh, B., Mandal, A., et al., 2018. Assessment of nanotoxicity (cadmium sulphide and copper oxide) using cytogenetical parameters in *Coriandrum sativum L.*(Apiaceae). Cytol. Genet. 52 (4), 299–308.

Prasanna, S.S., Balaji, K., Pandey, S., Rana, S., 2019. Metal oxide based nanomaterials and their polymer nanocomposites. Nanomaterials and Polymer Nanocomposites. Elsevier, pp. 123–144.

Praveen, A., Khan, E., Ngiimei D, S., Perwez, M., Sardar, M., Gupta, M., 2018. Iron oxide nanoparticles as nano-adsorbents: a possible way to reduce arsenic phytotoxicity in Indian mustard plant (*Brassica juncea L.*). J. Plant. Growth Regul. 37 (2), 612–624.

Rajput, V.D., Minkina, T., Fedorenko, A., Chernikova, N., Hassan, T., Mandzhieva, S., et al., 2021a. Effects of zinc oxide nanoparticles on physiological and anatomical indices in spring barley tissues. Nanomaterials 11 (7), 1722.

Rajput, V.D., Minkina, T., Feizi, M., Kumari, A., Khan, M., Mandzhieva, S., et al., 2021b. Effects of silicon and silicon-based nanoparticles on rhizosphere microbiome, plant stress and growth. Biology 10 (8), 791.

Rastogi, A., Zivcak, M., Sytar, O., Kalaji, H.M., He, X., Mbarki, S., et al., 2017. Impact of metal and metal oxide nanoparticles on plant: a critical review. Front. Chem. 5, 78.

Rawat, S., Pullagurala, V.L., Hernandez-Molina, M., Sun, Y., Niu, G., Hernandez-Viezcas, J.A., et al., 2018. Impacts of copper oxide nanoparticles on bell pepper (*Capsicum annum L.*) plants: a full life cycle study. Environ. Sci.: Nano 5 (1), 83–95.

Rodrigues, E.S., Montanha, G.S., de Almeida, E., Fantucci, H., Santos, R.M., de Carvalho, H. W., 2021. Effect of nano cerium oxide on soybean (*Glycine max L.* Merrill) crop exposed to environmentally relevant concentrations, Chemosphere, 273. p. 128492.

Saleh, A.M., Hassan, Y.M., Selim, S., AbdElgawad, H., 2019a. NiO-nanoparticles induce reduced phytotoxic hazards in wheat (*Triticum aestivum L.*) grown under future climate CO2. Chemosphere 220, 1047–1057.

Salehi, H., Miras-Moreno, B., Chehregani Rad, A., Pii, Y., Mimmo, T., Cesco, S., et al., 2019b. Relatively low dosages of CeO2 nanoparticles in the solid medium induce adjustments in the secondary metabolism and ionomic balance of bean (*Phaseolus vulgaris L.*) roots and leaves. J. Agric. Food Chem. 68 (1), 67–76.

Sarkar, J., Chakraborty, N., Chatterjee, A., Bhattacharjee, A., Dasgupta, D., Acharya, K., 2020. Green synthesized copper oxide nanoparticles ameliorate defence and antioxidant enzymes in *Lens culinaris*. Nanomaterials 10 (2), 312.

Shang, H., Ma, C., Li, C., Zhao, J., Elmer, W., White, J.C., et al., 2021. Copper oxide nanoparticle-embedded hydrogels enhance nutrient supply and growth of lettuce (*Lactuca sativa*) infected with *Fusarium oxysporum* f. sp. lactucae. Environ. Sci. Technol.

Sharan, A., Nara, S., 2019. Phytotoxic properties of zinc and cobalt oxide nanoparticles in algaes. Nanomaterials in Plants, Algae and Microorganisms. Academic Press, pp. 1–22.

Sharma, S., Uttam, K.N., 2019. Non-destructive and rapid interrogation of biochemical response of the leaves of wheat seedlings towards Al2O3 nanoparticles stress using attenuated total reflectance Fourier transform infrared spectroscopy. Vib. Spectrosc. 100, 142–151.

Shi, F., Qiao, H., 2021. Preparations, properties and applications of gallium oxide nanomaterials—a review. Nano Sel.

Silva, S., Ribeiro, T.P., Santos, C., Pinto, D.C., Silva, A.M., 2020. TiO2 nanoparticles induced sugar impairments and metabolic pathway shift towards amino acid metabolism in wheat. J. Hazard. Mater. 399, 122982.

Singh, D., Kumar, A., 2015. Effects of nano silver oxide and silver ions on growth of *Vigna radiata*. Bull. Environ. Contam. Toxicol. 95 (3), 379–384.

Singh, D., Kumar, A., 2018. Investigating long-term effect of nanoparticles on growth of Raphanus sativus plants: a trans-generational study. Ecotoxicology 27 (1), 23–31.

Souza, L.R.R., Bernardes, L.E., Barbetta, M.F.S., da Veiga, M.A.M.S., 2019. Iron oxide nanoparticle phytotoxicity to the aquatic plant *Lemna minor*: effect on reactive oxygen species (ROS) production and chlorophyll a/chlorophyll b ratio. Environ. Sci. Pollut. Res. 26 (23), 24121–24131.

Subpiramaniyam, S., Hong, S.C., Yi, P.I., Jang, S.H., Suh, J.M., Jung, E.S., et al., 2021. Influence of sawdust addition on the toxic effects of cadmium and copper oxide nanoparticles on *Vigna radiata* seeds. Environ. Pollut. 289, 117311.

Torbati, S., 2018. Phytotoxicological effects of bulk-nio and nio nanoparticles on lesser and giant duckweeds as model macrophytes: changes in the plants physiological responses. Iran. J. Toxicol. 12 (4), 31–39.

Tripathi, D.K., Mishra, R.K., Singh, S., Singh, S., Singh, V.P., Singh, P.K., et al., 2017. Nitric oxide ameliorates zinc oxide nanoparticles phytotoxicity in wheat seedlings: implication of the ascorbate-glutathione cycle. Front. Plant. Sci. 8, 1.

Wang, X.P., Li, Q.Q., Pei, Z.M., Wang, S.C., 2018. Effects of zinc oxide nanoparticles on the growth, photosynthetic traits, and antioxidative enzymes in tomato plants. Biol. Plant. 62 (4), 801–808.

Wu, H., Tito, N., Giraldo, J.P., 2017. Anionic cerium oxide nanoparticles protect plant photosynthesis from abiotic stress by scavenging reactive oxygen species. ACS Nano 11 (11), 11283–11297.

Wu, S.G., Huang, L., Head, J., Chen, D.R., Kong, I.C., Tang, Y.J., 2012. Phytotoxicity of metal oxide nanoparticles is related to both dissolved metals ions and adsorption of particles on seed surfaces. J. Pet. Env. Biotechnol. 3 (4), 126.

Xiao, Z., Yue, L., Wang, C., Chen, F., Ding, Y., Liu, Y., et al., 2021. Downregulation of the photosynthetic machinery and carbon storage signaling pathways mediate La2O3 nanoparticle toxicity on radish taproot formation. J. Hazard. Mater. 411, 124971.

Xiong, T., Dumat, C., Dappe, V., Vezin, H., Schreck, E., Shahid, M., et al., 2017. Copper oxide nanoparticle foliar uptake, phytotoxicity, and consequences for sustainable urban agriculture. Environ. Sci. Technol. 51 (9), 5242–5251.

Yassen, A., Abdallah, E., Gaballah, M., Zaghloul, S., 2017. Role of silicon dioxide nano fertilizer in mitigating salt stress on growth, yield and chemical composition of cucumber (*Cucumis sativus L.*). Int. J. Agric. Res. 22, 130–135.

Zhang, W., Huang, Y., Gong, H., Dang, F., Zhou, D., 2019. Different uptake of metal dioxide nanoparticles (ceria nanoparticles, zirconia nanoparticles and silica nanoparticles) by Wheat. Bull. Environ. Contam. Toxicol. 103 (1), 199–205.

Zhang, W., Long, J., Geng, J., Li, J., Wei, Z., 2020. Impact of titanium dioxide nanoparticles on Cd phytotoxicity and bioaccumulation in rice (*Oryza sativa L.*). Int. J. Environ. Res. Public Health 17 (9), 2979.

Zhao, X., Zhang, W., He, Y., Wang, L., Li, W., Yang, L., et al., 2021. Phytotoxicity of Y2O3 nanoparticles and Y3 + ions on rice seedlings under hydroponic culture. Chemosphere 263, 127943.

Zheng, H., Gu, Z., Pan, Y., Chen, J., Xie, Q., Xu, S., et al., 2021. Biotransformation of rare earth oxide nanoparticles eliciting microbiota imbalance. Part. Fibre Toxicol. 18 (1), 1–14.

Zhou, H., Wu, H., Zhang, F., Su, Y., Guan, W., Xie, Y., et al., 2021. Molecular basis of cerium oxide nanoparticle enhancement of rice salt tolerance and yield. Environ. Sci. Nano.

Chapter 7

Impact of metal oxide nanoparticles against biotic stress in plants

P. Akila Vaishnavi[1], N.K. Udaya Prakash[2], M. Jayanthi[2], M. Suganthi[2] and M. Thenmozhi[2]

[1]Department of Microbiology, Nadar Saraswathi College of Arts and Science, Theni, Tamil Nadu, India, [2]Department of Biotechnology, Vels Institute of Science, Technology and Advanced Studies (VISTAS), Chennai, Tamil Nadu, India

Chapter outline

7.1 Introduction

As the energy producer of the planet, plants are the basis on which all other life depends. They are the suppliers of food, maintain oxygen content, and are the primary source of accessory foods and vitamins. Without plants we would starve to death, die of suffocation, and succumb to a combination of deficiency diseases (William, 1944). Plants in their natural habitats are simultaneously and persistently confronted with a range of biotic components, which include biotrophic and necrotrophic bacteria, phytoplasmas, oomycetes, fungi, nematodes, and noncellular pathogens, that is, viruses and viroids (Mantri et al., 2010). Plant parasites are the cause of at least 25% of crop loss in the world's total gross agricultural production (Pandey et al., 2018). However, complex immune response pathways are provoked by the plant system against these pathogens (biotic stress) to survive. This is not

Nanometal Oxides in Horticulture and Agronomy. DOI: https://doi.org/10.1016/B978-0-323-91809-1.00006-8

specific to single stress but to the combination of stresses, both biotic and abiotic (Nejat and Mantri, 2017). The plant's form tends to change in every step. With every adverse condition and altered situation, a new plant starts up (John, 1853). If the plant fails on its own, the recent technology, that is, nanotechnology, comes in handy for the defense of plants against their biotic stress.

As nanoscience progressed, the use of engineered nanomaterials in agriculture became more popular in recent years. Delivery of genetic material, biosensor creation, and gradual release of nanomaterial-assisted fertilizers are among the recently developed applications (Acharya and Pal, 2020; Jatav and De, 2013). As metals proved their toxicity toward pathogens and pests, and as the nanoparticles (NPs) demonstrated their effectiveness toward their metallic counterparts, the metal oxide NPs (MONPs) play a major role in the fight against plant biotic stress. Nanotechnology has the potential to revolutionize the agriculture sector with new tools to promote molecular treatment of diseases, detect diseases in a rapid manner, and improve the ability of plants to absorb nutrients (Patolsky et al., 2006).

Metals are necessary nutrients for plants, and elements such as Co, Fe, Mn, Cu, Mo, Ni, and Zn help plants grow and develop normally (Singh et al., 2016). These metal nutrients are found in small amounts in the soil and are stably delivered to the plant by metal transporters (Clemens, 2001). Metals have long been used as antibacterial agents to help plants deal with biotic stress. Metals cause various types of injuries to microbial cells as a result of oxidative stress, protein dysfunction, and membrane damage (Lemire et al., 2013). There are different metal (silver, zinc, copper, magnesium, and titanium) NPs synthesized and studied for their antimicrobial activity against viral, bacterial, and fungal plant pathogens, (Dizaj et al., 2014), pests (Korghond et al., 2021), nematodes (Li et al., 2020), and oomycetes (Zabrieski et al., 2015).

7.2 Biotic stress in plants

The plants constantly face multiple challenges from other living organisms, especially pathogens, like viruses, bacteria, oomycetes, fungi, nematodes, and pests induce biotic stress.

In addition to general and specific stress reactions, redundant defensive systems also developed in response to biotic slaughter (Dreher and Callis, 2007). Disease resistance is associated with a plant's defense response that involves an integrated set of signal transduction pathways (Schenk et al., 2000). Plants have developed complex sensory mechanisms to identify biotic invasion and overcome the detriment of growth, yield, and survival. Plants have evolved a surfeit of responses to defend themselves against attacks by a broad spectrum of pests and pathogens, including viruses, nematodes, bacteria, fungi, and herbivorous insects (Hammond-Kosack and Jones, 2000).

Thus, plants tend to balance their response and biotic stress to combat the deleterious effect on their survival (Peck and Ron, 2020).

More than 1000 viruses are currently known to be capable of infecting plants. Despite a large number of possible combinations, the development of the disease is an exception in most outcomes and thus, in most cases, plants are capable of counteracting the harmful effects of viruses (Culver and Padmanabhan, 2007). Viruses are intracellular obligate parasites; they require host resources to complete their infection cycle, which causes local and systemic damage (Pallas and Garcia, 2011).

This resistance is owed to the absence of essential host susceptibility factors (passive resistance) or to the existence of several defense layers that the virus has to overcome. First, the virus needs to overcome a series of preexisting physical and chemical barriers in plants. Even though numerous possibilities result in viral infection, plants do not develop disease because they can resist their detrimental effects (Maule et al., 2002).

Pseudomonas aeruginosa clinical strains PAO1 and PA14 can infect the roots of two plant species, that is, Arabidopsis and sweet basil, provoking black necrotic regions in the roots of both plants and water-soaked lesions on leaves of sweet basils. However, plant mortality takes place 7 days after exposure (Walker et al., 2004). Zinc oxide (ZnO) NPs (25 Lg/mg) yielded high suppression of pathogenic bacteria *P. aeruginosa.*

Fungal parasites are the notorious pathogens feeding on the plant kingdom and they can be either necrotrophic (kill host cell by toxin secretion) or biotrophic (feed on living host cell). They cause vascular wilts, leaf spots, blotches, blights, cankers, and stunted growth in plants (Laluk and Mengiste, 2010). Airborne fungal pathogens that cause important pre and postharvest plant diseases, such as Alternaria tomato brown leaf spot, gray mold caused by *Botrytis cinerea*, Monilia stone fruit brown rot, and strawberry Anthracnose caused by *Colletotrichum* spp., require a large number of fungicide applications for their control. Their resistance to the fungicides is seriously impairing effective control (Malandrakis et al., 2013, 2018). Soilborne pathogens that cause wilting, such as *Verticillium dahliae* or root rots like FORL and *Fusarium solani*, are a wide range of pathogens that are difficult to control with conventional fungicides except for high-cost fumigants (Paplomatas et al., 2005).

Not all microbes are villainous. Some microbes are beneficial to plants and perform the same role as chemical fertilizers and pesticides, acting as biofertilizers and/or biopesticides. *Bacillus subtilis* exhibits both a direct and indirect biocontrol mechanism to suppress the diseases. The direct mechanism includes the synthesis of many secondary metabolites, hormones, cell-wall-degrading enzymes, and antioxidants that assist the plant in its defense against pathogen attack. The indirect mechanism includes the stimulation of plant growth and the induction of acquired systemic resistance. *B. subtilis* also plays a significant role in improving tolerance to biotic stresses (Hashem et al., 2019).

Nematodes are a costly burden in agriculture. More than 4,100 species of plant-parasitic nematodes are identified that can destroy all varieties of plants (Decraemer and Hunt, 2006). Ubiquitous in nature, they feed on economically important crops and represent a significant constraint on global food security. Nematodes cause soilborne diseases that in turn induce nutrient deficiency, stunted growth, and wilting. Root-knot nematodes (*Meloidogyne* spp.), cyst nematodes (*Heterodera* and *Globodera* spp.), and lesion nematodes (*Pratylenchus* spp.) are the most important species due to their intricate relationship with the host plants, wide host range, and the level of damage ensued by infection (Bernard et al., 2017). Around 15% of known nematode species directly target plant roots and prevent uptake of water and nutrition, intruding in the plant growth and development thereby reducing the overall quality and yield. Nematodes in the order *Tylenchida* are pathogens of plants (Sasser and Freckman, 1987).

Development in nanotechnology-based agricultural systems helps to manage nematodes by using NPs as drug carriers and a controllable drug targeting and releasing system. The NPs used as protectants and carriers against the nematodes are termed nanonematicides. Several phytopathogenic nematodes are very effectively managed with the means of NPs (Safeena and Zakeel, 2020).

On the contrary, mites and insects impair plants by either feeding (piercing or sucking) on them or laying eggs. The insects might also act as carriers of other viruses and bacteria (Schumann and D'Arcy, 2006). Plants have developed an elaborate immune system to combat such stresses (Taiz and Zeiger, 2006; Saijo and Loo, 2020). Plants have a passive first line of defense, which includes physical barriers, such as cuticles, wax, and trichomes, to avert pathogens and insects. Plants are also capable of producing chemical compounds to defend themselves from infecting pathogens (Taiz and Zeiger, 2006). Additionally, plants trigger defense against biotic agents by two levels of pathogen recognition (Dangl and McDowell, 2006).

7.3 Host–pathogen relation

The first level of pathogen recognition encompasses pattern recognition receptors (PRRs), which identify pathogen-associated molecular patterns (PAMPs). Such plant immunity is categorized as PAMP-triggered immunity (PTI) (Monaghan and Zipfel, 2012). Phytophagous pests respond by identification of herbivore-associated elicitors, herbivore-associated molecular patterns, or PRR herbivore effectors (Santamaria et al., 2013). The second level of pathogen recognition encircles plant resistance (R) proteins, which identify specific receptors from a pathogen (Avr proteins) (Dangl and McDowell, 2006; Gouveia et al., 2017; Abdul Malik et al., 2020). It is considered an effective mechanism of plant resistance to pests and involves effector-triggered immunity (ETI) (Kaloshian, 2004; Spoel and Dong, 2012). ETI stimulates hypersensitive responses and triggers programmed cell death in infected and surrounding cells. The proteins encoded by a majority of R genes have a specific domain with a conserved nucleotide-binding (NB) site. The second next important domain is a leucine-rich repeat (LRR). Pathogen effectors are recognized directly (physical association) or indirectly (association of an accessory protein) by NB-LRR receptors (Dodds and Rathjen, 2010). Sometimes, R gene-mediated plant response toward invading pathogen provokes a higher degree of defense, termed systemic acquired resistance (SAR). SAR generates whole-plant systemic resistance against a broad spectrum of pathogens. In SAR, a local encounter results in the stimulation of resistance to the other plant organs through interplant communication (Fu and Dong, 2013). Generally, both categories of plant immune responses induce the same reaction, but ETI is considered more rigorous to pathogen infection (Tao et al., 2003).

Perturbations in cytosolic calcium (Ca^{2+}) concentrations are the earliest signaling events occurring upon the exposure of plants to biotic stress. Ca^{2+} signals are the center to plant immune signaling pathways (Aldon et al., 2018). Rapid and transient perturbations in Ca^{2+} concentrations are crucial to gene reprogramming required to generate an adequate response (Reddy et al., 2011). The plant immune responses differ in their Ca^{2+} signatures, for example, Ca^{2+} transients upon PTI activation returns, to basic levels within a few minutes (Lecourieux et al., 2005), whereas ETI involves a prolonged increase in cytosolic Ca^{2+} levels lasting for several hours. Lanthanum is a well-known Ca^{2+} channel blocker that has been linked to both PTI and ETI immunological responses (Grant et al., 2000; Boudsocq et al., 2010). Precisely, in response to the biotic invasion, PTI and ETI activate the Ca^{2+} ion channels, increasing cytoplasmic Ca^{2+} concentrations.

7.4 Metal oxide nanoparticles

Researchers employ MONPs widely for numerous studies than other nanomaterials. The ionic metal oxides have unique physical and chemical properties due to their limited size and a high density of corner or edge surface sites (Chavali and Nikolova, 2019). The applications are aimed at health-related units due to their enticing antibacterial activity (Jangra et al., 2012). MONPs are used because of their high stability and ease of synthesis for achieving desired size, shape, and porosity. They are now potential tools because of their hydrophobic, hydrophilic, and negatively charged surfaces (Sanchez-Moreno et al., 2016). Unlike other NPs, metal oxide particles are significant as they are specific and show antimicrobial activity when administered in low concentrations (Fernandez-Garcia and Rodriguez, 2011). But still knowledge of MONP generation, deposition, and interaction is unknown for its application (Boddu et al., 2011).

The unlikely possibility might occur if the microbes undergo multiple mutations simultaneously (Zhu et al., 2014). Furthermore, the type, distribution, density, and charge on the NP surface determine the cytotoxic interactions between nanomaterial and microorganisms (Teske and Detweiler, 2015). Since the biological response is based on the various characteristics of the MONPs, the controlled synthesis techniques are focused on attaining the specification (Nikolova and Chavali, 2020). Engineered MONPs are significant to combat biotic stress in plants and are the potential to address micronutrient demand (Khan and Rizvi, 2014). The micronutrients play a vital role in activating some enzyme systems (Romheld and Marschner, 1991).

The MONPs handle various modes of action to achieve the target of antimicrobial activity; thus, it is difficult and almost unachievable for the microbes to develop resistance against them. The antibacterial mechanisms of MONPs generally include: (1) photocatalytic production of reactive oxygen species (ROS), which damages cellular components; (2) compromised integrity of bacterial membranes; (3) interruption of energy transduction and transport processes; and (4) inhibition of respiratory enzyme activity and DNA synthesis (Huang et al., 2008; Ovais et al., 2018).

MONPs, like antibiotics, can distinguish between microbial and host target cells. However, this is due to the cells deviating metal transport systems and metalloproteins (Jangra et al., 2012). Although MONPs were used to combat disease-resistant bacteria, their usage has been limited due to their toxicity (Cai et al., 2018). The challenges were overcome by using low-concentration MONPs, which are both safe and beneficial to agronomic crop productivity (Wang et al., 2016). Some studies have demonstrated that NP exposure improves free-radical scavenging potential and antioxidant enzymatic activities and alters microRNAs expression that regulates different morphological, physiological, and metabolic processes in plants (Siddiqi and Husen, 2016).

Also, the MONPs had to satisfy some requirements, like the kinetics of the MONPs should adhere to treating the target infections and they have to be biodegradable to eliminate further intervention (Nikolova and Chavali, 2020), considering their toxicity. The wide application of nanomaterials has raised many concerns about their potential risks to human health and the environment. Impact on the environment and NPs reaction on ecosystems, including their toxic effects on plants, soil, and microorganisms, are the important points considered and reviewed in recent studies (Hao et al., 2016; Mukherjee et al., 2016; Rui et al., 2016).

7.5 Mode of action

The high-efficiency antimicrobial activity of the MONPs is generally ascribed to their superior physicochemical properties, high surface-to-volume ratio, and unique nanoscale structural characteristics (Stoimenov et al., 2002). Nanomaterials have attracted considerable interest owing to their unique physicochemical properties. The MONPs have two main underlying toxicity mechanisms, that is, one with ROS-mediated toxicities and the other with non-ROS-mediated toxicities. ROS activates oxidative stress, which leads to lipid peroxidation (LPO) and cell membrane damage. In addition, ROS can trigger the apoptotic pathway by activating caspase-9 and caspase-3. Non-ROS-mediated toxicity mechanism includes the effect of released ions, excessive accumulation of NPs on the cell surface, and combination of NPs with specific death receptors (Yanli et al., 2017).

The NPs generally follow two major lethal pathways. They have related to each other and, in most cases simultaneously, as a cascade reaction (1) disruption of membrane potential and integrity and (2) production of ROS, also known as oxygen-free radicals (Huh and Kwon, 2011; Blecher et al., 2011). H_2O_2 is the most toxic ROS. At a neutral pH, H_2O_2 is formed by the Fenton reaction between H_2O_2 and $O_2^{\cdot}$, catalyzed by transition metals.

$$H_2O_2 + O_2^{\cdot -} \rightarrow OH^- + O_2 + OH^{\cdot}$$

The cells undergo LPO, protein damage, and membrane destruction. Since there is no scavenging system for these toxic radicals, excess accumulation of $OH^{\cdot}$ causes cell death (Pinto et al., 2003). The mode of action differs from one NP to other. Mainly, the mode of action involves the production of hydrogen peroxide (H_2O_2) from the surface of the NPs, as in ZnO NPs. The enzyme catalase present in the living cells helps convert H_2O_2 into water (H_2O) and oxygen (O_2), and, in turn, the oxygen (O_2) atom attracts electrons from the bacterial cell wall, collapsing the overall structure (Yoshikawa, 2004).

Cytotoxicity of copper oxide (CuO) NPs results not only from the small size of the particles, high specific surface value, and close interaction with microbial membranes but also from the formation of leached cuprum-peptide complexes leading to a several-fold increase in reactive oxygen intermediate generation, cell viability decrease, and general biomass growth suppression (Gunawan et al., 2011). The Cu binds with the phosphate groups that are part of the structural backbone of DNA molecules, causing the cell molecules damage. As a result, this mechanism may lead to the separation of the double helix (Meyer, 2001).

Magnesium oxide (MgO) NPs enhanced ultrasound-induced LPO in the liposomal membrane. It suggested that the mechanism of the antibacterial activity of the MgO NPs relied on the presence of defects or oxygen vacancy at the surface of the NP, which led to LPO and ROS generation (Karthikeyan et al., 2012). MONP also shows significant antimicrobial activity through connection to microbial DNA and proteins and prevents bacterial duplication, avoiding metabolic enzymes of the bacterial electron transport chain, leading to their inactivation (Holt and Bard, 2005). After reviewing all the antibacterial mechanisms, it is found that it is indeed a combinational and complex process that causes toxicity, and not likely that one single factor is responsible for the bacterial killing (Slavin et al., 2017). Antibiotic-resistant bacteria grow in a relatively short time, before introducing new antibiotics to the market. However, the hypothesis that the NPs have the potential to reduce or eliminate the growth of resistant bacteria because NPs target multiple biomolecules to avoid the development of resistant strains (Kittler et al., 2010).

The antifungal mechanism of the NPs involves three main strategies: mold inhibition, adsorption of mycotoxin, and reduction in toxic effect (Horky et al., 2018). The inhibitory effects of NPs on different fungi are demonstrated using various mechanisms. The opposite surface charge of the NPs and the fungi, on the other hand, paved the way for significant adhesion, showing modifications in harmful fungi: physical cutting, blocked water channels, and increased membrane permeability. In some conditions, an influx of NPs into the fungal cells happens, resulting in cell death (Wang et al., 2014). On the other hand, the NPs impairs the ROS-scavenging system, thereby accumulating the poisonous ROS in the cells (Liu et al., 2016). Another method shows that MONPs release metal ions that are toxic to fungal pathogens (Zarzuela et al., 2017).

Plant protection against viruses by NPs may occur by different mechanisms at the same time, including direct effects of NPs on the viral cell death, or by improving the capabilities of the plant defense by inducing resistance, taking into account the mechanics of occurrence and use in the field of plant protection. In insects, the NPs enter the cuticle, which blocks the skin pores and respiratory openings, resulting in suffocation and cell death. In other instances, the NPs enter the insect's bloodstream and poison it, causing it to lose its appetite. This action depends on the NP's ability to translocate through epithelial and endothelial cells of the insects (Yamanaka et al., 2008).

Seed treatment is one of the most advanced and demandable integrated pest management technologies for managing numerous crop pests (Taylor et al., 2001; Magalhaes et al., 2009). The mechanism may be the entry of the NPs into the insect's cuticle, which blocks the skin pores and respiratory openings and leads to suffocation and death. The toxic effect is due to the particles entering the insect blood, which leads to poisoning and the loss of appetite (Yamanaka et al., 2008). NPs may affect crops at biochemical, physiological, and molecular levels through changes in minerals, nutrition, and photosynthesis (Rizwan et al., 2017).

7.6 Zinc oxide nanoparticles

ZnO NPs are known to have multifunctional inorganic NPs. However, ZnO NPs have low toxicity, allowing them to successfully inhibit pathogen growth. They even furnish secondary benefits to soil fertility (Dimkpa et al., 2013), enhancing plant growth. ZnO NPs have unique characters, and they differ from their bulk counterparts in morphology, size, and distribution (Wang et al., 2009).

They are widely employed in various industries, including medicine and agriculture, because of the above-mentioned characteristics (El-Ansary, 2014). They also have superior durability, high selectivity, and heat resistance (Padmavathy and Vijayaraghavan, 2008). ZnO NPs and the bulk ZnO are both toxic—inhibiting growth and reproductive capability, mainly in eukaryotic organisms (Wang et al., 2009). Some studies suggest that dissociation of Zn^{+} is the cause of ZnO NPs toxicity (Franklin et al., 2007). However, it is safe for humans and animals since the toxicity is specific (Reddy et al., 2007; Berube, 2008). ZnO is the safest among the five zinc compounds; hence, they pass for the ideal antimicrobial reagent (Espitia et al., 2012).

The toxic nature of ZnO NPs ascribes to the production of ROS in the living cells (Li et al., 2016), which can occur in the protein, membrane, and DNA of host cells, and causes damage to biological systems (Nel et al., 2006; Sruthi et al., 2018). Among the various mechanisms involved in the antibacterial activity of ZnO NPs, the effective one would be the generation of hydrogen peroxide (H_2O_2) from the surface of ZnO. Most living cells have the enzyme catalase, which will convert H_2O_2 into water (H_2O) and oxygen (O_2), and, in turn, the oxygen (O_2) atom attracts electrons from the bacterial cell wall. As the pathogenic microbes carry a negative charge, this act collapses the entire basic molecular structure. The cell proteins and the layers of lipids on the cell surface are destroyed, resulting in the damaged or shattered cell wall (Ravishankar and Jamuna, 2011; Brayner et al., 2006; Amro et al., 2000), causing cytoplasm leakage and eventually the death of bacterial cells (Yoshikawa, 2004).

The shelf life of ZnO NPs depends on the stability of the NPs, provided by the stabilizer, whereas in the absence of the stabilizer, a dispersant is used. However, the combination could impact the antibacterial characteristics of the NPs (Wani and shah, 2012). The infection of Fusarium graminearum, the causative agent of Fusarium head blight (FHB) in wheat and barley, is reduced by 26% upon using ZnO NPs as fungicides. The ariel mycelial growth is inhibited and the hyphal spread was reduced in *F. graminearum* (Dimkpa et al., 2013).

Botrytis fungi are capable of causing a wide range of symptoms. Blights, spots, blotch, wilts, cankers, rots, and damping-off are typical symptoms among a range of hosts. The study shows that with ZnO NPs, *B. cinerea* growth is inhibited between 63% and 80%. The ZnO NPs affect the cellular functions of the fungi, resulting in deformation in fungal hyphae. *Penicillium expansum*, the pathogen responsible for the blue rot of apple and pear, is found to go down by 61%–91% while treated with ZnO NPs. ZnO NPs inhibited the conidial development and distorted the conidiophores of *P. expansum* (He et al., 2011). Similarly, ZnO suppressed the fungal pathogen *Aspergillus flavus*, which causes *Aspergillus* ear rot, one of the major diseases in corn (Jayaseelan et al., 2012).

Alternaria alternata, Fusarium oxysporum, Rhizopus stolonifer, and *Mucor plumbeus* on exposure to ZnO NPs and MgO NPs show a high inhibition rate in the germination of fungal spores. *R. stolonifer* and *M. plumbeus* demonstrated protein leakage. However, the maximum inhibition in the spore germination was found at the highest concentration (Wani and Shah, 2012). ZnO NPs are used to control the spread of *F. oxysporum*, which causes diseases like crown rot, head blight, and scab on cereal grains, and fusarium wilt in sweet potatoes, tomatoes, legumes, melons, and bananas (in which the infection is known as Panama disease). Furthermore, ZnO NPs enhance growth as Zn act as a precursor for the synthesis of auxin, cytokinin, and gibberellin, as well as inducing antioxidant enzyme activity, which aids pathogen defense (Gonzalez-Merino et al., 2021).

ZnO NPs and silica oxide NPs (SiO_2 NPs) are used to control tobacco mosaic virus (TMV) infection, which forms mosaic, mottling necrosis, stunting, leaf curling, and yellowing of plant tissues. Foliar spraying of NPs for 12 days inhibits the replication mechanism of TMV, activating the plant defense system and growth response. Furthermore, NP treatment induces the activity of ROS accumulation, peroxidase activity, catalase activity, and genes linked to systemic resistance (PR1 and PR2). Interestingly, NP treatment also increased the dry and fresh weight of tobacco plants (Abdelkhalek and Abdulaziz, 2020). Another study reported that ZnO and SiO_2 NPs interact with TMV envelope glycoproteins, which cause the direct injury of TMV shell proteins, TMV aggregation, and even fracture, thereby deactivating the virus (Cai et al., 2020).

Cucumber mosaic virus (CMV) is responsible for necrotic mosaic symptoms in Solanum melongena, and the disease severity is reduced by applying ZnO NPs (El-Sawy et al., 2017). The fungi-like oomycete pathogen *Peronospora tabacina* is the causative agent of Blue mold—a foliar disease in tobacco and certain other members of the Solanaceae family. Upon touching the leaves, the airborne spores enter the plant epidermis via germ tubes, then the hyphae that spread from a fruiting body, which in turn produces spores, thereby completing the life cycle (Gisi and Cohen, 1996). As the pathogen showed resistance to the current fungicide metalaxyl, the NPs were understudied to replace it (Lee et al., 1999). An experiment conducted on growing wheat plants on soil spiked with ZnO NPs showed potential pesticides (Du et al., 2011). ZnO NPs used against *P. tabacina*, and at the lower concentration of 8 mg/L, the ZnO NPs suppressed the spore germination and the spread of infection (Wagner et al., 2016).

Pythium ultimum, other oomycetes, is responsible for the damping-off and root rot diseases of hundreds of diverse plant hosts, including corn, soybean, potato, wheat, fir, and many ornamental species. *Pythium aphanidermatum* infects seeds, juvenile tissue, lower stems, fruit rot, root rot, and, in severe cases, stem rot of commercially produced poinsettia (*Euphorbia pulcherrima* Willd). Both CuO NPs and ZnO NPs inhibit the *Pythium* sp. infection by interfering in a process where Fe intake occurs. The NPs reduced the production of Fe-chelating siderophore-like metabolites. While CuO inhibits the ferric reductase activity on the mycelial surface, ZnO increased it, so both proved to be toxic to the *Pythium* sp. (Zabrieski et al., 2015).

In higher organisms, ZnO NPs may affect (i) the creatures by causing physical damage by interacting directly, (ii) the zinc ions dissolution, and (iii) the ROS production (Hou et al., 2018). The other mechanism in eukaryotic living cells involves pinocytosis (cell drinking), by which ZnO NPs enter the host cells, interact and damage the cell organelles, and promote organ failure (Neal, 2008). *Meloidogyne incognita* (root-knot nematode) is a plant-parasitic roundworm that typically incites large, usually irregular galls on roots as a result of parasitism. ZnO NPs when injected into the young mango plantlets and the nematode community went into depression significantly, in 72 h, the mortality rate was 72.86% in *M. incognita* was detected. The findings also demonstrate a disturbed nematode body 1 day after the application of ZnO NPs (Elansary et al., 2021).

Reticulitermes flavipes, the eastern subterranean termite, feeds on the woods, and is known as agricultural pests. The study on ZnO NPs as pesticides showed that 4% or less feeding happened on ZnO NPs impregnated wood, and upon the termites chewing the NP wood, 94%–99% of the mortality rate was observed (Clausen et al., 2011). *Trialeurodes vaporariorum* (greenhouse whitefly) is a pest for fruits, vegetables, and ornamental plants. The foliar sprayed ZnO NPs act on it causing maximum lethality (91.6%) at the highest concentration. The mode of action of ZnO NPs is likely to distort the insect's physiology (Khooshe-Bast et al., 2016).

In another study, the Bt–ZnO–NPs biopesticide effect was tested against the pulse beetle (*Callosobruchus maculatus*), which reduced the fecundity (eggs laid) and hatchability in a dose-dependent manner. Thus, Bt–ZnO–NPs were highly effective in the control of *C. maculatus* since they caused 100% mortality at 25 μg/mL (Malaikozhundan et al., 2017). *Bemisia tabaci* (Gennadius) (Hemiptera: Aleyrodidae), the whitefly, is a pest for cotton crops that feed on the yield and exponentially reduces the quality and quantity of the product. It is found that ZnO NPs are the most potent nanopesticides that brought down the count of whitefly (Shaker et al., 2020).

ZnO NPs posed high toxicity against *Spodoptera littoralis*, an Egyptian cotton leafworm. The experiments showed delayed effects by lowering pupation and adult emergence rates, by weight and enzyme activity in both larval and pupal, except phenoloxidase. Nanosilica does the same function on the pest next to ZnO NPs (Osman et al., 2015). Colloidal solutions of ZnO NPs act as the nanofertilizers, promoting seed germination, seedling vigor, plant and root growth, and boosting the yield. They are also used in the food industry as additives and packaging due to their antimicrobial properties (Gupta et al., 2015).

Unfortunately, ZnO NPs are toxic to beneficial microbes that help plant growth and development and fight pathogens. For instance, the antibacterial effect of ZnO NPs toward *Staphylococcus aureus* and *Proteus mirabilis* than *Serratia marcescens* and *Citrobacter freundii* involves the generation of a high rate of surface oxygen species, thereby disturbing the cell membrane, resulting in cell death (Gunalan et al., 2012). From various studies, it is understandable that ZnO NPs and their dissolution products, while being potentially toxic to plants and animals at high levels, may be tolerable in agricultural environments if the doses required for pathogen controls are low (Lin and Xing, 2008) (Table 7.1).

7.7 Titanium dioxide nanoparticle

Titanium dioxide (TiO_2) NP/titania is a highly active inorganic nanomaterial that is chemically stable, nontoxic, inexpensive, and generally recognized as a safe substance. They have an effective antimicrobial activity against the active plant pathogens by cell decomposition and superlipophilic (Pan et al., 2012; Zhao et al., 2005). One of the major factors in antimicrobial activity is light. TiO_2 is activated against pathogenic bacteria when exposed to ultraviolet (UV) light (Adams et al., 2006). The sterilizing rate is 99.00% in no light and 99.99% in UV light irradiation (Pablos et al., 2012).

TiO_2 NPs might not have an effective photocatalytic activity while working alone as they tend to aggregate (Heinlaan et al., 2008). The photocatalytic bactericidal action is strongly influenced by TiO_2 NP concentration. TiO_2 NP when comes into contact with the pathogenic bacteria coaggregates forming an emulsion. Strong irradiation of the aggregates releases metal ions slowly on the surface and activates photocatalysis. Under UV light, the electron holes in TiO_2 NP interact with water and oxygen to generate active oxygen species, especially hydroxyl radicals, thereby achieving antibacterial effects (Li et al., 2018). As the interior of coaggregates was not touched by the UV rays, the killing mechanism is slow and takes time. As soon as the light irradiation starts, a cascade of reactions takes place in the bacterial cells, such as potassium leakage and protein and RNA release. The slow decomposition of the coaggregates revealed that the bacterial cell wall structure is disturbed after 120 min of exposure. The cells have no chromosomal complex, their central portion is hollow, and the protoplasm is coarse in nature, evident as the cell is undergoing decomposition (Saito et al., 1992).

TiO_2 photocatalytic reaction causes the ROS-mediated LPO to take place and, as a result, other normal cellular functions associated with an intact membrane are lost. Due to the nature of the ROS, once initiated, the TiO_2-

TABLE 7.1 Application of ZnO NPs.

Microbe	Disease caused	Mode of action of MONP	Reference
Fusarium graminearum	*Fusarium* head blight in wheat and barley	The ariel mycelial growth was inhibited and the hyphal spread reduced	Dimkpa et al. (2013)
Botrytis cinerea	Blights, spots, blotches, wilts, cankers, rots, and damping-off in wide range of host	Deformation in fungal hyphae	He et al. (2011)
Penicillium expansum	Blue rot of apple and pear	Conidial development inhibition and distorted the conidiophores	He et al., 2011
Pseudomonas aeruginosa	Black necrotic regions in the roots	Reduce the necrosis	Walker et al. (2004)
Aspergillus flavus	*Aspergillus* ear rot	Rot is reduced	Jayaseelan et al. (2012)
Fusarium oxysporum	Crown rot, head blight, and scab on cereal grains, fusarium wilt in sweet potatoes, tomatoes, legumes, melons, and bananas	Improves the promoter effect related to the zinc activity as a precursor in auxins synthesis, cytokinins, and gibberellins biosynthesis, as well as the induction of the higher activity of antioxidant enzymes	Gonzalez-Merino et al. (2021)
Tobacco mosaic virus (TMV)	Mosaic, mottling necrosis, stunting, leaf curling, and yellowing of plant tissues	Interact with the TMV envelope glycoproteins causing the direct injury of TMV shell proteins, TMV aggregation, and even fracture, thereby deactivating the virus	Cai et al. (2020)
Cucumber mosaic virus	Necrotic mosaic symptoms in *Solanum melongena*	The disease is visibly reduced	El-Sawy et al. (2017)
Peronospora tabacina	Blue mold—a foliar disease in tobacco and other Solanaceae families	Suppressed the spore germination and the spread of infection	Wagner et al. (2016); Du et al. (2011)

(*Continued*)

TABLE 7.1 (Continued)

Microbe	Disease caused	Mode of action of MONP	Reference
Pythium ultimum	Damping-off and root rot diseases of hundreds of diverse plant hosts	Inhibit the infection by interfering in a process where Fe intake occurs	Zabrieski et al. (2015)
Pythium aphanidermatum	Seeds, juvenile tissue, lower stems, fruit rot, root rot, and, in severe cases, stem rot	Inhibit the infection by interfering in a process where Fe intake occurs	Zabrieski et al. (2015)
Meloidogyne incognita (root-knot nematode)	Incites large, usually irregular, galls on roots	Disturbed nematode body	Elansary et al. (2021)
Reticulitermes flavipes	Feed on the woods	Disoriented body	Clausen et al. (2011)
Trialeurodes vaporariorum (Greenhouse whitefly)	Pest for fruits, vegetables, and ornamental plants	Distort the insect's physiology	Khooshe-Bast et al. (2016)
Callosobruchus maculatus	–	Reduced the fecundity (eggs laid) and hatchability	Malaikozhundan et al. (2017)
Bemisia tabaci (Gennadius) (Hemiptera: Aleyrodidae)	Pest for cotton crops	Distort the insect's physiology	Shaker et al. (2020)
Spodoptera littoralis	Cotton leafworm	Latent effect via producing a high reduction in pupation and adult emergence rates, decreasing both larval and pupal weight of this pest and reducing estimated enzymes activity	Osman et al. (2015)
Staphylococcus aureus, P. mirabilis, S. marcescens, and *C. freundii*	Beneficial microbes	Generation of a high rate of surface oxygen species thereby disturbing the cell membrane, resulting in cell death	Gunalan et al. (2012)

mediated reaction cannot be terminated even by placing the reaction mixture on ice or in the dark (Maness et al., 1999). Another study found that the endotoxin, an integral component of the outer membrane, was destroyed under photocatalytic conditions when TiO_2 was used (Sunada et al., 1998). *Xanthomonas perforans* is an important pathogenic bacterium that causes the bacterial spot of tomato. Upon administration of TiO_2 NP, the photocatalysis sets in, resulting in antimicrobial effects against the bacterial spot disease. Interestingly, doping the TiO_2 NPs with Ag or Zn significantly increased the photocatalytic activity. Treatment of *X. perforans*-infected tomato plants with TiO_2/Zn NPs at approximately 500–800 ppm significantly reduced bacterial growth (Paret et al., 2013).

Pseudomonas syringae pv. lachrymans, a bacterium responsible for angular leaf spots, and *Pseudoperonospora cubensis*, an oomycete responsible for downy mildew on cucurbits, are destructive plant pathogens. Administration of TiO_2 NP reduced *P. syringae pv. lachrymans* and *P. cubensis* infection of cucumber by 69% and 91%, respectively, and also increased photosynthetic activity (30%) (Cui et al., 2009). The TiO_2 photocatalytic reaction reduced the conidial germination of the fungal pathogen. The photocatalytic reactions suppress *P. expansum* regardless of what fruit it may affect. The blue mold of apples, tomato fruit rot, brown lesions, and Penicillium rot in lemons are all controlled by the application of TiO_2 NP (Maneerat and Hayata, 2006)

Colletotrichum gloeosporioides, the fungi responsible for bitter rot, *Cladosporium oxysporum* which exhibits leaf spots, scab, postharvest rots, and other disease symptoms, and *P. steckii* experiment against CTS/nano-TiO_2 composite and the composite is found to be an effective fungicide. CTS can inhibit calcium and other nutrients, and thus, limit the growth of fungi. Furthermore, it has been believed that CTS directly destroys the function of the cell membrane, thus exhibiting antibacterial properties. Chitinase in pathogenic fungi can be overexpressed when the CTS concentration is high, which leads to the degradation of chitin, an important component of the fungal cell wall, thus destroying the cell wall of the pathogen and achieving the antifungal effect (Leuba and Stossel, 1996).

It may be that a CTS/nano-TiO_2 composite membrane solution acts on the cell wall of mold to cause changes in its components, leading to the leakage of cell contents and, ultimately, cell ablation, in which its enzymes catalyze the decomposition of chitin, and protein, nucleic acid, ammonia, free amino acids, organic phosphates, and organic sulfur compounds are released, resulting in the decrease of mycelium dry weight (Gao et al., 2020). The CTS/nano-TiO_2 composite coating showed a better antifungal effect, restraining the growth of the mycelium, destroying cell membrane integrity, inducing the extravasation of intracellular protein and nucleic acid, increasing the conductivity value of the fungal suspensions, and killing the molds (Xing et al., 2021).

The FHB is a destructive disease caused by *F. culmorum*. It is a reemerging disease in small grain cereals that impairs both the grain yield and the quality. Plant growth-promoting rhizobacteria (PGPR) such as *B. thuringiensis* AZP2, *Paenibacillus polymyxa* A26 (Timmusk et al., 2011, 2014b; Abd El-Daim et al., 2015), *P. polymyxa* A26Δsfp (Timmusk et al., 2013; Abd El-Daim et al., 2015), and *Alcaligenes faecalis* AF (Timmusk et al., 2018) along with TiO_2 NP is effective in boosting the potential of the root-colonizing microbiome to benefit the plant.

Broad bean stain virus (BBSV) is one of the viruses that infect seeds in the soil. The virus is seedborne in *Lens culinaris* (lentil), *Pisum sativum* (field pea), *Vicia faba* (faba bean, broad bean, tick bean), and *V. palaestina* (Makkouk et al., 1986). TiO_2 NP is provided to the faba bean plant by foliar spray or soil drench and the NP is found to be very effective against the virus, reducing the disease severity. In this case, the foliar spray is more efficient than the soil drench as the activity of BBSV is highly suppressed by TiO_2 NP sprayed on the leaves. This is because the spray provides direct contact between the TiO_2 NP and the virus, whereas in the soil drench, the TiO_2 NP is absorbed into soil particles and, subsequently, taken up by the plant and then reacted to the virus (Elsharkawy and Derbalah, 2019).

Turnip mosaic virus was allowed to infect *Nicotiana benthamiana* (tobacco plant) for the study purpose and treated with the engineered NPs. Upon infection, TuMV forms necrosis on the tobacco leaves. Fe_2O_3 or TiO_2 NPs are foliar sprayed on the plant for 21 days and then the plant is infected by TuMV. The administration of MONPs significantly increased the shoot biomass by 50%, providing micronutrients to the plant development. Both the MONPs are found to be accumulated in the chloroplast. The abundance of TuMV on the leaf surface suggests that viral proliferation is inhibited. In total, 15%–60% decrease in the relative amount of TuMV coat proteins proves the mode of action of the MONPs—by breaking down the viral outer envelope. Phytohormones are calculated to be increased by 40% with the MONPs treatment. This suggests that the Fe_2O_3 or TiO_2 NPs help in stimulating plant growth and activating defense mechanisms (Yi et al., 2018).

Sublethal doses of MONPs on bulb mite *Rhizoglyphus robini claparede* (Acari: Acaridae) extended their egg incubation time, larval period, and nymphal period. TiO_2 NP and MgO NP did the deed effectively compared to CuO NPs, ZnO NPs, and Fe_2O_3 NPs (Korghond et al., 2021). The larvicidal activity of TiO_2 NPs is demonstrated on *Rhynchophorus ferrugineus*, a snout beetle known as the red palm weevil. The weevil destroys the apical bud leading to the death of the plant itself. TiO_2 NPs are found to have increased the mortality rate of the pests in the larval stage, at concentration of 75 mg/L. Green synthesized TiO_2 NPs from *Moringa oleifera* are used for the study (Al-Bartya and Hamzab, 2015).

Like ZnO NPs, TiO_2 NPs are also effective against *S. littoralis*, an Egyptian cotton leafworm, the main pest of the cotton plant. TiO_2 NPs are effective against the survival of the second and fourth instar larvae of the pest, controlling them at the young stage. TiO_2 NPs + CuO NPs mixture has the same effect on *S. littoralis* (Shaker et al., 2017a). Onion thrips, *Thrips tabaci* Lindeman (Thysanoptera: Thripidae), is one of the most damaging insect pests of onions, *Allium cepa* L., which is an economically important agricultural crop cultivated worldwide. TiO_2, SiO_2, and CuO treatments in high concentration recorded the most potent effect by decreasing the thrips count to zero. Treatment with ZnO and FeO also caused a marked effect causing thrips to count. Jassid pest, *Empoasca lybica* (De Berg.) (Hemiptera: Cicadellidae), another pest, is removed from the equation with a high concentration of TiO_2 and CuO NPs. Red spider pest *Tetranychus telarius* (L.) (Acari: Tetranychidae) is also decreased by the treatment with TiO_2 NPs. TiO_2 and CuO NPs treatments induced the most potent against *Aphis gossypii* Glover (Hemiptera: Aphididae) (Shaker et al., 2020).

The impact of TiO_2 NP in the plant growth is also an important parameter to be considered since the protection of NPs against a virus is indirectly handled by improving plant health. NP improved germination in seeds and growth in treated plants. This boost may be because TiO_2 NP stimulates the production of carbohydrates, thereby increasing the process of photosynthesis due to its ability to absorb light (Chen et al., 2014). TiO_2 NP also regulates the activity of enzymes responsible for nitrogen metabolisms, such as glutamate dehydrogenase, nitrate reductase, glutamine synthase, and glutamic–pyruvic transaminase, which helps the plant to absorb nitrates. These enzymes help in the conversion of inorganic nitrogen to organic in the form of protein and chlorophyll, which may be responsible for the observed increase in fresh and dry weight (Jaberzadeh et al., 2013). This mechanism is similar to the study with TiO_2 NP in the wheat plant (Mahmoodzadeh and Aghili, 2014).

Caenorhabditis elegans is the free-living, nonparasitic nematode that is extensively used as a model organism in biology. Now, *C. elegans* is tested for TiO_2 NP exposure. It was found that administration of TiO_2 NP affected their neurons allegedly making shorter axons, which may be related to the impending worm locomotion. Even then, the application of TiO_2 did not alter the worm's body length; it certainly reduced the worm population by 50% within 72 h (Hu et al., 2018).

Bacillus amyloliquefaciens is a well-studied rhizosphere species that serves as PGPR and helps in the stress management of plants serving as biocontrol agents (Niazi et al., 2014). The bacteria have a net negative charge on the surface, whereas TiO_2 NP has a positive charge (Larsen et al., 2009). Phosphate and phosphonate ligands of MONPs function as a glue between bacteria and TiO_2 NP (Pązik et al., 2011). One can hypothesize that there is an increase in clustering that the bacteria cannot achieve by themselves or that there is increased growth. Not only did the NPs aid in the clustering of the bacteria but also the colonization of the roots was enhanced. The roots were covered to a higher extent with TiO_2 NP treated bacteria. The colonies are appreciably bigger and maybe even more highly ordered. In the presence of these bacteria, both the growth and the virulence of the fungi are suppressed. For instance, the improved attachment of *B. amyloliquifaciens* to oilseed rape roots in the presence of the NPs correlates with systemic protection against leaf infection by *A. brassicae* (Palmqvist et al., 2015).

Due to the photocatalytic activity of TiO_2 NP, almost all studies regarding them show positive effects on the plants (Raliya et al., 2015). Application of TiO_2 NP in *Spinacia oleracea* increases plant fresh and dry weight by more than 60%. The amount of Rubisco activase increased by 42%, whereas their activity increases 2.5 times (Gao et al., 2008). Sometimes the best things can add negative effects too. TiO_2 NP on *A. cepa* increases genotoxicity with the concentration (Demir et al., 2014). The application of NPs also may have nontarget effects (Ruotolo et al., 2018). At high concentrations, ZnO NPs are found to increase the cytotoxicity in root cells and DNA fragmentation is also noted. While the number of catalases was decreased, it elevated the ROS and glutathione peroxidase production (Ghosh et al., 2016). Exposure to Fe_3O_4 NPs did not change the germination process, plant growth, or chlorophyll content in the plant *Triticum aestivum*. The NP-indulged plants show favorable responses to prevent oxidative damage (Iannone et al., 2016) (Table 7.2).

7.8 Copper oxide nanoparticles

The CuO NPs are more physically and chemically stable than any known MONPs and they easily mix with polymers (Gawande et al., 2016). The first study to be made on CuO NPs as a bacteriocide/fungicide was a grand success, resulting in the suppression of the fungi-like oomycetes organism called *Phythopthora infestans*. This oomycete causes a serious late blight or potato blight disease in tomatoes and potatoes. The result demonstrated that administration of CuO NPs as low as 150–340 μg/L curbed the leaf lesions (Giannousi et al., 2013).

The antimicrobial effect of CuO NPs results from the release of Cu^{2+} by a Fenton-like chemistry reaction that may cause (i) LPO and protein damage (Dupont et al., 2011); (ii) the Cu ions formed produce ROS, which ruptures the cell wall of bacteria; and (iii) Cu ions have shown the ability to prohibit the expression of biofilm-forming enzymes (Fowler et al., 2019). On the whole, the inherent ability of copper to inflict damage upon multiple cellular functions has been harnessed by macrophages to kill and later digest bacteria and other microorganisms (Kanhed et al., 2014; Zarzuela et al., 2017; Dupont et al., 2011). *Trametes versicolor* is a fungus responsible for wood decay or white rot decay in higher trees such as almonds and peaches. This is a secondary infection spread at the point of injury. The antifungal activity of CuO NPs prevented the wood from undergoing fungal decay (Jo et al., 2009). Another

TABLE 7.2 Application of TiO_2 NPs.

Microbe	Disease caused	Mode of action of MONP	Reference
Xanthomonas perforans	Bacterial spot on tomato	Photocatalysis	Paret et al. (2013)
Pseudomonas syringae pv. Lachrymans	Angular leaf spot	Disease is suppressed	Cui et al. (2009)
Pseudoperonospora cubensis	Downy mildew on cucurbits	Disease is suppressed	Cui et al. (2009)
Penicillium expansum	The blue mold of apples, tomato fruit rot, brown lesions, and Penicillium rot in lemons	Reduces conidial germination	Maneerat and Hayata (2006)
Colletotrichum gloeosporioides	Bitter rot	CTS/nano-TiO_2 composite membrane solution acts on the cell wall of mold to cause changes in its components, leading to the leakage of cell contents and, ultimately, cell ablation	Leuba and Stossel (1996); Gao et al. (2020)
Cladosporium oxysporum	Leaf spots, scab, postharvest rots, and other disease symptoms		
Penicillium steckii	–		
Fusarium culmorum	Fusarium head blight in small grain cereals	Boost the potential of the root-colonizing microbiome	Timmusk et al. (2018)
Broad bean stain virus	Infect seeds in the soil	Highly suppressed	Elsharkawy and Derbalah (2019)
Bulb mite *Rhizoglyphus robini Claparede* (Acari: Acaridae)	Pest	Prolonged the egg incubation, larval period, and nymphal period	Korghond et al. (2021)
Turnip mosaic virus	Forms necrosis on the tobacco leaves	Break down the viral outer envelope and increase the phytohormones for plant growth	Yi et al. (2018)

(Continued)

TABLE 7.2 (Continued)

Microbe	Disease caused	Mode of action of MONP	Reference
Rhynchophorus ferrugineus	Destroys the apical bud leading to the death of the plant	Increased the mortality rate of the pests in the larval stage	Al-Bartya and Hamzab (2015)
Spodoptera littoralis an Egyptian cotton leafworm	The main pest of a cotton plant	Against the second and fourth instar larvae of the pest	Shaker et al. (2017a)
Onion thrips, *Thrips tabaci* Lindeman (Thysanoptera: Thripidae)	Pests of onions	Decreasing the thrips count to zero	Shaker et al. (2020)
Caenorhabditis elegans	Beneficial nematode	Affected their neurons, impeding worm locomotion and reducing the worm population	Hu et al. (2018)
Alternaria brassicae	Pathogen affecting broccoli, cabbage, and oilseed rape	Enhancing the attachment of *B. amyloliquifaciens* to oilseed rape roots and correlating with systemic protection against leaf infection	Palmqvist et al. (2015)

fungus, *Gloeophyllum trabeum*, which is known to cause brown rot decay on trees, is also suppressed by the CuO NPs (Terzi et al., 2016).

B. cinerea, another fungus attacking ornamental plants and economical vegetables, causes the gray mold on roses. When subjected to CuO NPs at the concentration of 50, 100, and 200 mg/L, suppression of the mycelial growth occurs, and the color of the rose petals covered with gray molds changes into dark red, indicating antifungal activity of the NPs. On examination, NPs on fungi turn them weak. They showed scarce growth and lacked viability. The hyphae were slim and delicate, which indicated the inhibitory effects of NPs on *B. cinerea* growth (Hao et al., 2017).

CuO NPs are found to affect the growth of four major pathogenic fungi: *F. oxysporum*, *Curvularia lunata*, *A. alternata*, and *Phoma destructiva* (severe disease-causing pathogen that attacks crop plants with leaf blight to root rot and, sometimes, even wilting of the plant). It is mainly due to their large surface area to volume ratio. It is concluded that they can be used as antifungal agents to protect crops from plant pathogenic fungi (Kanhed et al., 2014).

Foliar application of CuO NPs on tomato shoots inhibits the fusarium wilt disease progression caused by *F. oxysporum*. The Cu ions found in the roots served as a micronutrient for plant growth. This proves the hypothesis stating that the MONPs not only protect against pathogens by directly attacking them but also by indirectly affecting them by increasing the gross plant nutritional status (Ashraf et al., 2021). In the *Chrysanthemum* plant, the fusarium wilt had caused a 22% decrease in the yield and the application of CuO NPs increased the yield by 23%. Similar trends with plant height were observed. Horticultural quality was improved by 28% with nano CuO (Elmer et al., 2021). Similarly, in the watermelon plant, CuO NPs increased the production of fruit by 53%. It is concluded that CuO NP may be serving as a highly effective delivery agent for this micronutrient to suppress disease (Elmer et al., 2018).

V. dahliae is the causative agent of verticillium wilt in many plant species, causing leaves to curl and discolor. It may cause death in some plants. The seeds of eggplants were treated with CuO NPs, and the experiment was performed on the *V. dahliae* infested soil, resulting in a 17%–31% increase in the yield by suppressing the disease (Elmer et al., 2021). CuO NPs at a low concentration increased the expression of the exogenous gene producing Bt toxin protein in transgenic Bt cotton leaves and roots, boosting insect resistance (Van et al., 2016).

Sitophylus granarius and *Rhyzopertha dominica* are major stored wheat pests worldwide that cause the most quantitative and qualitative losses (Fields, 2006). CuO NPs exhibited high efficacy against *Sitophilus granarius* and *R. dominica* insects at lower concentrations where the concentration of NPs did not pose a toxicity threat to the plants or inhibit their growth. Plant morphology (lengths, fresh weight, and dry weight of root and shoot, as well as the number of leaves) and leaf pigments were improved (chlorophylls and carotenoids). CuO NPs also boosted soluble carbohydrates, proteins, and activities of antioxidant enzymes (Badawy et al., 2021).

Bulb mite *Rhizoglyphus robini Claparede* (Acari: Acaridae) impose biotic stress on a wide range of plants, such as ornamental plants, potato, carrots, etc. (Diaz et al., 2000). They enter the plant by tunneling through the corms, in the underground stem, and act as food storage, thereby paving access to fungi and viral infections. A study on MONPs against the bulb mite on *Crocus sativus L.* (Iridaceae), commonly known as saffron, shows that CuO NPs are top on the table to kill the mites effectively upon administration by lethal dosage (Korghond et al., 2021). CuO NPs showed multifunctional with significant application with great future implications in photocatalysis and biomedicines (Murthy et al., 2018).

CuO NPs when administered in *Lemna minor*, the common duckweed or lesser duckweed, shows a negative effect on the plants, but increased peroxidase, catalase, superoxide dismutase activity, and LPO, resulting in the inhibition of plant growth as the outcome (Song et al., 2016). The same CuO

NPs in *Elodea densa* stimulated photosynthesis when the concentration is less than around 0.25 mg/L level, whereas suppressing it above 1 mg/L concentration. The same NP produces different effects based on the concentration, and there was a constant rise in catalase and superoxide dismutase activities (Nekrasova et al., 2011) (Table 7.3).

7.9 Magnesium oxide nanoparticles

MgO NPs, with the advantages of nontoxicity, environmental friendliness, ease of availability, and biocompatibility with human cells, are recognized as safe disinfection agents by the US Food and Drug Administration without any harmful byproduct. Molecular toxicity investigation revealed that the toxic effects of MgO NP were mainly due to the dysregulation in oxidative stress leading to apoptosis. Accumulation and internalization of NPs and their interaction with cellular proteins affect the structural integrity and functionality (Verma et al., 2020). NP MgO has a significant antimicrobial activity due to strong interactions with the negative charge surfaces of bacterial membranes and spores. The generation of superoxide ions on the surface of MgO NP physically damages the peptide linkages in the bacterial cell wall (Huang et al., 2005). *Salmonella* sp. reach the plant kingdom through animal feces, and they move on leaves toward open stomata, colonize in veins, the bases of trichomes, and damaged leaf areas, which may provide shelter, and increase nutrient and water availability (Monier and Lindow, 2005).

Salmonella appear to be the successful secondary colonists, benefiting from the phytopathogen invasion, for example, suppression of plant defenses and tissue damage (lesions, water soaking, and soft rots). Studies have shown that soft-rot bacteria promote the proliferation of *Salmonella* in plants. Biotrophic plant pathogens, like *P. syringae* and *X. campestris*, can promote the growth or survival of *Salmonella* and enterohaemorrhagic *Escherichia coli* on plants (Barak and Liang, 2008; Aruscavage et al., 2010; Potnis et al., 2014). The antibacterial activity of MgO NP against these two bacteria increases with an increase in the administration concentration and the bacteria undergoes morphological changes. MgO NPs disrupt the cell membrane, causing intracellular leakage and cell death (Jin and He, 2011).

A highly pathogenic bacteria, *Ralstonia solanacearum*, is the causative agent of bacterial wilt, one of the most important vascular plant diseases, which affects more than 400 species of plants worldwide. These pathogens are soilborne and waterborne, penetrate the host through the roots, and cause wilting by massively colonizing the xylem vessels and producing vascular dysfunction. They mainly target Solanaceous plant species and other strategic crops for human supply such as tomato, pepper, eggplant, and Irish potato, and also in ornamentals of economic importance (Alvarez et al., 2019). No management method seems to be fully advisable against bacterial wilt, since crop protection chemicals do not provide enough control and

TABLE 7.3 Application of CuO NPs.

Microbe	Disease caused	Mode of action of MONP	References
Phythopthora infestans	Serious late blight or potato blight disease in tomato and potato	Curbed the leaf lesions	Giannousi et al. (2013)
Trametes versicolor	Wood decay or white rot decay in higher trees	Prevented the wood from undergoing fungal decay	Jo et al. (2009)
Gloeophyllum trabeum	Brown rot decay on trees	Suppressed the decay	Terzi et al. (2016)
B. cinerea	Gray molds on ornamental plants and economically vegetables	Suppression of the mycelial growth	Hao et al. (2017)
Fusarium oxysporum	Fusarium wilt on legumes, tomato, sweet potato, banana, etc.	Cu ions are found in the roots which served as a micronutrient for plant growth and indirectly help fight disease	Ashraf et al. (2021)
Verticillium dahliae	Verticillium wilt in many plant species, causing leaves to curl and discolor	Disease is suppressed	Elmer et al. (2021)
Sitophylus granarius and *Rhyzopertha dominica*	Pests	Enhanced leaf morphology (lengths, fresh weight, and dry weight of root and shoot, as well as leaves number) and leaf pigments (chlorophylls and carotenoids). Also boosted soluble carbohydrates, proteins, and activities of antioxidant enzymes	Badawy et al. (2021)
Bulb mite *Rhizoglyphus robini Claparede* (Acari: Acaridae)	Pest	Disfiguring and death	Korghond et al. (2021)

usually harm the environment and/or human health and favor the emergence of resistances and are expensive, physical treatments are quite ineffective, crop rotations are often impractical, and the pathogen displays high aggressiveness and endurance in adverse environmental conditions (Alvarez et al., 2019; Yuliar Nion and Toyota, 2015; Nicolopoulou-Stamati et al., 2016). Since no method of control worked on the vicious bacteria, a desperate novel tool was sought and found to be the MONPs. MgO NP is effective against the pathogen.

MgO NP attaches directly to the bacterial cell membrane causing physical injury to it, along with decreased motility and biofilm formation ability of *R. solanacearum*. ROS accumulation could also be a major reason for the antibacterial action, inducing DNA damage. MgO NPs had a significant impact on bacterial wilt, lowering of wilt index (Cai et al., 2018). Tobacco black shank and black root rot caused by *Phytophthora nicotianae* and *Thielaviopsis basicola* are widespread and significantly decrease crops' quality and yield. MgO NP was found to be an effective antifungal agent, proved by inhibiting spore germination, impeding mycelial growth and sporulation, and changing the zeta potential of fungal cells (Chen et al., 2020).

The antimycotic activity was found in a variety of plant pathogenic fungi. *A. alternata* is one of the most common fungi of the diseases, which causes leaf spots, rots, and blights on many parts of more than 380 host plants. *Fusarium* wilt is a widespread plant disease caused by the soil-inhabiting fungus *F. oxysporum*. Sweet potatoes, tomatoes, legumes, melons, and bananas (in which the infection is known as Panama disease) are susceptible to the fungi (Wani and Shah, 2012). Several postharvest storage rots are caused by *R. stolonifer* and *R. arrhizus*. In warm conditions, these fungi affect the soft tissues of harvested fruits, often causing watery leakage, spoiling them beyond edibility. The inhibitory effect of MgO NPs may be due to the release of extracellular enzymes and metabolites that serve as an agent for their survival (Perez-de-Luque and Diego, 2009) (Table 7.4).

7.9.1 Other metal oxides

Sitophilus oryzae (rice weevil), although a primary pest of stored foods, can also attack cereal plants in the field and infest them before harvest. Aluminum oxide NPs are effective against the pest compared to other NPs, killing all the insects in 7 days at the dosage of 0.1 g/kg. SiO_2 protects plants from biotic stress by increasing the expression of genes involved in disease resistance and photosynthetic protection (Luyckx et al., 2017; Manivannan and Ahn, 2017).

SiO_2 NP is valuable in seed production because when applied to the roots of 1-year-old seedlings by soaking for 6 h, it promoted lateral root growth, particularly root length and chlorophyll content (Bao-shan et al., 2004). SiO_2 NP is efficient against TMV, which causes mosaic, mottling necrosis,

TABLE 7.4 Application of MgO NPs.

Microbe	Disease caused	Mode of action of MONP	References
Salmonella sp.	Successful secondary colonists, benefiting from the phytopathogen invasion, for example, suppression of plant defenses and plant tissue damage	Distort the cell membrane causing intercellular leakage, thereby the death of the cell	Jin and He (2011)
Ralstonia solanacearum	Bacterial wilt in tomato, pepper, eggplant, and Irish potato, and also in ornamentals of economic importance	Attaches directly on the bacterial cell membrane causing physical injury to it, along with decreased motility and biofilm formation ability	Cai et al. (2018)
Phytophthora nicotianae	Tobacco black shank	Inhibiting spore germination, impeding mycelial growth and sporulation, and changing the zeta potential of fungal cells	Chen et al. (2020)
Thielaviopsis basicola	Black root rot	Inhibiting spore germination, impeding mycelial growth and sporulation, and changing the zeta potential of fungal cells	Chen et al. (2020)
Alternaria alternate	Leaf spots, rots, and blights on many parts	Release extracellular enzymes and metabolites that serve as an agent for their survival	Perez-de-Luque and Diego (2009)
Fusarium oxysporum	Fusarium wilt in sweet potatoes, tomatoes, legumes, melons, and bananas	Release extracellular enzymes and metabolites that serve as an agent for their survival	Perez-de-Luque and Diego (2009)
Rhizopus stolonifer and R. arrhizus	Storage rot	Release extracellular enzymes and metabolites that serve as an agent for their survival	Perez-de-Luque and Diego (2009)

stunting, leaf curling, and yellowing of plant tissues in plants, similar to ZnO NPs. The foliar spray of the NPs for 12 days consequently suppressed the virus and reduced the disease symptoms exponentially. SiO_2 NP can improve plant growth response in addition to its antiviral properties. The NPs treatment upregulated ROS, peroxidase activity, catalase activity, and systemic resistance related genes (PR1 and PR2). Revoking the plant defense not only by acting against the pathogen but also by reinforcing the development of a plant is an added advantage (Lin et al., 2019).

The Fe_3O_4 NPs treatment along with TMV caused aggregation and breakage. The direct interaction leads to larger particle sizes. The absorption and physiological effects of Fe_3O_4 NPs were examined using the foliar spraying technique to find resistance against the TMV. Fe_3O_4 NPs were transported and accumulated throughout the whole *N. benthamiana* plant. It activated plant antioxidants and upregulated salicylic acid (SA) synthesis, which leads to the expression of the SA-responsive PR gene (i.e., PR1 and PR2), thereby enhancing plant resistance against TMV (Cai et al., 2020).

Ag NPs attached with SiO_2 NPs and water-soluble polymer practically controlled powdery mildews of pumpkin and blazon *Rhizoctonia solani*. *B. cinerea* is a necrotrophic fungus most notable host may be wine grapes. *Magnaporthe grisea*, also known as rice blast fungus, causes a serious disease affecting rice. *P. ultimum* is a plant pathogen that causes damping-off and root rot disease in diverse plants. *C. gloeosporioides* bitter rot in various crops worldwide, particularly perennials in the tropical regions (El-Bendary and El-Helaly, 2013).

The efficacy of NiO nanostructures for controlling CMV in cucumber plants was evaluated by a reduction in disease severity, reduction in CMV accumulation, and expression of regulatory and defense-related genes. Cucumber plants treated with nickel oxide nanostructures (NONS) showed incredible suppression of CMV infection compared with nontreated plants. Treatment with nickel oxide nanostructures significantly improved cucumber fresh and dry weights, as well as the number of leaves. The induction of systemic resistance toward CMV by NONS is considered a novel strategy and the first report (Hamed Derbalah and Elsharkawy, 2019). Graphene oxide silver NP nanocomposites can suppress hyphae development, showing a significant effect in controlling leaf spot disease from *F. graminearum* (Chen et al., 2016b).

The physical and chemical parameters of the environment can change the character of the exposed NPs. This modification changes their reactivity, penetration, and translocation inside the plant, demonstrating the plant will elucidate different responses for the same NP. Thus, the unpredictability poses a threat to plants and other living organisms including human beings and requires extensive studies (Levard et al., 2012). However, when the MONPs are in low concentration or trace amounts, it acts as micronutrients

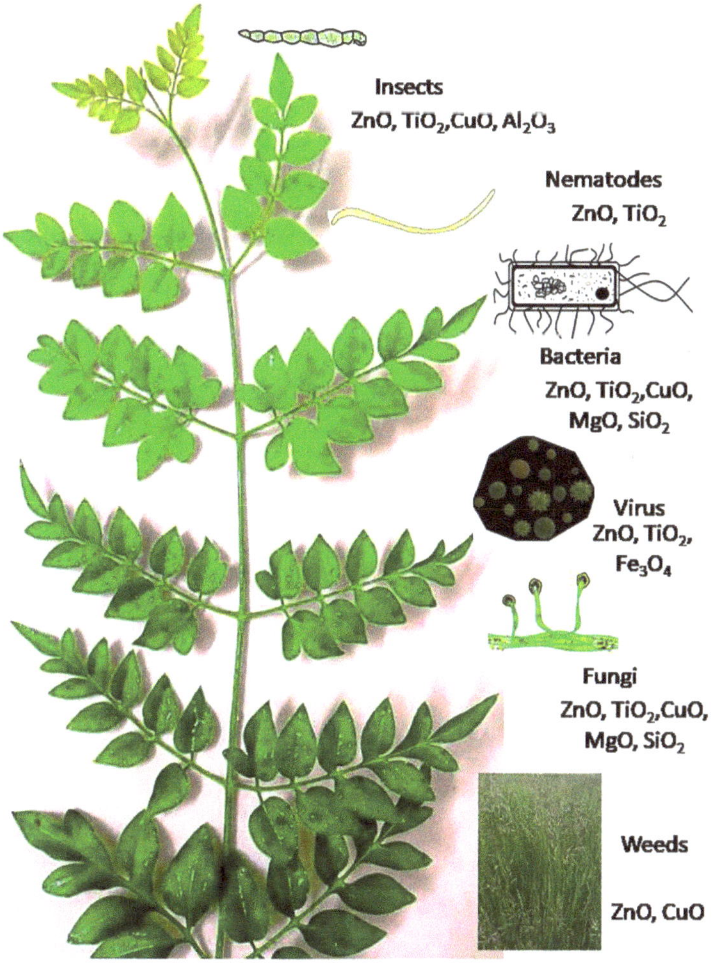

FIGURE 7.1 Different types of nanoparticles against biotic infection on plant.

or macronutrients for the plants, helping them in their growth and proliferation.

7.10 Conclusion

Based on the studies, it is evident that NPs come in handy in controlling the stress in plants whenever their fighting mechanism fails. The data obtained from this review reveal the impact of metal and MONPs on plants in controlling their stress due to biotic factors. Few studies show the positive impact of metal and MONPs in agriculture, but the mechanisms are largely unknown, and the research is still in its early stages. More research in this area is needed to fully comprehend the molecular mechanism of plant NP interaction. The study also reveals that MONPs cause morphological changes in plants to control stress. However, in employing NP to control the biotic stress in plants, care must be taken about the environmental toxicity too. Thus, the study recommends the usage of MONPs with utmost care in considering the environment (Fig. 7.1).

References

Abd El-Daim, I., Haggblom, P., Karlsson, M., Stenstrom, E., Timmusk, S., 2015. Paenibacillus polymyxa A26 Sfp-type PPTase inactivation limits bacterial antagonism against *Fusarium graminearum* but not of *F. culmorum*. Front. Plant. Sci. 1–8. Available from: https://doi.org/10.3389/fpls.2015.00368.

Abdul Malik, N.A., Kumar, I.S., Nadarajah, K., 2020. Elicitor and receptor molecules: orchestrators of plant defense and immunity. Intern. J. Mol. Sci. 21, 963. Available from: https://doi.org/10.3390/ijms21030963.

Acharya, A., Pal, P.B., 2020. Agriculture nanotechnology: translating research outcome to field applications by influencing environmental sustainability. Nano Impact 19, 100232.

Adams, L.K., Lyon, D.Y., Alvarez, P.J., 2006. Comparative eco-toxicity of nanoscale TiO2, SiO2, and ZnO water suspensions. Water Res. 40, 3527–3532.

Abdelkhalek, A., Al-Askar, A.A., 2020. Green synthesized ZnO nanoparticles mediated by *Mentha spicata* extract induce plant systemic resistance against tobacco mosaic virus. Appl. Sci 10, 5054. Available from: https://doi.org/10.3390/app10155054.

Al-Bartya, A.M., Hamzab, R.Z., 2015. Larvicidal, antioxidant activities and perturbation of transminases activities of titanium dioxide nanoparticles synthesized using *Moringa oleifera* leaves extract against the red palm weevil (*Rhynchophorus ferrugineus*). Eur. J. Pharm. Med. Res. 2 (6), 49–54.

Aldon, D., Mbengue, M., Mazars, C., Galaud, J.P., 2018. Calcium signalling in plant biotic interactions. Intern. J. Mol. Sci. 19, 665. Available from: https://doi.org/10.3390/ijms19030665.

Alvarez, B., Lopez, M.M., Biosca Elena, G., 2019. Biocontrol of the major plant pathogen *Ralstonia solanacearum* in irrigation water and host plants by novel waterborne lytic bacteriophages. Front. Microbiol. 10, 2813. Available from: https://doi.org/10.3389/fmicb.2019.02813.

Amro, N.A., Kotra, L.P., Wadu-Mesthrige, K., Bulychev, A., Mobashery, S., Liu, G., 2000. High-resolution atomic force microscopy studies of the *Escherichia coli* outer membrane: structural basis for permeability. Langmuir 16, 2789–2796.

Aruscavage, D., Phelan, P.L., Lee, K., LeJeune, J.T., 2010. Impact of changes in sugar exudate created by biological damage to tomato plants on the persistence of *Escherichia coli* O157: H7. J. Food Sci. 75 (4), M187–M192.

Ashraf, H., Anjum, T., Riaz, S., Ahmad, I.S., Irudayaraj, J., Javed, S., et al., 2021. Inhibition mechanism of green-synthesized copper oxide nanoparticles from *Cassia fistula* towards *Fusarium oxysporum* by boosting growth and defense response in tomatoes. Environ. Sci. Nano 8 (6), 1729–1748. Available from: https://doi.org/10.1039/d0en01281e.

Badawy, A.A., Abdelfattah, N.A.H., Salem, S.S., Awad, M.F., Fouda, A., 2021. Efficacy assessment of biosynthesized copper oxide nanoparticles (CuO-NPs) on stored grain insects and their impacts on morphological and physiological traits of wheat (*Triticum aestivum* L.) plant. Biology. 10 (3), 233. Available from: https://doi.org/10.3390/biology10030233.

Bao-shan, L., shao-qi, D., Chun-hui, L., Li-jun, F., Shu-chun, Q., Min, Y., 2004. Effect of TMS (nanostructured silicon dioxide) on growth of *Changbai larch* seedlings. J. Forest. Res. 15, 138. Available from: https://doi.org/10.1007/BF02856749.

Barak, J.D., Liang, A.S., 2008. Role of soil, crop debris, and a plant pathogen in Salmonella enterica contamination of tomato plants. PLoS One 3 (2), e1657.

Bernard, G.C., Egnin, M., Bonsi, C., 2017. The impact of plant-parasitic nematodes on agriculture and methods of control. In: Shah, M.M., Mahamood, M. (Eds.), Nematology-Concepts, Diagnosis and Control. IntechOpen, London, p. 121. Available from: http://doi.org/10.5772/intechopen.68958.

Berube, D.M., 2008. Rhetorical gamesmanship in the nano debates over sunscreens and nanoparticles. J. Nanopart. Res. 10, 23–27.

Blecher, K., Nasir, A., Friedman, A., 2011. The growing role of nanotechnology in combating infectious disease. Virulence 2 (5), 395–401.

Boddu, S.R., Gutti, V.R., Ghosh, T.K., Tompson, R.V., Loyalka, S.K., 2011. Gold, silver, and palladium nanoparticle/nano-agglomerate generation, collection, and characterization. J. Nanopart. Res. 13, 6591–6601. Available from: https://doi.org/10.1007/s11051-011-0566-x.

Boudsocq, M., Willmann, M.R., Mccormack, M., Lee, H., Shan, L., He, P., 2010. Differential innate immune signalling via Ca 2+ sensor protein kinases. Nature 464, 418–422. Available from: https://doi.org/10.1038/nature08794.

Brayner, R., Ferrari-Iliou, R., Brivois, N., Djediat, S., Benedetti, M.F., Fievet, F., 2006. Toxicological impact studies based on *Escherichia coli* bacteria in ultrafine ZnO nanoparticles colloidal medium. Nano Lett. 6, 866–870.

Cai, L., Chen, J.N., Liu, Z.W., Wang, H.C., Yang, H.K., Ding, W., 2018. Magnesium oxide nanoparticles: effective agricultural antibacterial agent against *Ralstonia solanacearum*. Front. Microbiol. 9, 790. Available from: https://doi.org/10.3389/fmicb.2018.00790.

Cai, L., Cai, L., Jia, H., Liu, C., Wang, D., Sun, X., 2020. Foliar exposure of Fe3O4 nanoparticles on *Nicotiana benthamiana*: evidence for nanoparticles uptake, plant growth promoter and defense response elicitor against plant virus. J. Hazard. Mater. 393, 122415.

Chavali, M.S., Nikolova, M.P., 2019. Metal oxide nanoparticles and their applications in nanotechnology. SN Appl. Sci. 1, 607. Available from: https://doi.org/10.1007/s42452-019-0592.

Chen, H., Seiber, J.N., Hotze, M., 2014. ACS select on nanotechnology in food and agriculture: a perspective on implications and applications. J. Agri Food Chem. 62, 1209–1212.

Chen, J., Sun, L., Cheng, Y., Lu, Z., Shao, K., Li, T., et al., 2016b. Graphene oxide-silver nanocomposite: novel agricultural antifungal agent against *Fusarium graminearum* for crop disease prevention. ACS Appl. Mater. Interfaces 8 (36), 24057–24070. Sep 14.

Chen, J., Wu, L., Lu, M., Lu, S., Li, Z., Ding, W., 2020. Comparative study on the fungicidal activity of metallic MgO nanoparticles and macroscale MgO against soil borne fungal phytopathogens. Front. Microbiol. 11, 365. Available from: https://doi.org/10.3389/fmicb.2020.00365.

Clausen, C.A., Kartal, S.N., Arango, R.A., Green, F., 2011. The role of particle size of particulate nano-zinc oxide wood preservatives on termite mortality and leach resistance. Nanoscale Res. Lett. 6, 42. Available from: https://doi.org/10.1186/1556-276X-6-427.

Clemens, S., 2001. Molecular mechanisms of plant metal tolerance and homeostasis. Planta 212 (475 486), 0032–0935.

Cui, H., Zhang, P., Gu, W., Jiang, J., 2009. Application of anatase TiO2 sol derived from peroxotitannic acid in crop diseases control and growth regulation. Technical Proceedings of the NSTI Nanotechnology Conference and Expo, NSTI-Nanotech 2. 286–289.

Culver, J.N., Padmanabhan, M.S., 2007. Virus-induced disease: altering host physiology one interaction at a time. Annu. Rev. Phytopathol. 45, 221–243.

Dangl, J.L., McDowell, J.M., 2006. Two modes of pathogen recognition by plants. Proc. Natl. Acad. Sci. U.S.A. 103, 8575–8576. Available from: https://doi.org/10.1073/pnas.0603183103.

Decraemer, W., Hunt, D., 2006. Structure and classification. In: Perry, R., Moens, M. (Eds.), Plant Nematology. CAB International, Oxfordshire, pp. 3–32. Available from: http://doi.org/10.1079/9781845930561.0000.

Demir, E., Kaya, N., Kaya, B., 2014. Genotoxic effects of zinc oxide and titanium dioxide nanoparticles on root meristem cells of *Allium cepa* by comet assay. Turk. J. Biol. 38, 31–39. Available from: https://doi.org/10.3906/biy-1306-11.

Diaz, A.O., Eckenrode, K., Villani, C., Oconnor, B.M., 2000. Biology, ecology, and management of the bulb mites of the genus Rhizoglyphus (Acari: Acaridae). Exp. Appl. Acarol. 24, 85–113. Available from: https://doi.org/10.1023/A:1006304300657.

Dimkpa, C.O., McLean, J.E., Britt, D.W., Anderson, A.J., 2013. Antifungal activity of ZnO nanoparticles and their interactive effect with a biocontrol bacterium on growth antagonism of the plant pathogen *Fusarium graminearum*. Biometals 26 (6), 913–924.

Dizaj, S.M., Farzaneh, L., Mohammad, B.-J., Mohammad Hossein, Z., Khosro, A., 2014. Antimicrobial activity of the metals and metal oxide nanoparticles. Mater. Sci. Eng.: C. 44, 278–284. Available from: https://doi.org/10.1016/j.msec.2014.08.031. ISSN 0928–4931.

Dodds, P.N., Rathjen, J.P., 2010. Plant immunity: towards an integrated view of plant-pathogen interactions. Nat. Rev. Genet. 11, 539–548. Available from: https://doi.org/10.1038/nrg2812.

Du, W., Sun, Y., Ji, R., Zhu, J., Wu, J., Guo, H., 2011. TiO2 and ZnO nanoparticles negatively affect wheat growth and soil enzyme activities in agricultrual soil. J. Environ. Monit. 13, 822–828.

Dupont, C.L., Grass, G., Rensing, C., 2011. Copper toxicity and the origin of bacterial resistance—new insights and applications. Metallomics 3 (11), 1109–1118. Available from: https://doi.org/10.1039/c1mt00107h. Epub 2011 Oct 10. PMID: 21984219.

Elansary, M., Ragaa, H., Maha, E., 2021. Using biosynthesized zinc oxide nanoparticles to alleviate the toxicity on banana parasitic-nematode. Research Square. https://doi.org/10.21203/rs.3.rs-186764/v1

El-Ansary, M.S.M., 2014. Current approaches of managing microbial diseases: focus on nanosciences technique. J. Coast. Life Med. 3, 658–662.

El-Bendary, H.M., El-Helaly, A.A., 2013. First record nanotechnology in agricultural: Silica nano-particles a potential new insecticide for pest control. App. Sci. Rep. 4 (3), 241–246.

Elmer, W., De La Torre-Roche, R., Pagano, L., Majumdar, S., Zuverza-Mena, N., Dimkpa, C., et al., 2018. Effect of metalloid and metal oxide nanoparticles on Fusarium wilt of watermelon. Plant. Dis. 102 (7), 1394–1401. Available from: https://doi.org/10.1094/PDIS-10-17-1621-RE. Epub. PMID: 30673561.

Elmer, W.H., de la Torre-Roche, R., Zuverza-Mena, N., Adisa, I.H., Dimkpa, C., Gardea-Torresdey, J., et al., 2021. Influence of single and combined mixtures of metal oxide nanoparticles on eggplant growth, yield, and *Verticillium wilt* severity. Plant. Dis. 105 (4), 1153–1161. Available from: https://doi.org/10.1094/PDIS-07-20-1636-RE.

El-Sawy, M., Elsharkawy, M., Abass, J., Kasem, M., 2017. Antiviral activity of 2-nitromethyl phenol, zinc nanoparticles and seaweed extract against cucumber mosaic virus (CMV) in *eggplant*. J. Virol. Antivir. Res. 6.

Elsharkawy, M.M., Derbalah, A., 2019. Antiviral activity of titanium dioxide nanostructures as a control strategy for broad bean strain virus in faba bean. Pest. Manag. Sci. 75 (3), 828–834. Available from: https://doi.org/10.1002/ps.5185.

Espitia, P.J.P., Soares, N.F.F., Coimbra, J.S.R., de Andrade, N.J., Cruz, R.S., 2012. Zinc oxide nanoparticles: synthesis, antimicrobial activity and food packaging applications. Food Bioprocess. Technol. 5, 1447–1464.

Fernandez-Garcia, M., Rodriguez, J.A., 2011. Metal oxide nanoparticles. In R.A. Scott (Ed.), Encyclopedia of Inorganic and Bioinorganic Chemistry. Available from: https://doi.org/10.1002/9781119951438.eibc0331.

Fields, P.G., 2006. Effect of *Pisum sativum* fractions on the mortality and progeny production of nine stored-grain beetles. J. Stored Prod. Res. 42, 86–96.

Fowler, L., Hakan, E., Caroline, O.-M., 2019. Effect of copper ion concentration on bacteria and cells. Materials 12, 3798. Available from: https://doi.org/10.3390/ma12223798.

Franklin, N., Rogers, N., Apte, S., Batey, G., Gadd, G., Casey, P., 2007. Comparative toxicity of nanoparticles ZnO, bulk ZnO, and $ZnCl_2$ to a fresh water microalgae (*Pseudokirchneriella subcapitata*): the importance of particle solubility. Env. Sci. Tech. 41, 8484–8490.

Fu, Z.Q., Dong, X., 2013. Systemic acquired resistance: turning local infection into global defense. Annu. Rev. Plant. Biol. 64, 839–863. Available from: https://doi.org/10.1146/annurev-arplant-042811-105606.

Gao, F., Liu, C., Qu, C., Zheng, L., Yang, F., Su, M., 2008. Improvement of spinach growth by nano-TiO(2) treatment related to the changes of *Rubisco activase*. Biometals 21, 211–217. Available from: https://doi.org/10.1007/s10534-007-9110-y.

Gao, Y., Wang, W., Su, F., Min, F., Liu, Z., Wei, Q., et al., 2020. Antifungal activity of dryocrassine on potato dry rot. Chin. Potato J. 34, 238–244.

Gawande, M.B., Goswami, A., Felpin, F.X., Asefa, T., Huang, T., Silva, R., et al., 2016. Cu and Cu-based nanoparticles: synthesis and applications in catalysis. Chem. Rev. 116, 3722–3811.

Ghosh, M., Jana, A., Sinha, S., Jothiramajayam, M., Nag, A., Chakraborty, A., 2016. Effects of ZnO nanoparticles in plants: cytotoxicity, genotoxicity, deregulation of antioxidant defenses, and cell-cycle arrest. Mutat. Res. Genet. Toxicol. Environ. Mutagen. 807, 25–32. Available from: https://doi.org/10.1016/j.mrgentox.2016.07.006.

Giannousi, K., Avramidis, I., Dendrinou-Samara, C., 2013. Synthesis, characterization and evaluation of copper based nanoparticles as agrochemicals against *Phytophthora infestans*. RSC Adv. 3 (44), 21743–21752.

Gisi, U., Cohen, Y., 1996. Resistance to phenylamide fungicides: a case study with Phytophthora infestans involving mating type and race structure. Ann. Rev. Phytopathol. 34, 549–572. Available from: https://doi.org/10.1146/annurev.phyto.34.1.549.

Gonzalez-Merino, A.M., Hernandez-Juarez, A., Betancourt-Galindo, R., Ochoa-Fuentes, Y.M., Valdez-Aguilar, L.A., Limon-Corona, M.L., 2021. Antifungal activity of zinc oxide nanoparticles in *Fusarium oxysporum—Solanum lycopersicum* pathosystem under controlled conditions. J. Phytopathol. 169, 533–544. Available from: https://doi.org/10.1111/jph.13023.

Gouveia, B.C., Calil, I.P., Machado, J.P.B., Santos, A.A., Fontes, E.P., 2017. Immune receptors and co-receptors in antiviral innate immunity in plants. Front. Microbiol. 7, 2139. Available from: https://doi.org/10.3389/fmicb.2016.02139.

Grant, M., Brown, I., Adams, S., Knight, M., Ainslie, A., Mansfield, J., 2000. The RPM1 plant disease resistance gene facilitates a rapid and sustained increase in cytosolic calcium that is necessary for the oxidative burst and hypersensitive cell death. Plant. J. 23, 441–450. Available from: https://doi.org/10.1046/j.1365-313x.2000.00804.x.

Gunalan, S., Sivaraj, R., Rajendran, V., 2012. Green synthesized ZnO nanoparticles against bacterial and fungal pathogens. Progr Nat. Sci. Mater. Int. 22, 693–700.

Gunawan, C., Teoh, W., Marquis, C., Amal, R., 2011. Cytotoxic origin of copper (II) oxide nanoparticles: comparative studies with micron-sized particles, leachate, and metal salts. ACS Nano 5 (9), 7214–7722.

Gupta, S., Tanuja, K., Ashutosh, V., Shweta, Y., 2015. Optimization of ZnO-NPs to investigate their safe application by assessing their effect on soil nematode *Caenorhabditis elegans*. Nanoscale Res. Lett. 10, 303. Available from: https://doi.org/10.1186/s11671-015-1010-4.

Hamed Derbalah, A.S., Elsharkawy, M.M., 2019. A new strategy to control *Cucumber mosaic* virus using fabricated NiO-nanostructures. J. Biotechnol. 20 (306), 134–141. Available from: https://doi.org/10.1016/j.jbiotec.2019.10.003. Epub 2019 Oct 5. PMID: 31593748.

Hammond-Kosack, K.E., Jones, J.D.G., 2000. Responses to plant pathogens. In: Buchanan, B.B., Gruissem, W., Jones, R.L. (Eds.), Biochemistry and Molecular Biology of Plants, American Society of Plant Physiology. Rockville, pp. 1102–1156.

Hao, Y., Cao, X., Ma, C., Zhang, Z., Zhao, N., Ali, A., et al., 2017. Potential applications and antifungal activities of engineered anomaterials against gray mold disease agent *Botrytis cinerea* on rose petals. Front. Plant. Sci. 8, 1332. Available from: https://doi.org/10.3389/fpls.2017.01332.

Hao, Y., Yu, F., Lv, R., Ma, C., Zhang, Z., Rui, Y., 2016. Carbon nanotubes filled with different ferromagnetic alloys affect the growth and development of rice seedlings by changing the C: N ratio and plant hormones concentrations. PLoS ONE 11, e0157264. Available from: https://doi.org/10.1371/journal.pone.0157264.

Hashem, A., Baby, T., Elsayed Fathi Abd, A., 2019. *Bacillus Subtilis*: a plant-growth promoting rhizobacterium that also impacts biotic stress. Saudi J. Biol. Sci. 26 (6), 1291–1297. ISSN 1319-562x.

He, L.L., Liu, Y., Mustapha, A., Lin, M.S., 2011. Antifungal activity of zinc oxide nanoparticles against *Botrytis cinerea* and *Penicillium expansum*. Microbiol. Res. 166 (3), 207–215.

Heinlaan, M., Ivask, A., Blinova, I., Dubourguier, H.C., Kahru, A., 2008. Toxicity of nanosized and bulk ZnO, CuO and TiO2 to bacteria *Vibrio fischeri* and crustaceans *Daphnia magna* and *Thamnocephalus platyurus*. Chemosphere 71, 1308–1316.

Holt, K.B., Bard, A.J., 2005. Interaction of silver (I) ions with the respiratory chain of *Escherichia coli*: an electrochemical and scanning electrochemical microscopy study of the antimicrobial mechanism of micromolar Ag+. Biochemistry 44, 13214–13223.

Horky, P., Skalickova, S., Baholet, D., Skladanka, J., 2018. Review, nanoparticles as a solution for eliminating the risk of mycotoxins. Nanomaterials 8, 727–748.

Hou, J., Wu, Y., Li, X., Wei, B., Li, S., Wang, X., 2018. Toxic effects of different types of zinc oxide nanoparticles on algae, plants, invertebrates, vertebrates and microorganisms. Chemosphere. 193, 852–860.

Hu, C.-C., Wu, G.-H., Hua, T.-E., Wagner, O.I., Yen, T.-J., 2018. Uptake of TiO_2 nanoparticles into *C. elegans* neurons negatively affects axonal growth and worm locomotion behavior. CS Appl. Mater. Interfaces 10 (10), 8485–8495.

Huang, L., Li, D.Q., Lin, Y.J., Wei, M., Evans, D.G., Duan, X.J., 2005. Controllable preparation of Nano-MgO and investigation of its bactericidal properties. Inorg. Biochem. 99 (5), 986–993.

Huang, Z., Zheng, X., Yan, D., Yin, G., Liao, X., Kang, Y., et al., 2008. Toxicological effect of ZnO nanoparticles based on bacteria. Apr 15; 24(8), 4140–41444.

Huh, J., Kwon, Y.J., 2011. Nanoantibiotics': a new paradigm for treating infectious diseases using nanomaterials in the antibiotics resistant era. J. Controlled Rel. 156 (2), 128–145.

Iannone, M.F., Groppa, M.D., de Sousa, M.E., van Raap, M.B.F., Benavides, M.P., 2016. Impact of magnetite iron oxide nanoparticles on wheat (*Triticum aestivum* L.) development: evaluation of oxidative damage. Environ. Exp. Bot. 131, 77–88. Available from: https://doi.org/10.1016/j.envexpbot.2016.07.004.

Jaberzadeh, A., Moaveni, P., Moghadam, H.R.T., Zahedi, H., 2013. Influence of bulk and nanoparticles titanium foliar application on some agronomic traits, seed gluten and starch contents of wheat subjected to water deficit stress. Not. Bot. Horti Agrobo 41, 201–207.

Jangra, S.L., Stalin, K., Dilbaghi, N., Kumar, S., Tawale, J., Singh, S.P., et al., 2012. Antimicrobial activity of zirconia (ZrO2) nanoparticles and zirconium complexes. J. Nanosci. Nanotechnol. Sep. 12 (9), 7105–7112. Available from: https://doi.org/10.1166/jnn.2012.6574.

Jatav, G.K., De, N., 2013. Application of nano-technology in soil-plant system. Asian J. Soil. Sci. 8 (1), 176–184.

Jayaseelan, C., Abdul Rahuman, A., Kirthi, A.V., Marimuthu, S., Santhoshkumar, T., Bagavan, A., et al., 2012. Novel microbial route to synthesize ZnO nanoparticles using *Aeromonas hydrophila* and their activity against pathogenic bacteria and fungi. Spectrochim. Acta A 90, 78–84.

Jin, T., He, Y., 2011. Antibacterial activities of magnesium oxide (MgO) nanoparticles against food borne pathogens. J. Nanopart. Res. 13, 6877–6885. Available from: https://doi.org/10.1007/s11051-011-0595-5.

Jo, Y., Kim, B.H., Jung, G., 2009. Antifungal activity of silver ions and nanoparticles on phytopathogenic fungi. Plant. Dis. 93, 1037–1043. Available from: https://doi.org/10.1094/PDIS-93-10-1037.

John, L., 1853. The Vegetable Kingdom; or, the Structure, Classification, and Uses of Plants, Illustrated upon the Natural System, third ed. Bradbury & Evans, London.

Kaloshian, I., 2004. Gene-for-gene disease resistance: bridging insect pest and pathogen defense. J. Chem. Ecol. 30, 2419–2438. Available from: https://doi.org/10.1007/s10886-004-7943-1.

Kanhed, P., Birla, S., Gaikwad, S., Gade, A., Seabra, A.B., Rubilar, O., 2014. In vitro antifungal efficacy of copper nanoparticles against selected crop pathogenic fungi. Mater. Lett. 115, 13–17. Available from: https://doi.org/10.1016/j.matlet.2013.10.011.

Karthikeyan, K., Govindasamy, M., Sang Jae, K., Kadarkaraithangam, J., Mariappan, P., 2012. Antibacterial activity of MgO nanoparticles based on lipid peroxidation by oxygen vacancy. J. Nanopart. Res. 14 (9), 1–10.

Kate, D., Judy, C., 2007. Ubiquitin, hormones and biotic stress in plants. Ann. Botany 99, 787–822.

Khan, M.R., Rizvi, T.F., 2014. Nanotechnology: scope and application in plant disease management. Plant. Pathol. J. 13, 214–231. Available from: https://doi.org/10.3923/ppj.2014.214.231.

Khooshe-Bast, Z., Sahebzadeh, N., Ghaffari-Moghaddam, M., Mirshekar, A., 2016. Insecticidal effects of zinc oxide nanoparticles and *Beauveria bassiana* TS11 on *Trialeurodes vaporariorum* (Westwood, 1856) (Hemiptera: Aleyrodidae). Acta Agri. Sloven. 107 (2), 299–309.

Kittler, S., Greulich, C., Diendorf, J., Koller, M., Epple, M., 2010. Toxicity of silver nanoparticles increases during storage because of slow dissolution under release of silver ions. Chem. Mater. 22, 4548–4554.

Korghond, G.T., Najmeh, S., Hossein, A., Sara, R., 2021. Acute toxicity and sublethal effects of metal oxide nanoparticles against the bulb mite. Syst. Appl. Acarol. 26 (4), 788–800. Available from: https://doi.org/10.11158/saa.26.4.9.

Laluk, K., Mengiste, T., 2010. Necrotroph attacks on plants: wanton destruction or covert extortion. Am. Soc. Plant. Biol. 8, e0136. Available from: https://doi.org/10.1199/tab.0136.

Larsen, M.U., Seward, M., Tripathi, A., Shapley, N.C., 2009. Biocompatible nanoparticles trigger rapid bacteria clustering. Biotechnol. Prog. 25 (4), 1094–1102.

Lecourieux, D., Lamotte, O., Bourque, S., Wendehenne, D., Mazars, C., Ranjeva, R., 2005. Proteinaceous and oligosaccharidic elicitors induce different calcium signatures in the nucleus of tobacco cells. Cell Calc. 38, 527–538. Available from: https://doi.org/10.1016/j.ceca.2005.06.036.

Lee, T.Y., Mizubuti, E., Fry, W.E., 1999. Genetics of metalaxyl resistance in Phytophthora infestans. Fungal Genet. Biol. 26, 118–130. Available from: https://doi.org/10.1006/fgbi.1998.1107.

Lemire, J., Harrison, J., Turner, R., 2013. Antimicrobial activity of metals: mechanisms, molecular targets and applications. Nat. Rev. Microbiol. 11, 371–384. Available from: https://doi.org/10.1038/nrmicro3028.

Leuba, J.L., Stossel, P., 1996. Chitosan and other polyamines: antifungal activity and interaction with biological membrace, Chitin in Nature and Technology, 279–286. Springer, Boston, MA, p. 43.

Levard, C., Hotze, E.M., Lowry, G.V., Brown Jr., G.E., 2012. Environmental transformations of silver nanoparticles: impact on stability and toxicity. Environ. Sci. Technol. 46, 6900–6914. Available from: https://doi.org/10.1021/es2037405.

Li, D., Jie, J., Yujie, Y., Dayong, W., 2020. Toxicity comparison of nanopolystyrene with three metal oxide nanoparticles in nematode *Caenorhabditis elegans*. Chemosphere 245, 125625. Available from: https://doi.org/10.1016/j.chemosphere.2019.125625. ISSN 0045–6535.

Li, L., Liu, Z., Aleksandr, M., Wang, T., 2018. Antibacterial mechanisms of a novel type picosecond laser-generated silver titanium nanoparticles and their toxicity to human cells. Int. J. Nanomed. 13, 89–101.

Li, Y., Niu, J., Shang, E., Crittenden, J.C., 2016. Influence of dissolved organic matter on photogenerated reactive oxygen species and metal-oxide nanoparticle toxicity. Water Res. 98, 9–18.

Lin, C., Changyun, L., Guangjin, F., Chaolong, L., Xianchao, S., 2019. Preventing viral disease by ZnONPs through directly deactivating TMV and activating plant immunity in *Nicotiana benthamiana*. Environ. Sci.: Nano 6, 3653–3669.

Lin, D., Xing, B., 2008. Root uptake and phytotoxicity of ZnO nanoparticles. Environ. Sci. Technol. 42, 5580–5585.

Liu, Z., Zhang, M., Han, X., Xu, H., Zhang, B., Yu, Q., 2016. TiO2 nanoparticles cause cell damage independent of apoptosis and autophagy by impairing the ROS-scavenging system in *Pichia pastoris*. Chem. Biol. Interact. 252, 9–18. Available from: https://doi.org/10.1016/j.cbi.2016.03.029.

Luyckx, M., Hausman, J.F., Lutts, S., Guerriero, G., 2017. Silicon and plants: current knowledge and technological perspectives. Front. Plant. Sci. 8, 411.

Magalhaes, L.C., Hunt, T.E., Siegfried, B.D., 2009. Efficacy of neonicotinoid seed treatments to reduce soybean aphid populations under field and controlled conditions in nebraska. J. Econ. Entomol. 102, 187–195.

Mahmoodzadeh, H., Aghili, R., 2014. Effect on germination and early growth characteristics in wheat plants (*Triticumae stivum* L.) seeds exposed to TiO_2 nanoparticles. J. Chem. Health Risks 4, 29–36.

Makkouk, K.M., Azzam, O.I., Katul, L., Rizkallah, A., Koumari, S., 1986. Seed transmission of broad bean stain in the wild legume *Vicia palaestina* Boiss. FABIS 16, 40–41.

Malaikozhundan, B., Vaseeharan, B., Vijayakumar, S., Thangaraj, M.P., 2017. *Bacillus thuringiensis* coated zinc oxide nanoparticle and its biopesticidal effects on the pulse beetle, *Callosobruchus maculatus*. J. Photochem. Photobiol. B Biol. 174, 306–314.

Malandrakis, A., Koukiasas, N., Veloukas, T., Karaoglanidis, G., Markoglou, A., 2013. Baseline sensitivity of *Monilinia laxa* from Greece to fenhexamid and analysis of fenhexamid-resistant mutants. Crop. Prot. 46, 13–17.

Malandrakis, A.A., Apostolidou, Z.A., Louka, D., Markoglou, A., Flouri, F., 2018. Biological and molecular characterization of field isolates of *Alternaria alternata* with single or double resistance to respiratory complex II and III inhibitors. Eur. J. Plant. Pathol. 152 (1), 199–211.

Maneerat, C., Hayata, Y., 2006. Antifungal activity of TiO2 photocatalysis against *Penicillium expansum* in vitro and in fruit tests. Int. J. Food Microbiol. 107, 99–103.

Maness, P.-C., Smolinski, S., Blake, D., Huang, Z., Wolfrum, E.J., Jacoby, W.A., 1999. Bactericidal activity of photocatalytic TiO2 reaction: toward an understanding of its killing mechanism. Appl. Environ. Microbiol. 65, 4094–4098. Available from: https://doi.org/10.1128/AEM.65.9.4094-4098.1999.

Manivannan, A., Ahn, Y.K., 2017. Silicon regulates potential genes involved in major physiological processes in plants to combat stress. Front. Plant Sci. 8, 1346.

Mantri, N.L., Ford, R., Coram, T.E., Pang, E.C.K., 2010. Evidence of unique and shared responses to major biotic and abiotic stresses in chickpea. Environ. Exp. Bot. 69, 286–292.

Maule, A., Leh, V., Lederer, C., 2002. The dialogue between viruses and hosts in compatible interactions. Curr. Opin. Plant. Biol. 5, 279–284.

Meyer, W., 2001. Coping with resistance to copper/silver disinfection. Water Eng. Manag. 148, 11.

Monaghan, J., Zipfel, C., 2012. Plant pattern recognition receptor complexes at the plasma membrane. Curr. Opin. Plant. Biol. 15, 349–357. Available from: https://doi.org/10.1016/j.pbi.2012.05.006.

Monier, J.M., Lindow, S.E., 2005. Aggregates of resident bacteria facilitate survival of immigrant bacteria on leaf surfaces. Microb. Ecol. 49 (3), 343–352.

Mukherjee, A., Majumdar, S., Servin, A.D., Pagano, L., Dhankher, O.P., White, J.C., 2016. Carbon nanomaterials in agriculture: a critical review. Front. Plant. Sci. 7, 172. Available from: https://doi.org/10.3389/fpls.2015.00172.

Murthy, H.C.A., Buzuayehu, A., Prakash, C.H., Shantaveerayya, K., 2018. A review on green synthesis and applications of Cu and CuO nanoparticles. Mater. Sci. Res. India 15, 279–295. Available from: https://doi.org/10.13005/msri/150311.

Neal, A.L., 2008. What can be inferred from bacterium-nanoparticle interactions about the potential consequences of environmental exposure to nanoparticles? Ecotoxicology. 17, 362–371.

Nejat, N., Mantri, N., 2017. Plant immune system: crosstalk between responses to biotic and abiotic stresses the missing link in understanding plant defense. Curr. Issues Mol. Biol. 23, 1–16.

Nekrasova, G.F., Ushakova, O.S., Ermakov, A.E., Uimin, M.A., 2011. Effects of copper (II) ions and copper oxide nanoparticles on *Elodea densa Planch.* Russ. J. Ecol. 42, 458–463. Available from: https://doi.org/10.1134/S1067413611060117.

Nel, A., Xia, T., Madler, L., Li, N., 2006. Toxic potential of materials at the nanolevel. Science. 311, 622–627.

Niazi, A., Manzoor, S., Bejai, S., Meijer, J., Bongcam-Rudloff, E., 2014. Complete genome sequence of a plant associated bacterium *Bacillus amyloliquefaciens* subsp. plantarum UCMB5033. Stand. Genom. Sci. 9 (3), 718–725.

Nicolopoulou-Stamati, P., Maipas, S., Kotampasi, C., Stamatis, P., Hens, L., 2016. Chemical pesticides and human health: the urgent need for a new concept in agriculture. Front. Public. Health 4, 148. Available from: https://doi.org/10.3389/fpubh.2016.00148.

Nikolova, M.P., Chavali, M.S., 2020. Metal oxide nanoparticles as biomedical materials. Biomimetics. 5 (2), 27. Available from: https://doi.org/10.3390/biomimetics5020027.

Osman, H.H., Abdel-Hafez, H.F., Khidr, A.A., 2015. Comparison between the efficacy of two nanoparticles and effective microorganisms on some biological and biochemical aspects of *Spodoptera littorals*. Intr. J. Agr. Innov. Res. 3 (6), 2319–2473.

Ovais, M., Zia, N., Khalil, A., Ayaz, M., Khalil, A., Ahmad, I., 2018. Nanoantibiotics: recent developments and future prospects. Front. Clin. Drug. Res. Anti-Infect. 5, 30–54. Available from: https://doi.org/10.2174/9781681086378119050006.

Pablos, C., van Grieken, R., Marugan, J., Munoz, A., 2012. Simultaneous photocatalytic oxidation of pharmaceuticals and inactivation of *Escherichia coli* in wastewater treatment plant effluents with suspended and immobilised TiO_2. Water Sci. Technol. 65, 2016–2023. Available from: https://doi.org/10.2166/wst.2012.868.

Padmavathy, N., Vijayaraghavan, R., 2008. Enhanced bioactivity of ZnO nanoparticles—an antimicrobial study. Sci. Technol. Adv. Mater. 9, 1–7.

Pallas, V., Garcia, J.A., 2011. How do plant viruses induce disease. Interactions and interference with host components. J. Gen. Virol. 92, 2691–2705. Available from: https://doi.org/10.1099/vir.0.034603-0.

Palmqvist, N.G., Bejai, S., Meijer, J., Seisenbaeva, G.A., Kessler, V.G., 2015. Nano Titania aided clustering and adhesion of beneficial bacteria to plant roots to enhance crop growth and stress management. Sci. Rep. 5, 10146.

Pan, X., Du, Y., Chen, W., Ji, M., Wang, L., Zhou, G., et al., 2012. Applied properties of Ag/TiO2 in antibacterial plastics. Eng. Plast. Appl. 40, 101–105.

Pandey, S., Giri, K., Kumar, R., Mishra, G., Raja Rishi, R., 2018. Nanopesticides: opportunities in crop protection and associated environmental risks. Proc. Natl. Acad. Sci. India, Sect. B Biol. Sci 88 (4), 1287–1308.

Paplomatas, E.J., Tjamos, S.E., Malandrakis, A.A., Kafka, A.L., Zouvelou, S.V., 2005. Evaluation of compost amendments for suppressiveness against *Verticillium wilt* of eggplant and study of mode of action using a novel *Arabidopsis patho* system. Eur. J. Plant. Pathol. 112 (2), 183–189.

Paret, M.L., Vallad, G.E., Averett, D.R., Jones, J.B., Olson, S.M., 2013. Photocatalysis: effect of light-activated nanoscale formulations of TiO(2) on *Xanthomonas perforans* and control of bacterial spot of tomato. Phytopathology. 103 (3), 228–236. Available from: https://doi.org/10.1094/PHYTO-08-12-0183-R.

Patolsky, F., Zheng, G., Lieber, C.M., 2006. Nanowire sensors for medicine and the life sciences. Nanomed. (Lond). Jun. 1 (1), 51–65. Available from: https://doi.org/10.2217/17435889.1.1.51.

Pązik, R., Andersson, R., Kepinski, L., Kessler, V.G., Seisenbaeva, G.A., 2011. Surface functionalization of the metal oxide nanoparticles with biologically active molecules containing phosphonate moieties. Case study of $BaTiO_3$. J. Phys. Chem. C. 115, 9850–9860. Available from: https://doi.org/10.1021/jp2000656.

Peck, S., Ron, M., 2020. Plant signaling in biotic and abiotic stress. J. Exp. Bot. 71 (5), 1649–1651. Available from: https://doi.org/10.1093/jxb/eraa051.

Perez-de-Luque, Diego, A.R., 2009. Nanotechnology for parasitic plant control. Pest. Manag. Sci. 65, 540–545.

Pinto, E., Sigaud-kutner, T., Leitao, M.A., Okamoto, O.K., Morse, D., Colepicolo, P., 2003. Heavy metal-induced oxidative stress in algae. J. Phycol. 39, 1008–1018. Available from: https://doi.org/10.1111/j.0022-3646.2003.02-193.x.

Potnis, N., Soto-Arias, J.P., Cowles, K.N., van Bruggen, A.H., Jones, J.B., Barak, J.D., 2014. *Xanthomonas perforans* colonization influences *Salmonella enterica* in the tomato phyllosphere. Appl. Env. Microbiol. 80 (10), 3173–3180.

Raliya, R., Nair, R., Chavalmane, S., Wang, W.N., Biswas, P., 2015. Mechanistic evaluation of translocation and physiological impact of titanium dioxide and zinc oxide nanoparticles on the tomato (*Solanum lycopersicum* L.) plant. Metallomics 7, 1584–1594. Available from: https://doi.org/10.1039/C5MT00168D.

Ravishankar, R.V., Jamuna, B.A., 2011. Nanoparticles and their potential application as antimicrobials, science against microbial pathogens: communicating current research and technological advances. In: Mendez-Vilas, A., (Ed.), Formatex, Microbiology Series, 3(1), 197–209.

Reddy, K.M., Kevin, F., Jason, B., Cory, H., Alex, P., 2007. Selective toxicity of zinc oxide nanoparticles to prokaryotic and eukaryotic systems. Appl. Phys. Lett. 90, 1–8.

Reddy, A.S., Ali, G.S., Celesnik, H., Day, I.S., 2011. Coping with stresses: roles of calcium-and calcium/calmodulin-regulated gene expression. Plant. Cell 23, 2010–2032. Available from: https://doi.org/10.1105/tpc.111.084988.

Rizwan, M., Ali, S., Qayyum, M., Ok, Y., Adrees, M., Ibrahim, M., et al., 2017. The effect of metal and metal oxide nanoparticles on the growth and physiology of globally important food crops: a critical review. J. Hazard. Mater. 322 (Pt A), 2–16.

Romheld, V., Marschner, H., 1991. Function of micronutrients in plants. In: Mortvedt, J.J., Cox, F.R., Shuman, L.M., Welch, R.M. (Eds.), Micronutrients in Agriculture. Soil Science Society of America, Inc, Madison, WI, pp. 297–328.

Rui, M., Ma, C., Hao, Y., Guo, J., Rui, Y., Tang, X., 2016. Iron oxide nanoparticles as a potential iron fertilizer for peanut (*Arachis hypogaea*). Front. Plant. Sci. 7, 815. Available from: https://doi.org/10.3389/fpls.2016.00815.

Ruotolo, R., Maestri, E., Pagano, L., Marmiroli, M., White, J.C., Marmiroli, N., 2018. Plant response to metal-containing engineered nanomaterials: an omics-based perspective. Environ. Sci. Technol. 52, 2451–2467.

Safeena, M.I.S., Zakeel, M.C.M., 2020. Nanobiotechnology-driven management of phytonematodes. In: Ansari, R., Rizvi, R., Mahmood, I. (Eds.), Management of Phytonematodes: Recent Advances and Future Challenges. Springer, Singapore. Available from: https://doi.org/10.1007/978-981-15-4087-5_1.

Saijo, Y., Loo, E.P.I., 2020. Plant immunity in signal integration between biotic and abiotic stress responses. N. Phytol. 225, 87–104. Available from: https://doi.org/10.1111/nph.15989.

Saito, T., Iwase, T., Horie, J., Morioka, T., 1992. Mode of photocatalytic bactericidal action of powdered semiconductor TiO2 on mutans streptococci. J. Photochem. Photobiol. B, Biol. 14, 369–379. Available from: https://doi.org/10.1016/1011-1344(92)85115-B.

Sanchez-Moreno, P., Ortega-Vinuesa, J.L., Peula-García, J.M., Marchal, J.A., Boulaiz, H., 2016. Smart drug-delivery systems for cancer nanotherapy. Curr. Drug. Targets 17, 339–359.

Santamaria, M.E., Martínez, M., Cambra, I., Grbic, V., Diaz, I., 2013. Understanding plant defence responses against herbivore attacks: an essential first step towards the development of sustainable resistance against pests. Transgen. Res. 22, 697–708. Available from: https://doi.org/10.1007/s11248-013-9725-4.

Sasser, J., Freckman, D., 1987. A world perspective on nematology: the role of the society. In: Veech, J., Dickson, D. (Eds.), Vistas on Nematology. Society of Nematologists, Inc., Hyattsville, MD, pp. 7–14.

Schenk, P.M., Kazan, K., Wilson, I., Anderson, J.P., Richmond, T., Somerville, S.C., et al., 2000. Coordinated plant defense responses in Arabidopsis revealed by microarray analysis. Proc Natl Acad Sci USA 97, 11655–11660.

Schumann, G.L., D'Arcy, C.J., 2006. Essential Plant Pathology. APS Press, London.

Shaker, A.M., Ibrahiem, M.A.A.E., Abdelrahim, F., Heba, Ahmed, Y., 2020. Impact of nanoparticle materials on the control of seedling pests in the Egyptian cotton. Egypt. J. Plant. Prot. Res. Inst. 3 (1), 138–147.

Shaker, A.M., Zaki, A.H., AbdelRahim, E.F., Khedr, M.H., 2017a. TiO2 nanoparticles as an effective nanopesticide for cotton leaf worm. Agric. Eng. Int: CIGR J. 61–68. Special issue.

Siddiqi, K.S., Husen, A., 2016. Engineered gold nanoparticles and plant adaptation potential. Nano Res. Lett. 11, 400.

Singh, R., Gautam, N., Mishra, A., Gupta, R., 2016. Heavy metals and living systems: an overview. Indian. J. Pharmacol. 43, 246. Available from: https://doi.org/10.4103/0253-7613.81505.

Slavin, Y.N., Asnis, J., Hafeli, U.O., 2017. Metal nanoparticles: understanding the mechanisms behind antibacterial activity. J. Nanobiotechnol 15, 65. Available from: https://doi.org/10.1186/s12951-017-0308-z.

Song, G., Hou, W., Gao, Y., Wang, Y., Lin, L., Zhang, Z., 2016. Effects of CuO nanoparticles on *Lemna minor*. Bot. Stud. 57, 3. Available from: https://doi.org/10.1186/s40529-016-0118-x.

Spoel, S.H., Dong, X., 2012. How do plants achieve immunity. Defence without specialized immune cells. Nat. Rev. Immunol. 12, 89–100. Available from: https://doi.org/10.1038/nri3141.

Sruthi, S., Ashtami, J., Mohanan, P.V., 2018. Biomedical application and hidden toxicity of zinc oxide nanoparticles. Mater. Today Chem. 10, 175–186.

Stoimenov, P.K., Klinger, R.L., Marchin, G.L., Klabunde, K.J., 2002. Metal oxide nanoparticles as bactericidal agents. Langmuir 18, 6679–6686. Available from: https://doi.org/10.1021/la0202374.

Sunada, K., Kikuchi, Y., Hashimoto, K., Fujishima, A., 1998. Bactericidal and detoxification effects of TiO2 thin film photocatalysts. Environ. Sci. Technol. 32, 726–728.

Taiz, L., Zeiger, E., 2006. Secondary metabolites and plant defense. Plant. Physiol. 4, 315–344.

Tao, Y., Xie, Z., Chen, W., Glazebrook, J., Chang, H.-S., Han, B., 2003. Quantitative nature of Arabidopsis responses during compatible and incompatible interactions with the bacterial pathogen *Pseudomonas syringae*. Plant Cell 15, 317–330. Available from: https://doi.org/10.1105/tpc.007591.

Taylor, A.G., Eckenrode, C.J., Straub, R.W., 2001. Seed coating technologies and treatments for onions: challenges and progress. Hort. Sci. 36, 199–205.

Terzi, E., Kartal, S.N., Yılgor, N., Rautkari, L., Yoshimura, T., 2016. Role of various nanoparticles in prevention of fungal decay, mold growth and termite attack in wood, and their effect on weathering properties and water repellency. Int. Biodeterior. Biodegrad. 107, 77–87. Available from: https://doi.org/10.1016/j.ibiod.2015.11.010.

Teske, S.S., Detweiler, C.S., 2015. The biomechanisms of metal and metal-oxide nanoparticles' interactions with cells. Int. J. Env. Res. Public. Health 12 (2), 1112–1134. Available from: https://doi.org/10.3390/ijerph120201112.

Timmusk, S., Abd El-Daim, I.A., Copolovici, L., Tanilas, T., Kännaste, A., Behers, L., 2014b. Drought-tolerance of wheat improved by rhizosphere bacteria from harsh environments: enhanced biomass production and reduced emissions of stress volatiles. PLoS ONE 9 (5), e96086. Available from: https://doi.org/10.1371/journal.pone.0096086.

Timmusk, S., Paalme, V., Pavlicek, T., Bergquist, J., Vangala, A., Danilas, T., 2011. Bacterial distribution in the rhizosphere of wild barley under contrasting microclimates. PLoS One 6 (3), e17968. Available from: https://doi.org/10.1371/journal.pone.0017968.

Timmusk, S., Gulaim, S., Lawrence, 2018. Titania (TiO2) nanoparticles enhance the performance of growth-promoting rhizobacteria. Sci. Rep. 8. Available from: https://doi.org/10.1038/s41598-017-18939-x.

Timmusk, S., Timmusk, K., Behers, L., 2013. Rhizobacterial plant drought stress tolerance enhancement. J. Food Sec. 11, 10–16.

Van, N.L., Ma, C., Shang, J., Rui, Y., Liu, S., Xing, B., 2016. Effects of CuO nanoparticles on insecticidal activity and phytotoxicity in conventional and transgenic cotton. Chemosphere 144, 661–670. Available from: https://doi.org/10.1016/j.chemosphere.2015.09.028. Epub 2015 Sep 25.

Verma, S.K., Nisha, K., Panda, P.K., Patel, P., Kumari, P., Mallick, M.A., et al., 2020. Green synthesized MgO nanoparticles infer biocompatibility by reducing in vivo molecular nanotoxicity in embryonic zebrafish through arginine interaction elicited apoptosis. Sci. Total. Env. 15 (713), 136521. Available from: https://doi.org/10.1016/j.scitotenv.2020.136521. Epub Jan 7. PMID: 31951838.

Wagner, G., Victor, K., Jonathan, D.J., Paul, M.B., 2016. Nanoparticles composed of Zn and ZnO inhibit *Peronospora tabacina* spore germination in vitro and *P. tabacina* infectivity on tobacco leaves. Nanomaterials 6, 50. Available from: https://doi.org/10.3390/nano6030050.

Walker, T.S., Bais, H.P., Déziel, E., 2004. *Pseudomonas aeruginosa*-plant root interactions. Pathogenicity, biofilm formation, root exudation. Plant. Physiol. 134 (1), 320–331. Available from: https://doi.org/10.1104/pp.103.027888.

Wang, H., Wick, R.L., Xing, B., 2009. Toxicity of nanoparticulate and bulk ZnO, Al2O3 and TiO2 to the nematode *Caenorhabditis elegans*. Env. Pollut. 157, 1171–1177.

Wang, X., Liu, X., Chen, J., Han, H., Yuan, Z., 2014. Evaluation and mechanism of antifungal effects of carbon nanomaterials in controlling plant fungal pathogen. Carbon 68, 798–806. Available from: https://doi.org/10.1016/j.carbon.2013.11.072.

Wang, X., Yang, X., Chen, S., Li, Q., Wang, W., Hou, C., 2016. Zinc oxide nanoparticles affect biomass accumulation and photosynthesis in Arabidopsis. Front. Plant. Sci. 6, 1243. Available from: https://doi.org/10.3389/fpls.2015.01243.

Wani, A.H., Shah, M.A., 2012. A unique and profound effect of MgO and ZnO nanoparticles on some plant pathogenic fungi. J. App Pharm. Sci. 2 (3), 40–44.

William, J.R., 1944. The importance of plants. Science 100 (2603), 440–443.

Xing, Y., Yi, R., Yang, H., Xu, Q., Huang, R., Tang, J., et al., 2021. Antifungal effect of chitosan/nano-TiO_2 composite coatings against *Colletotrichum gloeosporioides*, *Cladosporium oxysporum* and *Penicillium steckii*. Molecules. 26 (15), 4401. Available from: https://doi.org/10.3390/molecules26154401.

Yamanaka, K., Chun, S.J., Boillee, S., Fujimori-Tonou, N., Yamashita, H., Gutmann, D.H., et al., 2008. Astrocytes as determinants of disease progression in inherited amyotrophic lateral sclerosis. Nat. Neurosci. 11, 251–253.

Yanli, W., Lin, D., Chenjie, Y., Chenchen, L., Xiaojun, X., Yanan, H., et al., 2017. Toxic effects of metal oxide nanoparticles and their underlying mechanisms. Science China Materials 60 (2), 93–108. Available from: https://doi.org/10.1007/s40843-016-5157-0.

Yi, H., Wen, Y., Chuanxin, M., Jason, C.W., Zetian, Z., Muhammad, A., et al., 2018. Engineered nanomaterials suppress *Turnip mosaic virus* infection in tobacco (*Nicotiana benthamiana*). Environ. Sci.: Nano 5, 1685–1693. Available from: https://doi.org/10.1039/C8EN00014J.

Yoshikawa, S.J.T., 2004. Quantitative evaluation of antifungal activity of metallic oxide powders (MgO, CaO and ZnO) by an indirect conductimetric assay. J. Appl. Microbiol. 96 (4), 803–809.

Yuliar Nion, Y.A., Toyota, K., 2015. Recent trends in control methods for bacterial wilt diseases caused by *Ralstonia solanacearum*. Microbes Env. 30, 1–11. Available from: https://doi.org/10.1264/jsme2.ME14144.

Zabrieski, Z., Morrell, E., Hortin, J., Dimkpa, C., McLean, J., Britt, D., et al., 2015. Pesticidal activity of metal oxide nanoparticles on plant pathogenic isolates of pythium. Ecotoxicology 24 (6), 1305–1314. Available from: https://doi.org/10.1007/s10646-015-1505-x. Epub 2015 Jun 16. PMID: 26076749.

Zarzuela, R., Carb, M., Gil, M.L.A., Cantoral, J.M., Mosquera, M.J., 2017. CuO/SiO2 nanocomposites: a multifunctional coating for application on building stone. Mater. Des. 114, 364–372. Available from: https://doi.org/10.1016/j.matdes.2016.11.009.

Zhao, F., Zhou, X., Hong, Z., Zhou, S., Fan, X., 2005. Research progress of TiO2-based antibacterial material, Mater. Rep, 11. pp. 35–38.

Zhu, X., Radovic-Moreno, A.F., Wu, J., Langer, R., Shi, J., 2014. Nanomedicine in the management of microbial infection—overview and perspectives. Nano Today 9 (4), 478–498. Available from: https://doi.org/10.1016/j.nantod.2014.06.003. Aug 1.

Chapter 8

Impact of metal oxide nanoparticles against salinity stress in plants

Venkatachalam Vasudevan[1,2] and Vidhya Arumugam[1]
[1]*Department of Biotechnology, A.V.C. College (Autonomous), Mayiladuthurai, Tamil Nadu, India,* [2]*Department of Biotechnology, Bharathidasan University, Tiruchirappalli, Tamil Nadu, India*

Chapter outline

8.1 Introduction

According to the FAO, the global food supply must expand by 70% by 2050 to meet the rising demand of the world's growing population (FAO, 2013). Agriculture is one of the world's most important foundations of food production. Increasing land productivity in agriculture is difficult because the area under cultivation is expected to remain constant or even shrink as nonagricultural activities put more pressure on land. Furthermore, soil salinity is a substantial environmental constraint, limiting the amount of land available for sustainable agriculture and has emerged as a significant concern for global food security. Salinity affects approximately 62 million hectares (20% of cultivated land worldwide) of the world's irrigated land. The number is growing by the day, especially in arid and semiarid regions of the world (Etesami and Noori, 2019). Both glycophytic and halophytic plants are severely affected by salinity stress (Etesami and Beattie, 2018; Komaresofla et al., 2019). Salinity stress causes a slew of morphological, physiological,

Nanometal Oxides in Horticulture and Agronomy. DOI: https://doi.org/10.1016/B978-0-323-91809-1.00005-6

biochemical, and molecular alterations in plants affecting the plants' growth and productivity. One of the most typical effects of salinity stress on plants is a decrease in osmotic stress and an increase in ionic stress. Secondary stress caused by salinity include oxidative stress (the creation of reactive oxygen species, or ROS) and nutritional and hormonal abnormalities in plants (Kumar et al., 2019; Munns, 2005; Yang and Guo, 2018). Under salinity stress, these detrimental effects drastically reduce plant production (Parihar et al., 2015).

Plants respond to salt stress in various ways including CO_2 assimilation, osmotic regulation, ion compartmentation or exclusion, toxic ion uptake, chlorophyll content, chlorophyll fluorescence, ROS generation, photosynthetic electron transport, and antioxidant defenses (Munns and Tester, 2008; Tang et al., 2015), and it depends on a variety of factors including the solute type and concentration, the genetic potential of plant, and the severity of environmental stress (Arif et al., 2020).

Depending on the type and dosage of salt, plant cells have different ways of protecting themselves from the harmful effects of salinity stress. Plants develop defenses against salinity stress by modifying molecular, biochemical, and physiological processes at various levels. Ion hemostasis, modulation of enzymatic and nonenzymatic antioxidants, compatible solute buildup and osmotic protection, hormonal regulation, change in stress resistance gene expression, and nitric oxide generation are some of these mechanisms (Hanin et al., 2016; Van Zelm et al., 2020).

Breeding salt-tolerant plants and developing salt-resistant crops could be a useful technique to overcome the problem of decreasing global food production. However, conventional breeding strategies using interspecific or intergeneric hybridization have had minimal effectiveness in enhancing crop plant stress tolerance (Fita et al., 2015). Several recent studies have focused on innovative salinity management measures to reduce the harmful effects of salt on plants (Etesami and Glick, 2020; Etesami and Jeong, 2018; Liang et al., 2018).

In recent years, nanoparticles (NPs) have received much attention as one of the most promising strategies for improving plant growth and performance under salinity stress (Ahmad and Akhtar, 2019; Das and Das, 2019; Duhan et al., 2017). Nanomaterials consist of tiny components (at a scale of 100 nm or less), and these components affect the macro characteristics of materials. Nanomaterials have a higher surface area when compared to the same mass of material produced in a more extensive form. NPs can alter the chemical reactivity of materials, as well as their strength and electrical properties. NPs have higher reactivity and potential biological activity due to their high surface-to-volume ratio (Das and Das, 2019; Dubchak et al., 2010).

The administration of NPs has beneficial and adverse effects on plant growth at various developmental stages (Vishwakarma et al., 2018; Zulfiqar and Ashraf, 2021). The beneficial effects of utilizing NPs to overcome salt stress are demonstrated in diverse group of plants (Alabdallah and Alzahrani, 2020; Elsheery et al., 2020; Mahmoud et al., 2020a; Noohpisheh et al., 2021; Zulfiqar and Ashraf, 2021). Plants' morphological, physiological, and biochemical features are all profoundly affected by the application of NPs. NPs influence salt tolerance in plants by altering hormone concentrations, antioxidant enzyme activity, ion homeostasis, gene expression, and defense systems, according to previous research (Wahid et al., 2020; Zulfiqar and Ashraf, 2021). These beneficial effects may vary depending on the size, shape, and concentrations of NPs used, the different environmental conditions, and the plant species (Feizi et al., 2012; Mohamed et al., 2017; Wahid et al., 2020). Several investigations, for example, have described the negative and harmful consequences of high concentrations of NPs in plants, which differed depending on the quantities and characteristics of the NPs in plant tissues, growth phases, and species (Mattiello et al., 2015; Zulfiqar and Ashraf, 2021). As a result, 'concentration, size, treatment method, uptake by plants, characteristics, reactivity, and translocation of NPs into different tissues could all influence whether NPs interfere with various metabolic functions, resulting in harmful or beneficial effects (Mattiello et al., 2015; Paramo et al., 2020).

While there are some potential uses for NPs in agriculture, we still do not even fully understand the harmful consequences these compounds can have in the environment (especially in plants and living things) (Tripathi et al., 2017). In general, the extensive use of NPs in agriculture and plant research is still a point of contention (Nair et al., 2010). Although several studies have proven the positive/negative impacts of NPs on plants under salinity stress, only little is known about the linkage and interaction between NPs and intracellular processes in the plants under salinity stress.

Understanding the interactions between salinity-stressed plants and NPs opens up new opportunities for promoting crop growth in salt-affected soils.

8.2 Salinity stress in plants

With the increase in population and the deterioration of natural environments, soil saline alkalization has become an increasingly critical global problem' (Shabala, 2013). According to statistics, saline alkalization threatens over 7% of the world's land (more than 900 million hectares), and there are no adverse controls to prevent it (Li et al., 2014). As a result, soil saline alkalization is a critical abiotic stressor in, and it has become a major limiting factor for crop production in global agriculture (Wang et al., 2018). Plants face both salt stress and alkali stress as a result of soil saline alkalization.

According to the salt content and pH value, the degree of salt-alkali conditions is classified as mild (salt content less than 3 h, pH 7.1–8.5), moderate (salt content 3–6 h, pH 8.5–9.5), or severe (salt content >6 h, pH value >9.5) (Oster et al., 1999). NaCl, Na_2SO_4, and other neutral salts are the principal causes of salt stress. On the one hand, sodium ions will enter the cell directly via channel and carrier proteins, resulting in ion toxicity. The high ion concentration outside the cell, on the other hand, reduces the osmotic potential, driving water molecules out of the cell and causing physiological drought, that is, osmotic stress. Both of these factors cause metabolic problems in plants (Richards, 1947). $NaHCO_3$ and Na_2CO_3 cause alkali stress, which causes the pH to rise even more due to salt stress. Therefore, in addition to ionic toxicity and osmotic stress, high pH will disrupt cell pH stability, degrade cell membrane integrity, and reduce root vitality and photosynthetic performance (Zhang et al., 2017; Kaiwen et al., 2020). Many studies have revealed that combined salt-alkali stressors cause trophic ion imbalance, impaired osmotic adjustment ability, antioxidant system inhibition, and severe plant growth inhibition (Amirinejad et al., 2017; Chen et al., 2017; Wang et al., 2017, 2020). Plants that are moderately or severely alkali-stressed accumulate more carbohydrates and have much greater levels of ROS and Malondialdehyde (MDA) (Ye et al., 2021). In summary, osmotic stress, oxidative stress, ionic toxicity, and high-pH stress generated by mixed salt-alkali stress can degrade cell membrane structure, inactivate enzyme activity, and disturb ion balance in plant cells, among other things.

Depending on their ability to flourish in saltwater environments, plants are divided into two groups: halophytes and glycophytes. The former group possesses resistance mechanisms that allow it to survive higher salt concentrations, while the latter does not. The difference in behavior between them is attributed to the variation in their electron transport chain, assimilation of CO_2, photosynthetic pigment content, ROS generation, and sequestration. Salinity causes osmotic, ionic, and oxidative stress in plants, which affect their general metabolism and growth by altering their morphological, physiological, and molecular aspects. Increased water efflux due to increased Na + influx induces osmotic stress, damaging the photosynthetic machinery by disrupting the thylakoid membrane and Calvin-Benson cycle enzymes, resulting in a reduction in particular metabolites. The buildup of Na + and Cl ions causes a decrease in specific metabolites, resulting in nutritional insufficiency. The overproduction of ROS causes damage to nucleic acids, proteins, and lipids. They cause mutations, deletions, replication, transcription, and inhibition of signal transmission in DNA. They cause proteolysis susceptibility, amino acid profile change, chain fragmentation, and the buildup of cross-linked reaction products in proteins. They cause spontaneous oxidative chain reactions on unsaturated fatty acids in lipids. As a result, ROS destabilizes plasma membrane integrity by lipid peroxidation and breaks down protein.

8.3 Nanoparticles and metal oxide nanoparticles

Biosensing (Salata, 2004; Nehra et al., 2019), biofuel generation (Sekoai et al., 2019), organic and photochemical reactions (Song, 2015), and other applications have all made use of NPs. The main benefit of nanofertilizer is that it not only provides the highest micronutrients but also aids in soil reclamation. Several studies have found that specific amount of various NPs has significant favorable impacts on different plant types (Nair, 2016; Zhu et al., 2019; Kumar et al., 2019; Rastogi et al., 2017). The effect of a particular NP, on the other hand, is determined by the dose, kind, shape, structure, solubility, and duration of treatment (Aslani et al., 2014). Organic polymers (organic NPs) and inorganic components make up inorganic NPs. Inorganic NPs includes metals, metal oxides (SnO_2, Al_2O_3, CuO, ZrO_2, Cu_2O, MgO, La_2O_3, NiO, ZnO, TiO_2, CeO_2), and quantum dots (QDs). Liposomes, dendrimers, carbon nanomaterials, and polymeric micelles are examples of bioorganic NPs (Rajput et al., 2018; Nie et al., 2010; Kango et al., 2013). Plants absorb NPs at a rate of 15–20 times that of bulk nutrients (Lv et al., 2019; Srivastav et al., 2016). NPs have been used in agriculture to speed up seed germination and plant growth, as well as to protect plants from abiotic stresses like high or low temperature, salinity, drought, and flooding, as well as biotic challenges like fungi, bacteria, and insects (Vera-Reyes et al., 2018; Hao et al., 2018; Hao et al., 2017).

Silicon is beneficial for stressed plants, as it can help them to recover from drought, as well as from micronutrient and toxicity created due to other metals such as copper, aluminum, iron, zinc, etc. (Emamverdian et al., 2018). Nano-SiO_2 also improves maize seed germination by increasing the availability of nutrients and changing the pH and conductivity of the growth media (Suriyaprabha et al., 2012). Root growth in rice shows some impact while using QD and silica-coated QDs (Wang et al., 2014). Zinc oxide NPs (ZnONPs) have been used to study the seed germination in soybean, wheat, and onion and exhibits a beneficial effect at lower concentrations (Ali et al., 2021; Boonyanitipong et al., 2011; Sedghi et al., 2013; Raskar and Laware, 2014). Furthermore, specific tests on in vitro cultures and bananas enriched with ZnONPs revealed a beneficial effect on somatic embryogenesis and plantlet rebuilding. The observed tolerance to diverse biotic stresses was determined due to a higher amount of antioxidant enzymes [superoxide dismutase (SOD), peroxidase (POD), and catalase (CAT)] and biochemicals (such as proline) (Helaly et al., 2014).

ZnONPs significantly improved growth, photosynthetic efficacy, carbonic anhydrase, and free radical scavenging activity in tomato plants, according to a study (Faizan et al., 2018). Carbon nanotubes (CNTs) are carbon-based NPs have been widely employed for the promotion plant development. CNTs implanted in chloroplasts functioned as artificial antennae, were capable of capturing light of various wavelengths (ultraviolet, green, and near-infrared), enhancing seed germination, growth, and overall development in plants (Mukesh and Jha, 2017; Lahiani et al., 2013). The potential of single and multiwalled CNTs to penetrate the plant cell is also essential to design DNA and the biochemical delivery systems (Lara-Romero et al., 2017; Oloumi et al., 2018; Srinivasan and Saraswathi, 2010). Multiwalled CNTs were found to be a possible influencing factor for increased seed germination and plant growth in significant agricultural plants (barley, corn, and soybean) by inducing water and vital nutrient (Fe and Ca) uptake efficacy.

In soybean, barley, and corn, multiwalled CNTs have a gene-regulating influence on various water channel proteins (Lahiani et al., 2013). Several crops and noncrop plants can be treated with noble metal NPs such as Au and Ag (Dykman and Shchyogolev, 2018). In lettuce (*Lactuca sativa*) (Barrena et al., 2009), mustard (*Brassica juncea*) (Arora et al., 2012; Sharma et al., 2012), common bean (*Phaseolus vulgaris*), and corn (*Zea mays*), improved seed germination has been observed (Salama, 2012). In a study, *Arabidopsis thaliana* was treated with gold NPs and its remarkable effects seed germination and free radical scavenging capacity were noticed (Kumar et al., 2013). In a study where the morphology-based effects of NPs were analyzed, the decahedral-shaped silver NPs (AgNPs) had the most remarkable growth-stimulating effect in roots, while spherical-shaped AgNPs had the most significant anthocyanin accumulation in *Arabidopsis* seedlings, but had no influence on root growth (Syu et al., 2014). In terms of growth in *P. vulgaris*, the effects of conventionally synthesized and plant extract-based green AgNPs were examined. Administration of green AgNPs at a low dose significantly improved the leaf number, leaf area index, pod yield, and nitrate reductase activity. These NPs also successfully changed the pH of the soil from acidic to neutral, resulting in a significant improvement in cation exchange capacity, water retention capacity, and N/P concentration(Das et al., 2018).

8.4 Effects of metal oxide nanoparticles on salinity stress

NPs enter the plant system by various pathways, the most common of which are roots and leaves (Fig. 8.1). After entering the plant, NPs interact with it at the cellular and subcellular levels, affecting morphological, biochemical, physiological, and molecular changes (Khan et al., 2019). Depending on the nature of the NPs and the plant species, these interactions might be beneficial or harmful. The chemical nature, reactivity, size, and concentration of NPs determines the level of impact of NPs on a particular plant (Ma et al., 2010; Paramo et al., 2020; Tripathi et al., 2017; Zulfiqar and Ashraf, 2021). At concentrations below specified limitations, distinct NPs boost salinity-stressed plant growth and development (Ali et al., 2021; Aslani et al., 2014; Zulfiqar and Ashraf, 2021) through multiple recognized mechanisms (Fig. 8.1). Most of this research was conducted in artificial treatment conditions, such as plate growth medium and hydrophobic or pot conditions. To understand the impact of NPs on plant growth, we discuss the positive effects of NPs to enhance the salinity stress tolerance in plants.

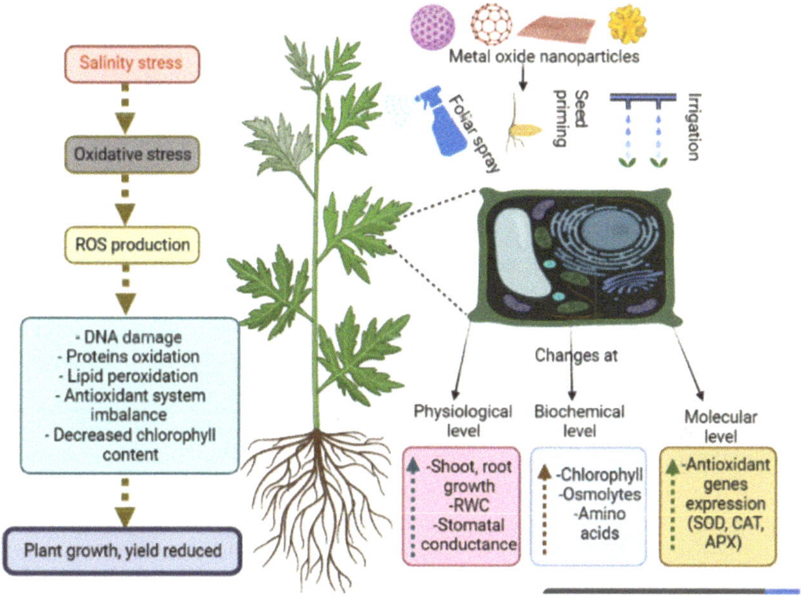

FIGURE 8.1 Effect of metal nanoparticles to mitigate the salinity stress in plants.

8.4.1 At the molecular level

The processes at the molecular level determine the biological functions of the plant. Research on the impact of NPs on plants at molecular level is a critical step in analyzing potential processes and plant impacts (Ali et al., 2021). Salinity stress affects gene expression and, as a result, regulates plant growth through changes in the expression of various genes involved in various regions of cells and their products. NPs cannot be efficient without interfering with cellular mechanisms and gene expression. The effects of ZnONPs (0, 20, and 40 mg/L) with less than 250 mM NaCl and plant growth-promoting bacteria (*Bacillus subtilis*, *Lactobacillus casei*, and *Bacillus pumilus*) in tomato plants (*Solanum lycopersicum*, *Lycopersicum linda*) were investigated (Hosseinpour et al., 2020). While salinity stress increased the polymorphism ratio, using NPs and bacteria together reduced the DNA hypermethylation caused by salinity stress. One of the epigenetic alterations in the cell that affects gene expression is DNA methylation, and the cell modifies gene expression during methylation and demethylation in salinity stress (Peng and Zhang, 2009). In another study, foliar application of ZnNPs on rapeseed plants (*Brassica napus* L.) under salinity stress caused a decrease in the expression of some genes (e.g., SKRD2, MYC, and MPK4) and an increase in the expression of other genes (e.g., ARP and MPK) (Hezaveh et al., 2019).

According to these studies, ZnNPs play a role in reducing salinity stress; Zn is involved in intra- and intercellular signaling, in DNA transcription (Caldelas and Weiss, 2017) and in lowering extreme environmental stresses like salinity stress (Caldelas and Weiss, 2017; Sofy et al., 2020). Future research should focus on molecular effects to better understand the NPs' mode of action under salinity conditions and determine the optimal concentration of ZnNPs. Low doses of ZnNPs have favorable effects, while excessive levels induced toxicity even in non-stress situations (Molnar et al., 2020). At a salinity of 100 mM NaCl, Ye et al. (2020) evaluated the effect of manganese (III) oxide NPs (0.1, 0.5, and 1 mg/L) on pepper plants. The researchers found that the NPs penetrated the seed coat, forming a corona-NPs complex. The critical function of manganese NPs (MnNPs), especially at a concentration of 1 mg/L, in increasing expression of SOD genes, and hence the production of SOD as one of the main enzymes for scavenging ROS under salinity stress was highlighted in this study. Molecular reports on MnNPs-induced improvement in salinity stress resistance are rare. As a result of Mn's role in bolstering the plant's defense system against various abiotic stresses (Ye et al., 2019), future research should concentrate on this area to determine the potential and unique roles of MnNPs in mediating salt stress tolerance.

In plants treated with cerium oxide NPs, there was a considerable drop in ROS levels and increased calcium content. The NPs alter signaling genes mediated by ROS and Ca^{2+}. Indeed, ROS and Ca^{2+} are significant and influential in plants' stress responses. Terpene synthetase genes (CAD1 and TPS) were similarly affected by cerium NPs (An et al., 2020). The study implies that the molecular impact of NPs is higher in seed tissue compared to seedling tissues. There has 'not been much research on cerium NPs inducing salt tolerance in plants. As a result, future research should focus on evaluating the method of action of cerium NPs on molecular pathways in plants facing the threat of climate change-related salt stress. Furthermore, cerium NP accumulation in several plant tissues is a dose-dependent phenomenon (Singh et al., 2019).

Under salinity stress conditions, silicon NPs had a favorable effect on *Cannabis sativa* L., resulting in improved growth and molecular alterations in this plant (Guerriero et al., 2021). Furthermore, a shift in the expression of genes including those which are involved in the manufacture of the hormone abscisic acid (ABA) such as OsZEP, OsNCED, and OsZEP reduced the effects of salinity stress by silicon (Tripathi et al., 2020). On raising the expression of the protein OsHMA3, more quantity of salt transfers into the vacuole and protects the plant from the detrimental effects of stress (Siddiqui et al., 2020). In summary, according to published studies (Etesami and Jeong, 2018; Etesami et al., 2020), silicon NPs-mediated growth improvement and salinity stress reduction require further investigation.

Copper as a micronutrient plays a significant role in mitigating the harmful effects of salinity on plants (Iqbal et al., 2018). Hernandez-Hernandez et al. (2018) investigated the effect of copper NPs (CuNPs) on tomato grown under salt stress. They found that they improved tomato growth by increasing the expression of SOD and jasmonic acid genes, reducing ionic and oxidative stress. CuNPs efficiently boost salinity tolerance by activating the antioxidant defense mechanism and the octadecanoid pathway of jasmonates. Although the application of CuNPs for reducing salt stress in plants is still in its early stages, findings reveal that this micronutrient in nanoforms has a significant potential for imparting salinity stress resistance in plants (Zulfiqar and Ashraf, 2021). As a result, more molecular research will provide necessary information on the mode of action of CuNPs to improve salinity resistance in plants. AgNPs are used in agriculture due to the benefits they provide in terms of growth. However, research on AgNPs as a viable option for reducing the harmful effects of salt on plants is limited (Zulfiqar and Ashraf, 2021). As a result, future research should elucidate their involvement in regulating salinity stress tolerance at the molecular level. Many studies recommend that AgNPs be used with caution and care since they can release silver ions (Ag^{+}) into the environment, which is very toxic in nature and can be harmful to species (Tortella et al., 2020). Therefore, it' is critical to understand how AgNPs affect plant growth and to what extent they concern the environment and creatures' health (Yan and Chen, 2019). Iron is well known for assisting plants in developing stress tolerance (Tripathi et al., 2018). Iron NPs (FeNPs) can alleviate salt stress in plants (Abdoli et al., 2020; Moradbeygi et al., 2020). However, the existing research lacks complete information on the exact metabolic pathways (at the molecular level) that they regulate (Zulfiqar and Ashraf, 2021).

8.4.2 At the biochemical level

8.4.2.1 Antioxidant modification

ROS generation is well known as one of the plants' reactions to abiotic challenges, such as salt stress. Plants have created antioxidant enzymes to deal with excessive ROS in salinity stressed plant cells (You and Chan, 2015). Many studies have shown that NPs increase the content of antioxidant enzymes (Abdoli et al., 2020; Gaafar et al., 2020; Gonzalez-García et al., 2021; Mushtaq et al., 2020). According to the researchers, some NPs exhibit antioxidant enzyme activities and hence aid the plant in overcoming the prevailing oxidative conditions within the plant. For example, cobalt, iron, and cesium NPs act as enzyme CAT, while cesium, manganese, copper, and iron NPs act as the enzyme POD (Rico et al., 2015). In a study, boosting the activity of ROS-activated nonselective cation channels (ROS-NSCC) increased ROS scavenging in *Arabidopsis* plants treated with cerium oxide NPs (Wu et al., 2018). Khan et al. (2020) examined the effect of seed priming with these AgNPs (10, 20, and 30 mM) on pearl millet (*Pennisetum glaucum* L.)

under salt stress. The presence of these NPs resulted in a considerable rise in growth features in this plant, which was related to an increase in antioxidant enzymes like SOD, CAT, and glutathione peroxidase (GPX) and a drop in the sodium to potassium ratio. The enhanced activity of antioxidant enzymes is the most important mechanism of AgNPs even at low concentrations (Sami et al., 2020). In an earlier study, TiO_2NPs induced an increase in antioxidant content and a decrease in H_2O_2 in *Dracocephalum moldavica* at doses of 0, 50, 100, and 200 mg/L under salinity stress (0, 50, and 100 mM NaCl) in a study (Gohari et al., 2020a). Cerium NPs also improved the activity of antioxidant enzymes in cotton plants, allowing them to eliminate ROS from inside the cell. Plant growth was also boosted by the NPs when exposed to salt (Liu et al., 2021). Gohari et al. (2020a) observed that low levels of multiwalled CNT had a salt ameliorating effect on *Ocimum basilicum* plants. Iron plays a role in salt tolerance by upregulating antioxidative enzymes (Singh and Bhatla, 2016). Moradbeygi et al. (2020) studied the effect of FeNPs on *D. moldavica* L. under salinity stress conditions. According to the findings, NPs improved the conditions for plant growth under salinity by increasing the level of phenolic and flavonoid chemicals, particularly in the roots, while decreasing the content of antioxidant enzyme activity. The nonenzymatic system is essential than the enzymatic system in this plant's ability to counteract stress detrimental consequences. As a result, when the plant's nonenzymatic defense mechanism is compromised, the enzymatic defense system kicks in to help eliminate free radicals and their damaging consequences. The evidence revealed above shows that different NPs can reduce oxidative stress in plants by altering antioxidant defense mechanisms (both enzymatic constituents and nonenzymatic constituents). Despite the previous findings, the interactions between the plant antioxidant enzyme system and NPs are still poorly known. Hence, a transcriptome analysis is needed to fully comprehend the processes governing NPs-mediated salinity stress response.

8.4.2.2 Phytohormones

Phytohormones play an essential part in plants' ability to adapt to varied environments through several processes. According to most research, these phytohormones assist in strengthening a plant's ability to withstand salinity stress (Fahad et al., 2015). Changes in the plant hormonal balance generated by NPs link with increased plant stress tolerance. The presence of AgNPs (Manickavasagam et al., 2019) considerably altered the levels of ABA, ethylene, and gibberellin in rice plants. Another study found that applying AgNPs to wheat plants under stressful conditions increased the content of IBA, NAA, and 6-benzyl amino purine (BAP) while decreasing the concentration of ABA. One of the primary mechanisms of AgNPs in boosting plant development under stress is due to these modifications in phytohormones (Abou-Zeid and Ismail, 2018). Selenium NP (SeNP) increases indole-3-acetic acid

(IAA) and ABA levels in plants. The SeNPs enhance root biomass and maintain the osmotic condition of the cells in salinity-stressed strawberry plants (Zahedi et al., 2019). In general, there is a lack of studies on the effects of NPs on growth hormones in salinity-stressed plants. The study of how NPs can modify the levels and their adaptive responses of phytohormones **can** be an **interesting**future research area in plant biology.

8.4.2.3 Compatible solutes

Plants produce more compatible solutes or osmolytes when exposed to high salinity conditions. The compatible solutes alleviate the limiting effect of high ion concentrations (Na^+ and Cl^-) and increase the activity of enzymes by stabilizing proteins and their complexes, as well as membranes under salinity stress (Etesami and Jeong, 2018). According to some reports, NPs can also improve plant tolerance to salt stress by altering the amounts of compatible solutes (Zulfiqar and Ashraf, 2021). NPs can improve plant tolerance to salinity stress by modifying the levels of solutes such as amino acids (e.g., proline) and total soluble sugars, which reduce the osmotic shock created due to ion toxicity (Na^+ and Cl^-) (Abdoli et al., 2020; Alabdallah and Alzahrani, 2020; An et al., 2020). ZnNP treatment to salinity-stressed Canola plants (*B. napus* L.), for example, relieved the salt-induced detrimental effects via osmolyte production and ionic control, according to Farouk and Al-Amri (2019). Wan et al. (2020) investigated the impact of carbon nanohorns on seedlings of *Sophora alopecuroides* under salt stress. The NPs foliar spray improved fresh root weight, leaf soluble sugar content, and leaf and root total protein levels, according to their findings. A little is known about the interaction between NPs and compatible solutes that necessitates future investigation.

8.4.3 At the physiological level

8.4.3.1 Photosynthesis

Photosynthesis is one of the essential processes impacted by salinity stress, depending on the kind of plant, salt dose, and other factors (Hnilickova et al., 2021). Several studies report that foliar application of NPs increases the concentration of chlorophyll in plants, allowing plants to synthesize more complexes for light harvesting, absorb more light energy and improve the rate of photosynthesis (Ali et al., 2021; Zulfiqar and Ashraf, 2021). Most of the NPs promote photosynthesis by raising the content of photosynthetic pigments in salinity-stressed plants (Abdoli et al., 2020; Aghdam et al., 2016; Alabdallah and Alzahrani, 2020; Hasanpour et al., 2015; Singh et al., 2021; Zulfiqar and Ashraf, 2021). Manganese is involved in the acceleration of photosynthetic electron transport rate and oxygen evolution in various cell areas (e.g., mitochondria, chloroplasts, the structure of specific enzymes, etc) (Pradhan et al., 2013). MnNPs also maintain an appropriate photosynthetic rate in plants under abiotic stress (Ye et al., 2020). Mn addition to salinity stressed *Vigna radiata* improved membrane stability index, chlorophyll content, and nitrate reductase activity (Shahi and Srivastava, 2018). Ullah et al. (2020) studied the impacts of TiO_2NPs on wheat growth in salinity-stressed conditions. The use of NPs increased wheat growth under saline stress, due to the NPs' ability to absorb light and aid photosynthesis and also due to a better water absorption by the wheat plant. Hong et al. (2005a,b) revealed the capacity of TiO_2NPs to preserve chloroplasts against aging due to their photocatalytic activities based on three different studies on the effect of TiO_2NPs on spinach plants. The foliar application of ZnNPs (10, 50, and 100 mg/L) to tomato plants under salinity stress (150 mM NaCl) significantly boosted chlorophyll content and photosynthetic rate, according to Faizan et al. (2021). Wang et al. (2020) studied the effect of ZnNP priming on wheat plant development under salinity stress, concluding that ZnNPs can promote wheat plant growth by altering the electron transfer chain and boosting enzymatic antioxidants. NPs have a variety of beneficial impacts, including protecting photosynthetic pigments and increasing photosynthetic capability. According to the findings of the studies stated above, NPs improve plant tolerance to salinity by increasing the photosynthetic efficiency. In future, more research into the mechanisms of different NPs-mediated control of photosynthesis-related aspects under salinity stress is required.

8.4.3.2 Regulation of ion homeostasis

Salinity stress may cause an extreme nutritional deficit in plants (Etesami and Maheshwari, 2018). For proper growth and development, all plants require an adequate number of vital plant nutrients. NPs have a significant impact on nutrient absorption, translocation, and eventual allocation in plant cells (Kopittke et al., 2019). The high potassium to sodium ratio disrupted by salinity stress is identified as one of the essential determinants for plant tolerance to salt stress. Owing to the application of cerium NPs in cotton plants, there is a raise in K:Na ratio that boosts the plant growth (Sytar et al., 2019; Tahjib-Ul-Arif et al., 2019). Cerium NPs, when compared to other existing materials, are capable of scavenging –OH for a long time. NPs accomplish this by enhancing the activity of plasma membrane K^+-selective (GORK) channels, which transport potassium into the cell (Wu et al., 2018). The study mentioned above demonstrates how NPs can help plants cope with salt stress by interfering with Na^+ absorption and altering the absorption of specific nutrients (via balancing nutrient uptake). However, to understand the effect of NPs in salt-stressed plants, it is essential to consider the balance of nutrients in the plants.

8.4.3.3 Maintenance of plant water balance

Salinity stress decreases the volume of uptake of water by plants. Increased osmotic stress in the soil solution limits water absorption by the plants. Osmotic stress affects water content in leaves, stomatal conductance, leaf growth (accelerated leaf senescence and leaf death), and photosynthesis (decreased chlorophyll concentrations), ultimately resulting in plant growth reduction (Munns and Tester, 2008). NPs increase water use efficiency in a variety of plant species and alleviate salt-induced osmotic stress (Mahmoud et al., 2020a; Zulfiqar and Ashraf, 2021).

Aquaporins are channel proteins members of the main intrinsic protein superfamily and are involved in plant water relations. Aquaporins are proteins that transport water and other tiny neutral molecules through biological membranes in plants (Kapilan et al., 2018). Higher expression of plasma membrane intrinsic protein aquaporins boost root hydraulic conductance, leading to increased water intake while reducing oxidative stress and membrane damage (Ali et al., 2021; Maurel et al., 2015; Zulfiqar and Ashraf, 2021). NPs treatments boost seed absorption and water retention, thus leading to increased seed germination (Khodakovskaya et al., 2009).

The aquaporins in the seed coats are regulated by NPs (Ali et al., 2021). One of the essential functions of silicon during salt stress is to increase aquaporin functioning (Rios et al., 2017). Although the processes by which NPs alter aquaporin expression and function are yet unknown, more research is needed to see if other mechanisms are also at work. Hence, we understand that oxidative stress, such as that caused by salt, damages the plasma membrane and alters the activity of aquaporins (negative regulation of plasma membrane aquaporin activities) and, as a result, root hydraulic conductivity (Liu et al., 2015). It suggests that NPs may improve plant water balance in salinity by upregulating essential antioxidative enzymes, which needs to be investigated more in the future.

8.5 Conclusion and perspectives

A thorough review of literature shows that MONPs increase plant growth and reduce salinity stress in plants under unfavorable conditions. According to the investigations, MONPs had considerable favorable impacts on salinity-stressed plants at the physiological, biochemical, and molecular levels (Fig. 8.1, Table 8.1), depending on the characteristics, application technique, and NP dosage. It appears that the judicious application of MONPs in agriculture will be a long-term strategy for alleviating salinity stress in the future. Furthermore, the application of MONPs (at optimal NP concentrations) may promote plant health and play a key role in low-input sustainable agriculture for both food and nonfood crops. The current understanding of how MONPs affect plants against salinity stress has advanced. The exact mechanism(s) by which they modulate plant physiology through the potentiation of host defense mechanisms still needs further investigation at the genomic, metabolomic, and proteomic levels. Other studies are required to explore the effect of various MONPs on plant gene expression and plant biochemistry. The comparative studies of interactions between multiple MONPs

TABLE 8.1 The positive impacts of various metal and metal oxide NPs on plants under salinity stress.

Metal NPs	Plant species	Concentration of NPs	Positive effect	References
Fe NPs	*Lycopersicon esculentum*	10, 50 and 100 mg/L	Improved photosynthesis, morphological trait,s and enzymes activities like CAT and SOD	Faizan et al. (2021)
FeNPs	*Pistacia vera*	2.9 mg/L	Chlorophyll degradation and decreased plasma membrane damage	Karimi et al. (2020)
ZnONPs	*Teriticum aestivum*	50 mg/L	Enhanced electron transport flux, activated the sucrose biosynthesis and antioxidant system	Wang et al. (2020)
ZnONPs	*Abelmoschus esculentus*	10 mg/L	Enhanced chlorophyll and antioxidant enzyme activity (SOD and CAT), decreased sugars and proline content	Alabdallah and Alzahrani (2020)
AgNPs	*Eustoma grandiflorum*	5, 10, and 20 mg/L	Decreased the POD activity and increased the activity of both SOD and CAT enzymes	Youssef et al. (2020)
MnNPs	*Capsicum annuum*	0.1, 0.5, and 1 mg/L	Improved morphological traits such as root development	Ye et al. (2020)
SiNPs	*Lycopersicon esculentum*	250 and 500 mg/L	Maintained concentration of chlorophylls, GSH, vitamin C, and PAL activity	Pinedo-Guerrero et al. (2020)
CuNPs	*Capsicum annuum*	100 and 500 mg/L Cu NPs	Increased chlorophylls, lycopene and glutathione POD activity in the leaves, and increased APX, POD, and CAT activity	Gonzalez-García et al. (2021)
Zn, Si, and B	*Solanum tuberosum L.*	20 mg/L ZnNPs, 15 mg/L Si and 1 mg/L B NPs	Improved morphological traits, increased the nutrient concentrations (N, P, K, Ca, Zn, and B)	Mahmoud et al. (2020b)

GSH, glutathione; *APX*, ascorbate peroxidase; *PAL*, phenylalanine ammonia lyase.

and plant would shed light on the mechanism governing salinity stress. There is a need of applied research to know the optimum MONPs concentration and suitable time and methods of the NPs application on the crop plants, especially when the plant is under salinity stress conditions. Apart from salinity stress, plants are exposed to multiple stresses simultaneously in nature. Most of the environmental stresses, especially salinity and drought, are present at the same time in agricultural fields. Hence, it is essential to study the effect of NPs on increasing resistance to salinity and drought stresses simultaneously. Although some types of NP transporters for NPs uptake (e.g., Si) is identified in several plant species, the transporter or channel protein responsible for loading various NPs remains unknown. At the subcellular level, most of the research related to NPs has concentrated on the cell wall. It would be interesting to investigate the distribution of NPs in the cell nucleus and organelles, which explain the biological roles of NPs in enhancing plant tolerance to salinity stress. Proteomic studies related to the effects of NPs would be interesting, and results may further contribute to the understanding of mechanisms of NPs-mediated impact on salinity stressed and nonstressed crop plants. It is needed to study the combined use of NPs and salinity-tolerant microorganisms that alleviate salinity stresses and compare their effects on the plant separately and dually. The underlying mechanisms by which MONPs regulate plant-salinity-tolerance have yet to be identified in plants. However, the essential dosage of NPs, exposure duration, translocation, accumulation, and mechanism of action on plants are all important factors to consider when developing an application strategy. Further, the buildup of soil, air, water, and creatures in the ecosystem are also essential in determining the precise consequences of MONPs. In general, a basic understanding of the interactions between plants and various MONPs would aid in a better understanding of mechanism of alleviating salinity stress by applying MONPs. Well-designed, large-scale, long-term field studies are necessary to assess the potential of MONPs in reducing salt stress, and to analyze the economic feasibility of various sources of NPs. There would be a growing interest in studying the mechanisms of MONP-mediated salinity stress tolerance in plants in future.

References

Abdoli, S., Ghassemi-Golezani, K., Alizadeh-Salteh, S., 2020. Responses of ajowan (*Trachyspermum ammi L.*) to exogenous salicylic acid and iron oxide nanoparticles under salt stress. Environ. Sci. Pollut. Res. 27, 36939–36953.

Abou-Zeid, H., Ismail, G., 2018. The role of priming with biosynthesized silver nanoparticles in the response of *Triticum aestivum L* to salt stress. Egypt. J. Bot. 58, 73–85.

Aghdam, M.T.B., Mohammadi, H., Ghorbanpour, M., 2016. Effects of nanoparticulate anatase titanium dioxide on physiological and biochemical performance of *Linum usitatissimum* (Linaceae) under well-watered and drought stress conditions. Braz. J. Bot. 39, 139–146.

Ahmad, I., Akhtar, M.S., 2019. Use of Nanoparticles in Alleviating Salt Stress. Salt Stress, Microbes, and Plant Interactions: Causes and Solution. Springer, pp. 199–215.

Alabdallah, N.M., Alzahrani, H.S., 2020. The potential mitigation effect of ZnO nanoparticles on [*Abelmoschus esculentus* L. Moench] metabolism under salt stress conditions. Saudi J. Biol. Sci. 27, 3132–3137.

Ali, S., Mehmood, A., Khan, N., 2021. Uptake, translocation, and consequences of nanomaterials on plant growth and stress adaptation. J. Nanomater 2021, 1–17.

Amirinejad, A.-A., Sayyari, M., Ghanbari, F., Kordi, S., 2017. Salicylic acid improves salinity-alkalinity tolerance in pepper (*Capsicum annuum L.*). Adv. Hort. Sci. 31, 157–163.

An, J., Hu, P., Li, F., Wu, H., Shen, Y., White, J.C., et al., 2020. Emerging investigator series: molecular mechanisms of plant salinity stress tolerance improvement by seed priming with cerium oxide nanoparticles. Environ. Sci. Nano 7, 2214–2228.

Arif, Y., Singh, P., Siddiqui, H., Bajguz, A., Hayat, S., 2020. Salinity induced physiological and biochemical changes in plants: an omic approach towards salt stress tolerance. Plant Physiol. Biochem. 156, 64–77.

Arora, S., Sharma, P., Kumar, S., Nayan, R., Khanna, P.K., Zaidi, M.G.H., 2012. Gold-nanoparticle induced enhancement in growth and seed yield of *Brassica juncea*. Plant Growth Regul. 66 (3), 303–310.

Aslani, F., Bagheri, S., Muhd Julkapli, N., Juraimi, A.S., Golestan Hashemi, F.S., Baghdadi, A., 2014. Effects of engineered nanomaterials on plants growth: an overview. Sci. World J. 2014, 1–28.

Barrena, R., Casals, E., Colon, J., Font, X., Sanchez, A., Puntes, V., 2009. Evaluation of the eco-toxicity of model nanoparticles. Chemosphere 75 (7), 850–857.

Boonyanitipong, P., Kositsup, B., Kumar, P., Baruah, S., Dutta, J., 2011. Toxicity of ZnO and TiO2 nanoparticles on germinating rice seed *Oryza sativa L.* Int. J. Biosci. Biochem. Bioinform. 1 (4), 282.

Caldelas, C., Weiss, D.J., 2017. Zinc homeostasis and isotopic fractionation in plants: a review. Plant Soil 411, 17–46.

Chen, Y., Li, Y., Sun, P., Chen, G., Xin, J., 2017. Interactive effects of salt and alkali stresses on growth, physiological responses and nutrient (N, P) removal performance of Ruppia maritima. Ecol. Eng. 104, 177–183.

Das, A., Das, B., 2019. Nanotechnology a Potential Tool to Mitigate Abiotic Stress in Crop Plants, Abiotic and Biotic Stress in Plants. IntechOpen.

Das, P., Barua, S., Sarkar, S., Karak, N., Bhattacharyya, P., Raza, N., et al., 2018. Plant extract–mediated green silver nanoparticles: efficacy as soil conditioner and plant growth promoter. J. Hazard. Mater. 346, 62–72.

Dubchak, S., Ogar, A., Mietelski, J.W., Turnau, K., 2010. Influence of silver and titanium nanoparticles on arbuscular mycorrhiza colonization and accumulation of radiocaesium in *Helianthus annuus*. Span. J. Agric. Res. 8, 103–108.

Duhan, J.S., Kumar, R., Kumar, N., Kaur, P., Nehra, K., Duhan, S., 2017. Nanotechnology: the new perspective in precision agriculture. Biotechnol. Rep. 15, 11–23.

Dykman, L., Shchyogolev, S., 2018. The effect of gold and silver nanoparticles on plant growth and development. In: Saylor, Y., Irby, V. (Eds.), Metal Nanoparticles. Nova, New York, pp. 63–300.

Elsheery, N.I., Helaly, M.N., El-Hoseiny, H.M., Alam-Eldein, S.M., 2020. Zinc oxide and silicone nanoparticles to improve the resistance mechanism and annual productivity of salt-stressed mango trees. Agronomy 10, 558.

Emamverdian, A., Ding, Y., Xie, Y., Sangari, S., 2018. Silicon mechanisms to ameliorate heavy metal stress in plants. Biomed. Res. Int. 2018, 8492898.

Etesami, H., Beattie, G.A., 2018. Mining halophytes for plant growth-promoting halotolerant bacteria to enhance the salinity tolerance of non-halophytic crops. Front. Microbiol. 9, 148.

Etesami, H., Jeong, B.R., 2018. Silicon (Si): review and future prospects on the action mechanisms in alleviating biotic and abiotic stresses in plants. Ecotoxicol. Environ. Saf. 147, 881–896.

Etesami, H., Maheshwari, D.K., 2018. Use of plant growth promoting rhizobacteria (PGPRs) with multiple plant growth promoting traits in stress agriculture: action mechanisms and future prospects. Ecotoxicol. Environ. Saf. 156, 225–246.

Etesami, H., Noori, F., 2019. Soil salinity as a Challenge for Sustainable Agriculture and Bacterial-Mediated Alleviation of Salinity Stress in Crop Plants, Saline Soil-Based Agriculture by Halotolerant Microorganisms. Springer, pp. 1–22.

Etesami, H., Glick, B.R., 2020. Halotolerant plant growth–promoting bacteria: prospects for alleviating salinity stress in plants. Environ. Exp. Bot. 178, 104124.

Etesami, H., Jeong, B.R., Rizwan, M., 2020. The use of silicon in stressed agriculture management: action mechanisms and future prospects. Metalloids in Plants: Advances and Future Prospects 381–431.

Fahad, S., Hussain, S., Matloob, A., Khan, F.A., Khaliq, A., Saud, S., et al., 2015. Phytohormones and plant responses to salinity stress: a review. Plant Growth Regul. 75, 391–404.

Faizan, M., Faraz, A., Yusuf, M., Khan, S.T., Hayat, S., 2018. Zinc oxide nanoparticle-mediated changes in photosynthetic efficiency and antioxidant system of tomato plants. Photosynthetica 56 (2), 678–686.

Faizan, M., Bhat, J.A., Chen, C., Alyemeni, M.N., Wijaya, L., Ahmad, P., et al., 2021. Zinc oxide nanoparticles (ZnO-NPs) induce salt tolerance by improving the antioxidant system and photosynthetic machinery in tomato. Plant Physiol. Biochem. 161, 122–130.

FAO, IFAD, and WFP, 2013. The State of Food Insecurity in the World: The Multiple Dimensions of Food Security 2013. FAO.

Farouk, S., Al-Amri, S.M., 2019. Exogenous zinc forms counteract NaCl-induced damage by regulating the antioxidant system, osmotic adjustment substances, and ions in canola (*Brassica napus* L. cv. Pactol) plants. J. Soil Sci. Plant Nutr. 19, 887–899.

Feizi, H., Moghaddam, P.R., Shahtahmassebi, N., Fotovat, A., 2012. Impact of bulk and nanosized titanium dioxide (TiO2) on wheat seed germination and seedling growth. Biol. Trace Elem. Res. 146, 101–106.

Fita, A., Rodríguez-Burruezo, A., Boscaiu, M., Prohens, J., Vicente, O., 2015. Breeding and domesticating crops adapted to drought and salinity: a new paradigm for increasing food production. Front. Plant Sci. 6, 978.

Gaafar, R., Diab, R., Halawa, M., Elshanshory, A., El-Shaer, A., Hamouda, M., 2020. Role of zinc oxide nanoparticles in ameliorating salt tolerance in soybean. Egypt. J. Bot. 60, 733–747.

Gohari, G., Mohammadi, A., Akbari, A., Panahirad, S., Dadpour, M.R., Fotopoulos, V., et al., 2020a. Titanium dioxide nanoparticles (TiO2 NPs) promote growth and ameliorate salinity stress effects on essential oil profile and biochemical attributes of *Dracocephalum moldavica*. Sci. Rep. 10, 1–14.

Gonzalez-García, Y., Cardenas Alvarez, C., Cadenas-Pliego, G., Benavides-Mendoza, A., Cabrera-de-la-Fuente, M., Sandoval-Rangel, A., et al., 2021. Effect of three nanoparticles (Se, Si and Cu) on the bioactive compounds of bell pepper fruits under saline stress. Plants 10, 217.

Guerriero, G., Sutera, F.M., Torabi-Pour, N., Renaut, J., Hausman, J.-F., Berni, R., et al., 2021. Phyto-courier, a silicon particle-based nano-biostimulant: evidence from cannabis sativa exposed to salinity. ACS Nano 15, 3061–3069.

Hanin, M., Ebel, C., Ngom, M., Laplaze, L., Masmoudi, K., 2016. New insights on plant salt tolerance mechanisms and their potential use for breeding. Front. Plant Sci. 7, 1787.

Hao, Y., Cao, X., Ma, C., Zhang, Z., Zhao, N., Ali, A., et al., 2017. Potential applications and antifungal activities of engineered nanomaterials against gray mold disease agent *Botrytis cinerea* on rose petals. Front. Plant Sci. 8, 1332.

Hao, Y., Yuan, W., Ma, C., White, J.C., Zhang, Z., Adeel, M., et al., 2018. Engineered nanomaterials suppress turnip mosaic virus infection in tobacco (*Nicotiana benthamiana*). Environ. Sci. Nano 5 (7), 1685–1693.

Hasanpour, H., Maali-Amir, R., Zeinali, H., 2015. Effect of TiO_2 nanoparticles on metabolic limitations to photosynthesis under cold in chickpea. Russ. J. Plant Physiol. 62, 779–787.

Helaly, M.N., El-Metwally, M.A., El-Hoseiny, H., Omar, S.A., El-Sheery, N.I., 2014. Effect of nanoparticles on biological contamination of in vitro cultures and organogenic regeneration of banana. Aust. J. Crop. Sci. 8 (4), 612.

Hernandez-Hernandez, H., Juarez-Maldonado, A., Benavides-Mendoza, A., OrtegaOrtiz, H., Cadenas-Pliego, G., Sanchez-Aspeytia, D., et al., 2018. Chitosan-PVA and copper nanoparticles improve growth and overexpress the SOD and JA genes in tomato plants under salt stress. Agronomy 8, 175.

Hezaveh, T.A., Pourakbar, L., Rahmani, F., Alipour, H., 2019. Interactive effects of salinity and ZnO nanoparticles on physiological and molecular parameters of rapeseed (*Brassica napus L.*). Commun. Soil Sci. Plant Anal. 50, 698–715.

Hnilickova, H., Kraus, K., Vachova, P., Hnilicka, F., 2021. Salinity stress affects photosynthesis, malondialdehyde formation, and proline content in *Portulaca oleracea L.* Plants 10, 845.

Hong, F., Yang, F., Liu, C., Gao, Q., Wan, Z., Gu, F., et al., 2005a. Influences of Nano-TiO2 on the chloroplast aging of spinach under light. Biol. Trace Elem. Res. 104, 249–260.

Hong, F., Zhou, J., Liu, C., Yang, F., Wu, C., Zheng, L., et al., 2005b. Effect of nanoTiO2 on photochemical reaction of chloroplasts of spinach. Biol. Trace Elem. Res. 105, 269–279.

Hosseinpour, A., Haliloglu, K., Tolga Cinisli, K., Ozkan, G., Ozturk, H.I., PourAboughadareh, A., et al., 2020. Application of zinc oxide nanoparticles and plant growth promoting bacteria reduces genetic impairment under salt stress in tomato (*Solanum lycopersicum L.* 'Linda'). Agriculture 10, 521.

Iqbal, M.N., Rasheed, R., Ashraf, M.Y., Ashraf, M.A., Hussain, I., 2018. Exogenously applied zinc and copper mitigate salinity effect in maize (*Zea mays L.*) by improving key physiological and biochemical attributes. Environ. Sci. Pollut. Res. 25, 23883–23896.

Kaiwen, G., Zisong, X., Yuze, H., Qi, S., Yue, W., Yanhui, C., et al., 2020. Effects of salt concentration, pH, and their interaction on plant growth, nutrient uptake, and photochemistry of alfalfa (*Medicago sativa*) leaves. Plant Signal. Behav. 15, 1832373.

Kango, S., Kalia, S., Celli, A., Njuguna, J., Habibi, Y., Kumar, R., 2013. Surface modification of inorganic nanoparticles for development of organic–inorganic nanocomposites: a review. Prog Pol Sci 38 (8), 1232–1261.

Kapilan, R., Vaziri, M., Zwiazek, J.J., 2018. Regulation of aquaporins in plants under stress. Biol. Res. 51, 1–11.

Karimi, S., Tavallali, V., Ferguson, L., Mirzaei, S., 2020. Developing a nano-Fe complex to supply iron and improve salinity tolerance of Pistachio under calcium bicarbonate stress. Commun. Soil Sci. Plant Anal. 51, 1835–1851.

Khan, M.R., Adam, V., Rizvi, T.F., Zhang, B., Ahamad, F., Jośko, I., et al., 2019. Nanoparticle–plant interactions: two-way traffic. Small 15, 1901794.

Khan, I., Raza, M.A., Awan, S.A., Shah, G.A., Rizwan, M., Ali, B., et al., 2020. Amelioration of salt induced toxicity in pearl millet by seed priming with silver nanoparticles (AgNPs): the oxidative damage, antioxidant enzymes and ions uptake are major determinants of salt tolerant capacity. Plant Physiol. Biochem. 156, 221–232.

Khodakovskaya, M., Dervishi, E., Mahmood, M., Xu, Y., Li, Z., Watanabe, F., et al., 2009. Carbon nanotubes are able to penetrate plant seed coat and dramatically affect seed germination and plant growth. ACS Nano 3, 3221–3227.

Komaresofla, B.R., Alikhani, H.A., Etesami, H., Khoshkholgh-Sima, N.A., 2019. Improved growth and salinity tolerance of the halophyte Salicornia sp. by co–inoculation with endophytic and rhizosphere bacteria. Appl. Soil Ecol. 138, 160–170.

Kopittke, P.M., Lombi, E., Wang, P., Schjoerring, J.K., Husted, S., 2019. Nanomaterials as fertilizers for improving plant mineral nutrition and environmental outcomes. Environ. Sci. Nano 6, 3513–3524.

Kumar, V., Guleria, P., Kumar, V., Yadav, S.K., 2013. Gold nanoparticle exposure induces growth and yield enhancement in *Arabidopsis thaliana*. Sci. Total Environ. 461–462, 462–468.

Kumar, M., Etesami, H., Kumar, V., 2019. Saline Soil-based Agriculture by Halotolerant Microorganisms. Springer.

Lahiani, M.H., Dervishi, E., Chen, J., Nima, Z., Gaume, A., Biris, A.S., et al., 2013. Impact of carbon nanotube exposure to seeds of valuable crops. ACS Appl. Mater. Interfaces. 5 (16), 7965–7973.

Lara-Romero, J., Campos-Garcia, J., Dasgupta-Schubert, N., Borjas-Garcia, S., Tiwari, D.K., Paraguay-Delgado, F., et al., 2017. Biological effects of carbon nanotubes generated in forest wildfire ecosystems rich in resinous trees on native plants. PeerJ 5, e3658.

Li, J., Pu, L., Han, M., Zhu, M., Zhang, R., Xiang, Y., 2014. Soil salinization research in China: advances and prospects. J. Geogr. Sci. 24, 943–960.

Liang, W., Ma, X., Wan, P., Liu, L., 2018. Plant salt-tolerance mechanism: a review. Biochem. Biophys. Res. Commun. 495, 286–291.

Liu, P., Yin, L., Wang, S., Zhang, M., Deng, X., Zhang, S., et al., 2015. Enhanced root hydraulic conductance by aquaporin regulation accounts for silicon alleviated saltinduced osmotic stress in *Sorghum bicolor L.* Environ. Exp. Bot. 111, 42–51.

Liu, J., Li, G., Chen, L., Gu, J., Wu, H., Li, Z., 2021. Cerium oxide nanoparticles improve cotton salt tolerance by enabling better ability to maintain cytosolic $K+/Na+$ ratio. J. Nanobiotechnol. 19, 1–16.

Lv, J., Christie, P., Zhang, S., 2019. Uptake, translocation, and transformation of metalbased nanoparticles in plants: recent advances and methodological challenges. Environ. Sci. Nano. 6 (1), 41–59.

Ma, Y., Kuang, L., He, X., Bai, W., Ding, Y., Zhang, Z., et al., 2010. Effects of rare earth oxide nanoparticles on root elongation of plants. Chemosphere 78, 273–279.

Mahmoud, A.W.M., Abdeldaym, E.A., Abdelaziz, S.M., El-Sawy, M.B.I., Mottaleb, S.A., 2020a. Synergetic effects of zinc, boron, silicon, and zeolite nanoparticles on confer tolerance in potato plants subjected to salinity. Agronomy 10, 19.

Mahmoud, L.M., Dutt, M., Shalan, A.M., El-Kady, M.E., El-Boray, M.S., Shabana, Y.M., et al., 2020b. Silicon nanoparticles mitigate oxidative stress of in vitroderived banana (Musa acuminata 'Grand Nain') under simulated water deficit or salinity stress. S. Afr. J. Bot. 132, 155–163.

Manickavasagam, M., Pavan, G., Vasudevan, V., 2019. A comprehensive study of the hormetic influence of biosynthesized AgNPs on regenerating rice calli of indica cv. IR64. Sci. Rep. 9, 8821.

Mattiello, A., Filippi, A., Poščić, F., Musetti, R., Salvatici, M.C., Giordano, C., et al., 2015. Evidence of phytotoxicity and genotoxicity in Hordeum vulgare L. exposed to CeO2 and TiO2 nanoparticles. Front. Plant Sci. 6, 1043.

Maurel, C., Boursiac, Y., Luu, D.-T., Santoni, V., Shahzad, Z., Verdoucq, L., 2015. Aquaporins in plants. Physiol. Rev. 95, 1321–1358.

Mohamed, A.K.S.H., Qayyum, M.F., Abdel-Hadi, A.M., Rehman, R.A., Ali, S., Rizwan, M., 2017. Interactive effect of salinity and silver nanoparticles on photosynthetic and biochemical parameters of wheat. Arch. Agron. Soil Sci. 63, 1736–1747.

Molnar, A., Papp, M., Zoltan Kovacs, D., B́elteky, P., Olah, D., Feigl, G., et al., 2020. Nitrooxidative signalling induced by chemically synthetized zinc oxide nanoparticles (ZnO NPs) in Brassica species. Chemosphere 251, 126419.

Moradbeygi, H., Jamei, R., Heidari, R., Darvishzadeh, R., 2020. Investigating the enzymatic and non-enzymatic antioxidant defense by applying iron oxide nanoparticles in *Dracocephalum moldavica* L. plant under salinity stress. Sci. Hortic. 272, 109537.

Mukesh, A.T., Jha, A., 2017. A review on: carbon nanotubes are vital for plant growth. Am J Agric Forestry. 5 (5–1), 1–9.

Munns, R., 2005. Genes and salt tolerance: bringing them together. N. Phytol. 167, 645–663.

Munns, R., Tester, M., 2008. Mechanisms of salinity tolerance. Annu. Rev. Plant Biol. 59, 651–681.

Mushtaq, A., Khan, Z., Khan, S., Rizwan, S., Jabeen, U., Bashir, F., et al., 2020. Effect of silicon on antioxidant enzymes of wheat (*Triticum aestivum L.*) grown under salt stress. Silicon 12, 2783–2788.

Nair, R., 2016. Effects of nanoparticles on plant growth and development. In: Kole, C., Kumar, D.S., Riya, V., Khodakovskaya, M.V. (Eds.), Plant Nanotechnology: Principles and Practices. Springer, New York, pp. 95–118.

Nair, R., Varghese, S.H., Nair, B.G., Maekawa, T., Yoshida, Y., Kumar, D.S., 2010. Nanoparticulate material delivery to plants. Plant Sci. 179, 154–163.

Nehra, A., Ahlawat, S., Singh, K.P., 2019. A biosensing expedition of nanopore: a review. Sens. Actuators B Chem. 284, 595–622.

Nie, Z., Petukhova, A., Kumacheva, E., 2010. Properties and emerging applications of selfassembled structures made from inorganic nanoparticles. Nat Nanotechnol. 5 (1), 15–25.

Noohpisheh, Z., Amiri, H., Mohammadi, A., Farhadi, S., 2021. Effect of the foliar application of zinc oxide nanoparticles on some biochemical and physiological parameters of *Trigonella foenum*-graecum under salinity stress. Plant Biosyst.—Int. J. Deal. Asp. Plant Biol. 155, 267–280.

Oloumi, H., Mousavi, E.A., Nejad, R.M., 2018. Multi-wall carbon nanotubes effects on plant seedlings growth and cadmium/lead uptake in vitro. Russ. J. Plant Physiol. 65 (2), 260–268.

Oster, J., Shainberg, I., Abrol, I., 1999. Reclamation of salt-affected soils. Agric. Drain. 38, 659–691.

Paramo, L.A., Feregrino-Ṕerez, A.A., Guevara, R., Mendoza, S., Esquivel, K., 2020. Nanoparticles in agroindustry: applications, toxicity, challenges, and trends. Nanomaterials 10, 1654.

Parihar, P., Singh, S., Singh, R., Singh, V.P., Prasad, S.M., 2015. Effect of salinity stress on plants and its tolerance strategies: a review. Environ. Sci. Pollut. Res. 22, 4056–4075.

Peng, H., Zhang, J., 2009. Plant genomic DNA methylation in response to stresses: potential applications and challenges in plant breeding. Prog. Nat. Sci. 19, 1037–1045.

Pinedo-Guerrero, Z.H., Cadenas-Pliego, G., Ortega-Ortiz, H., Gonzalez-Morales, S., Benavides-Mendoza, A., Valdés-Reyna, J., et al., 2020. Form of silica improves yield, fruit quality and antioxidant defense system of tomato plants under salt stress. Agriculture 10, 367.

Pradhan, S., Patra, P., Das, S., Chandra, S., Mitra, S., Dey, K.K., et al., 2013. Photochemical modulation of biosafe manganese nanoparticles on Vigna radiata: a detailed molecular, biochemical, and biophysical study. Environ. Sci. Technol. 47, 13122–13131.

Rajput, V., Minkina, T., Suskova, S., Mandzhieva, S.S., 2018. Effects of copper nanoparticles (CuO NPs) on crop plants: a mini review. BioNanoScience. 8 (1), 36–42.
Raskar, S., Laware, S., 2014. Effect of zinc oxide nanoparticles on cytology and seed germination in onion. Int. J. Curr. Microbiol. Appl. Sci. 3 (2), 467–473.
Rastogi, A., Zivcak, M., Sytar, O., Kalaji, H.M., He, X., Mbarki, S., et al., 2017. Impact of metal and metal oxide nanoparticles on plant: a critical review. Front. Chem. 5, 78.
Richards, L.A., 1947. Diagnosis and improvement of saline and alkaline soils. Soil Sci. 64, 432.
Rico, C.M., Peralta-Videa, J.R., Gardea-Torresdey, J.L., 2015. Chemistry, biochemistry of nanoparticles, and their role in antioxidant defense system in plants. In: Siddiqui, M., Al-Whaibi, M., Mohammad, F. (Eds.), Nanotechnology and Plant Sciences. Springer, Cham.
Rios, J.J., Martínez-Ballesta, M.C., Ruiz, J.M., Blasco, B., Carvajal, M., 2017. Siliconmediated improvement in plant salinity tolerance: the role of aquaporins. Front. Plant Sci. 8, 948.
Salama, H.M., 2012. Effects of silver nanoparticles in some crop plants, common bean (*Phaseolus vulgaris L.*) and corn (*Zea mays L.*). Int Res J Biotechnol. 3 (10), 190–197.
Salata, O.V., 2004. Applications of nanoparticles in biology and medicine. J Nanobiotech. 2 (1), 3.
Sami, F., Siddiqui, H., Hayat, S., 2020. Impact of silver nanoparticles on plant physiology: a critical review. In: Hayat, S., Pichtel, J., Faizan, M., Fariduddin, Q. (Eds.), Sustainable Agriculture Reviews 41. Sustainable Agriculture Reviews, 41. Springer, Cham.
Sedghi, M., Hadi, M., Toluie, S.G., 2013. Effect of nano zinc oxide on the germination parameters of soybean seeds under drought stress. Ann. West Univ. Timisoara Biol. 16 (2), 73.
Sekoai, P.T., Ouma, C.N.M., du Preez, S.P., Modisha, P., Engelbrecht, N., Bessarabov, D.G., et al., 2019. Application of nanoparticles in biofuels: an overview. Fuel. 237, 380–397.
Shabala, S., 2013. Learning from halophytes: physiological basis and strategies to improve abiotic stress tolerance in crops. Ann. Bot. 112, 1209–1221.
Shahi, S., Srivastava, M., 2018. Influence of foliar application of manganese on growth, pigment content, and nitrate reductase activity of *Vigna radiata* (L.) R. Wilczek under salinity. J. Plant Nutr. 41, 1397–1404.
Sharma, P., Bhatt, D., Zaidi, M.G.H., Saradhi, P.P., Khanna, P.K., Arora, S., 2012. Silver nanoparticle-mediated enhancement in growth and antioxidant status of *Brassica juncea*. Appl. Biochem. Biotechnol. 167 (8), 2225–2233.
Siddiqui, H., Ahmed, K.B.M., Sami, F., Hayat, S., 2020. Silicon nanoparticles and plants: current knowledge and future perspectives. Sustain. Agric. Rev. 41, 129–142.
Singh, N., Bhatla, S.C., 2016. Nitric oxide and iron modulate heme oxygenase activity as a long distance signaling response to salt stress in sunflower seedling cotyledons. Nitric Oxide 53, 54–64.
Singh, A., Hussain, I., Singh, N.B., Singh, H., 2019. Uptake, translocation and impact of green synthesized nanoceria on growth and antioxidant enzymes activity of *Solanum lycopersicum L.* Ecotoxicol. Environ. Saf. 182, 109410.
Singh, P., Arif, Y., Siddiqui, H., Sami, F., Zaidi, R., Azam, A., et al., 2021. Nanoparticles enhances the salinity toxicity tolerance in *Linum usitatissimum* L. by modulating the antioxidative enzymes, photosynthetic efficiency, redox status and cellular damage. Ecotoxicol. Environ. Saf. 213, 112020.
Sofy, M.R., Elhindi, K.M., Farouk, S., Alotaibi, M.A., 2020. Zinc and paclobutrazol mediated regulation of growth, upregulating antioxidant aptitude and plant productivity of pea plants under salinity. Plants 9, 1197.
Song, H., 2015. Metal hybrid nanoparticles for catalytic organic and photochemical transformations. Acc. Chem. Res. 48 (3), 491–499.
Srinivasan, C., Saraswathi, R., 2010. Nano-agriculture-carbon nanotubes enhance tomato seed germination and plant growth. Curr. Sci. 99 (3), 274–275.

Srivastav, A.K., Kumar, M., Ansari, N.G., Jain, A.K., Shankar, J., Arjaria, N., et al., 2016. A comprehensive toxicity study of zinc oxide nanoparticles vs their bulk in wistar rats: toxicity study of zinc oxide nanoparticles. Hum. Exp. Toxicol. 35 (12), 1286–1304.

Suriyaprabha, R., Karunakaran, G., Yuvakkumar, R., Rajendran, V., Kannan, N., 2012. Silica nanoparticles for increased silica availability in maize (Zea mays L.) seeds under hydroponic conditions. Curr Nanosci. 8 (6), 902–908.

Sytar, O., Kumari, P., Yadav, S., Brestic, M., Rastogi, A., 2019. Phytohormone priming: regulator for heavy metal stress in plants. J. Plant Growth Regul. 38, 739–752.

Syu, Y.Y., Hung, J.H., Chen, J.C., Chuang, H.W., 2014. Impacts of size and shape of silver nanoparticles on arabidopsis plant growth and gene expression. Plant Physiol. Biochem. 83, 57–64.

Tahjib-Ul-Arif, M., Sohag, A.A.M., Afrin, S., Bashar, K.K., Afrin, T., Mahamud, A.G.M.S.U., et al., 2019. Differential response of sugar beet to long-term mild to severe salinity in a soil–pot culture. Agriculture 9, 223.

Tang, X., Mu, X., Shao, H., Wang, H., Brestic, M., 2015. Global plant-responding mechanisms to salt stress: physiological and molecular levels and implications in biotechnology. Crit. Rev. Biotechnol. 35, 425–437.

Tortella, G.R., Rubilar, O., Durán, N., Diez, M.C., Martínez, M., Parada, J., et al., 2020. Silver nanoparticles: toxicity in model organisms as an overview of its hazard for human health and the environment. J. Hazard. Mater. 390, 121974.

Tripathi, D.K., Shweta, Singh, S., Singh, S., Pandey, R., Singh, V.P., et al., 2017. An overview on manufactured nanoparticles in plants: uptake, translocation, accumulation and phytotoxicity. Plant Physiol. Biochem. 110, 2–12.

Tripathi, D.K., Singh, S., Gaur, S., Singh, S., Yadav, V., Liu, S., et al., 2018. Acquisition and homeostasis of iron in higher plants and their probable role in abiotic stress tolerance. Front. Environ. Sci. 5, 86.

Tripathi, D.K., Vishwakarma, K., Singh, V.P., Prakash, V., Sharma, S., Muneer, S., et al., 2020. Silicon crosstalk with reactive oxygen species, phytohormones and other signaling molecules. J. Hazard. Mater. 124820.

Ullah, S., Adeel, M., Zain, M., Rizwan, M., Irshad, M.K., Jilani, G., et al., 2020. Physiological and biochemical response of wheat (*Triticum aestivum*) to TiO2 nanoparticles in phosphorous amended soil: a full life cycle study. J. Environ. Manag. 263, 110365.

Van Zelm, E., Zhang, Y., Testerink, C., 2020. Salt tolerance mechanisms of plants. Annu. Rev. Plant Biol. 71, 403–433.

Vera-Reyes, I., Vázquez-Núñez, E., Lira-Saldivar, R.H., Méndez-Argüello, B., 2018. Effects of nanoparticles on germination, growth, and plant crop development. In: López-Valdez, F., Fernández-Luqueño, F. (Eds.), Agricultural Nanobiotechnology: Modern Agriculture for a Sustainable Future. Springer, New York, pp. 77–110.

Vishwakarma, K., Upadhyay, N., Kumar, N., Tripathi, D.K., Chauhan, D.K., Sharma, S., et al., 2018. Potential applications and avenues of nanotechnology in sustainable agriculture. Nanomaterials in Plants, Algae, and Microorganisms. Elsevier, pp. 473–500.

Wahid, I., Kumari, S., Ahmad, R., Hussain, S.J., Alamri, S., Siddiqui, M.H., et al., 2020. Silver nanoparticle regulates salt tolerance in wheat through changes in ABA concentration, ion homeostasis, and defense systems. Biomolecules 10, 1506.

Wan, J., Wang, R., Bai, H., Wang, Y., Xu, J., 2020. Comparative physiological and metabolomics analysis reveals that single-walled carbon nanohorns and ZnO nanoparticles affect salt tolerance in *Sophora alopecuroides*. Environ. Sci. Nano 7, 2968–2981.

Wang, A., Zheng, Y., Peng, F., 2014. Thickness-controllable silica coating of CdTe QDs by reverse microemulsion method for the application in the growth of rice. J Spectrosc. 2014, 1–5.

Wang, X.-S., Ren, H.-L., Wei, Z.-W., Wang, Y.-W., Ren, W.-B., 2017. Effects of neutral salt and alkali on ion distributions in the roots, shoots, and leaves of two alfalfa cultivars with differing degrees of salt tolerance. J. Integr. Agric. 16, 1800–1807.

Wang, X., Wang, W., Huang, J., Peng, S., Xiong, D., 2018. Diffusional conductance to CO2 is the key limitation to photosynthesis in salt-stressed leaves of rice (*Oryza sativa*). Physiol. Plant. 163, 45–58.

Wang, Z., Li, H., Li, X., Xin, C., Si, J., Li, S., et al., 2020. Nano-ZnO priming induces salt tolerance by promoting photosynthetic carbon assimilation in wheat. Arch. Agron. Soil Sci. 66, 1259–1273.

Wu, H., Shabala, L., Shabala, S., Giraldo, J.P., 2018. Hydroxyl radical scavenging by cerium oxide nanoparticles improves Arabidopsis salinity tolerance by enhancing leaf mesophyll potassium retention. Environ. Sci. Nano 5, 1567–1583.

Yan, A., Chen, Z., 2019. Impacts of silver nanoparticles on plants: a focus on the phytotoxicity and underlying mechanism. Int. J. Mol. Sci. 20, 1003.

Yang, Y., Guo, Y., 2018. Elucidating the molecular mechanisms mediating plant saltstress responses. N. Phytol. 217, 523–539.

Ye, Y., Medina-Velo, I.A., Cota-Ruiz, K., Moreno-Olivas, F., Gardea-Torresdey, J.L., 2019. Can abiotic stresses in plants be alleviated by manganese nanoparticles or compounds? Ecotoxicol. Environ. Saf. 184, 109671.

Ye, Y., Cota-Ruiz, K., Hernandez-Viezcas, J.A., Valdés, C., Medina-Velo, I.A., Turley, R.S., et al., 2020. Manganese nanoparticles control salinity-modulated molecular responses in *Capsicum annuum L.* through priming: a sustainable approach for agriculture. ACS Sustain. Chem. Eng. 8, 1427–1436.

Ye, T., Wang, Y., Feng, Y.Q., Chan, Z., 2021. Physiological and metabolomic responses of bermudagrass (Cynodon dactylon) to alkali stress. Physiol. Plant. 171, 22–33.

You, J., Chan, Z., 2015. ROS regulation during abiotic stress responses in crop plants. Front. Plant Sci. 6, 1092.

Youssef, N.M., Ghareeb, Z.F., Ali, A.I., Taha, L.S., 2020. Mitigation of salinity stress on in vitro growth of Eustoma grandiflorum using zinc nanoparticles. Plant Arch. 2, 4547–4554.

Zahedi, S.M., Abdelrahman, M., Hosseini, M.S., Hoveizeh, N.F., Tran, L.-S.P., 2019. Alleviation of the effect of salinity on growth and yield of strawberry by foliar spray of selenium-nanoparticles. Environ. Pollut. 253, 246–258.

Zhang, H., Liu, X.-L., Zhang, R.-X., Yuan, H.-Y., Wang, M.-M., Yang, H.-Y., et al., 2017. Root damage under alkaline stress is associated with reactive oxygen species accumulation in rice (*Oryza sativa L.*). Front. Plant Sci. 8, 1580.

Zhu, Y., Xu, F., Liu, Q., Chen, M., Liu, X., Wang, Y., et al., 2019. Nanomaterials and plants: positive effects, toxicity and the remediation of metal and metalloid pollution in soil. Sci. Total Environ. 662, 414–421.

Zulfiqar, F., Ashraf, M., 2021. Nanoparticles potentially mediate salt stress tolerance in plants. Plant Physiol. Biochem. 160, 257–268.

Chapter 9

Impact of metal oxide nanoparticles against drought stress in plants

Hasna Abdul Salam[1] and Rajiv Periakaruppan[2,3]
[1]*Department of Biotechnology, Karpagam Academy of Higher Education, Coimbatore, Tamil Nadu, India,* [2]*College of Horticulture, Nanjing Agricultural University, Nanjing, Jiangsu Province, P.R. China,* [3]*Department of Biotechnology, PSG College of Arts & Science, Coimbatore, Tamil Nadu, India*

Chapter outline

9.1 Introduction

Agriculture is a significant part of the economy in many regions worldwide as it is a primary source of food and fabrics. It has a prominent role in population sustainability by producing food and raw materials. It involves not only crops but also livestock, and hence called the backbone of an economy. For example, in India, this sector constitutes about 17% of total GDP and 60% of employment opportunities (Arjun, 2013). The sustenance of agriculture is multifactorial. Advanced technology has brought innovations like conservation tillage that enhances carbon sequestration (Anthony, 2004). Growing food demand of rising population can be satisfied only by sustained agriculture; food crisis and food prices can be reduced by 40% and half, respectively.

Plant growth is affected by various factors like environmental conditions and stress throughout their life cycle. It exhibits variations among different

Nanometal Oxides in Horticulture and Agronomy. DOI: https://doi.org/10.1016/B978-0-323-91809-1.00002-0

species and even among plants of the same species. These are effectively combated by self-developed defense mechanisms (Ahmad et al., 2018; Kollist et al., 2019). The stress faced by plants adversely affecting their growth, development, and productivity can be broadly classified as biotic and abiotic. Abiotic stress may be physical or chemical, while biotic, as the name suggests, may be biological, for example, diseases (Verma et al., 2013). Pressure ranges from mild to severe. Severe stress can cause the death of crop plants by ceasing flowering and seed formation, and also induction of senescence (Zhu, 2002).

9.2 Abiotic stress faced by plants

Abiotic stress faced by plants include drought, waterlogging, cold, frost, heat, salinity, and mineral toxicity (Fig. 9.1).

The first line of defense in such conditions is the plant roots. Abscisic acid, a phytohormone, plays a prominent role in adapting to drought stress by modulation, moderation, and inhibition of growth-related and metabolic activities of plants (Seki et al., 2007). Drought affects nearly 45% of worldwide agricultural land (Dos Reis et al., 2016). It is a multidimensional abiotic stress, which, if prolonged, can lead to dehydration and death of plants (Takahashi et al., 2018). It affects stomatal closure, gaseous exchange, transpiration, and carbon assimilation during photosynthesis and is a significant force faced by plants in arid and semiarid regions (Chaves et al., 2003; Schulze et al., 2019). Nutrient mobility gets impeded, and uptake of nitrogen, phosphorus, and potassium, is worsened by drought (Baligar et al., 2001; Al-Kaisi et al., 2013). The water use efficiency (WUE) of plants gets severely affected (Gholami and Zahedi, 2019). Drought leads to the production of reactive oxygen species (ROS) that cause damages to tissues and macromolecules such as DNA, lipids,

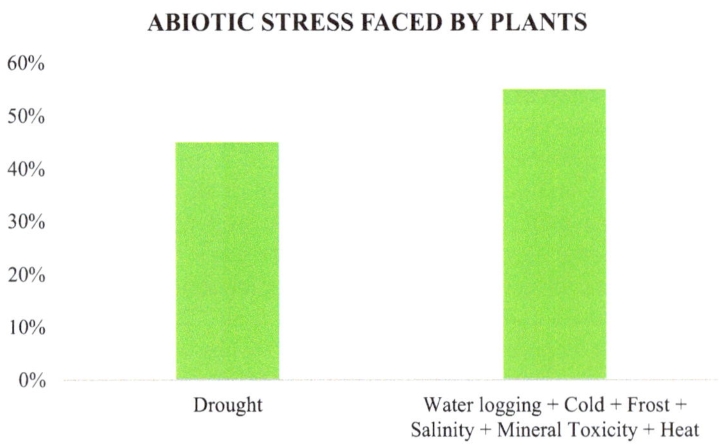

FIGURE 9.1 Abiotic stress faced by plants.

proteins, and carbohydrates (Reddy et al., 2004; Gill and Tuteja, 2010). Defense mechanisms include antioxidant defense systems and osmotic adjustments (Yin et al., 2005). Antioxidant enzymes are scavengers of ROS (Gill and Tuteja, 2010). Osmolytes or osmoprotectants lower the cell water potential, thereby enhancing water extraction capacity in environments with little water (Mashilo et al., 2018). Increasing tolerance to stress is essential for sustainable agriculture and crop production (Ahmad et al., 2020). Drought-tolerant crop varieties involve a lot of time for their development. But still, there are a few vigorous drought-tolerant species that are economically feasible (Hasan et al., 2018; Genc et al., 2019; Hasan et al., 2021a; Hasan et al., 2021b). Also, the safety of transgenic crops is a significant concern (De Lange et al., 2018).

Nanotechnology has gained attention as efficient fertilizer carriers and controlled-release vectors for preparing innovative fertilizers leading to an increase in nutrient use efficiency and decreased costs of environmental protection (Shallan et al., 2016). Nanostructures have physicochemical properties that are different from their native bulk compounds in aspects such as surface, optical, thermal, and electrical properties (Rastogi et al., 2017). Although nanoparticles are more commonly synthesized by physical and chemical methods, the application of toxic compounds during the process proves a demerit. Thus, the development of eco-friendly practices for biogenetic production generates more interest due to the versatility and simplicity of protocol (Popescu et al., 2010; Li et al., 2011). This has led to the emergence of nanobiotechnology that creates alternatives for the production of nanomaterials with novel properties that can be applied in the fields of biotechnology and agriculture (Chhipa, 2017; Giraldo et al., 2019). The potential use of nanoparticles in the context of their mechanism and role as stress mitigators in plants needs to be explored on a higher scale (Vishwakarma et al., 2018). Studies are also required to examine the mode of action of nanostructures, their interaction with biomolecules, and their impact on genetic levels under drought stress (Ahmad et al., 2020).

9.3 Metal and metal oxide nanostructures against drought stress

Nanofertilizers increase crop production, minimize the nutrient loss of fertilizers, and reduce soil toxicity caused by the application of chemical fertilizers and pesticides (León-Silva et al., 2018; Saranya et al., 2019; Pirzadah et al., 2019). The plant responses of engineered nanostructures are different from nonengineered/modified ones (Khan et al., 2017). Engineered nanoparticles offer compatibility of extra addition of surface molecules which makes them different from natural ones (Servin and White, 2016; Tan et al., 2017). Nanobiotechnology thus opens up new arenas for research in plant genetic engineering and the development of transgenic plants (Wang et al., 2016;

Verma et al., 2018). Nanoparticles are potential nanoherbicides and nanopesticides capable of the efficient release of content in desired quantity to target plant cell organelles (Cicek and Nadaroglu, 2015). Nanoparticles activate enzymatic and nonenzymatic antioxidants during abiotic stress (Ahmad et al., 2020). This depends on the size, shape, absorption, and dosage of the nanostructures involved though their molecular mechanisms are unclear (Ahmad et al., 2019). Three widely used metal oxide nanoparticles in agriculture include titanium dioxide (TiO_2), zinc oxide (ZnO), and iron oxide (Fe_3O_4) (Alabdallah et al., 2021). The potential of various metal oxide nanostructures to combat drought stress is summarized in Table 9.1.

9.3.1 Titanium dioxide nanoparticles

Titanium dioxide nanoparticles (TiO_2-NPs) have numerous effects on the morphology, biochemistry, and the physiology of plants (Djanaguiraman et al., 2018). Nano anatase TiO_2-NPs at 10 ppm concentration enhances the accumulation of proline and reduces H_2O_2 and malondialdehyde in *Dracocephalum moldavia* L. (Mohammadi et al., 2016). Malondialdehyde is known as a

TABLE 9.1 Metal oxide nanostructures to fight drought stress.

Metal oxide nanostructures	Functions
Titanium dioxide nanoparticles (TiO_2-NPs)	Enhances proline accumulation, shoot length, photosynthetic pigments Reduces H_2O_2 and malondialdehyde Increases gluten and starch content Impacts secondary metabolites
Zinc oxide nanoparticles (ZnO-NPs)	Interacts with root-colonizing microbes causing alteration in growth and function of plants Enhances photosynthetic activity, chlorophyll content, transpiration rate, stomatal conductance, water use efficiency, and enzymatic activity Maintains integrity of biological membranes
Iron oxide nanoparticles (Fe_3O_4-NPs)	Enhances root length, rate of photosynthesis, antioxidant enzyme activity, and nutrient uptake
Chitosan, silicon dioxide (silica) and selenium nanoparticles (SiO_2-NPs, Se-NPs)	Enhances relative water content, grain protein, proline content, fruit quality, expression of genes related to antioxidant defense Decreases lipid peroxidation

biomarker for lipid peroxidation (Denise et al., 2009). Drought stress in *Cuminum cyminum* L. increased carbohydrate, antioxidant enzymes activity, and proline contents (Mehdi et al., 2020). Proline accumulation is a general response during environmental stress (Liang et al., 2013). Proline in membrane structures acts as ROS scavengers under drought stress (Ashraf and Foolad, 2007). TiO_2-NPs are known to enhance the length of shoot and root, growth and yield of radish, corn, lettuce, cucumber, canola, and wheat (Lin and Xing, 2007; Mahmoodzadeh et al., 2013). Application of TiO_2-NPs under drought stress can substantially increase gluten and starch content (Raliya et al., 2015). The exogenous application of TiO_2-NPs under drought stress resulted in an increase in shoot length, and fresh and dry weight of wheat, as well as an increase in photosynthetic pigments of wheat and *Linum usitatissimum* (Aghdam et al., 2016; Faraji and Sepehri, 2019). TiO_2-NPs impact secondary metabolites considerably under drought stress (Sattari et al., 2020).

9.3.2 Zinc oxide nanoparticles

Zinc oxide nanoparticles (ZnO-NPs) interact with root-colonizing microbes in *Triticum aestivum* L, resulting in alteration of plant growth and function under drought (Yang et al., 2017). Zinc (Zn) has an active involvement in maintaining the integrity of biological membranes, which is critical for water absorption and utilization (Osakabe et al., 2014). It also influences the structure and function of enzymes. Treating *Hordeum vulgare* L. with 1, 3, and 5 mg Zn/kg for 4 weeks promoted growth, yield, and development, and fortified edible grains with crucial nutrients along with improvement in nitrogen acquisition (Dimkpa et al., 2019). Plants treated with ZnO-NPs or copper oxide nanoparticles (CuO-NPs) provided cross-protection during drought and metal stress (Yang et al., 2017). Applying Zn-NPs boosted root biomass and growth of above-ground tissues in the case of *Vigna radiata* seedlings during drought stress (Dhoke et al., 2013). Photosynthetic activity, chlorophyll content, transpiration rate, stomatal conductance, and WUE were found to get enhanced in maize seedlings, owing to an accumulation of osmolytes such as proline and sugars required to adjust osmotic functions (Sun et al., 2020). A significant increase to drought tolerance has been reported in *Moringa peregrina* species by applying ZnO-NPs that increased the activities of antioxidant polyphenol oxidase and peroxidase (Foroutan et al., 2018).

9.3.3 Iron oxide nanoparticles

Environmental iron levels affect the synthesis of chlorophyll content, photosynthetic activity, and metabolism (Mozafari et al., 2017). Foliar application of iron oxide nanoparticles (Fe_2O_3NPs) coated with citric acid enhances root length and photosynthetic rate (Dhoke et al., 2013). Fe_3O_4-NPs of size 40–53 nm improve plant growth, relative water content, and photosynthetic pigments of Fragaria × *ananassa* under drought stress conditions (Mozafari et al., 2017). *Oryza sativa* plants displayed an increase in their biomass, antioxidant enzyme activities, photosynthetic efficiency, and nutrient uptake under drought stress conditions (Ahmed et al., 2021).

9.3.4 Chitosan, silicon dioxide (silica), and selenium nanoparticles

Foliar application of 30, 60, and 90 ppm at the tillering stage of *Hordeum vulgare* L. enhanced the relative water content, grain protein, proline content, catalase, and superoxide dismutase (Behboudi et al., 2018). Chitosan containing nanocomposites promotes the slow release of nutrients, withholding reasonable quantities of water, thus facilitating salinity control and drought (Mushtaq et al., 2018). Chitosan nanoparticles formed by encapsulation of S-nitrosoglutathione increase NO-induced benefits for plants, thereby nullifying the negative impact of drought on the physiology and metabolism of plants (Silveira et al., 2019).

Spraying of solutions containing nanostructures of SiO_2, Se, and Se/SiO_2 promotes the growth and yield of strawberry plants under drought stress conditions due to the preservation of photosynthetic raw materials and produces more osmolytes like carbohydrate and proline. Exogenous spraying of Se/SiO_2 increases drought tolerance by increasing the activity of antioxidant enzymes and decreasing lipid peroxidation. Fruit quality and nutritional value are also enhanced (Seyed et al., 2020). Selenium and silicon improve the tolerance capacity of canola, wheat, and rice through increasing enzymatic antioxidant activities and maintaining photosynthetic apparatus performance (Bybordi, 2016; Sattar et al., 2017; Gao et al., 2018). Although selenium and silicon have not yet been identified as essential elements in plant growth, there have been reports of their favorable influences on yield and abiotic stress tolerance (Hasanuzzaman et al., 2014). They increase the production of soluble sugars via photosynthesis, which are effective osmolytes in preserving water equilibrium and growth stimulation (Jóźwiak and Politycka, 2019; Zhu et al., 2019). Silicon decreases transpiration, thereby postponing water loss (Mateos-Naranjo et al., 2013; Dehghanipoodeh et al., 2016). Exogenous application of Se and Si leads to expression of genes related to the antioxidant defense, which consequently increases the activity of SOD, CAT, and APX in maize and ramie plants, thereby helping to combat drought stress (Tang et al., 2015; Jiang et al., 2017). 100, 200, 300, and 400 mg/kg of nano silica have been shown to increase nutrient uptake and fruit yield of *Cucumis sativus* L. (Alsaeedi et al., 2019).

9.4 Conclusion

Drought stress is a significant problem that must be combated in order to sustain agriculture, and thus satisfy food demands of growing populations. It can be seen that environmental changes or abiotic stress must constantly be worked upon to avoid food crises and protect crop plants. The above review of various metal oxide nanoparticles shows that these nanostructures are magical bullets to increase tolerance to drought. This area poses a vast scope for the discovery and development of more novel particles to fight abiotic stress.

References

Aghdam, M.T.B., Mohammadi, H., Ghorbanpour, M., 2016. Effects of nanoparticulate anatase titanium dioxide on physiological and biochemical performance of *Linum usitatissimum* (Linaceae) under well-watered and drought stress conditions. Brazil. J. Bot. 39, 139–146.

Ahmad, J., Bagheri, R., Bashir, H., Baig, M.A., Al-Huqail, A., Ibrahim, M.M., et al., 2018. Organ-specific phytochemical profiling and antioxidant analysis of *Parthenium hysterophorus* L. BioMed. Res. Int. 2018, 1–10. Article ID 9535232.

Ahmad, J., Ali, A.A., Baig, M.A., Iqbal, M., Haq, I., Qureshi, M.I., 2019. Role of phytochelatins in cadmium stress tolerance in plants. Cadmium Toxicity and Tolerance in Plants. Academic Press, Cambridge, pp. 185–212.

Ahmad, J., Qamar, S., Kausar, N., Qureshi, M.I., 2020. Nanoparticles: the magic bullets in mitigating drought stress in plants. In: Hakeem, K.R., Pirzadah, T.B. (Eds.), Nanobiotechnology in Agriculture, Nanotechnology in the Life Sciences. pp. 145–161.

Ahmed, T., Noman, M., Manzoor, N., Shahid, M., Abdullah, M., Ali, L., et al., 2021. Nanoparticle-based amelioration of drought stress and cadmium toxicity in rice via triggering the stress responsive genetic mechanisms and nutrient acquisition. Ecotoxicol. Environ. Saf. 209, 111829.

Alabdallah, N.M., Hasan, M.M., Hammami, I., Alghamdi, A.I., Alshehri, D., Alatawi, H.A., 2021. Green synthesized metal oxide nanoparticles mediate growth regulation and physiology of crop plants under drought stress. Plants 10, 1730.

Al-Kaisi, M.M., Elmore, R.W., Guzman, J.G., Hanna, H.M., Hart, C.E., Helmers, M.J., et al., 2013. Drought impact on crop production and the soil: 2012 experiences from Iowa. J. Soil Water Conser. 68 (1), 19A–24A.

Alsaeedi, A., El-Ramady, H., Alshaal, T., El-Garawany, M., Elhawat, N., Al-Otaibi, A., 2019. Silica nanoparticles boost growth and productivity of cucumber under water deficit and salinity stresses by balancing nutrients uptake. Plant Physiol. Biochem. 139, 1–10.

Anthony, T., 2004. A critical assessment of organic farming-and-food assertions with particular respect to the UK and the potential environmental benefits of no-till agriculture. Crop Protect. 23 (9), 757–781.

Arjun, K.M., 2013. Indian agriculture- status, importance and role in Indian economy. Int. J. Agri. Food Sci. Technol. 4 (4), 343–346.

Ashraf, M., Foolad, M.R., 2007. Roles of glycine betaine and proline in improving plant abiotic stress resistance. Environ. Exp. Bot. 59 (2), 206–216.

Baligar, V.C., Fageria, N.K., He, Z.L., 2001. Nutrient use efficiency in plants. Commun. Soil Sci. Plant Anal. 32 (7–8), 921–950.

Behboudi, F., Tahmasebi Sarvestani, Z., Kassaee, M.Z., Sanavi, M., Mohamad, S.A., Sorooshzadeh, A., et al., 2018. Evaluation of chitosan nanoparticles effects on yield and yield components of barley (*Hordeum vulgare* L.) under late season drought stress. J. Water Environ. Nanotechnol. 3 (1), 22–39.

Bybordi, A., 2016. Infuence of zeolite, selenium and silicon upon some agronomic and physiologic characteristics of Canola grown under salinity. Commun. Soil Sci. Plant Anal. 47, 832–850.

Chaves, M.M., Maroco, J.P., Pereira, J.S., 2003. Understanding plant responses to drought—from genes to the whole plant. Funct. Plant Biol. 30 (3), 239–264.

Chhipa, H., 2017. Nanofertilizers and nanopesticides for agriculture. Environ. Chem. Lett. 15 (1), 15–22.

Cicek, S., Nadaroglu, H., 2015. The use of nanotechnology in the agriculture. Adv. Nano Res. 3 (4), 207–223.

De Lange, O., Klavins, E., Nemhauser, J., 2018. Synthetic genetic circuits in crop plants. Curr. Opin. Biotechnol. 49, 16–22.

Dehghanipoodeh, S., Ghobadi, C., Baninasab, B., Gheysari, M., Bidabadi, S.S., 2016. Effects of potassium silicate and nanosilica on quantitative and qualitative characteristics of a commercial strawberry (Fragaria × ananassa cv 'Camarosa'). J. Plant Nutr. 39, 502–507.

Denise, G., Lucas, S.M., Juliana, V., Clovis, P., Gabriella, S., Solange, C.G., et al., 2009. Importance of the lipid peroxidation biomarkers and methodological aspects FOR malondialdehyde quantification. Quim. Nova 32 (1), 169–174.

Dhoke, S.K., Mahajan, P., Kamble, R., Khanna, A., 2013. Effect of nanoparticles suspension on the growth of mung (*Vigna radiata*) seedlings by foliar spray method. Nanotechnol. Develop. 3, 1.

Dimkpa, C.O., Singh, U., Bindraban, P.S., Elmer, W.H., Gardea-Torresdey, J.L., White, J.C., 2019. Zinc oxide nanoparticles alleviate drought-induced alterations in sorghum performance, nutrient acquisition, and grain fortification. Sci. Total Environ. 688, 926–934.

Djanaguiraman, M., Belliraj, N., Bossmann, S.H., Prasad, P.V., 2018. High-temperature stress alleviation by selenium nanoparticle treatment in grain sorghum. ACS Omega 3 (3), 2479–2491.

Dos Reis, S.P., Marques, D.N., Lima, A.M., de Souza, C.R.B., 2016. Plant molecular adaptations and strategies under drought stress. Drought Stress Tolerance in Plants. Springer, Cham, pp. 91–122.

Faraji, J., Sepehri, A., 2019. Ameliorative effects of TiO_2 nanoparticles and sodium nitroprusside on seed germination and seedling growth of wheat under PEG-stimulated drought stress. J. Seed Sci. 41 (3), 309–317.

Foroutan, L., Solouki, M., Abdossi, A., Fakheri, B.A., 2018. The effects of zinc oxide nanoparticles on enzymatic and osmoprotectant alternations in different *Moringa peregrina* populations under drought stress. Int. J. Basic Sci. Med. 3 (4), 178–187.

Gao, M., Zhou, J., Liu, H., Zhang, W., Hu, Y., Liang, J., et al., 2018. Foliar spraying with silicon and selenium reduces cadmium uptake and mitigates cadmium toxicity in rice. Sci. Total Environ. 631–632, 1100–1108.

Genc, Y., Taylor, J., Lyons, G., Li, Y., Cheong, J., Appelbee, M., et al., 2019. Bread wheat with high salinity and sodicity tolerance. Front. Plant Sci. 10, 1280.

Gholami, R., Zahedi, S.M., 2019. Identifying superior drought-tolerant olive genotypes and their biochemical and some physiological responses to various irrigation levels. J. Plant Nutr. 42 (17), 2057–2069.

Gill, S.S., Tuteja, N., 2010. Reactive oxygen species and antioxidant machinery in abiotic stress tolerance in crop plants. Plant Physiol. Biochem. 48 (12), 909–930.

Giraldo, J.P., Wu, H., Newkirk, G.M., Kruss, S., 2019. Nanobiotechnology approaches for engineering smart plant sensors. Nat. Nanotechnol. 14 (6), 541–553.

Hasan, M.M., Hajar, A.S., Alharby, H.F., Hakeem, K.R., 2018. Effects of magnetized water on phenolic compounds, lipid peroxidation and antioxidant activity of Moringa species under drought stress. J. Animal Plant Sci. 28 (3), 803–810.

Hasan, M.M., Skalicky, M., Jahan, M.S., Hossain, M.N., Anwar, Z., Nie, Z.F., et al., 2021a. Spermine: its emerging role in regulating drought stress responses in plants. Cells 10 (2), 261.

Hasan, M.M., Gong, L., Nie, Z., Feng, L., Ahammed, G.J., Fang, X.W., 2021b. ABA-induced stomatal movements in vascular plants during dehydration vs rehydration. Environ. Exp. Bot. 186, 104436.

Hasanuzzaman, M., Nahar, K., Fujita, M., 2014. Silicon and selenium: two vital trace elements that confer abiotic stress tolerance to plants. Emerging Technologies and Management of Crop Stress Tolerance. pp. 377–422.

Jiang, C., Zu, C., Lu, D., Zheng, Q., Shen, J., Wang, H., et al., 2017. Effect of exogenous selenium supply on photosynthesis, Na + accumulation and antioxidative capacity of maize (*Zea mays* L.) under salinity stress. Sci. Rep. 7, 42039.

Jóźwiak, W., Politycka, B., 2019. Effect of selenium on alleviating oxidative stress caused by a water deficit in cucumber roots. Plants 8 (7), 217.

Khan, I., Saeed, K., Khan, I., 2017. Nanoparticles: properties, applications and toxicities. Arab. J. Chem. 12 (7), 908–931.

Kollist, H., Zandalinas, S.I., Sengupta, S., Nuhkat, M., Kangasjärvi, J., Mittler, R., 2019. Rapid responses to abiotic stress: priming the landscape for the signal transduction network. Trends Plant Sci. 24 (1), 25–37.

León-Silva, S., Arrieta-Cortes, R., Fernández-Luqueño, F., López-Valdez, F., 2018. Design and production of nanofertilizers. Agricultural Nanobiotechnology. Springer, Cham, pp. 17–31.

Li, X., Xu, H., Chen, Z., Chen, G., 2011. Biosynthesis of nanoparticles by microorganisms and their applications. J. Nanomater. 2011, 1–16. Article 8.

Liang, X., Zhang, L., Kumar, N.S., Becker, D.F., 2013. Proline mechanisms of stress survival. Antioxid. Redox Signal. 19 (9), 998–1011.

Lin, D., Xing, B., 2007. Phytotoxicity of nanoparticles, inhibition of seed germination and root growth. Environ. Pollut. 150 (2), 243–250.

Mahmoodzadeh, H., Nabavi, M., Kashefi, H., 2013. Effect of nanoscale titanium dioxide particles on the germination and growth of canola (*Brassica napus*). J. Ornam. Horticult. Plants 3 (1), 25–32.

Mashilo, J., Odindo, A.O., Shimelis, H.A., Musenge, P., Tesfay, S.Z., Magwaza, L.S., 2018. Photosynthetic response of bottle gourd [*Lagenaria siceraria* (Molina) Standl.] to drought stress: relationship between cucurbitacins accumulation and drought tolerance. Sci. Hortic. (Amsterdam) 231, 133–143.

Mateos-Naranjo, E., Andrades-Moreno, L., Davy, A.J., 2013. Silicon alleviates deleterious effects of high salinity on the halophytic grass *Spartina densifora*. Plant Physiol. Biochem. 63, 115–121.

Mehdi, S., Mohammadreza, Y., Arezo, P., Mohammad, M.A., 2020. Effects of titanium and silicon nanoparticles on antioxidant enzymes activity and some biochemical properties of *Cuminum cyminum* L. under drought stress. Banat's J. Biotechnol. 11 (21), 19–25.

Mohammadi, H., Esmailpour, M., Gheranpaye, A., 2016. Effects of TiO_2 nanoparticles and waterdeficit stress on morpho-physiological characteristics of dragonhead (*Dracocephalum moldavica* L.) plants. Acta Agri. Sloven. 107 (2), 385–396.

Mozafari, A., Havas, F., Ghaderi, N., 2017. Application of iron nanoparticles and salicylic acid in in vitro culture of strawberries (Fragaria × ananassa Duch.) to cope with drought stress. Plant Cell Tissue Organ Cult. 132 (3), 511–523.

Mushtaq, A., Jamil, N., Rizwan, S., Mandokhel, F., Riaz, M., Hornyak, G.L. et al., 2018. Engineered silica nanoparticles and silica nanoparticles containing controlled release fertilizer for drought and saline areas. In: IOP Conference Series: Materials Science and Engineering, vol. 414(1). IOP Publishing, Bristol, pp. 012–029.

Osakabe, Y., Osakabe, K., Shinozaki, K., Tran, L.S., 2014. Response of plants to water stress. Front. Plant Sci. 5, 86.

Pirzadah, T.B., Malik, B., Maqbool, T., Rehman, R.U., 2019. Development of nano-bioformulations of nutrients for sustainable agriculture. In: Prasad, R., Kumar, V., Kumar, M., Choudhary, D. (Eds.), Nanobiotechnology in Bioformulations: Nanotechnology in the Life Sciences. Springer, Cham, pp. 381–394.

Popescu, M., Velea, A., Lorinczi, A., 2010. Biogenic production of nanoparticles. Digest J. Nanomater. Biostruct. 5 (4), 1035–1040.

Raliya, R., Nair, R., Chavalmane, S., Wang, W.N., Biswas, P., 2015. Mechanistic evaluation of translocation and physiological impact of titanium dioxide and zinc oxide nanoparticles on the tomato (*Solanum lycopersicum* L.) plant. Metallomics 7 (12), 1584–1594.

Rastogi, A., Zivcak, M., Sytar, O., Kalaji, H.M., He, X., Mbarki, S., et al., 2017. Impact of metal and metal oxide nanoparticles on plant: a critical review. Front. Chem. 5, Article 78.

Reddy, A.R., Chaitanya, K.V., Vivekanandan, M., 2004. Drought–induced responses of photosynthesis and antioxidant metabolism in higher plants. J. Plant Physiol. 161 (11), 1189–1202.

Saranya, S., Aswani, R., Remakanthan, Radhakrishnan, E.K., 2019. Nanotechnology in agriculture. Nanotechnology for Agriculture. Springer, Singapore, pp. 1–17.

Sattar, A., Cheema, M.A., Abbas, T., Sher, A., Ijaz, M., Hussain, M., 2017. Separate and combined effects of silicon and selenium on salt tolerance of wheat plants. Russ. J. Plant Physiol. 64, 341–348.

Sattari, S.N., Jamei, R., Eslam, B.P., Lisar, S.Y.S., 2020. Titanium dioxide nanoparticles increase resistance of *L. iberica* to drought stress due to increased accumulation of protective antioxidants. Iran. J. Plant Physiol. 10, 3343–3354.

Schulze, E.D., Beck, E., Buchmann, N., Clemens, S., Müller-Hohenstein, K., Scherer-Lorenzen, M., 2019. Water deficiency (drought). Plant Ecology. Springer, Berlin, pp. 165–202.

Seki, S.K., Reddy, K.R., Li, J., 2007. Abscisic acid and abiotic stress tolerance in crop plants. Front. Plant Sci. 7, 571.

Servin, A.D., White, J.C., 2016. Nanotechnology in agriculture: next steps for understanding engineered nanoparticle exposure and risk. NanoImpact 1, 9–12.

Seyed, M.Z., Faezeh, M., Saadat, S., Mohsen, P., 2020. Selenium and silica nanostructure-based recovery of strawberry plants subjected to drought stress. Sci. Rep. 10, 17672.

Shallan, M.A., Hassan, H.M.M., Namich, A.A.M., Ibrahim, A.A., 2016. Biochemical and physiological effects of TiO_2 and SiO_2 nanoparticles on cotton plant under drought stress. Res. J. Pharm., Biol. Chem. Sci. 7 (4), 4–13.

Silveira, N.M., Seabra, A.B., Marcos, F.C., Pelegrino, M.T., Machado, E.C., Ribeiro, R.V., 2019. Encapsulation of Snitrosoglutathione into chitosan nanoparticles improves drought tolerance of sugarcane plants. Nitric Oxide 84, 38–44.

Sun, L., Song, F., Zhu, X., Liu, S., Liu, F., Wang, Y., et al., 2020. Nano-ZnO alleviates drought stress via modulating the plant water use and carbohydrate metabolism in maize. Arch. Agron. Soil Sci. 67 (2), 245–259.

Takahashi, F., Kuromori, T., Sato, H., Shinozaki, K., 2018. Regulatory gene networks in drought stress responses and resistance in plants. Survival Strategies in Extreme Cold and Desiccation. Springer, Singapore, pp. 189–214.

Tan, W., Du, W., Barrios, A.C., Armendariz Jr, R., Zuverza-Mena, N., Ji, Z., et al., 2017. Surface coating changes the physiological and biochemical impacts of nano-TiO_2 in basil (*Ocimum basilicum*) plants. Environ. Pollut. 222, 64–72.

Tang, H., Liu, Y., Gong, X., Zeng, G., Zheng, B., Wang, D., et al., 2015. Effects of selenium and silicon on enhancing antioxidative capacity in ramie (*Boehmeria nivea* (L.) Gaud) under cadmium stress. Environ. Sci. Pollut. Res. 22, 9999.

Verma, S., Nizam, S., Verma, P.K., 2013. Biotic and abiotic stress signaling in plants, Stress Signaling in Plants: Genomics and Proteomics Perspective, 1. pp. 25–49.

Verma, S.K., Das, A.K., Patel, M.K., Shah, A., Kumar, V., Gantait, S., 2018. Engineered nanomaterials for plant growth and development: a perspective analysis. Sci. Total Environ. 630, 1413–1435.

Vishwakarma, K., Upadhyay, N., Kumar, N., Tripathi, D.K., Chauhan, D.K., Sharma, S., et al., 2018. Potential applications and avenues of nanotechnology in sustainable agriculture. Nanomaterials in Plants, Algae, and Microorganisms. Academic Press, Cambridge, pp. 473–500.

Wang, P., Lombi, E., Zhao, F.J., Kopittke, P.M., 2016. Nanotechnology: a new opportunity in plant sciences. Trends Plant Sci. 21 (8), 699–712.

Yang, K.Y., Doxey, S., McLean, J.E., Britt, D., Watson, A., Al Qassy, D., et al., 2017. Remodeling of root morphology by CuO and ZnO nanoparticles: effects on drought tolerance for plants colonized by a beneficial pseudomonad. Botany 96 (3), 175–186.

Yin, C., Peng, Y., Zang, R., Zhua, Y., Li, C., 2005. Adaptive responses of *Populus kangdingensis* to drought stress. Physiol. Plant. 123 (4), 445–451.

Zhu, J.K., 2002. Salt and drought stress signal transduction in plants. Annu. Rev. Plant Biol. 53, 247–273.

Zhu, Y.X., Gong, H.J., Yin, J.L., 2019. Role of silicon in mediating salt tolerance in plants: a review. Plants 8 (6), 147.

Chapter 10

Impact of metal oxide nanoparticles against heavy metal stress in plants

Vidhya Arumugam[1,2] and Venkatachalam Vasudevan[1,3]
[1]*Department of Biotechnology, A.V.C. College (Autonomous), Mayiladuthurai, Tamil Nadu, India,* [2]*Department of Biotechnology, Bon Secours College for Women, Thanjavur, Tamil Nadu, India,* [3]*Department of Biotechnology, Bharathidasan University, Tiruchirappalli, Tamil Nadu, India*

Chapter outline

Nanometal Oxides in Horticulture and Agronomy. DOI: https://doi.org/10.1016/B978-0-323-91809-1.00015-9

10.1 Introduction

Heavy metal contamination has become a major issue in the environment as a result of rapid urbanization and industrialization. More than 10 million polluted sites exist worldwide, out of which heavy metal contamination accounts for more than half of them. Several studies have recently revealed that heavy metal pollution in farming soil is steadily growing. Heavy metal stress has a remarkable effect on plant development. Furthermore, an excess of heavy metals leads to systemic bone weakening, nervous system damage, reproductive failure, and even mortality. According to the Agency for Toxic Substances and Disease Registry and the US Environmental Protection Agency, Cd, Pb, As, and Hg are among the top 20 harmful compounds. To limit crop production, heavy metal primarily affects normal cell structure, the antioxidant system, and plant growth. Heavy metal contaminants such as lead (Pb), cadmium (Cd), and mercury (Hg) are difficult to remove from soil and water because chemical reactions do not quickly break down them into benign components. Contamination of agricultural land and water with Pb and Cd metal ions is one of the most severe global concerns due to its direct impact on ecosystems and human health (Malar et al., 2014b). Cd and Pb have been identified as the most common dangerous heavy metals found in the environment, owing to their negative effects on human health and significant crop production reduction (Ekmekci et al., 2009). Phytoremediation is a technique that employs plants to remediate soil or water by absorption, uptake, accumulation in the rhizosphere, and translocation into the aerial parts (Malar et al., 2014a). According to studies, certain hyperaccumulator plants could be used to successfully remediate harmful heavy metals from contaminated soil and water sources utilizing phytoremediation technology (Malar et al., 2014a,b; Manikandan et al., 2015). However, due to their resistance to different hazardous components, heavy metal deposition in biomass can range dramatically amongst plant species. Heavy metal toxicity has a significant impact on plant biomass growth, photosynthetic rate, and development (Sharma and Dubey, 2005). It also involves changes in physiological and biochemical activities such as changes in normal metabolic enzyme activities, water movement imbalances, altered cell membrane permeability, and mineral nutrition uptake inhibition. One of the most prevalent stress effects of heavy metal exposure in plants is the formation of reactive oxygen species (ROS), which can alter metabolism through oxidative cell damage (Choudhury et al., 2013). In cells, oxidative stress can cause lipid peroxidation, membrane damage from excessive hydrogen peroxide production, protein degradation, and DNA molecule harm, among other things (Malar et al., 2014a,b; Manikandan et al., 2015). Plants, on the other hand, contain effective heavy metal detoxification processes that activate a variety of antioxidative enzymes such as superoxide dismutase (SOD), catalase (CAT), peroxidase (POD), and ascorbate peroxidase (APX). Pb- and Cd-induced oxidative stress has previously been linked to physiochemical and molecular changes, including lower seed germination, reduced seedling development and photosynthetic pigment levels, water imbalance, and changes in genomic DNA (Gangwar and Singh, 2011). As a result, there has been a surge of interest in employing nanoparticles (NPs) as adsorbents in the medium to reduce heavy metal phytotoxicity. NPs are aggregates of molecules with dimensions of $>$ 100 nm (Roco, 2003). NPs of various particle sizes, geometries, and functionalities can be synthesized to meet specific requirements. NPs have a number of benefits over typical materials, including high surface activity, more surface reaction sites, high catalytic efficiency, and distinctive optical and magnetic features. NPs have been proven to be effective in agri cultural and environmental applications. NPs have been shown in several studies to promote plant seed germination, photosynthesis, oxidative stress resistance, rhizome growth and development, crop yield, and quality. When compared to typical fertilizers, NPs are used as nanofertilizers and nanopesticides since they are easily absorbed by plants and slowly released into the environment. NPs (such as CeO_2, TiO_2, and Mn_3O_4), on the other hand, can increase the activity of antioxidant enzymes, reducing the accumulation of ROS in plants, easing stress, and boosting quality and yield. Another method for increasing plant resistance to heavy metals is foliar spraying. Through the foliar application, selenium and silicon NPs, for example, relieved Cd and Pb stress in maize. Considering the significant levels of heavy metal stress that affect plant growth around the world, this chapter will focus on the stress management strategies used by plants through the application of various metal and metal oxide NPs (MNPs) and the possible ways of interaction between the NPs and the heavy metal that stressed the plant.

10.2 Heavy metal stress in plants

Heavy metals are nonbiodegradable, persistent inorganic chemical constituents with an atomic mass greater than 20 and a density greater than 5 g/cm^3 that cause cytotoxicity, genotoxicity, and mutagenesis in humans, animals, and plants. Essential micronutrients for normal plant growth (Fe, Mn, Zn, Cu, Mg, Mo, and Ni) and nonessential elements with uncertain biological and physiological functions (Fe, Mn, Zn, Cu, Mg, Mo, and Ni) are the two types of metals present in soils (Cd, Sb, Cr, Pb, As, Co, Ag, Se, and Hg). Heavy metals can be absorbed by plants on both their subsurface and aboveground surfaces. In the structure of enzymes and proteins, the essential elements are the key components that play a critical function. Plants require them in trace amounts for their growth, metabolism, and development; nevertheless, the concentration of both necessary and nonessential metals is a major determinant in the growing process of plants, and their presence in excess can result in growth inhibition. Bioactive metals are separated into two classes depending on their physicochemical properties: redox-active metals like Cr, Cu, Mn, and Fe, and nonredox active metals like Cd, Ni, Hg, Zn, and Al. By performing Haber–Weiss and Fenton reactions, redox metals can directly cause oxidative injury in plants, resulting in cell homeostasis disturbance, DNA-strand breakage, protein defragmentation, cell membrane damage, and damage to photosynthetic pigments, which can lead to cell death. Nonredox active metals, on the other hand, cause oxidative stress indirectly through a variety of processes, including glutathione depletion, protein sulfhydryl group binding, antioxidative enzyme inhibition, and induction of ROS-producing enzymes such as NADPH oxidases. When confronted with the stressful condition created by high concentrations of heavy metals, plants deploy a variety of intrinsic and extrinsic defense methods for tolerance or detoxification. Plants use an avoidance strategy to avert the development of stress by reducing or eliminating metal uptake from the soil, avoiding metal access into the plant root. This can be accomplished through a variety of processes, including metal immobilization through mycorrhizal association, metal sequestration, or complexation by the exudation of organic molecules from the root. If these strategies fail and heavy metals manage to enter plant tissues, tolerance mechanisms for detoxification are activated, which include metal sequestration and compartmentalization in various intracellular compartments (e.g., vacuole), metal ions trafficking, metal binding to the cell wall, biosynthesis or accumulation of osmolytes and osmoprotectants, and, finally, if all of these methods fail and plants become overwhelmed by heavy metal toxicity, antioxidant defense systems are activated.

10.2.1 Physiological effect of nanoparticles on plants

The physiology of plants is greatly influenced by NPs. They alter plant photosynthesis by affecting pigments, photosystems, and leaf protein contents (Laware and Raskar, 2014; Tripathi et al., 2015; Tayyab et al., 2016), and they affect physiological processes by altering ROS and antioxidant metabolism (Laware and Raskar, 2014; Tripathi et al., 2015; Rico et al., 2015; Shweta et al., 2016; Tayyab et al., 2016). In *Bacopa monnieri*, for example, AgNPs enhanced protein and carbohydrate content while decreasing CAT and POD activity and total phenol content (Krishnaraj et al., 2012). By removing ROS from the plant cells, Song et al. (2012) found that the application of TiO_2 NPs at 200 mg/mL increased the activities of POD and SOD and increased the content of chlorophyll, but the effect was reversed at 500 mg/mL and the cell membrane broke down. Upregulation of light absorbance, photosynthetic efficiency, and protection of chloroplast against aging are some of the other physiological functions of TiO_2 NPs (Yang et al., 2006). This may be because TiO_2 NPs increase the activity of antioxidant enzymes, protecting the chloroplast from extremely bright light (Hong et al., 2005). Similarly, foliar application of zinc oxide (ZnO) NPs at a concentration of 10 ppm improved leaf protein (soluble), phosphorus, and chlorophyll content in cluster beans (Raliya and Tarafdar, 2013). Zinc NPs have the tendency to improve the length of the root, its fresh and dry weight, stem lengths, and total biomass of plants. In *Oryza sativa* L., zinc NPs have been shown to regulate growth and physiological responses (Mankad et al., 2017; Sabir et al., 2014; Upadhyaya et al., 2016, 2017b). Application of various concentrations of Fe_2O_3 NPs increased seedling germination, growth, and physiological parameters of watermelon (Li et al., 2013); activities of CAT, SOD, and POD were also increased compared to control treatments; as a result, NPs play a critical role in the regulation of physiological processes in plants. Other NPs, such as carbon nanotubes, could act as artificial antennae for chloroplast, allowing it to capture UV, infrared, and green light (Giraldo et al., 2014; Shweta et al., 2017). According to studies, NPs changed the plant photosynthesis process by boosting the light absorption by photosystem I and drastically reducing energy transfer (Tayyab et al., 2016; Giraldo et al., 2014). By incorporating NPs into leaves, which can boost photosynthesis, leaves can be transformed into biochemical sensors. Because NPs have a significant impact on various physiological processes in plants based on their type and concentration, a thorough understanding of their physiochemical, electrical, biological, and other properties is critical in the development of nanomaterials that modulate these plant processes and find applications in improving crop productivity for long-term agricultural sustainability. Some of the examples of physiological impacts of NPs on plants are given in Table 10.1. Table 10.1 summarizes some of the physiological impacts of NPs on plants.

10.2.2 Effect of nanoparticles on reactive oxygen species and antioxidant system

According to the results of scientific studies, the NPs may interfere with many metabolic activities, causing influence on plants, dependent on their movement, characteristics, and reactivity. Alkoxy radical (RO), singlet oxygen ($^1O_2$), hydroxyl radical (OH), superoxide radical (O_2^-), and peroxy radical (ROO) are all particularly toxic to plant cells (Dismukes et al., 2001; Karuppanapandian et al., 2006a,b,c, 2008, 2009, 2011; Karuppanapandian and Manoharan, 2008; Vellosillo et al., 2010). ROS are produced by cellular metabolism or respiration, and in normal circumstances, antioxidative defense systems made up of enzymic and nonenzymic molecules scavenge the ROS (Navrot et al., 2007). However, the ROS generation is enhanced during stress and ROS are not removed properly, causing the plant cell to get disrupted. ROS attacks all of the cell's macromolecules, causing the metabolic system to malfunction and the cell to die (Karuppanapandian et al., 2011). Abiotic stress creates an overproduction of ROS, which causes substantial cell damage (Karuppanapandian et al., 2006a,b,c, Mittler, 2002, 2006). The accumulation of ROS causes the oxidation of proteins, lipids,

TABLE 10.1 Physiological effect of some nanoparticles in plants.

S. No.	Type of metal oxide nanoparticle	Effects on plants and underlying mechanism
1.	Silver	• Exerts intracellular ROS • Induces oxidative stress • Affects electron relay center • Induces secondary quinone production • Destroys photosynthetic pigments • Causes biochemical imbalance by depleting carbohydrate pool • Inhibits electron transfer • Causes genotoxicity and chromosome damage
2.	Gold	• Decreases chlorophyll production • Causes necrotic cell death
3.	Titanium	• Genotoxic mode of action • Disrupts antioxidant system in cells by generation of intracellular ROS
4.	Zinc	• Developmental gene expression affected • Cellular toxicity through intracellular ROS generation • Cell membrane destabilization
5.	Iron	• Damages aquaporins • Cytotoxicity in *Cucurbita pepo*
6.	Copper	• Causes DNA damage and lesion • Intracellular ROS generation
7.	Aluminum	• Damages mitochondrial protein • Expression of microRNA and transcription of the gene affected
8.	Magnesium	• Cytotoxic and genotoxic
9.	Calcium	• Enhances nutrition and affirms biotic stress response in plants
10.	Cadmium	• Destroys xylem structure • Affects internal cellular frameworks • Generates exorbitant intracellular ROS
11.	Manganese	• Enhances photosynthetic pathway
12.	Cobalt	• No effect on plant growth, catalase activity lowered in tomato • Root elongation inhibited in *Allium cepa*
13.	Nickel	• Phytotoxicity because of an increase in intracellular ROS level • Antioxidant profile hampered

(Continued)

TABLE 10.1 (Continued)

S. No.	Type of metal oxide nanoparticle	Effects on plants and underlying mechanism
		• Genotoxic and DNA breakage • P-53 cascade of cell death
14.	Cerium	• Intracellular ROS generation • Affects epigenetic inheritance • Scavenges superoxides • Electron transport chain efficiency increased

carbohydrates, and DNA (Arora et al., 2002; Demidchik, 2015), but NPs boost antioxidant enzyme activity. Nanoceria (cerium oxide NPs) have been proven to scavenge ROS in isolated chloroplasts and rescue plant photosynthesis from the negative effects of ROS accumulation during abiotic stress (Giraldo et al., 2014). Silicon has been shown to protect plants from abiotic stresses (Tripathi et al., 2014), and silicon NPs have been shown to improve abiotic stress tolerance by activating antioxidant enzymes and improving absorption mechanisms in plants (Tripathi et al., 2015, 2016, 2017). As a result, plants can withstand a variety of abiotic pressures (Liang et al., 2007). TiO_2 NPs protect chloroplast against intense light by boosting the activity of antioxidant enzymes (Hong et al., 2005). TiO_2 NPs are known to exhibit antioxidant and prooxidant effects. They also found that growth-promoting benefits in spinach were linked to higher amounts of chlorophyll b, soluble carbohydrates, and proline, as well as improved antioxidant enzyme activity. Broad beans can survive saline soil and grow better in these conditions after being treated with TiO_2 NPs, according to Latef et al. (2017b). Abiotic stress in *Spinacia oleracea* can be reduced by TiO_2 NPs, according to Lei et al. (2008), which lower the activities of different ROS and raise the activities of antioxidant enzymes. The effect of NPs on ROS and antioxidant metabolism is depending on the kind and concentration of NPs (Table 10.2).

10.3 Mechanism of nanoparticle uptake by plants

The NPs, after interacting with plant roots, translocate to aerial parts and accumulate in cellular or subcellular organelles. The initial step in bioaccumulation is the adsorption of NPs from the soil by plant roots. A variety of NPs showed that root adsorption led to nanoparticle accumulation, which was then followed by diffusion through plant tissues via various alterations such as crystal phase dissolution, biotransformation, and bioaccumulation. The size of the NPs is directly related to their absorption since it is a critical feature that allows them to pass through cell wall pores or plant stomata.

TABLE 10.2 Impact of some nanoparticles on ROS and antioxidant system in plants.

Serial no	Type of nanoparticle	Impact on ROS and antioxidant system	References
1.	Nanoceria (cerium oxide nanoparticles)	Scavenge ROS in isolated chloroplasts and protect plant photosynthesis from detrimental effects of ROS accumulation during abiotic stress	Giraldo et al. (2014)
2.	Si	Enhance antioxidant enzyme activation and increase plant capabilities to withstand abiotic stresses (salinity, drought, etc.)	Liang et al. (2007)
3.	TiO_2	Enhance antioxidant stress tolerance Increased antioxidant enzyme activities contribute to the reduction in hydrogen peroxide and malondialdehyde (MDA) contents under salinity	Lei et al. (2008), Latef et al. (2017a)
4.	Ag	Induce ROS (up to 10 mg/L); a significant increase in activity of SOD, CAT, and POD; a significant increase in the content of nonenzymatic antioxidants glutathione and MDA	Jiang et al. (2014 a,b)
5.	ZnO	Decreased MDA content and increased SOD, CAT, POD, and APX activities when applied under salinity stress	Latef et al. (2017a)

Furthermore, their size influences their transport processes inside plant cells [plasmodesmata (PD)] or organelles, altering their accumulation, toxicity, and transport kinetics. The type of the nanoparticle determines the surface area, aggregation, and reactivity on the cell surface or within plant structures. The NPs attached to the surface of the plant cell are directly dependent on the charge of the NPs, due to the evident negative charge of the cell wall. Following charge and particle size, understanding the fundamental structure of nanomaterials is necessary for assessing their impact on nanoparticle uptake, transport, and aggregation in plants. A careful investigation of the nature of NPs is required to understand and interpret the absorption, translocation, and accumulation processes. Monitoring and tracking NPs are critical for determining their mobility and localization to various structures and cell organelles within the plant. Small NPs (diameters of 3–5 nm) have been shown to penetrate plant roots via osmotic pressure, capillary pressures, or direct passage through root epidermal cells. The epidermal cells in the cell wall of roots are semipermeable containing tiny pores and restrict large NPs. Some NPs form new pores in the epidermal cell wall, making it easier for them to enter. NPs are apoplastically carried through extracellular gaps after crossing cell walls, allowing the xylem to flow unidirectionally upward. To enter the core vascular cylinder, NPs must symplastically cross the Casparian strip barrier. Endocytosis, pore creation, and transport are used to bind to the carrier proteins on the endodermal cell membrane. The NPs after getting internalized in the cytoplasm migrate from one cell to the next via PD. The NPs that are unable to internalize clump together on the Casparian strip, whereas those that reach the xylem are transported to the shoots and then returned to the roots via the phloem. Plants may absorb NPs through the epidermal cell wall, cortical cell cytoplasm, and nucleus. The NPs can enter the leaves via the stomata or cuticles. The cuticle acts as a major leaf barrier, limiting the size of NPs of size less than 5 nm. NPs larger than 10 nm enter the plant by stomata, and their cellular transport occurs via apoplastic and symplastic routes into the plant's vascular system. The cytoplasm of the neighboring cell is preferred for the transmission of NPs (between 10 and 50 nm) (symplastic route). As a result, bigger NPs (50–200 nm) are translocated across the cells (apoplastic route). Internalized NPs are carried through the phloem sieve tubes with the sugar flow. NPs that do not adhere to the root surface of the soil aggregate can alter nutrient absorption. Seeds can absorb NPs directly by entering the coat through parenchymatic intercellular gaps, followed by diffusion in the cotyledon. Because these organs act as effective sinks for the sap, NPs can flow bidirectionally and accumulate in roots, stems, fruits, grains, and young leaves to variable degrees as a result of vascular transport through phloem. Many water nutrients and nonessential metal complexes are thought to prefer the apoplastic pathway for translocation. The effective adsorption of NPs is achieved through the foliar application of NPs. The presence of trichrome, leaf exudates, and waxes, as well as the shape and chemical makeup of the leaf, are all important elements that influence the trapping of NPs on the leaf surface. Environmental factors and characteristics of NPs may influence the rate at which NPs accumulate at the root of plants.

10.4 Uptake and translocation pathways of nanoparticles in plants

10.4.1 Foliar uptake and top-down translocation of nanoparticles in plants

There are two different exposure modes of plants to NPs, namely foliar exposure and root exposure. The cuticle is the first natural barrier against NPs entering substrate tissue because the leaves of higher plants are protected by the waxy cuticle against water loss and uncontrolled exchange of other solutes. There are two routes for the uptake of solutes across the cuticle (cuticular pathway), namely nonpolar solutes via diffusion and permeation (lipophilic pathway) and polar solutes via polar aqueous pores (hydrophilic pathway) with estimated effective size ranging from about 0.6–4.8 nm. Accordingly, NPs below 4.8 nm in size may penetrate through the cuticle via the cuticular pathway directly, whereas many studies have reported foliar uptake and accumulation of NPs larger than 5 nm. In addition to the cuticular pathway, studies have demonstrated the uptake of hydrophilic substances through stomatal openings (stomatal pathway). The morphological size of stomatal apertures is approximately 25 μm in length and 3–10 μm in width. Eichert and Goldbach (2008), using an indirect calculation method, estimated the equivalent pore radius of this pathway to be larger than 20 nm. This stomatal pathway is the only confirmed pathway of foliar uptake of NPs from the leaf surface to the internal tissues. Numerous studies support this pathway of NP uptake, including the observation of various NPs or their aggregates in leaf stomata and the deeper tissues of different plant species including *Citrullus lanatus*, *Cucurbita pepo*, *Allium porrum*, *Lactuca sativa*, and *Arabidopsis thaliana* using Transmission Electron Microscope, Confocal Laser Scanning Microscope, or μ-XRF (X-ray Flourescence Spectrometer). In addition, plant species with different leaf morphology, stomatal size, and density are expected to have different capacities for foliar uptake of NPs. NPs may undergo long-distance transport via the vascular system after entering the leaf apoplast through the stomatal pathway. Traditionally, photosynthate, sugars, and macromolecules in the leaf, including small RNA and proteins, are able to transport downward via the phloem system to shoots and roots. In general, the long-distance transport of liquids in higher plants occurs via the vascular system, which is composed of the xylem and phloem conductive tissues. In the xylem system, the direction of flow is from bottom to top (from root to shoot), whereas, in the phloem system, the flow direction is from top to bottom (from shoot to root). This plant vascular system is noncirculatory in nature, which means that materials moving downward in the phloem do not circulate back to their original sites through the xylem. Therefore, the phloem system is the only possible foliar uptake pathway for NPs to translocate from leaf to root. Wang et al. studied the foliar uptake of four MNPs of the size range 24–47 nm by watermelon. They found that the small NPs could penetrate watermelon leaves following the stomatal pathway using TEM, and the metal elements were detected in the shoots and roots; therefore, they concluded that NPs passed through the shoots and finally reached the roots through the phloem sieve tubes. Thus, NPs can transport upward through the xylem and downward through the phloem. The NPs and their transformed products translocated in roots may subsequently be exuded into the rhizosphere, and the ones translocated in leaves may be exuded into the phyllosphere, which can further influence the microbial community in the rhizosphere or the phyllosphere.

10.4.2 Integrated root uptake and translocation pathways of nanoparticles in plants

After approaching plant roots, NPs are first adsorbed onto the root surface. Plant roots have a rough surface due to the presence of root hairs, which can secrete mucilage or small molecules such as organic acids, and therefore, the root surface generally exhibits a negative charge. These features lead to NPs with the positive surface charge being more likely to be adsorbed and accumulate on the root surface. For uptake and translocation to occur, NPs must traverse a series of physiological root barriers from the surface to the xylem vessels, including the root surface cuticle, epidermis, cortex, endodermis, Casparian strip, and finally transport upward to the shoots via the xylem. The role and composition of the root surface cuticle are similar to those of the leaf surface cuticle. However, the cuticle is underdeveloped on the surface of root hairs and developing root tips of main and secondary roots; thus, NPs may therefore be directly exposed to the epidermis in these regions. When approaching the root epidermis, there are two basic pathways for root uptake and transport of NPs in higher plants. The apoplastic pathway in which NPs first penetrate the pores of the cell wall and then diffuse into the space between the cell wall and the plasma membrane or pass through the intercellular space without crossing the cell membrane has been proposed. NPs smaller than 20 nm can pass through the cell walls. However, many NPs larger than 20 nm have been observed in intercellular spaces. One possible mechanism is that NPs may induce the destruction of the cell wall and enlarge the pore size. Another plausible hypothesis is that NPs enter the intercellular spaces or even the xylem through diseased roots or physical wounding as a consequence of belowground herbivores and mechanical injuries, such as accidents during seedling transplantation. By the apoplastic pathway, NPs can pass through the epidermis, cortex, and into the endodermis, but are prevented by the Casparian strip, a belt of specialized cell wall material sealed by lipophilic hydrocarbons located in adjacent cells of the endodermis around the vascular system, to prevent the entry of macromolecules and NPs into the vascular cylinder. However, under some special conditions, NPs may enter the vascular system and avoid the Casparian strip through the root apoplast, for example, through the root tip region where the Casparian strip has not yet formed, or the lateral root junction where the Casparian strip is disconnected. The lateral root junction may be an important apoplastic pathway by which NPs enter the vascular system. The other hypothesized pathway is the symplastic route, which is a cell-to-cell transport pathway. In addition to the traverse across the cell wall, there are two barriers by which NPs may be transported through the symplastic pathway. One is penetration by NPs of the cell membrane and entry into the cytoplasm. The other is the transport of NPs to the adjacent cells after entering the cell through PD. Several hypotheses have been proposed for the transmembrane transport of NPs by plant cells, e.g., through aquaporins, interconnected ion channels, endocytosis, and broken membrane intubation. However, ion channels work for specific ions, and although aquaporins are nonselective in nature, the protein channels of aquaporins are generally below 1 nm in diameter. The hypothesis of a water or ion channel pathway for NP cellular internalization must face challenges from the basic principles of plant physiology. Broken membranes have been observed in cellular internalization of carbon nanotubes but this is a destructive invasion that may induce membranolysis and cell death. The highest feasibility transmembrane pathway for NPs is through endocytosis. Plant cell endocytosis is poorly understood compared with that in animal systems. However, recent studies on nanoparticle internalization using isolated plant cells suggest that endocytosis pathways are involved. The surface properties of NPs play a crucial role in plant cell endocytosis. The internalized NPs remained in endocytotic vesicles in the cytoplasm, with sizes ranging from 0.2 to 3 μm. Cell walls separate individual plant cells. Symplastic transport takes place intercellularly through PD. PD are channels that span the plant cell wall and enable intercellular communication by linking the cytoplasm between adjacent cells.

10.5 Impact of nanoparticles and nanometal oxides against heavy metal stress in plants

Heavy metal stress has been proven to affect the plant's morphology, physiology, and biochemistry. To resist heavy metal stress in plants, a variety of techniques are used. For example, (1) lowering the quantity of accessible heavy metals in the soil, (2) regulating the expression of heavy metal transport genes in plants, (3) boosting the ability of plant antioxidant systems and physiological activities, and (4) promoting the production of protective agents (such as root exudates, phytochelatin, and organic acids). Heavy metals in the soil can be absorbed and converted by NPs, thereby lowering heavy metal mobility and bioavailability. Fe_3O_4 NPs, for example, reduced the mobility of Cd and other heavy metals in soil. After 3 years in the field, Wang et al. found that mercapto Si NPs changed Cd into a more stable component. Furthermore, some NPs, such as hydroxyapatite NPs, can release phosphate and raise the pH of the soil, minimizing the negative effects of heavy metals in the soil. The physiological role of apoplastic barriers in plant roots is to protect the plant and control the flow of water, ions, and oxygen. NPs can impact the establishment of apoplastic barriers, perhaps lowering the quantity of heavy metal in the root. However, the apoplastic barriers alone may not be an efficient technique to limit the damage caused by heavy metal stress since plant roots have numerous ion and protein transporters in the plasma membrane of root cells, which can transport heavy metal simultaneously. Furthermore, particular NPs can influence plant metal transport genes, strengthening the plant's extracellular barrier to heavy metal interception. The majority of NPs collect in cell walls, bind to heavy metals, and render them inaccessible by forming complexes. These complexes bind to the cell surface, preventing heavy metals from migrating through plants and lowering their biological activity. Furthermore, heavy metals can chelate organic acids deposited in the cell walls of plant roots and leaves, reducing the damage caused by heavy metal stress on plants. NPs have been demonstrated to boost the creation of structural protective agents, such as exogenous Si NPs that promoted the synthesis of organic acids and reduced Cd damage to the plant. Activating the plant's oxidation defense mechanism is another way to alleviate heavy metal stress. Plants produce ROS as a result of metabolic processes. Plants, for example, continuously create ROS in chloroplasts and other portions of cells during metabolic processes such as respiration and photosynthesis. ROS function as signal molecules in growth, development, and defense at low levels. Excessive ROS accumulation, on the other hand, is damaging to cell membranes, proteins, and other biological components under stress. Antioxidant enzymes such as SOD, CAT, APX, glutathione reductase, glutathione POD, and POD scavenge ROS in plants (POD). Nonenzymatic low-molecular-weight metabolites, such as Vitamin C, Vitamin E, and polyphenols, can also scavenge ROS. Under stress, metabolic pathways related to ROS elimination will be activated. For example, shikimate-phenylpropanoid biosynthesis and galactose, alanine, aspartic acid, and ascorbate metabolism can ameliorate the oxidative stress of the plant. As a result, using NPs with antioxidant enzyme activity (such as CeO_2 NPs, Fe_3O_4 NPs, Mn_3O_4 NPs, and C_{60}) can improve the ability of plants to reduce ROS and, as a result, reduce crop growth and production losses.

10.5.1 Impact of silver nanoparticles

AgNPs have attracted much attention among NPs because of their extensive application in nanobiology and so on (Rai et al., 2018; Wijnhoven et al., 2009). AgNPs can easily reach the plants through sludge and surface water. At high concentrations, AgNPs of 25 nm were reported to rupture the cell membrane and disrupt the vacuoles of root cells (Mazumdar and Ahmed, 2011) but when present in low concentrations (up to 30 g/mL), it was unable to penetrate the root cells (Mirzajani et al., 2013). Different sizes (2−50 nm) of NPs with different penetration capacities in the highest applied concentration (100 ppm), or the different chemical properties of biologically synthesized NP, induce protein and carbohydrate synthesis and decrease total phenol contents, which can be considered a positive effect. Plant growth was enhanced by AgNPs of size 200−800 nm, whereas root and shoot growth of different plants was positively influenced by AgNP of size 35−40 nm (Pallavi et al., 2016), which could be due to the inability of large NPs to penetrate at low concentrations, as reported by Mirzajani et al. (2013). Different sizes of AgNP were utilized in several investigations, and there was a definite association between size and tolerance against heavy metal toxicity.

Smaller NPs were always reported to have a greater impact on heavy metal toxicity than bigger NPs (Yin et al., 2012; Jiang et al., 2014a,b; Cvjetko et al., 2017). An increase in ROS was reported in response to AgNP stress, which also leads to increased synthesis of antioxidant enzymes and molecules as an adaptive mechanism (Jiang et al., 2014a,b; Cvjetko et al., 2017; Tripathi et al., 2017). AgNPs have also been found to have an effect on DNA and alter gene expression in a variety of plants (Patlolla et al., 2012; Qian et al., 2013; Cvjetko et al., 2017; Saha and Dutta Gupta, 2017). Physiological effects of AgNP include decreased transpiration, chlorophyll concentration (Mirzajani et al., 2013; Qian et al., 2013; Tripathi et al., 2017), and chlorophyll fluorescence (Mirzajani et al., 2013; Tripathi et al., 2017). When *Raphanus sativus* was treated with 2 nm AgNP at a concentration of 500 mg/L, significant changes in macromolecules, lipids, proteins, lignin, pectin, and cellulose were detected (Zuverza-Mena et al., 2016). AgNP was found to alter plant hormones such as cytokinin and auxin (Yin et al., 2012; Vinkovic et al., 2017). AgNP may have a distinct effect on plants when paired with different treatments/compounds. In *Zea mays*, AgNP treatment in combination with a magnetic field improved quantitative yields, whereas nitric oxide alleviated the influence of AgNP by modulating Ag absorption, an antioxidant system, oxidative stress, and anatomical structure (Berahmand et al., 2012; Tripathi et al., 2017).

10.5.2 Impact of copper nanoparticles

Copper is an essential micronutrient for normal plant metabolism and plays a role in several physiological processes such as photosynthesis, respiration, carbohydrate distribution, and protein metabolism. The copper and copper oxide NPs play a significant role in reducing stress caused by heavy metals like cadmium in plants. In particular, when the copper NPs are treated with cadmium-stressed wheat plants, the application of copper NPs had some impact on the growth of the plant with respect to an increase in length of root and shoot compared to nontreated plants and promoted the root biomass. The treatment of wheat plants with the biogenic copper NPs significantly promoted the uptake of N and P from soil under Cd stress. The N and P contents of the wheat plants treated with copper NPs have improved significantly as compared to nontreated plants. Moreover, the uptake of Na^+ decreased and the uptake of K^+ and Ca^{2+} increased in wheat plants treated with biogenic copper NPs. The application of copper NPs was also found to be effective in increasing the residual amount of Cd and decreasing plant-available Cd in soil. Another mechanism of retention of Cd in the soil leading to its reduced availability to wheat plants is the competition between copper NPs and Cd at active transport sites of root epidermal cells. This competition mechanism is achieved by increased in-planta Cu contents of wheat plants with an increase in the level of copper NPs. Plant roots uptake Cu from the soil and transport it to shoots via xylem transport channels, where water flow might be the key factor for this root to shoot translocation of Cu (Ali et al., 2019). Plants to uptake Cd from soil use the same transport channels, and hence there exists an antagonistic competition between Cu and Cd at active sites. Furthermore, Cu as an enzyme cofactor might have activated several vital enzymes involved in the growth and development of plants under the contaminated environment. The proposed mechanism elucidates an important understanding of the strategies used by biogenic NPs to mitigate heavy metal stress in plants. Moreover, biogenic copper NPs also modulated the cellular ionic balance to mitigate Cd toxicity. This increase in the biomass of wheat plants as well as reduced acropetal Cd translocation might be attributed to the immobilization of Cd onto the large surface areas of copper NPs due to the capping of diverse protein molecules. Consequently, the green copper nanoparticle-based approach could serve as the best alternative to remote techniques used in the alleviation of Cd stress in plants. However, plenty of research insights are required to explore the biochemical and molecular pathways driving the NP-based mitigation of metal stress in plants. The understanding of such mechanisms will help us develop integrated strategies to manage heavy metal stress in agricultural soils.

10.5.3 Impact of titanium nanoparticles

Cadmium is a nonessential and toxic trace element for plants, and Cd accumulation in plant parts can severally inhibit the growth of plants by impacting photosynthesis and arresting the normal chemical processes in plants (Bączek-Kwinta et al., 2011; Rizwan et al., 2017a). Titanium NPs have both positive and negative impacts on plants (Liu et al., 2014; Rizwan et al., 2017c). The titanium NPs diminish the Cd-induced phytotoxicity in most plants and predominantly in rice. The cadmium causes adverse impacts on growth (Rizwan et al., 2017a) and reduction in biomass. The higher Cd levels in plants may accelerate the EL and malondialdehyde (MDA) production, which ultimately alters the photosynthesis in plants. Cadmium may inhibit the growth of plants by disturbing the normal physiological and molecular mechanisms in plants through enhanced production of ROS (Arao et al., 2009; Rizwan et al., 2017a). The spray of titanium NPs on plants under Cd stress can significantly ameliorate the Cd-induced negative impacts on plant growth. Numerous studies have demonstrated that the application of NPs promoted crop growth under metal stress through several mechanisms such as increasing mineral nutrition or by alleviating metal stress in crops (Tripathi et al., 2015; Chen et al., 2018; Hussain et al., 2018). The increase in biomass might also be associated with the increase in mineral nutrients in nanoparticle-treated plants (Chen et al., 2018). TiO_2 NPs improved the photosynthetic rate in soybean under Cd toxicity (Singh and Lee, 2016) and chlorophyll contents in rice (Ji et al., 2017). The main toxicity of Cd to plants is the production of ROS, which may result in the loss of membrane integrity and also damages proteins and lipids (Rizwan et al., 2017b). The enzymes in plants play a major role in protecting the plants against oxidative stress. The Cd intake in rice tissues can be diminished by foliar use of both titanium and titanium oxide NPs. TiO_2 NPs reduce copper toxicity (Wang et al., 2015b) and Cd uptake in rice (Ji et al., 2017). In contrast, soil application of TiO_2 NPs (100–300 ppm) enhanced the Cd accumulation by plants (Singh and Lee, 2016). The higher Cd accumulation in soybean might be associated with the binding of Cd with NPs and could move from roots to shoots in plants (Singh and Lee, 2016). The foliar application of NPs may also be linked with the modification of the gene expression of the plants, which reduced the metal transfer from the roots to aerial parts (Liu et al., 2014). The impacts of the NPs on the plants depend upon the doses, shapes, types of NPs, and crop species (Rizwan et al., 2017c; Noman et al., 2018). Therefore, NPs should be applied carefully with the aim to improve crop growth and abiotic stress alleviation. The foliar application of NPs minimized the oxidative stress considerably and had positive effects on photosynthesis and gas exchange.

10.5.4 Impact of cadmium nanoparticles

Because of the advantages of green approaches over chemical methods, the utilization of plants for the manufacture of cadmium oxide NPs (CdO NPs) has attracted a lot of attention. In the laboratory, cadmium sulfide NPs (CdS NPs) can be synthesized by chemical methods and biological methods. Biosynthesis of CdS NPs using various plant products and microorganisms had gained much more importance than the other methods as it is less expensive and more environment friendly. Most of the chemical synthesis methods of CdS NPs use cadmium nitrate [Cd $(NO_3)^2$] and sodium sulfide (Na_2S) as precursors. In this case, cadmium nitrate acts as a source of Cd^{+2} and sodium sulfide as source of S^{-2}. For the bioreduction of metals, including CdO, various components of the plant, such as leaves, flowers, fruits, roots, and seeds, are used. The terpenoid and flavanone components of leaf extract considerably contribute to the stability of CdO NPs. Extracellular polymeric substances (EPS) can serve as binding sites for various metal ions and also act as a capping agent in the synthesis of NPs. The surface functionalization of EPS can enhance the adsorption of metal. Among the various functional groups, sulfur is mostly used because of its high stability constant, low solubility products, and easy preparation procedure. The sulfur group can easily bind with the cadmium ions in the aqueous medium and synthesizes CdS NPs. The CdS NPs incorporated into the functionalized EPS have a greater adsorption potential for metal ions than the functionalized EPS. The EPS with CdS NPs not only has better adsorption capacity but also has shown additional benefits like easy recovery, easy synthesis, absence of secondary pollutants, cost-effectiveness, and environmental-friendly. Thus, by further advancement of biotechnological and nanotechnological approaches, tolerance against stress created due to different heavy metals in plants could be improved by the phenomenon of enhanced adsorption and recovery carried out by cadmium and cadmium oxide NPs.

10.5.5 Impact of cerium nanoparticles

Rare-earth NPs, such as CeO_2 NPs, have a considerable impact on plant tissues and can help plants tolerate heavy metal stress. The lowering of heavy metal stress is based on the first oxidation of heavy metals such as arsenic. CeO_2 NPs can greatly reduce the amount of heavy metals that are translocated from plant roots. Application of CeO_2 NPs to plants under heavy metal stress might increase the formation of root exudates, which can contribute electrons to induce redox processes. This can lower heavy metal phytoavailability and improve plant development. Higher root exudate production is thought to be aided by the presence of CeO_2 NPs in the plant's roots. Although CeO_2 NPs can be taken up apoplastically by plants, dissolved CeO_2 NPs are transported considerably more efficiently in the shoots from roots. Heavy metals can also be adsorbent on the surfaces of NPs. The bioavailability of heavy metals will also be lowered as a result of this method.

10.5.6 Impact of aluminum nanoparticles

miRNA is considered to be one of the key factors in plants to develop stress responses against heavy metals. miRNAs are a newly discovered class of small regulatory RNAs that act in posttranscriptional gene regulation by either marking mRNAs for cleavage or by preventing ribosomal translation of the messenger RNA into a protein. miRNAs are highly conserved; therefore, they are thought to play a major role in biotic and abiotic stress responses. miRNAs have been shown to play key roles in plant response to many environmental stressors such as drought, salinity, and heavy metals. If the concentration of Al_2O_3 NPs is increased, nine miRNAs were significantly upregulated (miR159, miR162, miR167, miR169, miR395, miR396, miR397, miR398, and miR399). The great amount of upregulation in the expression of the nine miRNAs suggests that these miRNAs play an alternative role in mediating stress response to heavy metal accumulation in tobacco, or they mediate other defensive responses as a result of the Al_2O_3 NP treatment. The two common possibilities are either the NPs will adhere to the roots impeding any uptake of other toxic heavy metals or that the NPs are uptaken and translocated within the plant, therefore causing posttranscriptional gene regulation internally. One of the widespread used aluminum NPs is aluminum oxide (Al_2O_3) NPs. The efficiency of aluminum NPs in reducing heavy metal uptake depends on the concentration, particle size, surface area, and physicochemical properties of NPs (Keller et al., 2015; Miralles et al., 2012).

10.5.7 Impact of iron nanoparticles

Iron (Fe) is an essential element for humans, animals, and plants growth and development. It plays a vital role in physiological and biochemical reactions, such as cell metabolism, photosynthesis, and respiration. For example, the synthesis of specific chlorophyll–protein complexes in chloroplasts requires Fe, and Fe deficiency causes yellowing leaves and reduced photosynthetic capacity. Iron is also a cofactor for certain enzymes, closely related to enzyme activity. Fe NPs are particularly powerful adsorbents due to their unique structure and electronic properties. The alleviation mechanisms of iron NPs mainly include adsorption of heavy metal, promotion of the formation of iron film on the root surface, activation of the oxidation defense system, and scavenging the ROS. In addition, root growth will also be enhanced by presoaking the roots in Fe_2O_3 NPs at a lower dose. Nanoscale zero-valent iron has been shown to reduce the accumulation of heavy metals in sunflowers, increase SOD and POD in plant leaves, and promote plant growth. Hussain et al. found that soil and foliar application of Fe_2O_3 NPs (5, 10, 15, and 20 ppm) to wheat under Cd stress reduced both the leaf electrolyte leakage rate and the cadmium content in grains and increased the antioxidant enzyme activity and the dry weight of the wheat. They also found that the foliar spraying of Fe NPs is preferred to the soil application because the absorption of Fe in the soil may be affected by many factors, such as pH, and the combination with other minerals during the absorption process.

10.5.8 Impact of zinc nanoparticles

Zinc (Zn) is a trace element that is required by both plants and humans. ZnO NPs are commonly used MNPs in personal care products, paints, and coatings. It works by giving nutritional value, activating the oxidative stress system, and improving the physiological and biochemical processes of the plant. According to Ali et al., combining the use of ZnO NPs (foliar spraying) and biochar (soil application) significantly reduced Cd stress, which was more effective than using either ZnO NPs or biochar alone. It shows that combining the use of NPs with other materials could be a promising strategy for reducing heavy metals in plants. Both ZnO NPs and Zn^{2+} dramatically reduced arsenic concentration in rice roots and shoots. ZnO NPs, on the other hand, merely reduced Cd accumulation in rice shoots, whereas Zn^{2+} increased Cd accumulation in shoots. Wheat biomass, yield, and photosynthesis, which decreased under normal conditions due to Cd stress, were exacerbated by drought stress and the use of ZnO NPs relieved both Cd and drought stress (Khan et al.). Hussain et al. used foliar spray and soil application to apply different concentrations of ZnO NPs (25, 50, and 100 mg/L) to Cd-stressed wheat, and found that both applications enhanced wheat growth, photosynthesis, and grain yield under Cd stress. However, foliar treatment increased grain production and reduced Cd concentration in grains more effectively than soil application. Rizwan et al. (2017b) soaked wheat seeds in a medium containing ZnO NPs (25, 50, 75, and 100 mg/L) after 20 h in dark conditions at 25°C, and then selected seedlings with equivalent germination to be applied in cadmium-contaminated soil. When compared to the control group, wheat grain production was increased, Cd level in the grain was reduced, chlorophyll a and b content was increased, SOD and POD activities were increased dramatically, and the electrolyte leakage rate was decreased. These findings demonstrate that foliar application of ZnO NPs to wheat seeds at an adequate dose can improve heavy metal resistance (cadmium) in wheat plants.

10.5.9 Impact of nickel nanoparticles

Metal-based NPs are of two types: monometallic NPs and bimetallic NPs. Magnetic NPs contain iron (Fe), cobalt (Co), and nickel (Ni) and the 4s outermost electrons and the unsaturated 3d electron shell in magnetic metal elements are responsible for the magnetic and catalytic properties (Hyeon, 2003; Kang et al., 2011). The application of bimetallic NPs for alleviation of heavy metal stress is one of the latest advances in environmental nanotechnology. Ni–Fe NPs were utilized for the dechlorination of trichloroethylene, and the observed reactivity was found to be comparable to the Pd–Fe NPs. Reports also suggested that Ni–Fe bimetallic NPs immobilized on cation exchange resin were widely utilized for heavy metal reduction. The effects of different concentrations of nickel oxide NPs (NiO NPs) on plant development, oxidative stress, and antioxidant activities in *Nigella arvensis* L. hydroponically grown tissues were examined. The NiO NPs can influence an increase in the plant biomass at a minimum concentration (50 mg/L) of NiO NPs treatment. The NiO NPs trigger the enzymatic antioxidants such as APX, CAT, SOD, and POD in roots and shoots and exhibit a greater heavy metal reduction activity after exposure to 50–2500 mg/L of NiO NPs.

10.6 Summary and future prospective

Heavy metal stress has a major impact on plant morphology, physiology, and biochemistry. This chapter summarizes the research development in improving heavy metal stress resistance in plants via seed priming, soil application, foliar spraying, and solution application of various metal NPs and MNPs. Reducing the concentration of bioavailable heavy metal in the soil, regulating the expression of heavy metal transport genes in plants, improving the ability of the apoplastic barrier to intercept heavy metal, providing more

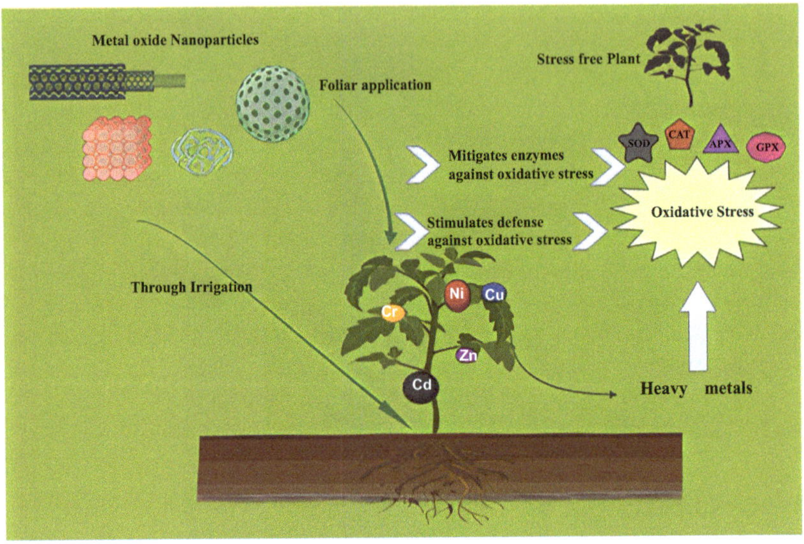

FIGURE 10.1 Impact of metal oxide nanoparticles against heavy metal stress in plants.

nutrition to plants, enhancing the ability of plant antioxidant systems, improving the function of organs, and increasing the production of structural protective agents are all important strategies for improving plant heavy metal resistance (such as root exudates, phytochelatin and organic acids). We also created a schematic diagram (Fig. 10.1) of probable NP pathways to reduce heavy metal stress and summarized the effect of different NPs on plants under different heavy metal stress (Table 10.1). With the growth of science and technology, a variety of NPs are becoming accessible on the market, including those that are surface coated or blended with other materials (such as surfactants) to improve or expand their activities. Furthermore, the physicochemical features of NPs (size and charge) that impact their interaction with heavy metal in agricultural soils need to be investigated further. Plant cell walls have a porosity of 15 nm, which allows some NPs to pass through. Furthermore, negatively charged NPs are more effective in penetrating the membrane than positively and neutrally charged NPs. In the research on alleviating heavy metal stress, the exposure level of NPs and the different application methods affect plants differently. Low dosage has no meaningful control impact, and high dosage would increase unneeded costs or have detrimental effects on plants, which is a serious problem. Furthermore, there are still questions about how to apply NPs in the soil environment. As a result, more research into the causes and treatment effects of various application modalities is required. Furthermore, combining various types of NPs with other materials is unquestionably a novel study direction. The use of NPs in combination with other alleviators such as hydrogen peroxide, nitric oxide, auxin, biochar, and heavy metal-resistant microbe strains, for example, is studied. Environmental science and other disciplines are already incorporating big data and artificial intelligence (AI) technology. Rossi et al., for example, used AI to forecast cadmium accumulation in *Brassica napus* based on physiological data of the plant. Based on the property of the soil, the variety of plants, and the property of NPs, it is expected that many physiological parameters of plants and the amount of heavy metals in plants may be predicted with the accumulation of experimental data and the advancement of technology. It is possible to choose the right types and concentrations of NPs for a certain application, cutting costs, and maximizing the benefits. Finally, one of the most important constraints limiting the use of NPs in agriculture is their high cost. As a result, producing efficient and cost-effective NPs is a crucial aspect of nanotechnology's future development. Researchers must undertake further in-depth research to understand how NPs can improve plant resistance and the impact of NPs on the terrestrial environment. Furthermore, combining NPs with computers and other disciplines will give us a more accurate application of NPs and environmental impact, allowing us to better promote nanoparticle technology for heavy metal stress control.

References

Ali, S., Rizwan, M., Noureen, S., Anwar, S., Ali, B., Naveed, M., et al., 2019. Combined use of biochar and zinc oxide nanoparticle foliar spray improved the plant growth and decreased the cadmium accumulation in rice (*Oryza sativa L.*) plant. Environ. Sci. Pollut. Res. 26, 11288–11299.

Arao, T., Kawasaki, A., Baba, K., Mori, S., Matsumoto, S., 2009. Effects of water management on cadmium and arsenic accumulation and dimethylarsinic acid concentrations in Japanese rice. Environ. Sci. Technol. 43, 9361–9367.

Arora, A., Sairam, R., Srivastava, G., 2002. Oxidative stress and antioxidative system in plants. Curr. Sci. 82, 1227–1238.

Bączek-Kwinta, R., Bartoszek, A., Kusznierewicz, B., Antonkiewicz, J., 2011. Physiological response of plants and cadmium accumulation in heads of two cultivars of white cabbage. J. Elem. 16, 355–364. Acta Physiologiae Plantarum (2019) 41:35. 1 3 Page 11 of 12 35.

Berahmand, A.A., Ghafariyan-Panahi, A., Sahabi, H., Feizi, H., Rezvani Moghaddam, P., Shahtahmassebi, N., et al., 2012. Effects silver nanoparticles and magnetic field on growth of fodder maize (*Zea mays* L.). Biol. Trace Elem. Res. 149, 419–424. Available from: https://doi.org/10.1007/s12011-012-9434-5.

Chen, R., Zhang, C., Zhao, Y., Huang, Y., Liu, Z., 2018. Foliar application with nano-silicon reduced cadmium accumulation in grains by inhibiting cadmium translocation in rice plants. Environ. Sci. Pollut. Res. 25, 2361–2368.

Choudhury, S., Panda, P., Sahoo, L., Panda, S.K., 2013. Reactive oxygen species signaling in plants under abiotic stress. Plant Signal. Behav. 8, e23681 Available from: https://doi.org/10.4161/psb.23681.

Cvjetko, P., Milošic, A., Domijan, A.M., Vrček, I.V., Tolić, S., Štefanić, P.P., et al., 2017. Toxicity of silver ions and differently coaté d silver nanoparticles in *Allium cepa* roots.

Ecotoxicol. Environ. Saf. 137, 18–28. Available from: https://doi.org/10.1016/j.ecoenv.2016.11.009.

Demidchik, V., 2015. Mechanisms of oxidative stress in plants: from classical chemistry to cell biology. Environ. Exp. Bot. 109, 212–228.

Dismukes, G.C., Klimov, V.V., Baranov, S.V., Kozlov, Y.N., DasGupta, J., Tyryshkin, A., 2001. The origin of atmospheric oxygen on Earth: the innovation of oxygenic photosynthesis. Proc. Natl. Acad. Sci. U. S. A. 98, 2170–2175.

Eichert, T., Goldbach, H.E., 2008. Equivalent pore radii of hydrophilic foliar uptake routes instomatous and astomatous leaf surfaces-further evidencefor a stomatal pathway. Physiol. Plant. 132, 491–502.

Ekmekci, Y., Tanyolac, D., Ayham, B., 2009. A crop tolerating oxidative stress induced by 706 excess pb:maize. Acta Physiol. Plant. 31, 319–330.

Gangwar, S., Singh, V.P., 2011. Indole acetic acid differently changes growth and nitrogen metabolism in *Pisum sativum* L. seedlings under chromium (VI) phytotoxicity: Implication of oxidative stress. Sci. Hortic. Amst. 129 (2), 321–328.

Giraldo, J.P., Landry, M.P., Faltermeier, S.M., McNicholas, T.P., Iverson, N.M., Boghossian, A. A., et al., 2014. Plant nanobionics approach to augment photosynthesis and biochemical sensing. Nat. Mater. 13, 400–408.

Hong, F., Yang, F., Liu, C., Gao, Q., Wan, Z., Gu, F., et al., 2005. Influence of nano TiO_2 on the chloroplast aging of spinach under light. Biol. Trace Elem. Res. 104, 249–260.

Hussain, A., Ali, S., Rizwan, M., Rehman, M.Z., Javed, M.R., Imran, M., et al., 2018. Zinc oxide nanoparticles alter the wheat physiological response and reduce the cadmium uptake by plants. Environ. Pollut. 242, 1518–1526.

Hyeon, T., 2003. Chemical synthesis of magnetic nanoparticles. Chem. Commun 8, 927–934.

Ji, Y., Zhou, Y., Ma, C., Feng, Y., Hao, Y., Rui, Y., et al., 2017. Jointed toxicity of TiO_2 NPs and Cd to rice seedlings: NPs alleviated Cd toxicity and Cd promoted NPs uptake. Plant. Physiol. Biochem. 110, 82–93.

Jiang, H.S., Qiu, X.N., Li, G.B., Li, W., Yin, L.Y., 2014a. Silver nanoparticles induced accumulation of reactive oxygen species and alteration of antioxidant systems in the aquatic plant Spirodela polyrhiza. Environ. Toxicol. Chem. 33, 1398–1405. Available from: https://doi.org/10.1002/etc.2577.

Jiang, H., Liu, J.K., Wang, J.D., Lu, Y., Zhang, M., Yang, X.H., et al., 2014b. The biotoxicity of hydroxyapatite nanoparticles to the plant growth. J. Hazard. Mater. 270–271.

Kang, K.-M., Kim, H.-W., Shim, I.-W., Kwak, H.-Y., 2011. Extracellular biosynthesis of silver nanoparticles by the culture supernatant of *Bacillus licheniformis*. Mater. Lett. 62, 4411–4413.

Karuppanapandian, T., Manoharan, K., 2008. Uptake and translocation of tri- and hexa-valent chromium and their effects on black gram (*Vigna mungo L.* Hepper cv. Co4) roots. J. Plant. Biol. 51, 192–201.

Karuppanapandian, T., Sinha, P.B., Kamarul, H.A., Manoharan, K., 2006a. Differential antioxidative responses of ascorbate-glutathione cycle enzymes and metabolites to chromium stress in green gram (*Vigna radiata L.* Wilczek) leaves. J. Plant. Biol. 49, 440–447.

Karuppanapandian, T., Sinha, P.B., Kamarul, H.A., Premkumar, G., Manoharan, K., 2006b. Aluminium-induced changes in antioxidative enzyme activities, hydrogen peroxide content and cell wall peroxidase activity in green gram (*Vigna radiata L.* cv. Wilczek) roots. J. Plant. Biol. 33, 241–246.

Karuppanapandian, T., Sinha, P.B., Premkumar, G., Manoharan, K., 2006c. Chromium toxicity: correlated with increased in degradation of photosynthetic pigments and total soluble protein and increased peroxidase activity in green gram (*Vigna radiata L.*) seedlings. J. Swamy Bot. Club 23, 117–122.

Karuppanapandian, T., Saranyadevi, A.R., Jeyalakshmi, K., Manoharan, K., 2008. Mechanism, control and regulation of leaf senescence in plants. J. Plant. Biol. 35, 141–155.

Karuppanapandian, T., Sinha, P.B., Kamarul, H.A., Manoharan, K., 2009. Chromium-induced accumulation of peroxide content, stimulation of antioxidative enzymes and lipid peroxidation in green gram (*Vigna radiata L.* cv. Wilczek) leaves. Afr. J. Biotechnol. 8, 475–479.

Karuppanapandian, T., Wang, H.W., Prabakaran, N., Jeyalakshmi, K., Kwon, M., Manoharan, K., et al., 2011. 2,4- dichlorophenoxyacetic acid-induced leaf senescence in mung bean (*Vigna radiata L.* Wilczek) and senescence inhibition by co-treatment with silver nanoparticles. Plant. Physiol. Biochem. 49, 168–177.

Keller, C., Rizwan, M., Davidian, J.C., Pokrovsky, O.S., Bovet, N., Chaurand, P., et al., 2015. Effect of silicon on wheat seedlings (*Triticum turgidum L.*) grown in hydroponics and exposed to 0 to 30 μM Cu. Planta 241, 847–860.

Krishnaraj, C., Jagan, E.G., Ramachandran, R., Abirami, S.M., Mohan, N., Kalaichelvan, P.T., 2012. Effect of biologically synthesized silver nanoparticles on Bacopamonnieri (Linn.) Wettst. plant growth metabolism. Process. Biochem. 47, 651–658.

Latef, A.A.H.A., Alhmad, M.F.A., Abdelfattah, K.E., 2017a. The possible roles of priming with ZnO nanoparticles in mitigation of salinity stress in lupine (*Lupinus termis*) plants. J. Plant. Growth Regul. 36 (1), 60–70.

Latef, A., Hamed, A.A., Srivastava, A.K., El-sadek, M.S.A., Kordrostami, M., Tran, L.S.P., 2017b. Titanium dioxide nanoparticles improve growth and enhance tolerance of broad bean plants under saline soil conditions. Land. Degrad. Dev. Available from: https://doi.org/10.1002/ldr.2780.

Laware, S.L., Raskar, S., 2014. Effect of titanium dioxide nanoparticles on hydrolytic and antioxidant enzymes during seed germination in onion. Int. J. Curr. Microbiol. Appl. Sci. 3 (7), 749–760.

Lei, Z., Su, M.Y., Wu, X., Liu, C., Qu, C.X., Chen, L., et al., 2008. Antioxidant stress is promoted by nano-anatase in spinach chloroplasts under UV-beta radiation. Biol. Trace Elem. Res. 121, 69–79.

Li, J., Chang, P.R., Huang, J., Wang, Y., Yuan, H., 2013. Physiological effects of magnetic iron oxide nanoparticles towards watermelon. J. Nanosci. Nanotechnol. 13, 5561–5567.

Liang, Y., Sun, W., Zhu, Y.G., Christie, P., 2007. Mechanisms of silicon mediated alleviation of abiotic stresses in higher plants: a review. Environ. Pollut. 147, 422–428.

Liu, C., Wei, L., Zhang, S., Xu, X., Li, F., 2014. Effects of nanoscale silica sol foliar application on arsenic uptake, distribution and oxidative damage defense in rice (*Oryza sativa* L.) under arsenic stress. RSC Adv 4, 57227–57234.

Malar, S., Favas, P.J.C., Sahi, S.V., Venkatachalam, P., 2014b. Effect of lead on 748 phytotoxicity, growth, biochemical alterations and its role on genomic template stability 749 in Sesbania grandiflora: A potential plant for phytoremediation. Ecotoxicol. Environ. Saf. 108, 249–257.

Malar, S., Sahi, S.V., Paulo, J.C.F., Venkatachalam, P., 2014a. Lead heavy metal toxicity induced changes on growth and antioxidative enzymes level in water hyacinths [Eichhornia crassipes (Mart.)]. Bot. Stud. 55, 54–65.

Mankad, M., Fougat, R.S., Patel, A., Mankad, P., Patil, G., Subhash, N., 2017. Physiological and Biochemical Effects of Zinc Oxide Nanoparticles on Rice. (Oryza sativa L.)

Manikandan, R., Sahi, S.V., Venkatachalam, P., 2015. Impact assessment of mercury accumulation and biochemical and molecular response of Mentha arvensis: a potential hyperaccumulator plant. ScientificWorldJournal 2015, 761, 715217. Available from: http://dx.doi.org/10.1155/2015/715217.

Mazumdar, H., Ahmed, G.U., 2011. Phytotoxicity effect of silver nanoparticles on *Oryza sativa*. Int. J. ChemTech. Res. 3, 1494–1500.

Miralles, P., Church, T.L., Harris, A.T., 2012. Toxicity, uptake, and translocation of engineered nanomaterials in vascular plants. Environ. Sci. Technol. 46 (17), 9224–9239.

Mirzajani, F., Askari, H., Hamzelou, S., Farzaneh, M., Ghassempour, A., 2013. Effect of silver nanoparticles on *Oryza sativa L.* and its rhizosphere bacteria. Ecotoxicol. Environ. Saf. 88, 48–54. Available from: https://doi.org/10.1016/j.ecoenv.2012.10.018.

Mittler, R., 2002. Oxidative stress, antioxidants and stress tolerance. Trends Plant. Sci. 7, 405–410.

Mittler, R., 2006. Abiotic stress, the field environment and stress combination. Trends Plant. Sci. 11 (1), 15–19.

Navrot, N., Roubier, N., Gelbaye, E., Jacquot, J.P., 2007. Reactive oxygen species generation and antioxidant systems in plant mitochondria. Physiol. Plant. 129, 185–195.

Noman, M.T., Ashraf, M.A., Ali, A., 2018. Synthesis and applications of nano-TiO_2: a review. Environ. Sci. Pollut. Res. Available from: https://doi.org/10.1007/s11356-018-3884-z.

Pallavi, Mehta, C.M., Srivastava, R., Arora, S., Sharma, A.K., 2016. Impact assessment of silver nanoparticles on plant growth and soil bacterial diversity. Biotech 6, 254. Available from: https://doi.org/10.1007/s13205-016-0567-7.

Patlolla, A.K., Berry, A., May, L., Tchounwou, P.B., 2012. Genotoxicity of silver nanoparticles in Vicia faba: a pilot study on the environmental monitoring of nanoparticles. Int. J. Environ. Res. Public Health 9, 1649–1662. Available from: https://doi.org/10.3390/ijerph9051649.

Qian, H., Peng, X., Han, X., Ren, J., Sun, L., Fu, Z., 2013. Comparison of the toxicity of silver nanoparticles and silver ions on the growth of terrestrial plant model *Arabidopsis thaliana*. J. Environ. Sci. 25, 1947–1955. Available from: https://doi.org/10.1016/S1001-0742(12)60301-5.

Rai, P.K., Kumar, V., Lee, S.S., Raza, N., Kim, K.H., Ok, Y.S., et al., 2018. Nanoparticle-plant interaction: implications in energy, environment, and agriculture. Environ. Int. 119, 1–19.

Raliya, R., Tarafdar, J.C., 2013. ZnO nanoparticle biosynthesis and its effect on phosphorous-mobilizing enzyme secretion and gum contents in cluster bean (*Cyamopsistetragonoloba L.*). Agric. Res. 2 (1), 48–57. 107.

Rico, C.M., Peralta-Videa, J.R., Gardea-Torresdey, J.L., 2015. Chemistry, biochemistry of nanoparticles, and their role in antioxidant defense system in plants. Nanotechnology and Plant Sciences. Springer International Publishing, pp. 1–17.

Rizwan, M., Ali, S., Adrees, M., Ibrahim, M., Tsang, D.C.W., Rehman, M.Z., et al., 2017a. A critical review on effects, tolerance mechanisms and management of cadmium in vegetables. Chemosphere 182, 90–105.

Rizwan, M., Ali, S., Hussain, A., Ali, Q., Shakoor, M.B., Rehman, M.Z., et al., 2017b. Effect of zinc-lysine on growth, yield and cadmium uptake in wheat (*Triticum aestivum L.*) and health risk assessment. Chemosphere 187, 35–42.

Rizwan, M., Ali, S., Qayyum, M.F., Ok, Y.S., Adrees, M., Ibrahim, M., et al., 2017c. Effect of metal and metal oxide nanoparticles on growth and physiology of globally important food crops: a critical review. J. Hazard. Mater. 322, 2–16.

Roco, M.C., 2003. Broader societal issue of nanotechnology. J. Nanopart. Res. 5, 181–189.

Sabir, S., Arshad, M., Chaudhari, S.K., 2014. Zinc oxide nanoparticles for revolutionizing agriculture: synthesis and applications. Sci. World J. 2014.

Saha, N., Dutta Gupta, S., 2017. Low-dose toxicity of biogenic silver nanoparticles fabricated by Swertia chirata on root tips and flower buds of *Allium cepa*. J. Hazard. Mater. 330, 18–28. Available from: https://doi.org/10.1016/j.jhazmat.2017.01.021.

Sharma, P., Dubey, R.S., 2005. Lead toxicity in plants. Braz. J. Plant Physiol. 17, 35–52.

Shweta, Tripathi, D.K., Singh, S., Singh, S., Dubey, N.K., Chauhan, D.K., 2016. Impact of nanoparticles on photosynthesis: challenges and opportunities. Mater. Focus. 5 (5), 405–411.

Shweta, Vishwakarma, K., Sharma, S., Narayan, R.P., Srivastava, P., Khan, A.S., et al., 2017. Plants and carbon nanotubes (CNTs) interface: present status and future prospects. Nanotechnology. Springer, Singapore, pp. 317–340.

Singh, J., Lee, B.K., 2016. Influence of nano-TiO_2 particles on the bioaccumulation of Cd in soybean plants (*Glycine max*): a possible mechanism for the removal of Cd from the contaminated soil. J. Environ. Manage. 170, 88–96.

Song, G., Gao, Y., Wu, H., Hou, W., Zhang, C., 2012. Physiological effect of anatase TiO_2 nanoparticle on Lemna minor. Environ. Toxicol. Chem. 31, 2147e2152.

Tayyab, Q.M., Almas, M.H., Jilani, G., Razzaq, A., 2016. Nanoparticles and plant growth dynamics: a review. J. Appl. Agric. Biotechnol. 1 (2), 14–22.

Tripathi, D.K., Singh, V.P., Gangwar, S., Prasad, S.M., Maurya, J.N., Chauhan, D.K., 2014. Role of silicon in enrichment of plant nutrients and protection from biotic and abiotic stresses. Improvement of Crops in the Era of Climatic. Springer, New York, pp. 39–56.

Tripathi, D.K., Singh, V.P., Prasad, S.M., Chauhan, D.K., Dubey, N.K., 2015. Silicon nanoparticles (SiNp) alleviate chromium (VI) phytotoxicity in Pisum sativum (L.) seedlings. Plant. Physiol. Biochem. 96, 189–198.

Tripathi, D.K., Singh, S., Singh, V.P., Prasad, S.M., Chauhan, D.K., Dubey, N.K., 2016. Silicon nanoparticles more efficiently alleviate arsenate toxicity than silicon in maize cultivar and hybrid differing in arsenate tolerance. Front. Environ. Sci. 4, 1–14.

Tripathi, D.K., Singh, S., Singh, S., Srivastava, P.K., Singh, V.P., Singh, S., et al., 2017. Nitric oxide alleviates silver nanoparticles (AgNps)-induced phytotoxicity in Pisum sativum seedlings. Plant. Physiol. Biochem. 110, 167–177. Available from: https://doi.org/10.1016/j.plaphy.2016.06.015.

Upadhyaya, H., Roy, H., Shome, S., Tewari, S., Bhattacharya, M.K., Panda, S.K., 2017b. Physiological impact of zinc nanoparticle on germination of rice (*Oryza sativa* L.) seed. J. Plant Sci. Phytopathol. 1, 062e070.

Upadhyaya, H., Shome, S., Tewari, S., Bhattacharya, M.K., Panda, S.K., 2016. Zinc nanoparticles induced comparative growth responses in rice (*Oryza sativa* L.) cultivars. In: Paul, S., Tewari, S. (Eds.), Frontiers of Research in Physical Sciences pp. 71e77.

Vellosillo, T., Vicente, J., Kulasekaran, S., Hamberg, M., Castresana, C., 2010. Emerging complexity in reactive oxygen species production and signaling during the response of plants to pathogens. Plant. Physiol. 154, 444–448.

Vinkovic, T., Novák, O., Strnad, M., Goessler, W., Jurašin, D.D., Parađiković, N., et al., 2017. Cytokinin response in pepper plants (*Capsicum annuum L.*) exposed to silver nanoparticles. Environ. Res. 156, 10–18. Available from: https://doi.org/10.1016/j.envres.2017.03.015.

Wang, Y., Peng, C., Fang, H., Sun, L., Zhang, H., Feng, J., et al., 2015b. Mitigation of Cu(II) phytotoxicity to rice (*Oryza sativa*) in the presence of TiO_2 and CeO_2 nanoparticles combined with humic acid. Environ. Toxicol. Chem. 34, 1588–1596.

Wijnhoven, S.W.P., Peijnenburg, W.J.G.M., Herberts, C.A., Hagens, W.I., Oomen, A.G., Heugens, E.H.W., et al., 2009. Nano-silver—a review of available data and knowledge gaps in human and environmental risk assessment. Nanotoxicology 3, 109–138. Available from: https://doi.org/10.1080/17435390902725914.

Yang, F., Hong, F., You, W., Liu, C., Gao, F., Wu, C., Yang, P., 2006. Influence of nano-anatase TiO_2 on the nitrogen metabolism of growing spinach. Biol. Trace Elem. Res. 110 (2), 179e190.

Yin, L., Colman, B.P., McGill, B.M., Wright, J.P., Bernhardt, E.S., 2012. Effects of silver nanoparticle exposure on germination and early growth of eleven wetland plants. PLoS One 7, e47674. Available from: https://doi.org/10.1371/journal.pone.0047674.

Zuverza-Mena, N., Armendariz, R., Peralta-Videa, J.R., Gardea-Torresdey, J.L., 2016. Effects of silver nanoparticles on radish sprouts: root growth reduction and modifications in the nutritional value. Front. Plant. Sci. 7, 90. Available from: https://doi.org/10.3389/fpls.2016.00090.

Chapter 11

Role of metal oxide nanoparticles in the plant/rhizo microbiome

Jeyapragash Danaraj[1], Rajiv Periakaruppan[2,3], Yosuva Mariasingarayan[4] and Subasankari Kumaresan[5]

[1]*Centre for Ocean Research (DST-FIST Sponsored Centre), MoES—Earth Science and Technology Cell (Marine Biotechnological Studies), Col. Dr. Jeppiaar Research Park, Sathyabama Institute of Science and Technology, Chennai, Tamil Nadu, India,* [2]*College of Horticulture, Nanjing Agricultural University, Nanjing, Jiangsu Province, P.R. China,* [3]*Department of Biotechnology, PSG College of Arts & Science, Coimbatore, Tamil Nadu, India,* [4]*Centre for Advanced Study in Marine Biology, Annamalai University, Parangipettai, Tamil Nadu, India,* [5]*Department of Biochemistry, Dhanalakshmi Srinivasan College of Arts and Science for Women, Perambalur, Tamil Nadu, India*

Chapter outline

11.1 Introduction

In recent years, nanoparticles (NPs) and nanostructured materials represent some of the dynamic research arenas among scientific community with techno-economic input in many application domains. This prominence is due to the technological advancements in tuning up the physicochemical characteristics

Nanometal Oxides in Horticulture and Agronomy. DOI: https://doi.org/10.1016/B978-0-323-91809-1.00011-1

including melting point, thermal and electrical conductivity, wettability, light absorption, and catalytic activity resulting in enhanced performance than their bulk counterparts (Jeevanandam et al., 2018). A nanometer (nm) is a unit that represents 10^9 m in length and nanomaterials are described as materials with a length of 1–100 nm in atleast one dimension. NPs are not new to the science and environment, as human beings already exploited the reinforcement of ceramic mixtures including nanofibres and natural absetos, more than 4500 years ago (Heiligtag and Niederberger, 2013). Around the third century (BC), Egyptians prepared the first synthetic pigment named "Egyptian blue" which is a sintered mixture nano-sized glass and quartz used in some countries like Egypt, Mesopotomia, and Greece for decorative purpose that has been observed during archeological explorations. However, in the 13th and 14th century, the synthesis of metallic NPs was started using chemical methods during which the Egyptians and Mesopotamians started making glass using the metals, which was considered as the beginning of NP era (Schaming and Remita, 2015). From the late Bronze Age (1200–1000 BC), red glass has been found in Italy that was colored by the surface plasmon excitation of Cu NPs. In addition, Celtic enamels originated during the period of 400–100 BC have been reported to contain Cu NPs and cuprous oxide (Brun et al., 1991). From ancient times, nanomaterials have been used as colorant for materials, but the synthetic design and NP engineering for various applications has started only in the last few decades. Engineered NPs are offered in different size and shapes, which suit with the surface molecules and make them different from naturally occurring materials (Radad et al., 2012). In recent years, engineered NPs have been employed in the cosmetics industries, electronic devices, antimicrobial paints, and textile industries, and these NPs are designed to be highly reactive, and they have advanced features, not like the natural counterparts, which may be harmful to different life forms including microorganisms and animals (Luechinger et al., 2009).

11.2 Sources of metal oxide nanoparticles

Metal and metal oxide NPs reveal various physiochemical properties and are different from the native compounds in terms of their surface, optical, thermal, and electrical properties, which are synthesized by the addition of oxidizing/reducing agents (Sanchez-Dominguez et al., 2009). The source from which these metal oxide NPs are synthesized are classified into three main categories based on their origin: (i) incidental nanomaterials, which are produced incidentally as an industrial processe byproduct from vehicles engine exhaust, welding fumes, burning processes, and natural processes such as forest fires; (ii) engineered NPs, which are synthesized by human beings in order to obtain the required properties for desired applications. Engineered NPs can be categorized based on the size, structure, toxicity, and chemical composition and these NPs includes CuO, TiO_2, ZnO, $CuZnFe_2O_4$, Fe_3O_4, and Fe_2O_3, which are used primarily in a variety of applications, despite the fact that few studies reported their hazardous consequence to the environment (Li et al., 2012); (iii) nanomaterials from the natural sources such as insects, plants, animals, and human bodies. NPs have been utilized in various household, industrial, and medicinal products since the last few decades and owing to this increased significance, different industries are involved in synthesizing novel nanomaterials for the services and product improvements. The use of NPs in industries may cause an intensive effect during its release into the environment. Although many of them are useful, some are hazardous to the plants, animals, algae, and microorganisms. They may, therefore, pose threat to the environment. NPs are frequently used in foods, cosmetics, drug delivery, photonic crystals, paints, and bioremediation (Obernosterer et al., 2005; Lead and Wilkinson, 2006). It was reported that about 3000 tons of titanium oxide NP are produced every year and about 50% of which are used in personal care products (Keller and Lazareva, 2014; Cornelis et al., 2014). Manjunatha et al. (2016) also reported that copper oxide NPs induce membrane damage to *Escherichia coli* as was proved by $K+$ leakage. A study conducted by Servin et al. (2015) showed that the increased day's exposure of Zebra fish (*Danio rerio*) to titanium oxide NPs decreases the viable embryos. In contrast to the above studies, a few NPs produced in large scale have the potential benefits by making its way into the environment. Most of the NPs are susceptible to the environmental conditions such as the variation in the physical and chemical parameters during which they will change their state of aggregation, oxidation, and precipitation during the secondary phase (Levard et al., 2012). In addition, the NPs' composition may also change their properties and consequently their reactivity, penetration, and translaocation inside the plants, which in turn exhibit different responses in plants to the same NP.

11.3 Plant-soil microbiome and nanoparticles

Microorganisms are abundant and ubiquitous in nature that play diverse and vital roles in the soil and plants, which was recently proved by microbiome studies of various habitats (Cho and Blaser, 2012; Cryan and Dinan, 2019). Soil accumulates higher density of microorganisms, with up to 10^8 bacterial cells per gram of soil, which contribute considerably to the soil biomass (Islam et al., 2020). The type of soil, climatic change and edaphic factors, nutrients, and oxygen availability influence the abundance and diversity of microorganisms, because these parameters vary among the different habitats including the microbial communities as well. Literature studies on microbiomes have revealed that the archaeal and bacterial species found in various soil types and are unique (Fierer, 2017). The biomass of bacteria and fungi in the soil is reported to be 10^2-10^4 times more than the microorganisms including viruses, Archaea, and Protists (Kandel et al., 2017; Eisenhauer et al., 2017). The microbiome data showed the predominance of fungi including Agaricomycetes, Archaeorhizomycetes, Chytridiomycota, Dothideomycetes, Eurotiomycetes, Glomeromycota, and Zygomycota (Fierer, 2017) isolated from a variety of 66 soil samples of different habitats, whereas the predominant bacterial population found including Acidobacteria, Alphaproteobacteria, Betaproteobacteria, Bacteroidetes, Deltaproteobacteria, Gammaproteobacteria, Verrucomicrobia, and Proteobacteria. In the Archaea, the members of Crenarchaeota and Euryarchaeota were found higher in the samples studied and these microbes in soil play very important roles such as nutrient solubilization like zinc and phosphate solubilization, mineralization, nutrient fixation (Prakash et al., 2015; Jacoby et al., 2017), and nutrient cycling (Sahu et al., 2017).

In contrast, plant microbiomes are the microorganisms that live on various plant parts and habitate in the aerial parts of leaf, external root surface, and in the rhizosphere region (Gupta et al., 2021). Plant–microbiome interactions are more complex and are influenced by numerous factors including the plant nature, age and health, secretions, physicochemical conditions, as well the existing environment. It was reported that the plant microbiomes are species-specific and colonize in the rhizosphere region. Furthermore, microorganisms residing in the endosphere, phyllosphere, and rhizospere region may also vary considerably even under the same set of conditions (Trivedi et al., 2020). Hence, numerous studies on plant microbiome have focused on their relationship, plant-specific interaction, and vertical transmission of core microbiomes, plant endophyte behavior, and their metabolic interactions. In addition, other types of interactions occur between the plants and their microbiomes; that is, some microorganisms are harmful to the plants by causing diseases, while others are beneficial for the plants (Mansfield et al., 2012). The plant-associated microorganisms show symbiotic relationship and support their growth and development. Further, the plants insist the associated microorganisms to perform specific functions by discharging the plant exudates (Berendsen et al., 2018). In addition, the bacteria residing in the plant support nitrogen fixation, nutrient mobilization, effector molecule production (Lucke et al., 2020), micronutrients sequestration (Khan et al., 2022), stress tolerance, and defensive mechanisms against plant pathogens. Interestingly, Panke-Buisse et al. (2014) reported that the flowering time and the biomass of *Arabidopsis thaliana* are controlled by its microbiome which reside in the plants, while another study reported that the microbiome present in the rhizosphere region of *A. thaliana* plays a crucial role in the defense mechanisms when it is exposed to diseases (Berendsen et al., 2012). Pérez-Jaramillo et al. (2017) similarly observed that the microbial community associated with the *Nicotiana attenuata* plays a protective role against the wilt disease. The rhizosphere region of the wild plants is found to possess a high population of Bacteroidates, but due to the domestication, decline in the Bacteroidate populations with increased population of Actinobacteria and Proteobacteria is observed (Pérez-Jaramillo et al., 2017).

In the plant rhizosphere region, diversified form of the natural nanomaterials are found including the clay, organic materials, and other metal oxide NPs, apart from several NPs released in the soil and water from various industries to the environment. According to the reports, several million tons of metal oxide NPs including the Ag, Au, SiO_2, ZnO, TiO_2, and CNT (carbon nanotube) are synthesized at high quantity worldwide and released into the environment either during their production or during their use. Some of the nanomaterials exhibit significant usage in coatings, paints, electronic, cosmetics, and optics (Mandlik et al., 2020), and it is reported that roughly 0.1%–2.0% is released into the environment during the production phase.

11.4 Impact of metal oxide nanoparticles on soil properties

In soil, nanomaterials are usually found in the form of Si, Al, Na, Ca, Fe, Ba, Sr, Rb, as well as carbonates, silicates, sulfates, oxides, hydroxides, and phosphates (Buzea and Pacheco, 2017). The presence/accumulation of NPs in the soil are controlled by certain characteristics such as pH, soil salinity, soil structure, soil organic matter, cation exchange properties, etc., and these parameters control the behavior of any nanoelement in the soil rhizosphere (Jilling et al., 2018). Furthermore, the rhizosphere region regulates the movement of nanoelements from soil to the plants and three main types of silica NPs that occur in the soil in solid state: amorphous, crystalline, and microcrystalline. It is reported that the crystalline type of silica are used for the silicates and silica materials. The Si subsists in the form of poorly crystalline or microcrystalline types, while the amorphous forms are lithogenic or biogenic in detail and are available at small quantities in the range upto 30 mg/g of the total soil. The Si solubility in the solid stage significantly reduces its concentration in the soil solution, and its solubility depends on the density and range of silica tetrahedral (Sommer et al., 2006). Apart from the naturally occurring NPs, a variety of engineered metal and metal oxide NPs are also discarded in the soil and water due to the worldwide applications in different nanomaterials based products. Piccinno et al. (2011) estimated that a few million tons of various metal and metal oxide NPs including, Ag, Au, SiO_2, ZnO, TiO_2, and CNT are under production worldwide annually. Garner et al. (2017) developed a model to understand the dynamic fate and transport of nanomaterials and predicted that the accumulation varies in a time-dependent manner under various environmental conditions. The concentration of nanomaterials in the soil varies from one place to another due to the direct use of biosolids, irrigation with engineered nanomaterials in waste water, and nanomaterials based pestisides and fertilizers. In general, the applications of different engineered nanomaterials decide their fate in different environments, such as TiO_2 which is used in the pigments and reaches the landfills directly (Sun et al., 2014). Gottschalk et al. (2009) reported that the concentration of engineered nanomaterials was found in the soil treated with sludge, and in Europe it was estimated that ~0.3–1.3 μg/kg TiO_2 reached the soil every year, whereas in United States and Europe ~0.05 and 0.09 μg/kg/Y of ZnO, respectively. Nanomaterials at higher concentrations discharged in the soil are estimated to affect the soil characteristics, and hence the life of various nanomaterials in different environments including the soil varies with the inherent properties of the nanomaterials and with environmental characteristics.

11.5 Effect of nanoparticles in plant health and microbiome

Nanomaterials cover a diverse range of materials; however, few of them are widely used and in recent years, the environment is at risk due to the exposure of nanomaterials. They directly impact the plants and animals exist in the soil, thereby reducing the growth and development of the organisms. Metal and metal oxide NPs of TiO_2, silver, zinc oxide, cerium dioxide, copper, copper oxide, aluminum, nickel, and iron are most commonly used in industries and hence are mostly studied for their effects on different plants. Recently, nanomaterials are commonly used in the agriculture as nanofertilizers and nanopesticides (Wang et al., 2016). For example, the chitosan NPs were used to encapsulate the herbicide, which efficiently improved the growth and development of plants at higher rate (Maruyama et al., 2016). In addition, the silicon NPs have been used to deliver the DNA, proteins, and other biomolecules in plant systems (Martin-Ortigosa et al., 2014). In agriculture, nanomaterials such as nanozeolites and hydrogels help in the soil quality improvement and nanosensors are used to monitor the plant and soil health (Fraceto et al., 2016). Slomberg and Schoenfisch (2012) reported that the silica NPs are not toxic to the plants. However, other authors reported that some are toxic to the plants due to the decrease of pH. To understand the toxicity of metal and metal oxide nanomaterials to the plants and the environment, information on the analytical instruments is required with respect to size, concentration, zeta potential, and its plant uptake along with their effects. The impact of nanomaterials on plants at the physiological, morphological, and genotoxic levels needs to be understood and hence, it is of much significance to focus on the effective use of NPs in agriculture. The effects of different metal and metal oxide nanoprticles on plants are tabulated in Table 11.1.

11.5.1 Effect of silver nanoparticles

Silver NPs have gained much attention among the different NPs due to their significant use in various products including their uses in antimicrobial agents, soaps, shampoo, toothpaste, waste water treatment, and other food packaging materials (Wijnhoven et al., 2009). Due to the increasing use, the synthesis of these NPs is increasing rapidly with a range upto 150–2500

TABLE 11.1 Impact of different nanoparticles on plant systems.

Nanomaterials	Tested organisms	Dose	Impacts	References
AgNP	*Oryza sativum*	50, 500, 1000 mg/L	Rupture of cell wall and damaged vacuoles in the root cells	Mazumdar and Ahmed (2011)
AgNP	Zucchini	100–500 mg/L	Decreased phytohormone induction	Stampoulis et al. (2009)
AgNP	Zeamays	40 g/h	Improved quantitative yields	Berendsen et al. (2012)
AgNP	Ryegrass and flax	40 mg/L	Reduced germination	Yin et al. (2011)
AgNP	*Cucurbita pepo*	250, 750 mg/L	Decreased plant biomass and transpiration	Hawthorne et al. (2012)
AgNP	Barley	1.5 g/L	Reduced germination	El-Temsah and Joner (2012)
Zno	*Arabidopsis thaliana*	200–300 mg/L	Reduced photosynthesis	Wu et al. (2017)
CuONP	*Phaseolus radiatesTriticum aestivum*	200, 400, 600, 800, 1000 mg/L	Reduced seedlings; diminished growth with increased NP concentration;occurrence of root damage	Lee et al. (2008)
CuONP	*Raphanus sativusLolium perenneLolium rigidum*	10, 100, 50, 1,000 mg/L	Increased concentration resulted in DNA damage	Atha et al. (2012)
CuONP	*Lemna gibba*	600 mg/L	Decreased growth rate; reduced PSII activity and electron transport; inactivation of PSII reaction centers	Perreault et al. (2014)

(*Continued*)

TABLE 11.1 (Continued)

Nanomaterials	Tested organisms	Dose	Impacts	References
CuONP	*Oryza sativum*	2.5, 10, 50, 100, 1000 mg/L	Decreased number of thylakoids; higher accumulation of naoparticles; photosynthetic pigments' contents	Da Costa and Sharma (2016)
CuNP	Wheat	3, 10, 30, 300 mg/kg	Decreased root elongation	Adams et al. (2017)
TiO_2	*Hydrilla verticillata*	0.01, 0.1, 1, 10 mg/L	Increased catalase and glutathione reductase activity	Okupnik and Pflugmacher (2016)
TiO_2	*Allium cepa*	10, 100, 1000 mg/L	Increased genotoxicity	Demir et al. (2014)
TiO_2	*Linum usitatissimum*	100 mg/mL	Decreased root biomass and root length; reduced germination after 24 h	Clément et al. (2013)
TiO_2	*Triticum aestivum*	100, 200, 300 mg/L	Increased agronomic traits and starch condition	Jaberzadeh et al. (2013)
TiO_2	*Alium cepa*	12.5, 25, 50, 100 mg/L	Increased ROS concentration	Pakrashi et al. (2014)

tons/year (El-Temsah and Joner, 2012). It was reported that the AsNP was unable to enter the root cells of *Oryza sativa* and destroy the cell structure (Mazumdar and Ahamed, 2011; Mirzajani et al., 2013).

11.5.2 Effect of copper and copper oxide nanoparticle

Copper is one of the essential micronutrients, which is used in many enzymes and proteins, therefore playing a significant role in the plant health and nutrition. It has wide range of applications including antimicrobial agents, catalysis, gas sensors, batteries, etc. (Kandel et al., 2017). However copper NPs lead to toxicity in microbial and plant communities that resulted in 75% of growth inhibition of zuchhini. Furthermore, the morphology of the roots was found to be affected and with higher concentration it directly affected the plants with reduced growth inhibition. Vale et al. (2016) observed that copper NPs showed altered response and influenced the genetic diversity of the plants studied.

11.5.3 Influence of nanoparticles on plant and soil microbiome

Literature documented that the NPs influence the plant growth directly or indirectly by mitigating the microbial community of the soil or plant (Table 11.2). Moll et al. (2017) reported that titanium oxide and zinc oxide NPs affect the microbial communities in the soil. In addition, these NPs reduce the nitrogen fixing and methane oxidizing bacteria following the NP exposure. Grun et al. (2018) documented that titanium oxide NP is most commonly found in the soil and studied its effect on the wheat microbiome. Exposure of plants with silver NPs resulted in decrease in the nitrogen fixers,

TABLE 11.2 Effect of plant microbiome and microbes mediated interaction with the nanoparticles.

Nanomaterials	Tested organisms	Dose	Impacts	References
AgNP	Nitrogen fixing and methane oxidizing bacteria	1 mg/kg soil	Decreased population	Grun et al. (2018)
TiO_2	Rhizospheric bacteria	100 μg/mL	Decreased population	Kibbey and Strevett (2019)
CuO	Denitrifiers and nitrifiers	500 mg/kg	Growth inhibition	Zhao et al. (2020)
CeO_2	Nitrogen fixing bacteria	1 g/kg soil	Growth inhibition	Priester et al. (2012)

soil microbial biomass, and the leucine aminopeptidase activity (Grun et al., 2018). Kibbey and Strevett (2019) reported that the addition of titanium oxide and polystyrene NPs accumulated in the rhizosphere region resulted in decreased population of rhizospheric bacteria by blocking the shoot and root growth. Other toxic effects due to NP exposures involve the purine metabolism, disruption of cell membrane, DNA damage, denaturation of proteins, free radicals formation, oxidative stress induction, respiration inhibition, mutagenesis, etc. (Khan et al., 2016). Furthermore, the organisms in the rhizosphere region were found to be sensitive to various nanomaterials, and the rate of microbial growth and the bacteria capability to produce sensitivity influence the bacterial sensitivity to the NPs (McGee, 2020).

11.6 Metal and metal oxide nanoparticle interaction with plants

It is evident from the literature that the NPs are toxic to the plants at higher concentration, which is uptaken by the plants by means of translocation (Fig. 11.1). Moreover, based on their transportation, properties, and reactivity, the metal and metal oxide NPs may interfere with different metabolic activity to produce an impact on plants. Mirzajani et al. (2013) reported that NPs present at higher concentration damage the plant cell wall and plasma membrane, thereby penetrating and diffusing with the different plant processes. The NPs present in the rhizosphere region enter via the root junctions and wounds. Plant cell walls are a cellulose-based structure that allows small

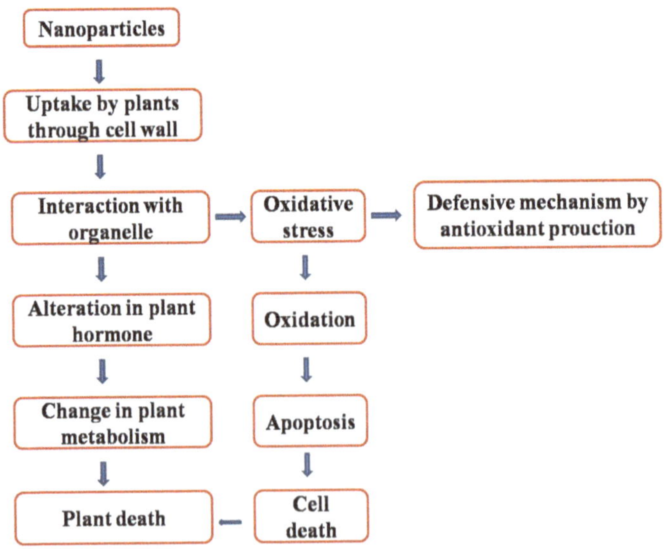

FIGURE 11.1 Mechanism of metal and metal oxide nanoparticle uptake by plants.

particles to enter while restricting larger ones; thus, smaller NPs can pass through this layer more easily than larger NPs. The plant cell wall has a size exclusion limit of 5 to 20 nm. NPs may interfere with plant metabolism in several ways, including providing micronutrients, regulating genes, and interfering with various oxidative processes in plants, which results in oxidative burst (Hossain et al., 2015). The NPs that enter the body can disrupt the electron transport chain of mitochondria and chloroplasts, resulting in an oxidative burst as measured by rise in ROS concentration (Pakrashi et al., 2014; Cvjetko et al., 2017). It has been reported previously that under the influence of various stress factors, the rate of carbon fixation is limited, causing an increase in photoinhibition, and potentially leading to overproduction of superoxide anion radicals and H_2O. (Foyer and Noctor, 2005). Once produced as a result of NP interaction, ROS is known to interact with almost all cellular components, causing protein modifications, lipid peroxidation, and DNA damage (Van Breusegem and Dat, 2006). Several studies have found that lipid peroxidation and DNA damage are increasing in plants with interaction of NPs. A high concentration of NPs may have a negative impact on photosynthesis, resulting in plant growth suppression or death. Several studies have found that NP exposure causes a significant decrease in plant growth. In some plants, the root is the primary organ for uptake of NPs from soil/water, while the shoot is unaffected (Vinkovic et al., 2017). It has previously been reported that the influence of reduced carbon fixation rate due to stress factors increases photoinhibition, potentially steering the photosystem in favor of superoxide anion radical overproduction as well as H_2O (Foyer and Noctor, 2005). Once the ROS is produced as the result of NP interaction, it is known that it interacts with nearly all cellular components, resulting in protein production alterations, lipid peroxidation, and DNA damage. Several reports have revealed that plant–NP interactions cause an increase in lipid peroxidation and DNA damage (Atha et al., 2012; Sahu et al., 2017; Cvjetko et al., 2017), which confirm that plant interaction with NPs results in lipid formation peroxidation and increased ROS production. The increase in cytokinin levels in *Capsicum annuum* in response to AgNP stress and decreases in IAA and ABA in a cotton plant in exposure to CuONP indicate that NPs influence plant hormonal balance, thereby affecting plant metabolism. As a result, it can be stated that hazardous nature and effect of NPs in plants is primarily mediated by ROS.

11.7 Conclusion

Nanomaterials are nanoscale versions of various organic and inorganic compounds with highly desirable properties. These nanomaterials can be produced by a variety of industries and sectors, including agriculture, using chemical, physical, or biological methods. The benefits of metal oxide NPs used in a wide range of crops have long been recognized, implying the significance of Si fertilization as a long-term option for sustainable agriculture. To ensure productivity, agricultural lands with strict cultivation systems, particularly those with naturally low soluble Si soils, can be modified. NPs are increasingly being used in modern life and are becoming a part of it. However, due to the demands of modern life, the environment cannot be ignored. According to the studies, metal and metal oxide NPs in excess are harmful to plants, whereas when present in trace amounts, they can be beneficial. As a result, the increasing concentration of NPs in the environment may have serious consequences for agriculture in the future. Because these nanomaterials are produced on an industrial scale, they are also released to various environments, including soil with no risk assessment. Some of the nanomaterials discarded into the soil in significant quantities include silver, SiO_2, titania, iron, iron oxides, zinc oxide, alumina, and others. Tons of these nanomaterials have accumulated in the soil as a result of their continuous and unregulated release. There have been a few studies that show the beneficial role of metal and metal oxide NPs in agriculture, but the mechanisms are not well understood, and the studies are in their early stages. As a result, extensive research is required before introducing NPs into the field. The majority of the research shows that metal and metal oxide NPs cause morphological changes in plants.

Acknowledgment

Our sincere gratitude to the Sathyabama Institute of Science and Technology (Deemed to be University) and Karpagam Academy of Higher Education (Deemed to be University) for providing the lab facility.

References

Adams, J., Wright, M., Wagner, H., Valiente, J., Britt, D., Anderson, A., 2017. Cu from dissolution of CuO nanoparticles signals changes in root morphology. Plant Physiology and Biochemistry. 110, 108–117.

Atha, D.H., Wang, H., Petersen, E.J., Cleveland, D., Holbrook, R.D., Jaruga, P., 2012. Copper oxide nanoparticle mediated DNA damage in terrestrial plant models. Environ. Sci. Technol. 46, 1819–1827.

Berendsen, R.L., Pieterse, C.M.J., Bakker, P.A.H.M., 2012. The rhizosphere microbiome and plant health. Trends Plant. Sci. 17, 478–486.

Berendsen, R.L., Vismans, G., Yu, K., Song, Y., de Jonge, R., Burgman, W.P., et al., 2018. Disease-induced assemblage of a plant-beneficial bacterial consortium. ISME J. 12, 1496–1507.

Brun, N., Mazerolles, L., Pernot, M., 1991. Microstructure of opaque red glass containing copper. J. Mater. Sci. Lett. 10, 1418–1420.

Buzea, C., Pacheco, I., 2017. Nanomaterials and their classification. In: Shukla, A.K. (Ed.), In EMR/ESR/EPR Spectroscopy for Characterization of Nanomaterials. Springer India, New Delhi, India, pp. 3–45.

Cho, I., Blaser, M.J., 2012. The human microbiome: at the interface of health and disease. Nat. Rev. Genet. 13, 260–270.

Clément, L., Hurel, C., Marmier, N., 2013. Toxicity of TiO2 nanoparticles to cladocerans, algae, rotifers and plants – Effects of size and crystalline structure. Chemosphere 90 (3), 1083–1090.

Cornelis, G., Hund-Rinke, K., Kuhlbusch, T., Brink, N.V.D., Nickel, C., 2014. Fate and bioavailability of engineered nanoparticles in soils: a review. Crit. Rev. Environ. Sci. Technol. 44, 2720–2764.

Cryan, J.F., Dinan, T.G., 2019. Talking about a microbiome revolution. Nat. Microbiol. 4, 552–553.

Cvjetko, P., A. Milošić, A., Domijan, A.M., Vinković Vrček, I., Tolić, S., Peharec Štefanić, P., Letofsky-Papst, I., Tkalec, M., Balen, B., 2017. Toxicity of silver ions and differently coated silver nanoparticles in Allium cepa roots. Ecotoxicol Environ Saf 137, 18–28.

Da Costa, M.V.J., Sharma, P.K., 2016. Effect of copper oxide nanoparticles on growth, morphology, photosynthesis, and antioxidant response in *Oryza sativa*. Photosynthetica. 54, 110–119.

Demir, A., Baykal, A., Sözeri, H., 2014. Green synthesis of Fe_3O_4 nanoparticles by one-pot saccharide-assisted hydrothermal method. Turk. J. Chem 38, 825–836.

Eisenhauer, N., Lanoue, A., Strecker, T., Scheu, S., Steinauer, K., Thakur, M.P., et al., 2017. Root biomass and exudates link plant diversity with soil bacterial and fungal biomass. Sci. Rep. 7, 44641.

El-Temsah, Y.S., Joner, E.J., 2012. Impact of Fe and Ag nanoparticles on seed germination and differences in bioavailability during exposure in aqueous suspension and soil. Environ. Toxicol. 27, 42–49.

Fierer, N., 2017. Embracing the unknown: disentangling the complexities of the soil microbiome. Nat. Rev. Microbiol. 15, 579.

Foyer, C.H., Noctor, G., 2005. Oxidant and antioxidant signalling in plants: a re-evaluation of the concept of oxidative stress in a physiological context. Plant, Cell & Environment. 28 (8), 1–12.

Fraceto, L.F., Grillo, R., de Medeiros, G.A., Scognamiglio, V., Rea, G., Bartolucci, C., 2016. Nanotechnology in agriculture: which innovation potential does it have? Front. Environ. Sci. 4, 20.

Garner, K.L., Suh, S., Keller, A.A., 2017. Assessing the risk of engineered nanomaterials in the environment: development and application of the nanofate model. Environ. Sci. Technol. 51, 5541–5551.

Gottschalk, F., Sonderer, T., Scholz, R.W., Nowack, B., 2009. Modeled Environmental Concentrations of Engineered Nanomaterials (TiO2, ZnO, Ag, CNT, Fullerenes) for Different Regions. Environ. Sci. Technol. 43, 9216–9222.

Grun, A.L., Straskraba, S., Schulz, S., Schloter, M., Emmerling, C., 2018. Long-term effects of environmentally relevant concentrations of silver nanoparticles on microbial biomass, enzyme activity, and functional genes involved in the nitrogen cycle of loamy soil. J. Environ. Sci. 69, 12–22.

Gupta, R., Anand, G., Gaur, R., Yadav, D., 2021. Plant-microbiome interactions for sustainable agriculture: a review. Physiol. Mol. Biol. Plants 27, 165–179.

Hawthorne, J., Musante, C., Sinha, S.K., White, J.C., 2012. Accumulation and phytotoxicity of engineered nanoparticles to Cucurbita Pepo. Int. J. Phytoremediation 14, 429–442.

Heiligtag, F.J., Niederberger, M., 2013. The fascinating world of nanoparticle research. Mater. Today 16, 262–271.

Hossain, Z., Mustafa, G., Komatsu, S., 2015. Plant responses to nanoparticle stress. Int. J. Mol. Sci. 16, 26644–26653.

Islam, W., Noman, A., Naveed, H., Huang, Z., 2020. Role of environmental factors in shaping the soil microbiome. Environ. Sci. Pollut. Res. 27, 41225–41247.

Jaberzadeh, A., Moaveni, P., Moghadam, H.R.T., Zahedi, H., 2013. Influence of bulk and nanoparticles titanium foliar application on some agronomic traits, seed gluten and starch contents of wheat subjected to water deficit stress. Not. Bot. Horti Agrobo 41, 201–207.

Jacoby, R., Peukert, M., Succurro, A., Koprivova, A., Kopriva, S., 2017. The role of soil microorganisms in plant mineral nutritioncurrent knowledge and future directions. Front. Plant. Sci. 8, 1617.

Jeevanandam, J., Barhoum, A., Chan, Y.S., Dufresne, A., Danquah, M.K., 2018. Review on nanoparticles and nanostructured materials: history, sources, toxicity and regulations. Beilstein J. Nanotechnol. 2018 (9), 1050–1074.

Jilling, A., Keiluweit, M., Contosta, A.R., Frey, S., Schimel, J., Schnecker, J., et al., 2018. Minerals in the rhizosphere: overlooked mediators of soil nitrogen availability to plants and microbes. Biogeochemistry 139, 103–122.

Kandel, S.L., Joubert, P.M., Doty, S.L., 2017. Bacterial endophyte colonization and distribution within plants. Microorganisms 5, 77.

Keller, A.A., Lazareva, A., 2014. Predicted releases of engineered nanomaterials: from global to regional to local. Environ. Sci. Technol. Lett. 1, 65–70.

Khan, S.T., Malik, A., Ahmad, F., 2022. Role of zinc homeostasis in plant growth. In Microbial Biofertilizers and Micronutrient Availability. Springer, New York, NY, pp. 179–195.

Khan, S.T., Musarrat, J., Al-Khedhairy, A.A., 2016. Countering drug resistance, infectious diseases, and sepsis using metal and metal oxides nanoparticles: current status. Colloids Surf., B 146, 70–83.

Kibbey, T.C.G., Strevett, K.A., 2019. The effect of nanoparticles on soil and rhizosphere bacteria and plant growth in lettuce seedlings. Chemosphere 221, 703–707.

Lead, J.R., Wilkinson, K.J., 2006. Aquatic colloids and nanoparticles: current knowledge and future trends. Env. Chem. 3, 159–171.

Lee, J., Feng, J., Campbell, K.B., Scheffler, B.E., Garrett, W.M., Thibivilliers, S., Stacey, G., Naiman, D.Q., Tucker, M.L., Pastor-Corrales, M.A., Cooper, B., 2009. Quantitative proteomic analysis of bean plants infected by a virulent and avirulent obligate rust fungus. Mol. Cell. Proteomics. 8 (1), 19–31.

Levard, C., Hotze, E.M., Lowry, G.V., Brown Jr., G.E., 2012. Environmental transformations of silver nanoparticles: impact on stability and toxicity. Environ. Sci. Technol. 46, 6900–6914.

Li, C., Adamcik, J., Mezzenga, R., 2012. Biodegradable nanocomposites of amyloid fibrils and graphene with shape-memory and enzyme-sensing properties. Nat. Nanotechnol. 7, 421.

Lucke, M., Correa, M.G., Levy, A., 2020. The role of secretion systems, effectors, and secondary metabolites of beneficial rhizobacteria in interactions with plants and microbes. Front. Plant. Sci. 11, 1718.

Luechinger, N.A., Grass, R.N., Athanassiou, E.K., Stark, W.J., 2009. Bottom-up fabrication of metal/metal nanocomposites from nanoparticles of immiscible metals. Chem. Mater. 22 (1), 155–160.

Mandlik, R., Thakral, V., Raturi, G., Shinde, S., Nikolíc, M., Tripathi, D.K., et al., 2020. Significance of silicon uptake, transport, and deposition in plants. J. Exp. Bot. 71, 6703–6718.

Manjunatha, S.B., Biradar, D.P., Aladakatti, Y.R., 2016. Nanotechnology and its applications in agriculture: a review. J. Farm. Sci. 29, 1–13.

Mansfield, J., Genin, S., Magori, S., Citovsky, V., Sriariyanum, M., Ronald, P., et al., 2012. Top 10 plant pathogenic bacteria in molecular plant pathology. Mol. Plant. Pathol. 13, 614–629.

Martin-Ortigosa, S., Peterson, D.J., Valenstein, J.S., Lin, V.S., Trewyn, B.G., Lyznik, L.A., et al., 2014. Mesoporous silica nanoparticle-mediated intracellular CRE protein delivery for maize genome editing via loxP site excision. Plant. Physiol. 164, 537–547.

Maruyama, C.R., Guilger, M., Pascoli, M., Bileshy-José, N., Abhilash, P.C., Fraceto, L.F., et al., 2016. Nanoparticles based on chitosan as carriers for the combined herbicides imazapic and imazapyr. Sci. Rep. 6, 19768.

Mazumdar, H., Ahmed, G., 2011. Phytotoxicity effect of Silver nanoparticles on *Oryza sativa*. J Chemtech. Res 3, 1494–1500.

McGee, C.F., 2020. The effects of silver nanoparticles on the microbial nitrogen cycle: a review of the known risks. Environ. Sci. Pollut. Res. 27, 31061–31073.

Mirzajani, F., Askari, H., Hamzelou, S., Farzaneh, M., Ghassempour, A., 2013. Effect of silver nanoparticles on Oryza sativa L. and its rhizosphere bacteria. Ecotoxicol. Environ. Saf. 88, 48–54.

Moll, J., Klingenfuss, F., Widmer, F., Gogos, A., Bucheli, T.D., Hartmann, M., et al., 2017. Effects of titanium dioxide nanoparticles on soil microbial communities and wheat biomass. Soil. Biol. Biochem. 111, 85–93.

Obernosterer, I., Catala, P., Reinthaler, T., Herndl, G.J., Lebaron, P., 2005. Enhanced heterotrophic activity in the surface microlayer of the Mediterranean Sea. Aquat. Microb. Ecol. 39, 293–302.

Okupnik, A., Pflugmacher, S., 2016. Oxidative stress response of the aquatic macrophyte Hydrilla verticillata exposed to TiO_2 nanoparticles. Environ Toxicol Chem. 35 (11), 2859–2866.

Pakrashi, S., Jain, N., Dalai, S., Jayakumar, J., Chandrasekaran, P.T., Raichur, A.M., 2014. In Vivo Genotoxicity Assessment of Titanium Dioxide Nanoparticles by Allium cepa Root Tip Assay at High Exposure Concentrations. PLoS ONE 9 (2), e87789.

Panke-Buisse, K., Poole, A.C., Goodrich, J.K., Ley, R.E., Kao-Kniffin, J., 2014. Selection on soil microbiomes reveals reproducible impacts on plant function. ISME J. 9, 980.

Pérez-Jaramillo, J.E., Carrión, V.J., Bosse, M., Ferrão, L.F.V., de Hollander, M., Garcia, A.A.F., et al., 2017. Linking rhizosphere microbiome composition of wild and domesticated Phaseolus vulgaris to genotypic and root phenotypic traits. ISME J. 11, 2244.

Perreault, F., Oukarroum, A., Melegari, S.P., Matias, W.G., Popovic, R., 2012. Polymer coating of copper oxide nanoparticles increases nanoparticles uptake and toxicity in the green alga Chlamydomonas reinhardtii. Chemosphere. 87 (11), 1388–1394.

Piccinno, F., Gottschalk, F., Seeger, S., Nowack, B., 2011. Industrial production quantities and uses of ten engineered nanomaterials in Europe and the world. J. Nanopart. Res. 14, 1109.

Prakash, O., Sharma, R., Rahi, P., Karthikeyan, N., 2015. Role of Microorganisms in plant nutrition and health. In: Rakshit, A., Singh, H.B., Sen, A. (Eds.), Nutrient Use Efficiency: From Basics to Advances. Springer, New Delhi, India, pp. 125–161.

Priester, J.H., Ge, Y., Mielke, R.E., Horst, A.M., Moritz, S.C., Espinosa, K., et al., 2012. Soybean susceptibility to manufactured nanomaterials with evidence for food quality and soil fertility interruption. Proc. Natl. Acad. Sci. USA 109, E2451–E2456.

Radad, K., Al-Shraim, M., Moldzio, R., Rausch, W.D., 2012. Recent advances in benefits and hazards of engineered nanoparticles. Environ. Toxicol. Pharmacol. 34, 661–672.

Sahu, N., Vasu, D., Sahu, A., Lal, N., Singh, S., 2017. Strength of microbes in nutrient cycling: a key to soil health. Agriculturally Important Microbes for Sustainable Agriculture. Springer, Singapore, pp. 69–86.

Sanchez-Dominguez, M., Boutonnet, M., Solans, C., 2009. A novel approach to metal and metal oxide nanoparticle synthesis: the oilin-water microemulsion reaction method. J. Nanopart. Res. 11, 1823.

Schaming, D., Remita, H., 2015. Nanotechnology: from the ancient time to nowadays. Found. Chem. 17, 187–205.

Servin, A., Elmer, W., Mukherjee, A., De la Torre-Roche, R., Hamdi, H., White, J.C., et al., 2015. Review of the use of engineered nanomaterials to suppress plant disease and enhance crop yield. J. Nanopart. Res. 17, 92.

Slomberg, D.L., Schoenfisch, M.H., 2012. Silica nanoparticle phytotoxicity to *Arabidopsis thaliana*. Environ. Sci. Technol. 46, 10247–10254.

Sommer, M., Kaczorek, D., Kuzyakov, Y., Breuer, J., 2006. Silicon pools and fluxes in soils and landscapes—a review. J. Plant. Nutr. Soil. Sci. 169, 310–329.

Stampoulis, D., Sinha, S., White, J.C., 2009. Assay-dependent phytotoxicity of nanoparticles to plants. Environ. Sci. Technol. 43, 9473–9479.

Sun, T.Y., Gottschalk, F., Hungerbühler, K., Nowack, B., 2014. Comprehensive probabilistic modelling of environmental emissions of engineered nanomaterials. Environ. Pollut. 185, 69–76.

Trivedi, P., Leach, J.E., Tringe, S.G., Sa, T., Singh, B.K., 2020. Plant–microbiome interactions: from community assembly to plant health. Nat. Rev. Microbiol. 18, 607–621.

Vale, G., Mehennaoui, K., Cambier, S., Libralato, G., Jomini, S., Domingos, R.F., 2016. Manufactured nanoparticles in the aquatic environment-biochemical responses on freshwater organisms: A critical overview. Aquatic Toxicology 170, 162–174.

Van Breusegem, F., Dat, J.F., 2006. Reactive oxygen species in plant cell death. Plant Physiol. 141 (2), 384–390.

Vinković, T., Novák, O., Strnad, M., Goessler, W., Jurašin, D.D., Parađiković, N., Vrček, I.V., 2017. Cytokinin response in pepper plants (*Capsicum annuum* L.) exposed to silver nanoparticles. Environmental Research 156, 10–18.

Wang, P., Lombi, E., Zhao, F.J., Kopittke, P.M., 2016. Nanotechnology: a new opportunity in plant sciences. Trends Plant. Sci. 21, 699–712.

Wijnhoven, S.W.P., Peijnenburg, W.J.G.M., Herberts, C.A., Hagens, W.I., Oomen, A.G., Heugens, E.H.W., Roszek, B., Bisschops, J., Gosens, I., Van De Meent, D., Dekkers, S., De Jong, W.H., van Zijverden, M., Sips, A.J.A.M., Geertsma, R.E., 2009. Nano-silver – a review of available data and knowledge gaps in human and environmental risk assessment. Nanotoxicology 3 (1), 109–138.

Wu, H., Tito, N., Giraldo, J.P., 2017. Anionic cerium oxide nanoparticles protect plant photosynthesis from abiotic stress by scavenging reactive oxygen species. ACS Nano 11, 11283–11297.

Yin, L., Cheng, Y., Espinasse, B., Colman, B.P., Auffan, M., Wiesner, M., et al., 2011. More than the ions: the effects of silver nanoparticles on *Lolium multiflorum*. Environ. Sci. Tecnol. 45, 2360–2367.

Zhao, S., Su, X., Wang, Y., Yang, X., Bi, M., He, Q., et al., 2020. Copper oxide nanoparticles inhibited denitrifying enzymes and electron transport system activities to influence soil denitrification and N2O emission. Chemosphere 245, 125394.

Chapter 12

"Metal oxide nanoparticle" vs. "released ions": in soil and plants

Selvam Sathish[1], Anil Kumar Moola[2,3] and Venkatachalam Vasudevan[1,4]

[1]Department of Biotechnology, Bharathidasan University, Tiruchirappalli, Tamil Nadu, India, [2]Department of Biotechnology, Aditya Degree and PG College, Kakinada, Andhra Pradesh, India, [3]Department of Entomology, College of Agriculture, Food and Environment, Agriculture Science Center North, University of Kentucky, Lexington, KY, United States, [4]Department of Biotechnology, A.V.C. College (Autonomous), Mayiladuthurai, Tamil Nadu, India

Chapter outline

12.1 Introduction

Nanotechnology is fetching a new field of research that may revolutionize the agricultural sector with novel tools to maximize crop productivity. In recent years, the fast growth of nanotechnology has opened up new avenues for manufacturing nanomaterials with desired particle sizes and shapes for effective applications in agriculture, industry, and biomedicine. It allows us to maximize the benefits of managing these resources through efficient products in the form of nanopesticides (Ahmed et al., 2021; Chen, 2018). Nanoparticles (NPs) are materials widely used in various industries, including textiles, paints, cosmetics, medicine, and agriculture. Because of their

Nanometal Oxides in Horticulture and Agronomy. DOI: https://doi.org/10.1016/B978-0-323-91809-1.00007-X

unique characteristics, such as size, surface area, tunable physical/chemical properties, molecular, mechanical, electric, thermal, catalytic, and shape, their introduction into ecosystems poses a threat to human health and food security (Ameen et al., 2021; Rastogi et al., 2017; Shafey, 2020). The application of NPs in agriculture is explored as a key tool for increasing plant productivity, controlling numerous diseases, improving plant growth, regulating metabolic processes, and improving the nutritional content of crop species. However, massive production of nano-based goods and leaching of NPs either from industrial discharge or from nano-enabled household products into diverse environments pose their sustainability, adding a vast amount of NPs to both aquatic and terrestrial environments and other biospheres (Ma et al., 2014; Eduok et al., 2015; Brown, 2017; Kulizhskiy et al., 2017). Because of the tremendous application of NPs in commercial applications, numerous industries are developing new characteristic NPs to strengthen their products and services. Certain companies that manufacture extensive usage of NPs could expect NPs to be released into the atmosphere. NPs can contaminate natural systems through various activities, including inefficient industrial waste management and disposal of goods by the consumer (Rajput et al., 2018a; Sun et al., 2021).

In the past two decades, a significant amount of research has been done to test the potential of NPs under various environmental conditions (Rossi et al., 2017; Awad et al., 2019; Rizwan et al., 2021). Among the different NPs reported, metallic NPs are still charming the scientists to explore new magnitudes for their respective worth, which is attributed to their corresponding small sizes. These NPs are easy to synthesize because of their distinctive shape, size, structure, composition, novel physicochemical properties, and encapsulation (Subbenaik, 2016). With the rapid development of nanotechnology, metal oxide NPs (MONPs) constitute a significant class of NPs that are highly used in a wide range of products. MONPs are widely explored in agricultural applications because they provide multiple actions like supply of micronutrients to plants as fertilizer, plant defense mechanism, and control of plant pathogens as pesticides. The production and use of MONPs have been extensively applied in the past decade due to the different commercial applications, including medical treatment, cosmetics, sunscreens, wastewater treatment, solar cells, electronics, semiconductors, and catalysts (Song et al., 2019; Servin et al., 2017). Earlier several authors highlighted the positive effect of MONPs on diverse plant species, including Coriandrum *sativum* (Zuverza-Mena et al., 2015), tomato and eggplant (Elmer and White, 2016), and cotton (Le Van et al., 2016a,b). In recent literature, MONPs have been reported as one of the most toxic NPs to soil biota (Parada et al., 2019). These toxic effects forced people to concern about soil health, fertility, and security issues. The overproduction and use of NPs ultimately disturb the soil and plant system, leading to adverse effects on plants and animals (Ameen et al., 2021; Shemawar et al., 2021). The toxic or beneficial impact of MONPs on seed germination, growth, and plant morphology depends on the fate and behavior of these NPs in the plant growth medium. Both are governed by intrinsic factors like chemical composition, size, shape, surface characteristics, reactivity, and efficacious doses, and environmental or external factors like growth media, the form of application, aging, plant species, and localization of MONPs in plant tissues (Ameen et al., 2021). The chapter will contribute to a better understanding of the fate and behavior of NPs in the soil, their uptake and distribution within plants, the impact of soil microbial communities, and their interactions with other pollutants, which is an essential prerequisite to environmentally realistic studies of NP ecotoxicology.

12.2 Impact of metal oxide nanoparticles on plants

The NPs widespread in the atmosphere, water, and soil interact with plants. After uptake of NPs in plant roots, they are translocated to the various plant parts and enter the food chain, which could ultimately affect the ecosystem and human health. MONPs with multifunctional properties are used extensively in various industries and are released into the natural systems as industrial effluents and waste nanoproducts. During the last two decades, a considerable effort has been conducted on metal NPs and MONPs, particularly from an agricultural perspective, because these NPs can easily slip into the plant system (Arif et al., 2016; Cartwright et al., 2020; Rizwan et al., 2017). The plants uptake NPs via mechanisms such as passive diffusion, ion or water molecular pathways, aquaporins, carrier proteins, endocytosis, and new pore formation, binding with organic matter, which depends on the type of metal ions and plant species. Plants grown on soils enriched with MONPs or derived metal molecules would possibly deposit these in their roots and eventually translocate to aerial parts, leading to transfer to the food web. The underlying mechanisms of plant response to MONPs are yet to be fully understood. The extent of NPs determines the impact of NPs on plants' (1) uptake, (2) accumulation in plant organs, and (3) subsequent translocation to various sites. These three processes also depend upon physicochemical features of NPs, genotypes, and the anatomy of plants (Landa et al., 2016; Rastogi et al., 2017).

Several MONPs contain essential nutrients for plants, potentially replacing conventional nutrient sources as slow-releasing fertilizers. For instance, zinc oxide NPs (ZnONPs) have been explored as an effective, slow-releasing Zn fertilizer to alleviate the Zn deficiency in agricultural soils. Similarly, Servin et al. (2015) demonstrated that copper oxide NPs (CuONPs) are the slow-releasing source for Cu need in long-time crops. Silicon oxide NPs (SiO_2NPs) act as a novel carrier for essential nutrients for plants to alleviate salinity stress (Mushtaq et al., 2018). Stefani and Vélez (2021) reported that MONPs could hasten seed germination, trigger aerial and radical biomass, and upsurge crop yield. ZnONPs are also widely used as antimicrobial and antigenic properties, including veterinary sciences, food storage, and killing microbes to enhance the shelf life of food products (Prasanna et al., 2019; Parada et al., 2019). Further, MONPs are also serving as effective pesticides and pathogen controlling agents due to their antimicrobial activities (Naseem and Durrani, 2021). On the other hand, plants grown under metal stress reduced the uptake of trace elements which cause increased biomass and photosynthesis due to the positive effect of NPs present in the soil (Mustafa and Komatsu, 2016; Rizwan et al., 2021). Soil is an important source and the ultimate sink for the deposition of NPs in the ecosystem. The concentration of NPs in soil has been reported to be higher than that in the atmospheric and aqueous environment (Ma et al., 2017; Nair, 2018). The accumulation of MONPs in agricultural soil may provide the most significant exposure pathway for the entry of MONPs into the food chain (Du et al., 2017). MONPs inevitably can enter the ecosystem via various manufacturing, transportation, consumer usage, and improper disposal activities. Once MONPs are discharged into the soil, they may undergo a series of transformations, which eventually determine the bioavailability and toxicity of NPs. MONPs can cause biological toxicity by releasing metal ions and generating reactive oxygen species (ROS) and can be absorbed by plants, posing a potential threat to human health via transferring in the food chain (Liu et al., 2020). Although the positive or negative effect of NPs voraciously studied in soil and plants, the functional mechanism behind the scene remains unclear and requires extensive investigation.

12.2.1 Zinc oxide nanoparticles

Zinc (Zn) and zinc oxide (ZnO) are generally categorized as commonly used metal/metal oxide engineered nanomaterials. Zn is an essential micronutrient for all ecosystems, predominantly plants, but excess load can cause necrotic spots, chlorosis, leaf discoloration, and plant growth arrest (Rajput et al., 2018b). Zn/ZnO deficiency was found to be the most common micronutrient deficiency and most important yield-limiting nutrient after nitrogen, phosphorus, and potassium (Aslani et al., 2014). Earlier several researchers documented the positive and negative role of ZnO nanomaterials on plant growth and its yield. For example, increased biomass accumulation was recorded on cluster bean leaves after spraying with ZnONPs compared to ZnO (Raliya and Tarafdar, 2013). Soliman et al. (2015) indicated that after foliar application of ZnO and Fe_3O NPs on Moringa peregrine plants exhibited improved growth parameters in contrast to control treatment. Zhao et al. (2014) reported that ZnONPs treated cucumber plants showed an increase in micronutrient uptake. However, in pea and soybean, beyond the optimum dose ZnONPs significantly hampered the content of chlorophyll (Mukherjee et al., 2014; Rao and Shekhawat, 2014). This toxic effect on the photosynthesis system is highly relying on the concentration and type of NPs. Asli and Neumann (2009) observed that maize seedlings treated with TiO_2NPs revealed a reduction in leaf growth and leaf functions via disturbing water uptake. Peanut plants exposed to different concentrations of ZnONPs exhibited increased pod biomass and kernel and shelling percentage (Prasad et al., 2012). The possible reason for the positive effect can be reduced Zn concentrations and enhanced uptake of other nutrients in peanut plants. Under hydroponic conditions, ZnONPs with 20 nm in diameter are uptaken by ryegrass roots, leading to the accumulation of Zn content in roots at 140 mg Zn/kg dry weight upon 1 g/L of ZnONPs exposure. In contrast, the accumulation of Zn content is very low in leaves which are 1.4 mg Zn/kg dry weight. This accumulation caused phytotoxicity resulting in shrinkage of the root tip, biomass reduction, vacuolization, and collapse of cortical cells and root epidermal, which are partially attributed to NP dissolution and subsequent release of toxic Zn^{2+} ions (Lin and Xing, 2008). The anatomical and ultrastructural changes of rapeseed leaves and roots exposed with ZnONPs were investigated by Mousavi kouhi et al. (2015). Root tip diameter as well as epidermal and pericycle cells size gradually decreased owing to ZnONPs. Further, these NPs also reduced the mitochondria size, plastoglobuli, and chloroplasts number. These modifications could be ascribed to the discharge of Zn ions from the surplus NPs and their interaction with plant root systems.

12.2.2 CuO nanoparticles

Copper is an essential micronutrient incorporated in many proteins and enzymes, therefore playing a significant role in plant health and nutrition. At genomic level analysis, CuONPs concentration at 0.2 mg/L had no impact on the expression of genes associated with oxidative stress, glutathione, sulfur assimilation, and proline biosynthesis in *Arabidopsis thaliana*, whereas at higher concentrations, upregulated gene expression was detected (Nair and Chung, 2014). Atha et al. (2012) noticed a massive deposition of oxidatively changed and mutagenic DNA lesions in various plants, implying DNA damage due to CuONPs treatment. CuONPs were also found to reduce electron transport, thylakoid number per grana, photosynthetic rate, transpiration rate, photosynthetic pigments, and stomatal conductance by inactivating PS II reaction centers (Perreault et al., 2014; Costa and Sharma, 2016). Hong et al. (2015) found an increase in the uptake of various micronutrients including P, Cu, and S in *Medicago sativa* shoots. P and Fe uptake were drastically reduced in *Lactuca sativa* shoot after being treated with CuONPs at 5, 10, and 20 mg/L doses. However, in another study, Adhikari et al. (2012) demonstrated that the application of CuONPs severely hindered the soybean and chickpea root growth and length of wheat seedlings. In addition, under CuONPs stress, cucumber seed germination significantly decreased to 23.3% at 600 mg/L (Moon et al., 2014). Application of CuONPs for 14 days on wheat seedlings reduced the length of shoot and root by 13% and 59%, respectively, and revealed necrotic spots on roots, and the roots were thinner and more brittle than the control. Phytotoxicity is the release of toxic Cu ions from CuONPs (Dimkpa et al., 2012). Similarly, Wang et al. (2015) found that the application of 5 mg/L CuONPs 72 h exposure inhibited biomass and root in rice plants after 72 h exposure. Barley plants exposed to 0.5, 1.0, and 1.5 mM concentrations of CuONPs suppressed the root length and shoot weight at 10 and 20 days compared to control plants (Shaw et al., 2014). The above-said literature revealed that the Cu and CuONPs at elevated concentrations are toxic to plants, and that they influence the growth, physiology, and biochemistry of plants.

12.2.3 TiO_2 nanoparticles

MONPs have been investigated for their prospective interaction in plants and soil. TiO_2NPs have been studied for their possible implications on the growth, development, antioxidant defense system, and other nutritional factors in diverse crops. TiO_2NPs are widely used in different applications and have both positive and negative effects on photosynthesis. Earlier studies said that TiO_2NPs improved the photosynthesis rate in spinach, whereas, in tomato plants, photosynthesis was adversely affected (Mingyu et al., 2007; Qi et al., 2013). Similarly, TiO_2NPs reduced the hydrogen peroxide, malondialdehyde, and electrolyte leakage in chickpea (*Cicer arietinum* L.) compared with nontreated plants. The reason for these adverse effects was the accumulation of TiO_2 on the surface of the cell wall. Administration of NPs over the leaf surface usually enters via stomatal openings, ion channels, stigma, wounds, and base of trichrome and are then translocated to diverse parts of plant tissues. However, NP accumulation on photosynthetic areas causes foliar chamber heating, which leads to gas exchange alterations owing to the stomatal barrier, which affects changes in plants' various physiological and cellular activities (Uzu et al., 2010; Fernández and Eichert, 2009). According to an earlier report, most metals can enter the plant through leaf stomata and are transported to the root through the xylem and phloem (Avellan et al., 2019).

TiO_2NPs are used as additives in pesticides and other complex formulations as active ingredients or carriers. They increase adhesion or accelerate the degradation of organic pesticides. Chand and Siddiqui (2012) stated that nano-anatase TiO_2 enhances photosynthesis by triggering the rubisco activase transcripts, which could eventually help plants grow fast. Moreover, bulk use of MONPs also creates oxidative bursts in plants by discharging their metal ions. This oxidative stress may cause ROS production, hindering numerous metabolic reactions and impairing plant photosynthesis and gas exchange capability (Rizwan et al., 2016; Das et al., 2015). However, the occurrence of TiO_2 also certainly impacts plants growth. For instance, TiO_2NPs stimulated the growth of spinach by increasing nitrogen metabolism and photosynthetic rate. Haghighi and da Silva (2014) recorded 100%

seed germination on onion and tomato upon treatment of 100 mg/L concentrations of TiO_2NPs, and 400 mg/L concentrations revealed 100% germination in radish. Treatment of TiO_2NPs enhances fennel seeds vigor index, growth, and yield of wheat plants under drought conditions compared to the control and bulk TiO_2 exposed plants (Jaberzadeh et al., 2013; Verma et al., 2018). Exposure of TiO_2NPs increased nitrate reductase, enhanced water absorption capacity, and triggered the antioxidant mechanism in soybean. The plant produced from TiO_2NPs treated seeds exhibited high dry weight, elevated photosynthetic mechanism, and higher production chlorophyll-a compared to control (Mingfang et al., 2013). Zheng et al. (2005) and Yang et al. (2006) confirmed that TiO_2NPs might have increased the absorption of the inorganic nutrient, hastened organic substances breakdown, and triggered oxygen-free radicals quenching during the photosynthetic action, subsequently resulting in a higher photosynthetic rate.

NPs can be uptaken and accumulated in any part of the plant, such as leaves, stems, and roots. After the uptake of NPs, these can be translocated in the plants through the vascular system (Raliya et al., 2015). When NPs enter the leaves, they can be transported by phloem through bidirectional pathways. When NPs are introduced into the root, xylem transports them with the help of water. These NPs can create new pores in the roots by the reaction of NPs. Accumulated metal-based NPs in roots can be transported via symplastic or apoplastic pathways to shoots (Castiglione et al., 2011). Application of 30 mg/mL TiO_2NPs increased parsley seed germination to 92% compared to the untreated control (Song et al., 2013). TiO_2NPs increased seed germination in wheat plants at 2 to 10 ppm concentration, whereas poor germination was noted on higher concentrations of 100 and 500 ppm (Feizi et al., 2012). It can be concluded that TiO_2NPs at a lesser dose could act as a seed priming agent for crops. However, researchers outlined that higher concentrations of NPs were toxic to plants and decreased the germination of seeds. However, positive results of TiO_2NPs on seed germination are highly influenced by size, concentration, exposure time, growth environments, and plant species. Thus, further studies are needed to emphasize the effect of TiO_2NPs application in agriculture crops.

12.3 Impact of metal oxide nanoparticles and ions on soil

Since NPs are biologically nondestructive, they exist in the soil system for extended periods. Their alone or combined action modifies the soil fertility, soil microflora, and the physiology and metabolism of critical plants (Fayiga and Saha, 2017; Yang et al., 2017). The release of soluble metal ions from NPs (a process of dissolution of NPs) is highly governed by the physicochemical properties of NPs and the chemistry of the ecological system. The influence of soil pH, natural organic content, ionic strength, and temperature are critical environmental factors for releasing metals in the soil. Specific

NPs are vulnerable to dissolution, that is, the release of ions from the NPs surface throughout their life cycle (Ahmed et al., 2021; Cartwright et al., 2020). NPs must be stable in a soil-like atmosphere while delivering metals or metal ions at a controlled harmless rate to target the roots zones of plants and function as effective delivery vehicles (Ahmed et al., 2021; Cartwright et al., 2020). For instance, the dissolution of ZnO and CuO nanoparticles was higher at acidic pH and lower at alkaline pH (Omar et al., 2014; Odzak et al., 2017).

Furthermore, Dimkpa et al. (2013) investigated the interactions between ZnONPs and wheat. It has been observed to cause Zn phosphate accumulation in shoots, which might be related to the dissolution of ZnONPs and internalization of Zn^{2+} ions. Gelabert et al. (2014) found that ZnONPs coating with organic material hindered the dissolution process, which took 7 days to reach its maximum, while uncoated ZnONPs displayed maximum dissolution in 1 h. It is ascribed that the surface chemistry of NPs is considered to be a significant factor in the increase or decrease rate of dissolution. A low concentration of sulfide less than 1 mg/L can initiate AgNPs to release Ag^{+} ions, which subsequently react with sulfide to produce silver sulfide (Ag_2S) nanolinkages with adjacent NPs (Liu et al., 2011).

NP deposition in soil systems may sink to the ground via soil and can directly impact human health (Mahdi et al., 2018). To eliminate organic and inorganic contaminants from groundwater, NPs have been utilized successfully (Matlochova et al., 2013). Such NPs could be an additional source of nano-pollution to groundwater. In addition to soil system and groundwater, a vast amount of engineered NPs are discharged into the atmosphere by several industrial activities and agricultural amendments, including sewage treatment plants, waste incineration, landfills, manufacturing units, power plants, and during transportation (Fig. 12.1) (Jacob et al., 2013; Mukherjee and Acharya, 2018; Gottschalk and Nowack, 2011). Once the NP dissolution is activated, the pool of reactive ions upsurges; the NPs act as reservoirs for the constant release of ions. Most NPs displayed solubility, implying that they could release the constituent metal ions. For instance, foliar application of AgNPs in plants could release silver ions and trigger toxic reactions, which include mutilation of electron transport, deactivation of key enzymes involved in the respiratory system, cytolysis, membrane permeability, and gene mutation (Hong et al., 2021; Li et al., 2017). Organic acids emitted by maize roots in the rhizosphere layer, such as glutamic and malic acids, might facilitate the discharge of Zn ions from ZnONPs bound to the soil matrix (Zhao et al., 2012). Some MONPs cause toxicity by dissolution and releasing a large concentration of essential ions such as Zn^{2+} and Cu^{2+} or by generating excess ROS (Anjum et al., 2015; Minkina et al., 2018). This ROS affects mitochondrial respiration and apoptosis, lipid peroxidation in the cell membrane, and cause oxidative stress in various plant species (Dimkpa et al., 2012). At

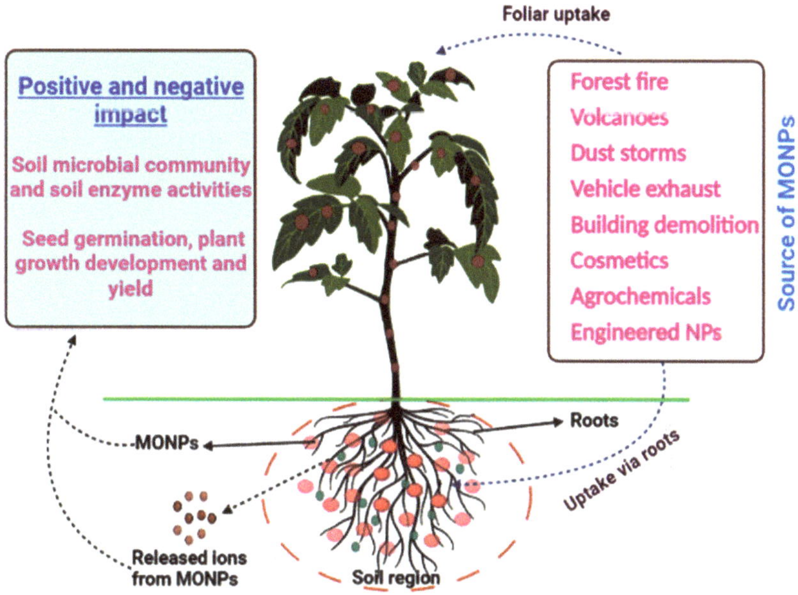

FIGURE 12.1 Interaction of metal oxide NPs and ions in plant–soil system.

present, inadequate information is available on how NPs and released ions disturb the plant and soil microbial community.

In comparison to water and air, soil ecosystems were projected as the primary sink for the deposition of engineered NPs. Numerous studies have shown that the incorporation of NPs into the soil can affect soil microbial community and soil enzyme activities (You et al., 2018; Ameen et al., 2021; Zhang et al., 2020). Indeed, some MONPs were reported to have positive effects on plants (Table 12.1). According to Shrestha et al. (2015), multiwalled carbon nanotubes considerably reduced the cytotoxicity and bioavailability of polycyclic aromatic hydrocarbons in the soil microbial biomass, and the effects were reliant on the soil organic quantity. He et al. (2011) found that metal NPs have stimulated the invertase and soil urease activities which could be a reason for the modifications in the bacterial community. Du et al. (2011) reported that TiO_2NPs significantly promoted the urease activity in soil than the control, while ZnONPs reduced the protease and catalase activities. It may be hypothesized that the decrease of these enzyme activities is attributed to the release of dissolved ions in soil from ZnONPs, which could reduce the nitrogen fixation in soil microorganisms. However, these NPs also discovered some undesirable impacts in the soil system. CuONPs at 10 mg/g reduced bacterial hydrolytic activity, oxidative potential, and changes in community composition in sandy loam soil, particularly for the Bacilli class abundance, and decreased up to 38% compared to the control (Frenk et al., 2013). Ge et al. (2011) evidenced

TABLE 12.1 Impact of metal oxide nanoparticles in plants.

NPs	Size	Plant species	Effect (positive/negative)	References
CuO	10–100	Lettuce	Increased shoot biomass	Wang et al. (2019)
ZnO	Not specified	*Triticum aestivum*	Biomass, grain yield, and Zn concentration in grain increased when compared with plants treated with bulk $ZnSO_4$	Du et al. (2019)
ZnO	10	*Cucumis sativus*	The concentration of sugars and glutelin in the fruit was higher than these in the control treatment	Zhao et al. (2014)
ZnO	20	*Cicer arietinum*	The aerial biomass, radical, and root length increased 27%, 37%, and 53% respectively, compared to the control treatment	Mahajan et al. (2011)
ZnO	Not specified	*Cicer arietinum*	The dry weight of leaves increased	Burman et al. (2013)
ZnO	1.2–6.8	*Cyamopsis tetragonoloba*	Significant improvement in plant length, biomass accumulation, photosynthetic pigments, and proteins soluble in leaves, rhizospheric microbial population, and enzymatic activity of acid and alkaline phosphatase increased	Raliya and Tarafdar (2013)
ZnO	Not specified	*Solanum lycopersicum*	The germination rate, seedling vigor, the pigment level, proteins, and sugar increased	Singh et al. (2016)
ZnO	20	*Triticum aestivum*	Enhancement in Zn concentration in grain	Zhang et al. (2018)
ZnO	18	*Triticum aestivum*	15% decrease in grain yield with simultaneous internalization of Zn in grain by 229% and 186% under fresh and weathered ZnO nanoparticle exposure	Dimkpa et al. (2018)
ZnO		*Arachis hypogaea*	The germination, seedling vigor, stem and root growth, as well as pod yield, were elevated compared to the control treatment and the treatment with bulk $ZnSO_4$	Prasad et al. (2012)

ZnO	18	*Sorghum bicolor*	Enhanced accumulation of nitrogen and potassium	Dimkpa et al. (2017)
TiO_2		*Brassica napus*	The germination and seedling index increased 75% and 1.6, respectively, in comparison with plants without the addition of NPs	Mahmoodzadeh et al. (2013)
ZnO	20 ± 5	*Lolium perenne*	Low biomass and root length and decrease in seedling biomass	Lin and Xing (2008)
ZnO	80	*Vicia faba*	Root cells indicated chromosomal aberrations and alterations in the cell cycle. Enzyme systems displayed an altered expressed pattern	Youssef and Elamawi (2020)
TiO_2	>50	*Allium cepa*	Reduced the mitotic index by 60%	Debnath et al. (2020)
TiO_2	25 ± 0.6	*Solanum lycopersicum*	Enhanced plant growth, root length, and biomass and chlorophyll content	Raliya et al. (2015)
TiO_2	100	*Triticum aestivum*	Reduction in biomass, NPs found on root surface	Du et al. (2011)
TiO_2	90–98	*Allium cepa*	Increase in ROS activity and genotoxic effects	Pakrashi et al. (2014)
TiO_2	27 ± 4	*Cucumis sativus*	Enhanced K and P availability in fruit	Servin et al. (2013)
ZnO	16–30	*Vigna* sp	At low concentration of NPs higher nitrogenase activity in the four legumes tested	Kumar et al. (2015)
ZnO	<10	*Zea mays*	Reduction in root length, and alteration in root anatomy	Pokhrel and Dubey (2013)
ZnO	10–125	*Pisum sativum*	Decreased chlorophyll content by 77%	Mukherjee et al. (2014)
ZnO	<50	*Brassica napus*	Decreased root, shoot length, and dry massed of plant	Mousavi Kouhi et al. (2014)

(*Continued*)

TABLE 12.1 (Continued)

NPs	Size	Plant species	Effect (positive/negative)	References
ZnO	<100	*Brassica juncea*	Decreased root, shoot length, and biomass, ROS and antioxidant enzymes activity increased	Rao and Shekhawat (2014)
ZnO	2–54	*Gossypium hirsutum*	Improved photosynthetic pigment levels, biomass, and protein contents; improved the activity of antioxidant defense enzymes; enhanced SOD and POX isoenzymes expression levels	Venkatachalam et al. (2017)
CuO	<50	*Oryza sativa*	Reduced percentage of germination, loss of viability of root cells, and oxidative stress	Shaw and Hossain (2013)
CuO	Not specified	*Triticum aestivum*	Enhanced biomass accumulation	Dimkpa et al. (2012)
CuO	<50	*Glycine max L.*	Reduced root, shoot growth, weight, and chlorophyll and fresh weights were drastically reduced at all concentrations	Nair and Chung (2014)
CuO	<30	*Gossypium hirsutum*	Negative effect on plant height, number of root hairs, root length, and shoot biomass with higher dose	Nhan et al. (2015)
Cu_2O	Not specified	*Lycopersicm esculentum*	Improved seed germination, leaf pigment, and increased root/shoot elongation	Ananda et al. (2019)
ZnO	<50	*Arabidopsis thaliana*	Growth inhibition, low chlorophyll content, reduction of photosynthetic rate, stomatal conductance, and intracellular CO_2	Wang et al. (2016)

the effect of TiO_2 and ZnONPs soil bacterial communities over 60 days and observed decreased biomass and altered the composition of the soil bacterial community. You et al. (2018) described that ZnONPs, when released into the soil, could harm the microbes and significantly decrease the microbial community size in saline-alkali soil after 30-day incubation. Further study should focus on the characterization of the dissolution of the NPs in soil and its influence on soil micronutrient bioavailability.

12.4 The mechanism behind metal oxide nanoparticles uptake by plants and soil interaction cycle

The recent rapid development of nanotechnologies raises questions regarding the uptake and translocations of NPs in the natural systems and their possible health and environmental threats. MONPs are nontoxic at particular concentrations, and it contains several essential elements like Zn, Cu, Fe, Ti, and Mn NPs. Metal and MONPs can convert or disintegrate in soil owing to the solubility of ZnO and CuO. Penetration of NPs into plant tissues seems to be influenced by particle size, and the maximum size, usually 40–50 nm, is permitted to enter and accumulate inside the cells (Pérez-de-Luque, 2017; Tripathi et al., 2017). For instance, small MONP size range 24–47 nm could penetrate watermelon leaves following the stomatal pathway. Therefore, authors concluded that NPs entered via shoots touched the roots through phloem sieve tubes (Wang et al., 2013). The uptake of NPs in plants is influenced by a series of factors such as particle size, surface functionalization, morphology, exposure conditions, plant species, plant growth stage, root integrity, and rhizosphere processes. Roots are in direct interaction with NPs, and they can acquire them from the soil and deliver them to aerial parts of plants. This uptake is facilitated by permeable and thinner cuticles of roots and the cell wall of root hairs (Banerjee et al., 2019). Earlier several researchers explored the uptake and accumulation of several NPs by plant root cells (Li et al., 2017; Vithanage et al., 2017). The uptake of NPs could be accelerated by transpiration, with a positive connection between the rate of water consumed and the uptake of NPs (Rico et al., 2013). For instance, along with water uptake through the xylem, the migration of CuONPs from roots to shoot of maize is observed under transmission electron microscopy (TEM) and energy-dispersive X-ray of xylem sap (Wang et al., 2013).

On the other hand, CeO_2NPs of $>$ 7 nm in diameter are uptaken by other crops, including alfalfa, tomato, cucumber, and corn (Lopez-Moreno et al., 2017). The CeO_2NPs having a diameter of more than 7 nm up to 25 nm are absorbed by cucumber roots and travel to their shoots. These studies suggest that the difference in NP uptake mode could be attributed to NP size and root pore diameter. Foliar spraying of NPs is deposited on the surface of the leaf and is able to permeate immediately inside the plant system owing to nano size and gaseous absorption by plants (Hong et al., 2021; Wang et al., 2013).

Xylem and phloem act as vehicle distribution and translocation of NPs from one organ to other plant tissues. The mechanism behind the uptake of NPs is usually considered an active-transport mechanism that includes other cellular operations such as recycling, signaling, and plasma membrane regulation (Lv et al., 2019; Etxeberria et al., 2016). However, the uptake and translocation process varies from NP to NP. For instance, TiO_2NPs in cucumber roots are translocated to leaves and fruits without bioconversion (Servin et al., 2013). The translocation of NPs is generally influenced by four facets: (1) size of NP, (2) surface area and charge, (3) growth phase of the plant, and (4) plant cell inside environment. In general, the NP translocation process ensues by symplastic or apoplastic pathways. In the symplastic pathway, NP transportation occurs intracellularly after moving the adjacent cell through plasmodesmata and sieve plates. In apoplastic pathways, initially, NPs pass through the pores of the cell wall and then diffuse into the space between the cell wall and the plasma membrane or enter longitudinal channels in the cell wall, extracellular space beyond touching the cell membrane (Lv et al., 2019; Pérez-de-Luque, 2017). Extensive investigations indicated that NPs could cross through plant cell walls and are localized inside cytosol or cell organelles. For example, Servin et al. (2012, 2013) proved that TiO_2NPs were translocated to cucumber leaf trichomes, and fruits were confirmed by micro-X-ray near edge spectroscopy analysis. Another study by Nhan et al. (2015) detected the presence of CeO_2NPs in various organelles, including vacuole, chloroplast, and plasma membrane of cotton plants grown under a hydroponic system.

12.5 Conclusion and future prospective

Metal ions and MONPs reduced or improved plant growth and development of various plants species. Plant response varied dramatically across NPs and was partially connected to concentration and the size of the NPs. Although NPs have a positive or negative effect on plants growth, the mechanisms underlying this are still unexplored, exclusively in soil. More studies are needed to clarify the aspect of bioavailability of NPs and metal ions in soils and their distribution and migration. This will deliver a better understanding of the toxic or benign accurate soil concentrations of various metal-based NPs. However, future research should emphasize understanding the molecular mechanism of cellular pathways, and therefore, further efforts need to be directed on the impact of MONPs on plant physiology and molecular biology. Considering the structural and functional toxicity of metal NPs to beneficial soil microbes, the strategies for safe disposal of metal NPs in soil agro-ecosystems are endorsed, and safe by design approaches need to be developed to minimize the contact of NPs with beneficial soil microbes.

References

Adhikari, T., Kundu, S., Biswas, A.K., Tarafdar, J.C., Rao, A.S., 2012. Effect of copper oxide nano particle on seed germination of selected crops. J. Agric. Sci. Technol. 2, 815.

Ahmed, B., Rizvi, A., Ali, K., Lee, J., Zaidi, A., Khan, M.S., et al., 2021. Nanoparticles in the soil–plant system: a review. Environ. Chem. Lett.

Ameen, F., Alsamhary, K., Alabdullatif, J.A., ALNadhari, S., 2021. A review on metal-based nanoparticles and their toxicity to beneficial soil bacteria and fungi. Ecotoxicol. Environ. Saf. 213, 112027.

Ananda, S., Shobha, G., Shashidhara, K.S., Mahadimane, V., 2019. Nano- cuprous oxide enhances seed germination and seedling growth in *Lycopersicum esculentum* plants. J. Drug. Deliv. Ther. 9, 296–302.

Anjum, N.A., Adam, V., Kizek, R., Duarte, A.C., Pereira, E., Iqbal, M., et al., 2015. Nanoscale copper in the soil-plant system toxicity and underlying potential mechanisms. Environ. Res. 138, 306–325.

Arif, N., Yadav, V., Singh, S., Singh, S., Ahmad, P., Mishra, R.K., et al., 2016. Influence of high and low levels of plant-beneficial heavy metal ions on plant growth and development. Front. Environ. Sci. 4.

Aslani, F., Bagheri, S., Muhd Julkapli, N., Juraimi, A.S., Hashemi, F.S.G., Baghdadi, A., 2014. Effects of engineered nanomaterials on plants growth: an overview. Sci. World J. 2014.

Asli, S., Neumann, P.M., 2009. Colloidal suspensions of clay or titanium dioxide nanoparticles can inhibit leaf growth and transpiration via physical effects on root water transport. Plant, Cell Environ. 32, 577–584.

Atha, D.H., Wang, H., Petersen, E.J., et al., 2012. Copper oxide nanoparticle mediated DNA damage in terrestrial plant models. Environ. Sci. Technol. 46, 1819–1827.

Avellan, A., Yun, J., Zhang, Y., Spielman-Sun, E., Unrine, J.M., Thieme, J., et al., 2019. Nanoparticle size and coating chemistry control foliar uptake pathways, translocation, and leaf-to-rhizosphere transport in wheat. ACS Nano 13, 5291–5305.

Awad, Y.M., Vithanage, M., Niazi, N.K., Rizwan, M., Rinklebe, J., Yang, J.E., et al., 2019. Potential toxicity of trace elements and nanomaterials to Chinese cabbage in arsenic-and lead-contaminated soil amended with biochars. Environ. Geochem. Health. 41, 1777–1791.

Banerjee, K., Pramanik, P., Maity, A., Joshi, D.C., Wani, S.H., Krishnan, P., 2019. Methods of using nanomaterials to plant systems and their delivery to plants (Mode of entry, uptake, translocation, accumulation, biotransformation and barriers). Advances in Phytonanotechnology. Academic Press, pp. 123–152.

Brown, J., 2017. Impact of silver nanoparticles on wastewater treatment. Nanotechnologies for Environmental Remediation. Springer, Cham, pp. 255–267.

Burman, U., Saini, M., Kumar, P., 2013. Effect of zinc oxide nanoparticles on growth and antioxidant system of chickpea seedlings. Toxicol. Environ. Chem. 95, 605–612.

Cartwright, A., Jackson, K., Morgan, C., Anderson, A., Britt, D.W., 2020. A review of metal and metal-oxide nanoparticle coating technologies to inhibit agglomeration and increase bioactivity for agricultural applications. Agronomy 10, 1–20.

Castiglione, M.R., Giorgetti, L., Geri, C., Cremonini, R., 2011. The effects of nano-TiO_2 on seed germination, development and mitosis of root tip cells of *Vicia narbonensis* L. and Zea mays L. J. Nanopart. Res. 13, 2443–2449.

Chand, N., Siddiqui, N., 2012. Improvement in thermo mechanical and optical properties of in situ synthesized PMMA/TiO_2 nanocomposite. Compos. Interfaces. 19, 51–58.

Chen, H., 2018. Metal based nanoparticles in agricultural system: behavior, transport, and interaction with plants. Chem. Speciat. Bioavailab. 30, 123–134.

Costa, M.V.J., Sharma, P.K., 2016. Effect of copper oxide nanoparticles on growth, morphology, photosynthesis, and antioxidant response in *Oryza sativa*. Photosynthetica 54, 110–119.

Das, S., Wolfson, B.P., Tetard, L., Tharkur, J., Bazata, J., Santra, S., 2015. Effect of N-acetyl cysteine coated CdS: Mn/ZnS quantum dots on seed germination and seedling growth of snow pea (*Pisum sativum* L.): imaging and spectroscopic studies. Environ. Sci.: Nano. 2, 203–212.

Debnath, P., Mondal, A., Sen, K., Mishra, D., Mondal, N.K., 2020. Genotoxicity study of nano Al_2O_3, TiO_2 and ZnO along with UV-B exposure: an *Allium cepa* root tip assay. Sci. Total. Environ. 713, 136592.

Dimkpa, C.O., McLean, J.E., Latta, D.E., Manangón, E., Britt, D.W., Johnson, W.P., et al., 2012. CuO and ZnO nanoparticles: phytotoxicity, Met. speciation, induction Oxid. Stress. sand-grown wheat. J. Nanopart. Res. 14, 1–15.

Dimkpa, C.O., Latta, D.E., McLean, J.E., Britt, D.W., Boyanov, M.I., Anderson, A.J., 2013. Fate of CuO and ZnO nano-and microparticles in the plant environment. Environ. Sci. Technol. 47, 4734–4742.

Dimkpa, C.O., White, J.C., Elmer, W.H., Gardea-Torresdey, J., 2017. Nano-particle and ionic Zn promote nutrient loading of sorghum grain under low NPK fertilization. J. Agric. Food. Chem. 65, 8552–8559.

Dimkpa, C.O., Singh, U., Bindraban, P.S., et al., 2018. Exposure to weathered and fresh nanoparticle and ionic Zn in soil promotes grain yield and modulates nutrient acquisition in wheat (*Triticum aestivum* L.). J. Agric. Food Chem. 66, 9645–9656.

Du, W., Sun, Y., Ji, R., et al., 2011. TiO_2 and ZnO nanoparticles negatively affect wheat growth and soil enzyme activities in agricultural soil. J. Environ. Monit. 13, 822–828.

Du, W., Gardea-Torresdey, J.L., Xie, Y., et al., 2017. Elevated CO_2 levels modify TiO_2 nanoparticle effects on rice and soil microbial communities. Sci. Total. Environ. 578, 408–416.

Du, W., Yang, J., Peng, Q., Liang, X., Mao, H., 2019. Comparison study of zinc nanoparticles and zinc sulphate on wheat growth: from toxicity and zinc biofortification. Chemosphere. 227 (10), 9–116.

Eduok, S., Hendry, C., Ferguson, R., et al., 2015. Insights into the effect of mixed engineered nanoparticles on activated sludge performance. FEMS Microbiol. Ecol.

Elmer, W.H., White, J.C., 2016. The use of metallic oxide nanoparticles to enhance growth of tomatoes and eggplants in disease infested soil or soilless medium. Environ. Sci. Nano 3, 1072–1079.

Etxeberria, E., Gonzalez, P., Bhattacharya, P., Sharma, P., Ke, P.C., 2016. Determining the size exclusion for nanoparticles in citrus leaves. Hort. Sci. 51, 732–737.

Fayiga, A.O., Saha, U.K., 2017. Nanoparticles in biosolids: effect on soil health and crop growth. Ann. Environ. Sci. Toxicol. 2, 059–067.

Feizi, H., Moghaddam, P.R., Shahtahmassebi, N., Fotovat, A., 2012. Impact of bulk and nanosized titanium dioxide (TiO_2) on wheat seed germination and seedling growth. Biol. Trace Elem. Res. 146, 101–106.

Fernández, V., Eichert, T., 2009. Uptake of hydrophilic solutes through plant leaves: current state of knowledge and perspectives of foliar fertilization. Crit. Rev. Plant. Sci. 28, 36–68.

Frenk, S., Ben-Moshe, T., Dror, I., Berkowitz, B., Minz, D., 2013. Effect of metal oxide nanoparticles on microbial community structure and function in two different soil types. PLoS One 8, e84441.

Ge, Y., Schimel, J.P., Holden, P.A., 2011. Evidence for negative effects of TiO_2 and ZnO nanoparticles on soil bacterial communities. Environ. Sci. Technol. 45, 1659–1664.

Gelabert, A., Sivry, Y., Ferrari, R., Akrout, A., Cordier, L., Nowak, S., et al., 2014. Uncoated and coated ZnO nanoparticle life cycle in synthetic seawater. Environ. Toxicol. Chem. 33, 341–349.

Gottschalk, F., Nowack, B., 2011. The release of engineered nanomaterials to the environment. J. Environ. Monit. 13, 1145–1155.

Haghighi, M., da Silva, J.A.T., 2014. The effect of N-TiO_2 on tomato, onion, and radish seed germination. J. Crop. Sci. Biotechnol. 17, 221–227.

He, S., Feng, Y., Ren, H., Zhang, Y., Gu, N., Lin, X., 2011. The impact of iron oxide magnetic nanoparticles on the soil bacterial community. J. Soils Sediment. 11, 1408–1417.

Hong, J., Rico, C.M., Zhao, L., Adeleye, A.S., Keller, A.A., Peralta-Videa, J.R., et al., 2015. Toxic effects of copper-based nanoparticles or compounds to lettuce (*Lactuca sativa*) and alfalfa (*Medicago sativa*). Environ. Sci. Process Impacts 17, 177–185.

Hong, J., Wang, C., Wagner, D.C., Gardea-Torresdey, J.L., He, F., Rico, C.M., 2021. Foliar application of nanoparticles: mechanisms of absorption, transfer, and multiple impacts. Environ. Sci. Nano 8, 1196–1210.

Jaberzadeh, A., Moaveni, P., Moghadam, H.R.T., Zahedi, H., 2013. Influence of bulk and nanoparticles titanium foliar application on some agronomic traits, seed gluten and starch contents of wheat subjected to water deficit stress. Not. Bot. Horti Agrobot. 41, 201–207.

Jacob, D.L., Borchardt, J.D., Navaratnam, L., Otte, M.L., Bezbaruah, A.N., 2013. Uptake and translocation of Ti from nanoparticles in crops and wetland plants. Int. J. Phytoremediat. 15, 142–153.

Kulizhskiy, S.P., Loiko, S.V., Morgalev, Y.N., et al., 2017. Investigation of platinum and nickel nanoparticles migration and accumulation in soils within the south eastern part of west siberia. Nano. Hybrids. Compos. 13, 115–122.

Kumar, P., Burman, U., Santra, P., 2015. Effect ofnano-zinc oxide on nitrogenase activity in legumes: an interplay of concentration and exposure time. Int. Nano Lett. 5, 191–198.

Landa, P., Cyrusova, T., Jerabkova, J., et al., 2016. Effect of metal oxides on plant germination: phytotoxicity of nanoparticles, bulk materials, and metal ions. Water Air Soil. Pollut. 227, 448.

Le Van, N., Ma, C., Shang, J., Rui, Y., Liu, S., Xing, B., 2016a. Effects of CuO nanoparticles on insecticidal activity and phytotoxicity in conventional and transgenic cotton. Chemosphere 144 (66), 1–670.

Le Van, N., Rui, Y., Cao, W., Shang, J., Liu, S., Nguyen Quang, T., et al., 2016b. Toxicity and bio-effects of CuO nanoparticles on transgenic Ipt-cotton. J. Plant. Interact. 11, 108–116.

Li, C.C., Dang, F., Li, M., Zhu, M., Zhong, H., Hintelmann, H., et al., 2017. Effects of exposure pathways on the accumulation and phytotoxicity of silver nanoparticles in soybean and rice. Nanotoxicol 11, 699–709.

Lin, D., Xing, B., 2008. Root uptake and phytotoxicity of ZnO nanoparticles. Environ. Sci. Technol. 42, 5580–5585.

Liu, J., Pennell, K.G., Hurt, R.H., 2011. Kinetics and mechanisms of nanosilver oxysulfidation. Environ. Sci. Technol. 45, 7345–7353.

Liu, W., Zeb, A., Lian, J., Wu, J., Xiong, H., Tang, J., et al., 2020. Interactions of metal-based and metal-oxide-based nanoparticles (MBNPs and MONPs) with crop plants: a critical review of research progress and prospects. Environ. Rev. 28, 294–310.

Lopez-Moreno, M.L., de la Rosa, G., Cruz-Jiménez, G., Castellano, L., Peralta-Videa, J.R., Gardea-Torresdey, J.L., 2017. Effect of ZnO nanoparticles on corn seedlings at different temperatures; X-ray absorption spectroscopy and ICP/OES studies. Microchem. J. 134, 54–61.

Lv, J., Christie, P., Zhang, S., 2019. Uptake, translocation, and transformation of metal-based nanoparticles in plants: recent advances and methodological challenges. Environ. Sci. Nano 6, 41–59.

Mahajan, P., Dhoke, S.K., Khanna, A.S., 2011. Effect of nano-ZnO particle suspension on growth of mung (*Vigna radiata*) and gram (*Cicer arietinum*) seedlings using plant agar method. J. Nanotechnol. 2011.

Mahdi, K.N., Peters, R., van der Ploeg, M., Ritsema, C., Geissen, V., 2018. Tracking the transport of silver nanoparticles in soil: a saturated column experiment. Water Air Soil. Pollut. 229, 334.

Mahmoodzadeh, H., Nabavi, M., Kashefi, H., 2013. Effect of nanoscale titanium dioxide particles on the germination and growth of canola (*Brassica napus*). J. Ornament. Plant. 25–32.

Matlochova, A., Placha, D., Rapantová, N., 2013. The application of nanoscale materials in groundwater remediation. Pol. J. Environ. stud. 22.

Ma, R., Levard, C., Judy, J.D., et al., 2014. Fate of zinc oxide and silver nanoparticles in a pilot wastewater treatment plant and in processed biosolids. Environ. Sci. Technol. 48, 104–112.

Ma, C., Liu, H., Chen, G.C., Zhao, Q., Eitzer, B., Wang, Z.H., et al., 2017. Effects of titanium oxide nanoparticles on tetracycline accumulation and toxicity in *Oryza sativa* (L.). Environ. Sci. Nano. 4, 1827–1839.

Mingfang, Q., Yufeng, L., Tianlai, L., 2013. Nano-TiO_2 improve the photosynthesis of tomato leaves under mild heat stress, biological trace element research. Biol. Trace Elem. Res. 156, 323–328.

Mingyu, S., Xiao, W., Chao, L., Chunxiang, Q., Xiaoqing, L., Liang, C., et al., 2007. Promotion of energy transfer and oxygen evolution in spinach photosystem II by nano-anatase TiO_2. Biol. Trace Elem. Res. 119, 183–192.

Minkina, T.M., Nevidomskaya, D.G., Shuvaeva, V.A., Soldatov, A.V., Tsitsuashvili, V.S., Zubavichus, Y.V., et al., 2018. Studying the transformation of Cu^{2+} ions in soils and mineral phases by the XRD, XANES, and sequential fractionation methods. J. Geochem. Explor. 184, 365–371.

Moon, Y.S., Park, E.S., Kim, T.O., Lee, H.S., Lee, S.E., 2014. SELDI-TOF MS-based discovery of a biomarker in *Cucumis sativus* seeds exposed to CuO nanoparticles. Environ. Toxicol. Pharmacol. 38, 922–931.

Mousavi Kouhi, S.M., Lahouti, M., Ganjeali, A., Entezari, M.H., 2014. Comparative phytotoxicity of ZnO nanoparticles, ZnO microparticles, and Zn^{2+} on rapeseed (*Brassica napus* L.): investigating a wide range of concentrations. Toxicol. Environ. Chem. 96, 861–868.

Mukherjee, K., Acharya, K., 2018. Toxicological effect of metal oxide nanoparticles on Soil and aquatic habitats. Arch. Environ. Contam. Toxicol. 75, 175–186.

Mukherjee, A., Pokhrel, S., Bandyopadhyay, S., Mädler, L., Peralta-Videa, J.R., Gardea-Torresdey, J.L., 2014. A soil mediated phyto-toxicological study of iron doped zinc oxide nanoparticles (Fe@ ZnO) in green peas (*Pisum sativum* L.). Chem. Eng. J. 258, 394–401.

Mushtaq, A., Jamil, N., Rizwan, S., Mandokhel, F., Riaz, M., Hornyak, G.L., et al., 2018. Engineered silica nanoparticles and silica nanoparticles containing controlled release fertilizer for drought and saline areas. IOP Conference Series: Materials Science and Engineering. IOP Publishing.

Mustafa, G., Komatsu, S., 2016. Toxicity of heavy metals and metal-containing nanoparticles on plants. Biochim. Biophys. Acta—Proteins Proteom. 1864, 932–944.

Nair, R., 2018. Plant response strategies to engineered metal oxide nanoparticles: a review. Phytotoxicity Nanopart. 377–393.

Nair, P.M.G., Chung, I.M., 2014. A mechanistic study on the toxic effect of copper oxide nanoparticles in soybean (*Glycine max* L.) root development and lignification of root cells. Biol. Trace Elem. Res. 162 (1), 342–352.

Naseem, T., Durrani, T., 2021. The role of some important metal oxide nanoparticles for wastewater and antibacterial applications: a review. Environ. Chem. Ecotoxicol.

Nhan, L.V., Ma, C., Rui, Y., Liu, S., Li, X., Xing, B., et al., 2015. Phytotoxic mechanism of nanoparticles: destruction of chloroplasts and vascular bundles and alteration of nutrient absorption. Sci. Rep. 5, 11618.

Odzak, N., Kistler, D., Sigg, L., 2017. Influence of daylight on the fate of silver and zinc oxide nanoparticles in natural aquatic environments. Environ. Pollut. 226, 1–11.

Omar, F.M., Aziz, H.A., Stoll, S., 2014. Aggregation and disaggregation of ZnO nanoparticles: influence of pH and adsorption of Suwannee River humic acid. Sci. Total. Environ. 468–469, 195–201.

Pakrashi, S., Jain, N., Dalai, S., Jayakumar, J., Chandrasekaran, P.T., Raichur, A.M., et al., 2014. In vivo genotoxicity assessment of titanium dioxide nanoparticles by *Allium cepa* root tip assay at high exposure concentrations. PLoS One 9 (2), e87789.

Parada, J., Rubilar, O., Fernández-Baldo, M.A., Bertolino, F.A., Durán, N., Seabra, A.B., et al., 2019. The nanotechnology among US: are metal and metal oxides nanoparticles a nano or mega risk for soil microbial communities? Crit. Rev. Biotechnol. 39, 157–172.

Perreault, F., Samadani, M., Dewez, D., 2014. Effect of soluble copper released from copper oxide nanoparticles solubilisation on growth and photosynthetic processes of *Lemna gibba* L. Nanotoxicol 8, 374–382.

Pokhrel, L.R., Dubey, B., 2013. Evaluation of developmental responses of two crop plants exposed to silver and zinc oxide nanoparticles. Sci. Total. Environ. 452, 321–332.

Prasad, T.N.V.K.V., Sudhakar, P., Sreenivasulu, Y., Latha, P., Munaswamy, V., Raja Reddy, K., et al., 2012. Effect of nanoscale zinc oxide particles on the germination, growth and yield of peanut. J. Plant. Nutr. 35, 906–927.

Prasanna, S.S., Balaji, K., Pandey, S., Rana, S., 2019. Metal oxide based nanomaterials and their polymer nanocomposites. Nanomaterials and Polymer Nanocomposites. Elsevier, pp. 123–144.

Pérez-de-Luque, A., 2017. Interaction of nanomaterials with plants: what do we need for real applications in agriculture? Front. Environ. Sci. 5, 1–7.

Qi, M., Liu, Y., Li, T., 2013. Nano-TiO_2 improve the photosynthesis of tomato leaves under mild heat stress. Biol. Trace Elem. Res. 156, 323–328.

Rajput, V.D., Minkina, T., Fedorenko, A., Tsitsuashvili, V., Mandzhieva, S., Sushkova, S., et al., 2018a. Metal oxide nanoparticles: applications and effects on soil ecosystems. Soil Contamination, Risk Assessment and Remediation. 81–106.

Rajput, V.D., Minkina, T.M., Behal, A., Sushkova, S.N., Mandzhieva, S., Singh, R., et al., 2018b. Effects of zinc-oxide nanoparticles on soil, plants, animals and soil organisms: a review. Environ. Nanotechnol. Monit. Manag. 9, 76–84.

Raliya, R., Nair, R., Chavalmane, S., Wang, W.N., Biswas, P., 2015. Mechanistic evaluation of translocation and physiological impact of titanium dioxide and zinc oxide nanoparticles on the tomato (*Solanum lycopersicum* L.) plant. Metallomics 7, 1584–1594.

Raliya, R., Tarafdar, J.C., 2013. ZnO nanoparticle biosynthesis and its effect on phosphorous-mobilizing enzyme secretion and gum contents in Clusterbean (*Cyamopsis tetragonoloba* L.). Agri Res. 2, 48–57.

Rao, S., Shekhawat, G.S., 2014. Toxicity of ZnO engineered nanoparticles and evaluation of their effect on growth, metabolism and tissue specific accumulation in *Brassica juncea.* J. Environ. Chem. Eng. 2, 105–114.

Rastogi, A., Zivcak, M., Sytar, O., Kalaji, H.M., He, X., Mbarki, S., et al., 2017. Impact of metal and metal oxide nanoparticles on plant: a critical review. Front. Chem. 5, 1–16.

Rico, C.M., Morales, M.I., Barrios, A.C., McCreary, R., Hong, J., Lee, W.Y., et al., 2013. Effect of cerium oxide nanoparticles on the quality of rice (*Oryza sativa* L.) grains. J. Agric. Food. Chem. 67, 11278–11285.

Rizwan, M., Ali, S., Adrees, M., Rizvi, H., Zia-ur-Rehman, M., Hannan, F., et al., 2016. Cadmium stress in rice: toxic effects, tolerance mechanisms, and management: a critical review. Environ. Sci. Pollut. Res. 23, 17859–17879.

Rizwan, M., Ali, S., Qayyum, M.F., Ok, Y.S., Adrees, M., Ibrahim, M., et al., 2017. Effect of metal and metal oxide nanoparticles on growth and physiology of globally important food crops: a critical review. J. Hazard. Mater. 322, 2–16.

Rizwan, M., Ali, S., Rehman, M.Z., Riaz, M., Adrees, M., Hussain, A., et al., 2021. Effects of nanoparticles on trace element uptake and toxicity in plants: a review. Ecotoxicol. Environ. Saf. 221.

Rossi, L., Zhang, W., Schwab, A.P., Ma, X., 2017. Uptake, accumulation, and in planta distribution of coexisting cerium oxide nanoparticles and cadmium in *Glycine max* (L.). Merr. Environ. Sci. Technol. 51, 12815–12824.

Servin, A.D., Morales, M.I., Castillo-Michel, H., Hernandez-Viezcas, J.A., Munoz, B., Zhao, L., et al., 2013. Synchrotron verification of TiO_2 accumulation in cucumber fruit: a possible pathway of TiO_2 nanoparticle transfer from soil into the food chain. Environ. Sci. Technol. 47, 11592–11598.

Servin, A., Elmer, W., Mukherjee, A., De la Torre-Roche, R., Hamdi, H., White, J.C., et al., 2015. A review of the use of engineered nanomaterials to suppress plant disease and enhance crop yield. J. Nanopart. Res. 17, 92.

Servin, A.D., Pagano, L., Castillo-Michel, H., et al., 2017. Weathering in soil increases nanoparticle CuO bioaccumulation within a terrestrial food chain. Nanotoxicol 11, 98–111.

Shafey, A.M.E., 2020. Green synthesis of metal and metal oxide nanoparticles from plant leaf extracts and their applications: a review. Green. Process. Synth. 9, 304–339.

Shaw, A.K., Hossain, Z., 2013. Impact of nano-CuO stress on rice (*Oryza sativa* L.) seedlings. Chemosphere 93, 906–915.

Shaw, A.K., Ghosh, S., Kalaji, H.M., Bosa, K., Brestic, M., Zivcak, M., et al., 2014. Nano-CuO stress induced modulation of antioxidative defense and photosynthetic performance of Syrian barley (*Hordeum vulgare* L.). Environ. Exp. Bot. 102, 37–47.

Shemawar, M.A., Hussain, S., Mahmood, F., Iqbal, M., Shahid, M., Ibrahim, M., et al., 2021. Toxicity of biogenic zinc oxide nanoparticles to soil organic matter cycling and their interaction with rice-straw derived biochar. Sci. Rep. 11, 1–12.

Shrestha, B., Anderson, T.A., Acosta-Martinez, V., Payton, P., Cañas-Carrell, J.E., 2015. The influence of multiwalled carbon nanotubes on Polycyclic Aromatic Hydrocarbon (PAH) bioavailability and toxicity to soil microbial communities in alfalfa rhizosphere. Ecotoxicol. Environ. Saf. 116, 143–149.

Singh, A., Singh, N.B., Hussain, I., Singh, H., Yadav, V., Singh, S.C., 2016. Green synthesis of nano zinc oxide and evaluation of its impact on germination and metabolic activity of *Solanum lycopersicum*. J. Biotechnol 233, 84–94.

Soliman, A.S., El-feky, S.A., Darwish, E., 2015. Alleviation of salt stress on *Moringa peregrina* using foliar application of nanofertilizers. J. Hort. For. 7, 36–47.

Song, U., Shin, M., Lee, G., Roh, J., Kim, Y., Lee, E.J., 2013. Functional analysis of TiO_2 nanoparticle toxicity in three plant species. Biol. Trace Elem. Res. 155, 93–103.

Song, W., Zhao, B., Wang, C., Ozaki, Y., Lu, X., 2019. Functional nanomaterials with unique enzyme-like characteristics for sensing applications. J. Mater. Chem. B 7, 850–875.

Stefani, Y., Vélez, P., 2021. Interaction of metal nanoparticles – plants – microorganisms in agriculture and soil remediation. J. Nanopart. Res.

Subbenaik, S.C., 2016. Physical and chemical nature of nanoparticles. Plant Nanotechnology. Springer, Cham, pp. 15–27.

Sun, W., Dou, F., Li, C., Ma, X., Ma, L.Q., 2021. Impacts of metallic nanoparticles and transformed products on soil health. Crit. Rev. Environ. Sci. Technol. 51, 973–1002.

Tripathi, D.K., Tripathi, A., Shweta, Singh, S., Singh, Y., Vishwakarma, K., et al., 2017. Uptake, accumulation and toxicity of silver nanoparticle in autotrophic plants, and heterotrophic microbes: a concentric review. Front. Microbiol. 8, 1–16.

Uzu, G., Sobanska, S., Sarret, G., Munoz, M., Dumat, C., 2010. Foliar lead uptake by lettuce exposed to atmospheric fallouts. Environ. Sci. Technol. 44, 1036–1042.

Venkatachalam, P., Priyanka, N., Manikandan, K., Ganeshbabu, I., Indiraarulselvi, P., Geetha, N., et al., 2017. Enhanced plant growth promoting role of phycomolecules coated zinc oxide nanoparticles with P supplementation in cotton (*Gossypium hirsutum* L.). Plant. Physiol. Biochem. 110, 118–127.

Verma, S.K., Das, A.K., Patel, M.K., Shah, A., Kumar, V., Gantait, S., 2018. Engineered nanomaterials for plant growth and development: a perspective analysis. Sci. Total. Environ. 630, 1413–1435.

Vithanage, M., Herath, I., Almaroai, Y.A., Rajapaksha, A.U., Huang, L., Sung, J.K., et al., 2017. Effects of carbon nanotube and biochar on bioavailability of Pb, Cu and Sb in multi-metal contaminated soil. Environ. Geochem. Health 39, 1409–1420.

Wang, W.N., Tarafdar, J.C., Biswas, P., 2013. Nanoparticle synthesis and delivery by an aerosol route for watermelon plant foliar uptake. J. Nanopart. Res. 15, 1–13.

Wang, S., Liu, H., Zhang, Y., Xin, H., 2015. The effect of CuO NPs on reactive oxygen species and cell cycle gene expression in roots of rice. Environ. Toxicol. Chem. 34, 554–561.

Wang, X., Yang, X., Chen, S., Li, Q., Wang, W., Hou, C., et al., 2016. Zinc oxide nanoparticles affect biomass accumulation and photosynthesis in *Arabidopsis*. Front. Plant. Sci. 6, 1243.

Wang, Y., Lin, Y., Xu, Y., Yin, Y., Guo, H., Du, W., 2019. Divergence in response of lettuce (var. ramosa Hort.) to copper oxide nanoparticles/microparticles as potential agricultural fertilizer. Environ. Pollut. Bioavail. 31, 80–84.

Yang, F., Hong, F., You, W., Liu, C., Gao, F., Wu, C., et al., 2006. Influence of nano-anatase TiO_2 on the nitrogen metabolism of growing spinach. Biol. Trace Elem. Res. 110, 179–190.

Yang, J., Cao, W., Rui, Y., 2017. Interactions between nanoparticles and plants: phytotoxicity and defense mechanisms. J. Plant. Interact 12, 158–169.

Youssef, M.S., Elamawi, R.M., 2020. Evaluation of phytotoxicity, cytotoxicity, and genotoxicity of ZnO nanoparticles in *Vicia faba*. Environ. Sci. Pollut. Res. 27, 18972–18984.

You, T., Liu, D., Chen, J., Yang, Z., Dou, R., Gao, X., et al., 2018. Effects of metal oxide nanoparticles on soil enzyme activities and bacterial communities in two different soil types. J. Soils Sediments. 18, 211–221.

Zhang, T., Sun, H., Lv, Z., et al., 2018. Using synchrotron-based approaches to examine the foliar application of $ZnSO_4$ and ZnO nanoparticles for field-grown winter wheat. J. Agric. Food Chem. 66, 2572–2579.

Zhang, W., Long, J., Li, J., Zhang, M., Ye, X., Chang, W., et al., 2020. Effect of metal oxide nanoparticles on the chemical speciation of heavy metals and micronutrient bioavailability in paddy soil. Int. J. Environ. Res. Public. Health 17 (7), 2482.

Zhao, L., Peralta-Videa, J.R., Ren, M., Varela-Ramirez, A., Li, C., Hernandez-Viezcas, J.A., et al., 2012. Transport of Zn in a sandy loam soil treated with ZnO NPs and uptake by corn plants: electron microprobe and confocal microscopy studies. Chem. Eng. J. 184, 1–8.

Zhao, L., Peralta-Videa, J.R., Rico, C.M., Hernandez-Viezcas, J.A., Sun, Y., Niu, G., et al., 2014. CeO_2 and ZnO nanoparticles change the nutritional qualities of cucumber (*Cucumis sativus*). J. Agric. Food. Chem. 62, 2752–2759.

Zheng, L., Hong, F., Lu, S., Liu, C., 2005. Effect of nano-TiO(2) on strength of naturally aged seeds and growth of spinach. Biol. Trace Elem. Res. 104, 83–91.

Zuverza-Mena, N., Medina-Velo, I.A., Barrios, A.C., Tan, W., Peralta-Videa, J.R., Gardea-Torresdey, J.L., 2015. Copper nanoparticles/compounds impact agronomic and physiological parameters in cilantro (*Coriandrum sativum*). Environ. Sci: Process. Impacts 17, 1783–1793.

Chapter 13

Toxicity in plants by metal oxide nanoparticles

Rajiv Periakaruppan[1], P. Vanathi[2], G. Priyanka[2] and D. Vidhya[3]
[1]Department of Biotechnology, PSG College of Arts & Science, Coimbatore, Tamil Nadu, India, [2]Department of Biotechnology, Sri Ramakrishna College of Arts & Science, Coimbatore, Tamil Nadu, India, [3]TRM Biotech Private Limited, Namakkal, Tamil Nadu, India

Chapter outline

Nanometal Oxides in Horticulture and Agronomy. DOI: https://doi.org/10.1016/B978-0-323-91809-1.00012-3

13.1 Introduction

Nanotechnology is a promising tool for addressing any global challenge in a long-term manner. Nanomaterials are seen as a major platform to bring about the agri-nanotechnology revolution. In agronomy, nanomaterials intention is to diminish the usage of toxic chemical fertilizers; reduce the nutrient loss that occurs in conception; and increase productivity over pests, diseases, and nutrient conservation (Prasad et al., 2017). Several nanomaterials have been produced to test the impact of nanomaterials on crop protection and productivity, including metal, metal oxide based, carbon-based single-walled, and multiwalled nanotubes (Kah et al., 2013; Kah, 2015; Rizwan et al., 2017).

The metal oxide nanoparticles (NPs) possess distinct chemical, electrical, and optical properties and thereby become a natural choice for an inclusive array of applications and are extensively utilized in many fields, including agriculture. Ag, ZnO, CuO, TiO_2, and Fe_3O_4 are metal oxide based nanomaterials that are frequently used in agriculture. They could be used to improve agricultural productivity by using metals and metal oxides as nanopesticides, nanofertilizers, or nanosensors (Wang et al., 2016a,b).

The florae are directly exhibited to the NPs as we make progress toward sustainable agriculture. Interactions between these NPs and crops cause their accumulation in various food sources. Several researchers have looked at the accumulation of metal (Ag) and other metal oxides (like CuO, CeO_2, and ZnO) in the edible tissues of various crops (Rico et al., 2011; Ebbs et al., 2016; Ma et al., 2010a,b). As the metal oxide NPs are found to be accumulated in many edible parts of food sources, the contact between metal oxide NPs and living beings is becoming closer. Furthermore, some metal oxides are soluble and produce dissolved metal ions, which would increase their environmental consequences. The toxicity of the NPs is considered to depend on the concentration, element size, shape, zeta potential, and bioavailability, in addition to the plant species under study (Ma et al., 2010a,b). Therefore, a critical analysis of the information regarding the concentration, size, zeta potential, uptake by a plant, and impact on a plant is required to decide whether the metal oxide NPs pose a concern to plant species and the environment. At larger dosages, metal and metal oxide NPs are harmful to plants, and the toxicity mechanism in plants is primarily mediated by the generation of reactive oxygen species (ROS) (Rastogi et al., 2017).

The plants' interaction with metal oxide NPs, as well as their toxicity mechanism, is still unclear. This chapter aims to highlight metal oxide NP stress, phytotoxicity, and genotoxicity mechanisms in plants.

13.2 Applications of metal oxide nanoparticles in agriculture

Nanotechnology is widely adopted in agriculture as it can enhance productivity, simulate plant growth, improve the quality of soil, and reduce postharvest loss (Fraceto et al., 2016). NPs are being explored for their potential to be used as nanofertilizers, nanopesticides, growth enhancers, oxidative stress reducers, and nanosensors (Sohail et al., 2021). Over the last decade, metal oxide NPs are being increasingly investigated and some of the major applications of the metal oxide NPs are as follows.

Micronutrients are little amounts of nutrients that are required for the plant to grow. NPs show an energetic part in the delivery of micronutrients and are utilized to transport micronutrients efficiently (Monreal et al., 2016; Kah et al., 2018). An increase in biomass, yield, and nutrient accumulation was observed in crops like cucumber (Zhao et al., 2013), ground nut, and tomato (Prasad et al., 2012; Singh et al., 2013) when they are treated with ZnO NPs. The NPs could therefore be used for biofortification as well. Wheat seeds treated with iron oxide NPs showed increased iron content (Sundaria et al., 2019), whereas zinc content increased in maize seeds treated with ZnO NPs (Subbaiah et al., 2016). This has the potential to help people who are deficient in zinc and iron.

The metal oxide NPs are also being utilized for the control of pests and diseases. The toxic effects of titanium oxide NPs on insect pests were reported (Goswami et al., 2010). The ZnO nanomaterials are found to have natural antifungal activity (Kairyte et al., 2013). The nanopesticides formed by metal oxides of copper, magnesium, cerium, and zinc are shown to control the pests at the field level and also the storage pests (Malaikozhundan and Vinodhini, 2018; Pandey et al., 2018; Tiwari et al., 2020). Hence, they could be widely exploited for the control of diseases and pests. The ZnO NPs are also used in the nanosensors, to monitor soil conditions, pesticide detection, and pathogen attack (Fraceto et al., 2018; Kaushal and Wani, 2017).

Plants' resilience toward abiotic anxieties such as deficiency of nutrients, water scarcity, salinity, and other oxidative damage might be improved by the use of metal oxide NPs. Anionic cerium oxide improved the tolerance of *Arabidopsis* to abiotic challenges (Wu et al., 2017) by boosting photosynthetic performance, and CeO_2 NPs were shown to relieve salt stress in *Brassica napus* (Rossi et al., 2017). Researchers found that treating wheat and bamboo with ZnO reduced oxidative stress and heavy metal noxiousness (Rizwan et al., 2019; Emamverdian et al., 2020).

Although metal oxide NPs can improve crop yield and make agriculture sustainable, as proven by several research publications and patents, they are still in the development stage. Not only metal oxide NPs but also the majority of all other NPs with promising agricultural applications are yet to make their way into the market (Parisi et al., 2015). One explanation for this could be that plant nanomaterial interaction, toxicity, and environmental and human implications are still unknown.

13.3 Nanotoxicology

The toxicity of nanomaterials is studied in nanotoxicology, and it explains the interface methods of tiny material through active organisms. Some characteristics of standard toxification are not applicable to the perception of NPs. Conferring to the standard toxicology "the dosage of the material only makes poison." NPs toxicity not only depends on the dosage of the particles but also depends on parameters like aggregation, size, and surface area also determines the poisonous level of the NPs (Hobson, 2016; Gautam and Chattopadhyaya, 2016). The increase in demand for products facilitates the development of the product at nanoscale level in improving the agricultural field that may offer more recompenses but then there is nonexistence in acquaintance in the toxicological effect of nanomaterials on the biotic system (Pacheco et al., 2010). The nanomaterial toxicity is due to the physiochemical properties of NPs like size distribution of NPs, surface area, morphology, absorption, and superficial charge of the nanomaterials (Corredor et al., 2009). The toxicity of NPs also depends on the method of synthesis of NPs and the developmental phase when NPs come in contact with plants either through root or by foliar spray method. It also depends on the exposure time of the NPs over plants (Tarafdar et al., 2012).

13.4 Plant response to metal oxide nanoparticles

Once the NP has come in contact with the plant system either through soil application or through the foliar application, they have to be taken up by the plants. The plants have developed two mechanisms to uptake the nanomaterials: root origin to leaflet and leaf and pome/fruit to root base (Ma et al., 2015).

Plant mechanisms can be altered by the NPs in a variety of ways like providing micronutrients, regulating stress-related genes, and interfering with oxidative processes (Liu and Lal, 2015; Nair and Chung, 2014; Hossain et al., 2015). The metal oxide NPs when absorbed could either dissolve or stimulate the generation of ROS. In addition, the dose and size of the NPs will impact the plant's response to NPs (Da Rosa, 2013). Metal oxide NPs have a positive effect on plant growth and development when they are used at an optimal concentration (Siddiqi and Husen, 2016; Venkatachalam et al., 2017). Plants react physiologically and biochemically to the stress that metal oxide NPs impose on them.

13.5 Physiological response of plants to metal oxide nanoparticles

The physiological response of the plant is the major response of the plant to any abiotic stress, including the stress induced by the metal oxide NPs.

Ma et al. (2015) showed that when plants were treated with metal and metal oxide NPs, they often showed an increase or decrease in biomass.

Treatment with ZnO NPs amended chlorophyll and proteins in cluster beans (Raliya and Tarafdar, 2013) and Fe_2O_3-treated soybean showed an increased photosynthetic rate (Alidoust and Isoda, 2013). Similarly, the total nitrogen, protein, oxygen, and chlorophyll content of spinach plants treated with TiO_2 NPs increased, promoting photosynthesis in the plants (Yang et al., 2007).

Other experiments utilizing TiO_2 NPs on tomatoes and duckweed found no improvement in the plant's chlorophyll level (Song et al., 2013). Linseed florae have higher quantities of seed oil when treated with TiO_2 NPs (Baiazidi-Aghdam et al., 2016). Accumulation of Fe_2O_3 NPs to the soil was found to upsurge the chlorophyll content, biomass, and entire iron constituents of peanut plants. The improvement in the chlorophyll content and the functioning of the photosystem would result in the production of primary metabolites, which positively correlated with the increased biomass and yield in the metal oxide NP treated plants.

Soybean plants treated with ZnO at 500 mg/kg showed negative effects on the developmental stages. Flowering was delayed by a week in wheat florae medicated with CeO_2 nanomaterials (Du et al., 2015). The metal oxide NPs also caused phytotoxic effects when exposed to a higher concentration. ZnO and CuO NPs (1000 mg/kg) caused negative impacts on potato tuber biomass (Bradfield et al., 2017). The lower doses of the metal oxide NPs stimulated growth, whereas the higher dose of NPs would induce toxicity in plants.

Pollen abortion was seen in *Arabidopsis* plant treated with 3 mg/L of Fe_2O_3 NPs, without any changes to the plant morphology (Bombin et al., 2015). This is of concern as the NP might not show any visible morphological changes, but still affect the developmental process. Therefore, a thorough understanding of the biochemical and molecular responses of metal oxide NPs on plants should be studied.

13.6 Biochemical and molecular responses of plant to the metal oxide nanoparticles

The alterations instigated by the interactivity of the NPs with the plant system would result in an oxidative burst. When a NP interferes with the electron transport system of plants, a rise in ROS level is seen (Pakrashi et al., 2014; Siddiqi and Husen, 2017; Cvjetko et al., 2017). NP stress in plants results in production of ROS, which includes singlet oxygen, superoxide, hydrogen peroxide, and hydroxyl radicals (Iqbal et al., 2020; Siddiqi and Husen, 2017). The ROS generated would interact with various biological components, causing harm to plant cells. However, the mechanism underlying plant responses to metal oxide NPs is not completely figured out.

There are two methods by which oxidative stress is handled by the plant: enzymatic and nonenzymatic methods. In enzymatic methods, we could see the upregulation of enzymes, ascorbate peroxidase (APX), catalase (CAT), superoxide dismutase (SOD), monodehydroascorbate reductase, guaiacol peroxidase, glutathione reductase, and dehydroascorbate reductase intricate in scavenging the ROS and reactive nitrogen species, that is, oxidants collectively called as free radicals. The consequence of nonenzymatic methods describes based on the activities of ascorbic acid, reduced glutathione (GSH), α-tocopherol, carotenoids, flavonoids, and proline (Das and Roychoudhury, 2014). The outcome of various common metallic oxide NPs on ROS mechanism besides response to oxidative stress is summarized in Tables 13.1–13.4.

Various biochemical pathways are altered upon treatment with metal oxide NPs. CuO NPs caused a significant rise in the amounts of several antioxidant molecules, implying that plants had activated the protective mechanism (Song et al., 2016). CeO_2 NPs increased the root soluble proteins in the bean after prolonged exposure (500 mg/kg) to which the plants presumably produced elevated levels of peroxidase enzyme to sustain cellular homeostasis. The cerium (IV) oxide (CeO_2) treated wheat plants showed a modified amino acid composition when compared to the control plants (Rico et al., 2014).

miRNAs are involved in plant developmental processes and tolerance to abiotic stress in plants. An elevated level of miRNA was observed in the tobacco plants treated with TiO_2 (Frazier et al., 2014). TiO_2 NPs boosted lycopene content in tomatoes (Raliya et al., 2015). Al_2O_3 NPs were found to induce programmed cell death in wheat roots (Yanık and Vardar, 2018). Altered phytohormone levels were also seen in the plants treated with CuO NPs (Nair and Chung, 2014). The heat shock protein 70 (hsp70) was also involved in the stress response to the exposure of CeO_2 NPs (400 or 800 mg/kg) in corn, besides upregulation of CAT and APX genes (Zhao et al., 2012). Yue et al. (2017) showed significant downregulation of AQP genes, showing that reduced water intake could be one of the plant defensive strategies utilized to limit La_2O_3 uptake in maize.

As there are many cellular events associated with a mechanism, there is a probable potential for cross talk. Hence, the proteome and transcriptome analysis is being used for a detailed study of the molecular responses induced in plants by the NPs (Landa et al., 2015).

13.7 Metabolomics effects in plants due to metal oxide nanoparticle exposure

A majority of studies have focused on just one biological reaction to investigate the mechanism of nanotoxicity; other key biological processes in the plant system may have been overlooked. Understanding metabolic profiles

TABLE 13.1 Mechanism of plant response to zinc oxide nanoparticle.

Plant species	Part used in study	Type of culture	Size of NP (nm)	Concentration of the nanoparticles	Effect of nanoparticles on plants	References
Allium cepa	Bulb	Hydroponic	<100	25, 50, 75, 100 μg/mL	Dose-dependent increase of TBARS content indicating, higher lipid peroxidation, upsurge in chromosomal aberrations, reduction in mitotic index	Kumari et al. (2011)
Fagopyrumesculentum	Root	Hydroponic	<50	1, 5, 10, 100, 1000, and 2000 mg/L	Increased ROS enzyme activity, increased CAT and GSH, increased antioxidants	Lee et al. (2013)
Zea mays	Root, stem, and leaves	Soil	10	100–800 mg/kg	Reduced APX and CAT levels	Zhao et al. (2013)
Solanum lycopersicum	Root, leaves, fruit	Amended soil	<100	300, 600, and 1000 mg/kg	Increased H_2O_2, reduced APX and SOD activity, increased CAT activity in roots and coupled with lipid peroxidation in roots Inhibitory effect on nonenzymatic antioxidants	Akanbi-Gada et al. (2019)
Oryza sativa spp. *japonica*	Shoot and root tissues	Hydroponic	< 50	25, 50, and 100 mg/L	Decreased CAT, APX and SOD Enhanced oxidative damage	Chen et al. (2018)
Arabidopsis thaliana	Leaves	Soil	<50	50, 100, 200, 250, and 300 mg/L	Inhibition of chlorophyll synthesis genes and photosynthesis Increased expression of carotenoid synthesis genes	Wang et al. (2016a,b)

TABLE 13.2 Mechanism of plant response to copper oxide nanoparticles.

Plant species	Part used in study	Type of culture	Size of NP (nm)	Concentration of the nanoparticles	Effect of nanoparticles on plants	References
Elodea densa	Leaf	Hydropoics	30	0.025, 0.25, 0.5, 1.0, and 5.0 mg Cu/L	Increased SOD and CAT at lower doses and decreased levels at higher doses Increased TBARS content till 1 mg/L and diminished by 5 mg/L Reduced photosynthesis at higher concentrations (at 5.0 mg decreased SOD and CAT)	Nekrasova et al. (2011)
Oryza sativa	Leaf	Hydroponics	<50	0.5 mM, 1.0 mM, and 1.5 mM	Reduced seed germination, increased H_2O_2 levels, increased proline, MDA, and reduced carotenoids. Increased APX, SOD, glutathione reductase, and MDAR. Reduced dehydroascorbate reductase	Shaw and Hossain (2013)
Glycine max	Roots	½ Ms	50	50–500 mg/L	Increase in hydrogen peroxide formation and lignin contents, altered phytohormones, reduced the development of shoot, fresh weight, and total chlorophyll a and b constitution	Nair and Chung (2014)
Brassica juncea	Roots	½ MS	<50	20–500 mg/L	Reduced APX and CAT in roots Reduced carotenoids after 100 mg/L Decrease in total chlorophyll Increased SOD Increased in H_2O_2 levels, MDA levels, peroxidase enzyme, and upsurge in lignin of shoots and roots	Nair and Chung (2015)
Brassica rapa	Seedling	Hydroponics	25–55	50, 250, and 500 mg/L	Decreased chlorophyll a and b content, carotenoid, and sugar content Increased MDA, H_2O_2 and ROS DNA damage and lipid peroxidation	Chung et al. (2019)

TABLE 13.3 Mechanism of plant response to cerium oxide nanoparticle.

Plant species	Part used in study	Type of culture	Size of NP (nm)	Concentration of the nanoparticles	Effect of nanoparticles on plants	References
Zea mays	Shoot	Soil	10 ± 1	400–800 mg/kg	Increased catalase and ascorbate peroxidase No changes in MDA content Upregulation of HSP70	Zhao et al. (2012)
Oryza sativa	Shoot and root	Hydroponics (filter paper on Petri dishes)	8 ± 1	62.5, 125, 250, and 500 mg/L	Amplification of enzyme activities and rise in levels of ascorbate and thiols, high MDA levels (500 mg/L) Reduced H_2O_2 levels in shoots and roots at 62.5 and 125 mg/L Altered oxidative enzymes	Rico et al. (2013)
Coriandrum sativum	Root and shoot	Soil	8	62.5–500 mg/kg	Alterations in APX and CAT activity	Morales et al. (2013)
Raphanus sativus	Leaf and root	Soil	<25 nm	10–100 mg/kg	Improved chlorophyll content Increased antioxidant enzyme activity	Gui et al. (2017)

TABLE 13.4 Mechanism of plant response to iron oxide nanoparticles.

NPs	Plant species	Part used in study	Type of culture	Size of NP (nm)	Concentration of the nanoparticles	Effect of nanoparticles on plants	References
Fe_2O_3	*Helianthus annuus*	Root and leaves	Hydroponics	20–100	50 and 100 mg/L	Inhibition of hydraulic conductivity Reduced micronutrient uptake Reduced chlorophyll concentration Yield parameters unchanged	Martínez-Fernández et al. (2016)
Fe_3O_4	*Lolium perenne* *Cucurbita mixta*	Root and shoots	Hydroponics	25	30 and 100 mg/L	Enhanced oxidative stress Increased SOD and CAT levels Increased MDA level in roots and shoots	Wang et al. (2011)

will aid in a more thorough knowledge of the mechanisms of interaction. Thereby, to fully exploit the potential of NPs, a thorough understanding of the various biological networks at different levels is required. Metabolomics has looked into the phytotoxicity caused by engineered nanomaterials, but it is still in a primitive state (Li et al., 2018). Numerous research studies have been undertaken to examine the metabolic impact of metal oxide NPs on plants and to investigate the mechanism of stress and toxicity (Table 13.5).

Overall, the metal oxide NPs impact mainly the carbohydrate, amino acid chemical processes, energy metabolism, oxidative stress response, lipid peroxidation, and secondary metabolism, according to research publications. This is consistent with the findings of Li et al. (2018), wherein they have comprehensively summarized the effect of NPs on four basic categories: carbohydrate, amino acid, lipid, and secondary metabolite metabolism. Studying the underlying molecular mechanisms will help us design the NPs designated for particular applications without making them toxic to the plants or the human system.

13.8 Determination of metal oxide nanoparticles toxicity

The toxicity metabolism of metallic oxide NPs might be explored by combining metabolomic, proteomic, transcriptomic, and other biochemical research. A schematic representation of the toxicity metabolic processes is seen in Fig. 13.1.

A reduction in biomass accumulation during photosynthesis indicated that the carbon metabolism and energy metabolism might be disturbed. Disturbed photosynthetic machinery in the *Elodea densa* was thought to be due to improper functioning of the pigment complex or to inactivation of RuBP carboxylase (Nekrasova et al., 2011). Enhanced accumulation of proline failed to protect the CuO NP treated rice plant at higher concentrations (Shaw and Hossain, 2013). Reduced levels of prolamin were initiated in the plants treated with 500 mg/kg of CeO_2 NPs (Rico et al., 2013). Seed storage proteins involved in the nutrient storing and carbohydrate metabolic process diminished in the dosage-reliant method when treated with CeO_2 NPs. Reduced production of the osmolite proline or other amino acids indicates a malfunctioning amino acid metabolic machinery, which leads to a reduction in precursors needed for the formation of antioxidant enzymes, resulting in the accumulation of ROS. The ROS produced would cause oxidative damage to the cells. The concentration of the antioxidant enzymes was higher, still the free radicals produced were also found to be higher in rice plants treated with CuO NPs (Shaw and Hossain, 2013). A possible explanation could be that the elevated levels of enzymes were still not sufficient for scavenging the free radicals produced. Similarly, in all the research articles mentioned in Tables 13.1–13.4, the increased activity of the enzymes might indicate higher production of antioxidant enzymes that are still not sufficient to combat

TABLE 13.5 Metabolomic effects in plants due to metal oxide nanoparticle exposure.

NP	Size (nm)	Plant	Route of exposure	Concentration	Investigated tissue	Pathways that show alterations	References
ZnO	50	*Cucumis sativus*	Foliar	100 mg/L	Leaf	Photosynthesis—increased chlorophyll contents and PSII activity Oxidative stress response through increased CAT and SOD activities Amino acid metabolism Flavone and flavonol glycosides biosynthesis Secondary metabolism Indole alkaloid biosynthesis	Li et al. (2021)
					Root	Isoquinoline alkaloid biosynthesis Amino acid metabolism—Tyrosine metabolism	
TiO_2	20	*Oryza sativa*	Soil	100, 250, 500 mg/L	Leaf	TCA cycle Pentose phosphate pathway Carbohydrate synthesis metabolism Fatty acid biosynthesis Amino acid biosynthesis Secondary metabolism	Wu et al. (2017)
TiO_2	5–10	*Zea mays*	Soil	100 mg/kg	Leaf	Amino acid metabolism Secondary metabolism Nonprotein amino acid accumulation Putrescine biosynthesis Glycerophospholipid, glyoxylate, and dicarboxylate metabolism	Zhao et al. (2019)

Fe_3O_4	30	*Zea mays*	Soil	100 mg/kg	Leaves	Nitrogen metabolism TCA cycle Glycolysis Gluconeogenesis Plant hormone biosynthesis Pyrimidine metabolism	Zhao et al. (2019)
CeO_2	4	*Spinaciaoleracea*	Foliar	0.3 mg/ plant	Leaves	Amino acid metabolism Amino sugar and nucleotide sugar metabolism Glycolysis Gluconeogenesis	Zhang et al. (2019)
				3 mg/ plant		Amino sugar and nucleotide sugar metabolism	
				0.3 mg/ plant	Root	Isoquinoline alkaloid biosynthesis Tyrosine metabolism	
				3 mg/ plant		Amino acid metabolism Sulfur metabolism	
CeO_2	10–30	*Phaseolus vulgaris*	Foliar	250–2000 mg/L	Leaf	Photosynthesis and electron transport chain Protein turnover and folding Oxidative and osmotic stress Biosynthesis of lipids and fatty acids	Salehi et al. (2018)
			Root		Leaf	Photosynthesis and electron transport chain Protein turnover and folding Oxidative stress	

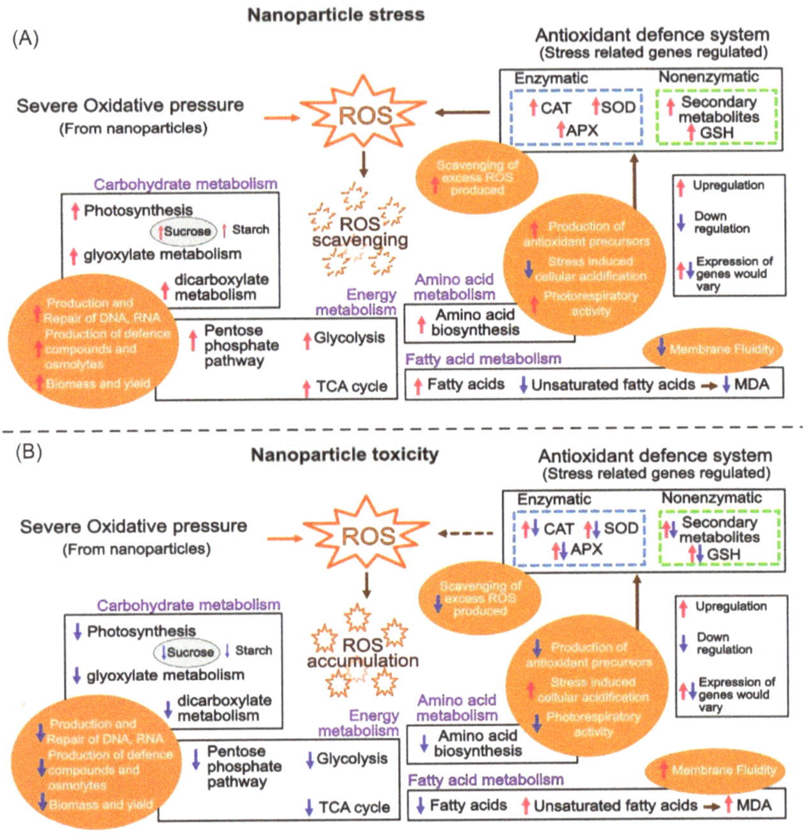

FIGURE 13.1 Overview of a possible mechanism of nanoparticle stress and toxicity in plants. (A) Plants response to nanoparticle stress. (B) Plant response to nanoparticle toxicity. The effect of nanoparticles on carbon, amino acid, lipid, and secondary metabolism, which would eventually lead to stress response (protection of cells) or toxicity (cellular damage) in florae (Li et al., 2018).

the stress induced. Prolonged exposure to CeO_2 (500 mg/kg) in beans resulted in decreased antioxidant activities. The reduction of CAT, SOD, APX, and GSH might indicate that at concentrations that are toxic to the plant, the entire pathway might be affected leading to the reduction of antioxidant enzymes and secondary metabolites. In studies, where the antioxidant enzymes were found to be higher, the malondialdehyde (MDA) levels were also found to be higher (Tables 13.1–13.3). Fatty acids form the integral component of the cell membranes. During times of stress, changes in fatty acid content may cause damage to cell membranes. Increased production of membrane lipids as a result of the ZnO NP stress in rice plants protected the cell membranes (Wu et al., 2017). There was increased activity of thiobarbituric acid reactive species in onion (Kumari et al., 2011) and MDA levels in tomatoes (Akanbi-Gada et al., 2019) in a dosage-reliant method upon acquaintance with ZnO NPs. These parameters indicate the levels of polyunsaturated fatty acid, which directly indicates membrane fluidity. Increased membrane fluidity causes the electrolytic leakage of the cell membranes and eventually cell damage (Fig. 13.1).

Mangalampalli et al. (2018) found that treatment with MgO NPs resulted in a dosage-reliant increase in ROS of *Allium cepa* root tips. Lipid peroxidation and MDA levels in the cells rose in tandem with the increasing superoxide and hydrogen peroxide radicals generated. The results also showed that as the dose of MgO NPs (up to 100 μg/mL) was raised, DNA damage and chromosomal abnormalities were increased and the mitotic index was decreased (Mangalampalli et al., 2018). Similarly, Liman et al. (2019) found that CeO_2 NPs at a concentration of reduced mitotic index increased chromosomal aberrations and DNA impairment in *A. cepa* root cells. In all studies, low doses of NP exposure were found to be favorable to the plant and did not cause any toxic effect on them.

13.9 Phytotoxicity of nanoparticles

NPs interact with plants to enter the plant cell and get accumulated in plant cells, which affect the fate and transport of NPs to the environment. NPs attach to the root of the NPs and accumulate in plant roots. It causes physiochemical toxicity to plants and the microorganisms present in the soil. Phytotoxicity is observed in plants due to metallic NPs by proclamation of higher absorption of essential ions and release of an excess of ROS over redox reaction.

13.10 Diverse metallic nanoparticles cause phytotoxicity

Different metal oxide NPs such as zinc oxide, copper oxide, gold, silver, titanium oxide, aluminum oxide, cerium oxide, and iron oxide were synthesized via topdown and bottomup planning approaches. Nanomaterials on exposure to plants as fertilizers, herbicides, and pesticides show both inhibitory and stimulatory effects. The responses of metallic NPs to major crops are listed in Tables 13.6 and 13.7.

13.10.1 Silver (Ag) nanoparticles

Ag NPs are one among the widely used NPs for commercial importance. The NP surface charge may contribute to the phytotoxicity in soil. It affects the internal cells of the plant by generating more ROS. Ag NPs also interrupt the electron transport chain in plants by obstructing the transport of electrons to affected tissues. Long time exposure to these NPs causes nutrient deficiency in plants (Zuverza-Mena et al., 2016). In *Oryza sativa*, Ag NPs $<30\,\mu g/L$ concentration destroys the biochemical integrity and structure of the cell and inhibits shoot and root growth. In the exemplar plant *Arabidopsis thaliana*, Ag NPs reduce the seed propagation of the plant. Due to Ag NP stress, photosynthetic pigment chlorophyll production was declined and biochemical parameters like starch, protein, and carbohydrate contents were reduced. The harmfulness of Ag NPs to plants is mainly due to the magnitude, surface extent, and surface charge of NPs.

13.10.2 Gold (Au) nanoparticles

Au NPs are commercially imperative and extensively utilized in various fields such as drug delivery, catalysis, and sensors. In plants, Au NPs increase light absorption which leads to an increase in the plasmon effect and reduce the photosynthetic pigment production by decreasing the electron transport to the cells (Falco et al., 2011). The translocation of Au NPs in plants toward different parts showed the necrosis of leaves and cells in plants. NPs activate the cell death cascade through the apoptosis mechanisms. When the size of the NPs increases, it causes toxicity in plants.

13.10.3 Cerium nanoparticles

CeO NPs are mainly used in fuel and electric-powered industries. Although cerium is intermittent ground metal, it enters the food chain and causes various effects on living organisms. CeO NPs affect the root length and growth of the plants. These NPs are less toxic than compared to other NPs. CeO NPs increase the efficacy of electron transport in plants. In low dosage, it enhances the growth of the plants. Even though it has positive effects, it shows some toxicity in plants like it correspondingly affects the epigenetic reminiscence of the plant (Morales et al., 2013).

13.10.4 Titanium nanoparticles

TiO_2 NPs are predominantly used in various applications like cosmetic industries, consumer goods, medical industries, and hospitality. The phytotoxicity of TiO_2 NPs in plants is mainly due to the size of the NPs. The NPs attached to the root of the plant inhibit the elongation of the root in a dose-reliant method. The dimensions of the NP interior of plant cells remain constant. The genome of the plant gets affected when a large quantity of TiO_2 NPs accumulate in plant cell and cause genotoxicity in plants. The phytotoxicity is mainly due to the production of ROS, consequential in disturbances of cellular free radicals.

TABLE 13.6 Response of crops on exposure to nanoparticles.

Crops	Nanoparticles	Concentration of nanoparticles	Impacts	Key references
Oryza sativa	AgNP < 20 nm	0.2, 0.5 and 1 μg/mL	Decreased shoot and root elongation, total chlorophyll content, and fresh weight of root.	Nair and Chung (2014)
Oryza sativa	AgNP < 100 nm	30, and 60 μg/mL	More than 60 μg/mL condensed the shoot and root development. The morphology of the root cells damage	Mirzajani et al. (2013)
Lycopersicon esculentum	AgNP = 10–15 nm	0, 50, 100, 1000, 2500, and 5000 μg/mL	Reduced elongation of root and shoot, increased ROS production, increased SOD, and decreased the fruit productivity	Song et al. (2013)
Arachis hypogaea	AgNP = 20 nm	50–2000 mg/kg	Reduction in herbal biomass, tallness, seed mass, and crop profit	Rui et al. (2017)
Hordeum vulgare	AuNP = 2–21 nm	0, 1, 3, 5, 8, and 10 μg/mL	Irrelevant discrepancy detected in seed germination, chlorosis of leaves, blackening of roots, and biomass reduction	Feichtmeier et al. (2015)
Triticum aestivum	CuO = 50 nm ZnO = 100 nm	500 mg/kg for each	Reduced shoot and root height, diminished biomass production, and condensed chlorophyll content	Dimkpa et al. (2012)
Viciafaba	Ag NP = 60 nm	12.5–100 mg/L	Genotoxic effect increased and chromosomal aberration is more. Decreases the mitotic index	Patlolla et al. (2012)
Arabidopsis thaliana	CuO = 30 nm	0.5, 1, 2, 5, 10, 20, 50, 100 mg/L	Dosage-reliant reduction in root length, fresh weightiness and chlorophyll constitution	Nair and Chung (2014)

(*Continued*)

TABLE 13.6 (Continued)

Crops	Nanoparticles	Concentration of nanoparticles	Impacts	Key references
Lemna minor	CuO = 40 nm	10–200 mg/L	Increases in peroxidase, CAT, SOD, and lipid peroxidation activity that impede plant development	Song et al. (2016)
Glycine max	ZnO NP <50 nm	500 ppm	Decreases root length, fresh weight of root, losses cell viability of root, decreases the fresh weight of shoot, downregulation of gene, and accumulation of more SOD	Yoon et al. (2014)
Zea mays	ZnO NP <50 nm	500–2000 mg/L	Inhibit the root prolongation	
Triticumaestivum	ZnO NP < 100 nm	500 mg/kg	Decreased root length, increased the oxidized glutathione and lipid peroxidation in roots, and reduced chlorophyll constituents in shoots.	Dimkpa et al. (2012)
Lactuca sativa	CuO NP = 5–15 nm	0.002–8 mg/L	Decreased seed propagation and root elongation	Hong et al. (2015)
Raphanussativus, Loliumrigidum	CuO NP = 6 nm and CuO NP bulk <200 nm	10–1000 mg/L	Inhibit shoot and root development, increase free radical damage to plant nucleic acid	Zuverza-Mena et al. (2016)
Zea mays	ZnO NP <410 nm	20 mg/L	The NPs accumulate and enter the root cells, root epidermis, and cortex	
Glycine max	ZnO = 10 nm	50–500 mg/kg soil	This NP accumulates in the entire tissues, particularly in leaves	Yoon et al. (2014)

Helianthus annuus	$Fe_2O_3 < 100$ nm	50 and 100 mg/L	It reduces the root hydraulic conductivity equipped 26% at 100 mg/L concentration of NPs	Martínez-Fernández et al. (2016)
Oryza sativa	$Fe_2O_3 = 7–13$ nm	2–200 mg/L	The root hormones ABA (abscisic acid) and IAA (indole-3-acetic acid) are inhibited by iron NP concentration	Yang et al. (2015)
Triticumaestivum	Cuo NP < 50 nm, ZnO NP < 100 nm	500 mg/kg soil	Reduce shoot and root dimensions, biomass development, decreased chlorophyll content, and increase peroxidase and catalase in roots	Savi et al. (2015)
Gossypiumhirsutum	SiO_2 NP < 30 nm	10–2000 mg/L	Reduced shoot and root length, and decreased root biomass production	Brar et al. (2021)
Gossypiumhirsutum	CeO_2 NPs	500–2000 mg/L	Decreased herbal height, shoot, and root biomass production, and reduced nutrients elements in roots	López-Moreno et al. (2010)

TABLE 13.7 Metal oxide nanoparticles and their primary mechanisms that affects the plants.

Metal oxide nanoparticles	Primary mechanisms effects in plants	References
Ag NPs (silver)	Degrades photosynthetic pigment (chlorophyll) and decreases the chlorophyll production Increases the oxidative stress Induces the intracellular ROS generation Inhibits the electron transport chain process Causes necrosis/apoptosis of cell Causes the genotoxicity and chromosomal damage	Falco et al. (2011) Oukarroum et al. (2013) Oukarroum et al. (2013) Vannini et al. (2013) Sabo-Attwood et al. (2012)
Au NPs (gold)	Causes necrosis of plant cell Causes genotoxicity in plants	Sabo-Attwood et al. (2012)
Zn/ZnO (zinc/ zinc oxide)	Expression of the gene affected Cellular toxicity increased due to a large amount of ROS generation	Lee et al. (2010) Pokhrel and Dubey (2013)

13.10.5 Zinc oxide nanoparticles

ZnO nanomaterials are sphere-shaped metal NPs with a large surface extent. ZnO NPs are used in various applications, such as antimicrobial, antifungal, and antibiotic agents. It is also used in various fields like cosmetics, medical, paintings, dressing materials, plastics, buildings, nanofibers, textiles, and solar cells (Pokhrel and Dubey, 2013). ZnO NPs are selectively permeable to the seed coat of certain plants. The seed germination and root length are affected in plants due to the large amount of usage of ZnO NPs. At 20 mg/L concentration of ZnO NPs, the root elongation and seed germination are affected in rape and ryegrass (Raskar and Laware, 2014). The cytological aberrations are seen in *Allium* sps. on exposure to ZnO NPs. On long time exposure of ZnO NPs (500 mg/L) on *Glycine max* plant affects the cellular mechanisms of excess ROS production in the plant due to oxidative stress and a greater number of free radicals were formed.

13.10.6 Iron nanoparticles

Iron NPs range between 1 and 100 nm. It is used in various fields. The NPs are superparamagnetic in nature. Owing to their superparamagnetic property, nanomaterials have extensive applications in electrical, catalysis, biomedical, etc. In plants, a large amount dosage of iron NPs causes cell damage and blocks the normal functioning of the cell and the aqua channel sextant in the cell membrane. Iron NPs enter the plant tissues through the root hairs and root cells, which affect the root length and reduce the leaves count and leaf width (Lee et al., 2010). Water uptake by iron NPs is affected by the dosage and size of the iron NPs (50–100 mg/L). It mainly affects the root surface, root hydraulic conductivity, and nutrient uptake. It also affects the macronutrient level in the shoot region and lowers the content of chlorophyll pigmentation (Martínez-Fernández et al., 2016).

13.11 Physiochemical stress due to nanoparticle exposure

Physiochemical process is the regular intracellular and extracellular process that takes place in normal plant cells. These processes are essential for the plants to get energy and for survival. Any aberration from this normal process causes physiochemical stress in plants. The physiochemical stress is caused by some intrinsic and extrinsic factors. Extrinsic factors like biotic and abiotic stress affect the normal functions of cells that cascade many intrinsic factors and affect cellular function and cause necrosis of cell or cell death. The physiochemical properties and their interaction comprehend monarchies of biochemistry, molecular biology, genetics, etc. These all fields are interrelated and deal with biological molecules and their chemical interactions and possessions that rely on physiochemical stress (Knorre et al., 2013).

Xenobiotic compounds' intrusion in the plants' propinquity disturbs the internal functions of plants, which alters the behavior of plants, and this drastic change in plants causes physiochemical stress in them (Knorre et al., 2013). These stresses affect the rate of germination of seed, root elongation, and the morphology of xylem; reduce leaf development; increase ROS production; reduce pigment production; and cause DNA damage and cell death. After exposure to a high dosage of NPs, few physiochemical stresses are observed and tabulated (Table 13.8).

NPs enter plants through various parts mainly through roots and leaves. NPs after perturbation interact with cellular and subcellular cells in plants and cause morphological changes that affect the biochemical reactions and physiological and molecular states of the plants (Ali et al., 2021). The diverse nanomaterials endorse stress in plants either by biotic or abiotic factors on plant evolution and progress at a concentration below dosage level by different mechanisms (Zulfiqar and Ashraf, 2021).

13.12 Genotoxicity effect in plants due to nanoparticles

The plant functions depend on the mechanisms that occur at a microscopic level. Nevertheless, evolution made at the molecular level is influenced by the NPs evaluating mechanisms and effects in plants. NPs are not effective

TABLE 13.8 Physiochemical stress in plants due to nanoparticle exposure.

Nanoparticles	Plants name	Symptoms
Titanium dioxide, NiO, Al_2O_3, MnO_2	*Allium cepa*, ryegrass, lettuce, corn, cucumber, *O. sativa*, *Nicotiana tabacum*	Damage the mitochondria, photosynthetic pigment production, electron transport chain, genotoxicity, and DNA
CdO, iron oxide, zinc oxide, NiO	*Lycopersicon esculentum*, *O. sativa*, *C. pepo*	Damage root elongation, harm xylem development, impair cell membrane, and increase intracellular ROS generation
Silver oxide, ZnO, NiO, CdO, iron oxide, cerium oxide, gold NPs, titanium dioxide, manganese dioxide, magnesium NPs	*Allium cepa*, *L. esculentum*, *Nicotiana tabacum*, ryegrass, *C. pepo*, *Cucumis sativus*	Limit the germination rate of plants, reduce the root length, seed development, decrease leaves development, damage the intracellular cells, and necrosis of plant cells

without interaction with cellular mechanisms. When the NPs enter the plant cells, it interacts with the cellular mechanisms and affects the gene expression, which shows effects on plant growth involved in diverse portions of the plant cells (Ali et al., 2021). There are some mechanisms involved in genotoxicity of plants. First mechanisms are involved in the release of lethal constituents from NPs to the plants. For illustration, from AgNPs, the free Ag^+ ions are released and some other toxic ions are released from NPs and cause DNA damage to plants. Fe^{2+}, Ag^+, Mn^+, Cu^+, and Zn^+ are transition ions released from iron oxide, silver, manganese, copper, and zinc oxide NPs that are involved in the generation of a large amount of ROS through a Fenton-type reaction. These released transition ions bind with DNA bases. The second mechanism is ROS generation via surface interactions. For instance, Ag^+ ions interact with H_2O_2 and produce more ROS. The third reaction is straight physiological interactions with NPs. For example, the silver NPs interact with cell membrane or DNA or they also interact with mitochondria and chloroplast which disrupts the cellular functions and interrupt the ATP synthesis resulting in DNA impairment and cell demise (Kruszewski et al., 2011). Pulate et al. (2011) reported that roots of *A. cepa* treated with AgNPs increase the chromosomal aberrations, decrease mitotic indices, increase mitotic abnormalities like C- metaphase, anaphasic bridge, and disturbed anaphase. Table 13.9 and Fig. 13.2 show the genotoxicity in plants.

TABLE 13.9 Genotoxicity effects of various nanoparticles on plants.

Plants name	Nanoparticles	Size (nm)	Genotoxicity	References
Oedogonium, Pitophora, and *Chlorella vulgaris*	Ag	10–15	Genetic damage, chromosomal abnormalities, and mitotic disturbances	Dash et al. (2012)
Loliumperenne	CuO	100	DNA impairment	Atha et al. (2012)
Allium cepa	TiO_2	< 25	Chromosomal abnormalities as well as chromosomal interruptions, besides sticky shirker chromosome	Pakrashi et al. (2014)
Glycine max	ZnO, CeO_2	20	Reduce mitotic indices, mutation, DNA damage, and chromosomal aberrations	Raskar and Laware (2014), López-Moreno et al. (2010)
Allium cepa	CuO	20–25	Destruction in mitotic phases in chromosomes with laggard disturbed telophase and vacuolated nucleus	Nagaonkar et al. (2015)

13.13 Interaction of genotoxic substances and their mechanisms

The genotoxicity mechanism divided into direct and indirect mechanisms is shown in Fig. 13.2.

13.13.1 Direct mechanism of genotoxicity

It involves the physical interactions with DNA that influence the damage in the DNA base, which causes phosphorylation and alters the expression of genes. The surface area and surface charge of the NPs are important factors for causing genotoxicity in plants (Tae Kim et al., 2013). The NPs that cross the cell membrane and nuclear membrane interact directly with DNA that is organized in chromosomes or chromatin depending on the cell cycle through

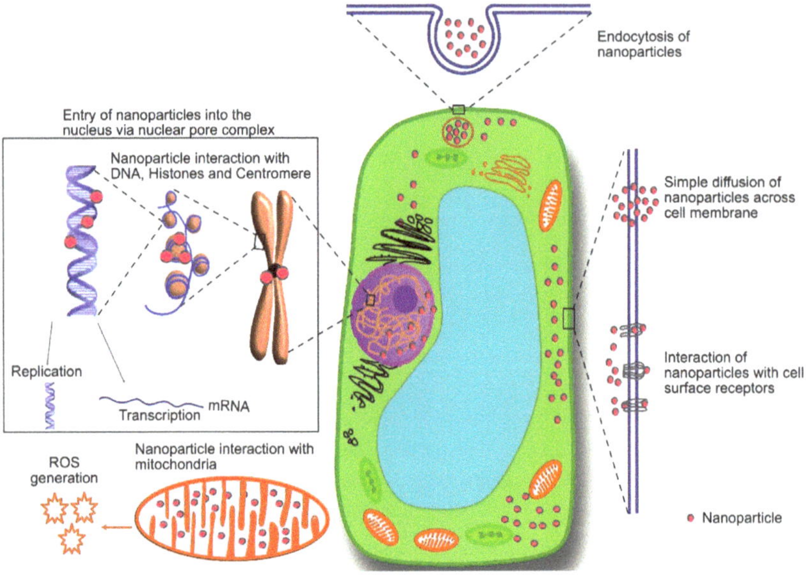

FIGURE 13.2 Interaction of nanoparticles with genetic material: chromosomal abnormalities and cytotoxic effects.

endocytosis or diffusion mechanisms. This causes effects on replication, and transcription in DNA structures that affects the mitotic indices, causing chromosomal breakage and loss of chromosomes (Wang et al., 2013).

13.13.2 Indirect mechanism of genotoxicity

In this indirect mechanism, the NPs interact with nuclear protein (proteins involved in transcription, replication, and repair mechanisms) not with DNA or mitotic spindles. The NPs also interrupt the cell cycle checkpoints, come into contact with enzymes used in antioxidant mechanisms, induce ROS generation, and defect the cell cycle mechanisms. It also reduces the DNA repair mechanisms, increases oxidative stress, and alters gene expressions (Wang et al., 2013). For occurrence, the interface of NPs with glutathione (antioxidant components) decreases glutathione and superoxide dismutase (antioxidant enzymes) and generation of more ROS causes DNA damage (Barillet et al., 2010).

13.14 Chromosome interaction with genetic material

13.14.1 Interactions of nanoparticles with chromosome and genetic material

The plants exposed to NPs show chromosome breakage, chromosome stickiness, translocation, fragmentations, gaps, and micronuclei formation and predominantly cause chromosomal aberrations (Kumari et al., 2011). The NPs like Ag NPs, CuO, and ZnO cause inhibition of DNA synthesis and M phase of the cell cycle in mitosis, and interrupt the transcription process in DNA to RNA and the replication of DNA. For instance, AgNPs induce chromosomal stickiness, mixture of chromatin fibers, and loss of genetic material in plants. Chromosomal stickiness occurs in plants mainly in metaphase (physical adhesion) and makes chromosomes sticky together. Induction of aberrations causes chromosomal bridges or breaks leading to forfeiture of genetic substantial, that is, DNA. Apart from DNA interactions, it also interrelates with nuclear protein intricate in DNA replications, transcription, and repair mechanisms (Magdolenova et al., 2014). When these types of interactions occur, the loss of protein takes place leading to the formation of a gap in chromosomes and loss of chromatin. The nuclear proteins will be inactivated by a large amount of ROS generation (Fig. 13.2) (Magdolenova et al., 2014).

13.14.2 Effects of nanoparticles on spindle fiber formation

NPs are one of the aneugenic agents interacting through mitotic spindle assembly checkpoints, centrioles otherwise allied proteins lead to gain or loss of daughter cells. The high mitotic anomalies, namely chromosome sheathing, nuclear fragmentation, genetic material collapse or break down, alteration of anaphase, and C-metaphase, cause effects in mitotic spindle fiber formation. The plant on exposure to AgNPs causing interaction with the sulfhydryl (SH) group in tubulin induces impairment in spindle fiber formation and functions and alters the chromosomal orientations at different phases of the cell-division cycle (Tanti et al., 2012).

13.14.3 Nanoparticle interactions on micronuclei formation and DNA damage

During the cell division process, DNA breaks and induces micronuclei formation, which results in fragmentation and laggards from the nucleus and leads to disruption of spindle fiber formation, leading to the loss of genetic material. The plant cells on exposure to NPs like Ag, Cu, Zn, and Au induce micronuclei development, which causes alteration in genetic material and inhibition of S-phase that damage the DNA (Hackenberg et al., 2011b).

13.14.4 Interaction of nanoparticle toxicity on cell cycle

The plants on exposure to NPs could also interact with protein kinase enzymes that are involved in signal transduction and regulation of cell division and cell cycle causing disturbances in cell cycle checkpoints and functions. The NPs disrupt the mitotic checkpoint and enzymes involved in the cell cycle like protein kinase and cytokinesis (Magdolenova et al., 2014).

13.15 Conclusion and future prospective

NPs are predominantly used in agricultural fields for nutrient management and to increase the growth and yield of crops. As NPs move to various parts of the plants and the plant cells, it interacts with the plant's physiochemical properties and biological functions and also with the genes of the plants. The toxicity of the NPs relies upon surface area, superficial charge, and size in addition to NP concentration. When the dosage of the NPs increases, the cellular mechanisms of the plant are affected and cause an upsurge in the creation of ROS and a decline in the activity of antioxidant enzymes like SOD that cause genotoxicity in plants and physiological changes in plants. Chemical composition, surface charge, size, agglomeration, and structure of NPs have an impact on genotoxicity in plants. After all types of effects, lastly, when DNA repair mechanisms are incomplete, it leads to the loss of hereditary material that causes gene variants. Nanomaterials have adverse effects on physiological, cellular, and molecular mechanisms in plants that may indirectly affect humans and microorganisms, which is not good for the future generation. Extreme care should be taken to prevent the entry of NPs into the food chain due to its nanotoxicological effects seen in plants because plants are the primary producers in the food chain where every organism depends on them for their energy production. Therefore, the increasing concentration of NPs causes a serious influence on the environment and agriculture in the future. Before NPs come out as products, a hefty expanse of field imprints is desired to appraise the likelihood of the NPs.

References

Aghdam, M.T.B., Mohammadi, H., Ghorbanpour, M., 2016. Effects of nanoparticulateanatase titanium dioxide on physiological and biochemical performance of Linumusitatissimum (Linaceae) under well-watered and drought stress conditions. Brazil. J. Bot. 39 (1), 139–146.

Akanbi-Gada, M.A., Ogunkunle, C.O., Vishwakarma, V., Viswanathan, K., Fatoba, P.O., 2019. Phytotoxicity of nano-zinc oxide to tomato plant (*Solanumlycopersicum L.*): Zn uptake, stress enzymes response and influence on non-enzymatic antioxidants in fruits. Environ. Technol. Innov. 14, 100325.

Ali, S., Iqbal, M., Naseer, A., Yaseen, M., Bibi, I., Nazir, A., et al., 2021. State of the art of gold (Au) nanoparticles synthesis via green routs and applications: a review. Environ. Nanotechnol., Monitor. Manage. 100511.

Alidoust, D., Isoda, A., 2013. Effect of γFe2O3 nanoparticles on photosynthetic characteristic of soybean (*Glycine max* (L.) Merr.): foliar spray vs soil amendment. Acta Physiol. Plant. 35 (12), 3365–3375.

Atha, D.H., Wang, H., Petersen, E.J., Cleveland, D., Holbrook, R.D., Jaruga, P., et al., 2012. Copper oxide nanoparticle mediated DNA damage in terrestrial plant models. Environ. Sci. Technol. 46 (3), 1819–1827.

Barillet, S., Jugan, M.L., Laye, M., Leconte, Y., Herlin-Boime, N., Reynaud, C., et al., 2010. In vitro evaluation of SiC nanoparticles impact on A549 pulmonary cells, cyto-genotoxicity and oxidative stress. Toxicol. Lett. 198, 324–330.

Bombin, S., LeFebvre, M., Sherwood, J., Xu, Y., Bao, Y., Ramonell, K.M., 2015. Developmental and reproductive effects of iron oxide nanoparticles in *Arabidopsis thaliana*. Int. J. Mol. Sci. 16 (10), 24174–24193.

Bradfield, S.J., Kumar, P., White, J.C., Ebbs, S.D., 2017. Zinc, copper, or cerium accumulation from metal oxide nanoparticles or ions in sweet potato: yield effects and projected dietary intake from consumption. Plant Physiol. Biochem. 110, 128–137.

Chen, J., Dou, R., Yang, Z., You, T., Gao, X., Wang, L., 2018. Phytotoxicity and bioaccumulation of zinc oxide nanoparticles in rice (*Oryza sativa L.*). Plant Physiol. Biochem. 130, 604–612.

Chung, I.M., Rekha, K., Venkidasamy, B., Thiruvengadam, M., 2019. Effect of copper oxide nanoparticles on the physiology, bioactive molecules, and transcriptional changes in *Brassica rapa* ssp. rapa seedlings. Water Air Soil Pollut. 230 (2), 1–14.

Corredor, E., Testillano, P.S., Coronado, M.J., González-Melendi, P., Fernández-Pacheco, R., Marquina, C., et al., 2009. Nanoparticle penetration and transport in living pumpkin plants: in situ subcellular identification. BMC Plant Biol. 9 (1), 45.

Cvjetko, P., Milošić, A., Domijan, A.M., Vrček, I.V., Tolić, S., Štefanić, P.P., et al., 2017. Toxicity of silver ions and differently coated silver nanoparticles in *Allium cepa* roots. Ecotoxicol. Environ. Saf. 137, 18–28.

Da Rosa, E.L.S., 2013. Kinetic effects of TiO2 fine particles and nanoparticles aggregates on the nanomechanical properties of human neutrophils assessed by force spectroscopy. BMC Biophys. 6 (1), 1–10.

Das, K., Roychoudhury, A., 2014. Reactive oxygen species (ROS) and response of antioxidants as ROS-scavengers during environmental stress in plants. Front. Environ. Sci. 2, 53.

Dash, A., Singh, A.P., Chaudhary, B.R., Singh, S.K., Dash, D., 2012. Effect of silver nanoparticles on growth of eukaryotic green algae. Nano Micro Lett. 4, 158–165.

Dimkpa, C.O., McLean, J.E., Latta, D.E., Manangon, E., Britt, D.W., Johnson, W.P., et al., 2012. CuO and ZnO nanoparticles: phytotoxicity, metal speciation, and induction of oxidative stress in sand-grown wheat. J. Nanopart. Res. 14, 1125.

Du, W., Gardea-Torresdey, J.L., Ji, R., Yin, Y., Zhu, J., Peralta-Videa, J.R., et al., 2015. Physiological and biochemical changes imposed by CeO2 nanoparticles on wheat: a life cycle field study. Environ. Sci. Technol. 49 (19), 11884–11893.

Ebbs, S.D., Bradfield, S.J., Kumar, P., White, J.C., Musante, C., Ma, X., 2016. Accumulation of zinc, copper, or cerium in carrot (Daucuscarota) exposed to metal oxide nanoparticles and metal ions. Environ. Sci.: Nano 3 (1), 114–126.

Emamverdian, A., Ding, Y., Mokhberdoran, F., Ahmad, Z., Xie, Y., 2020. Determination of heavy metal tolerance threshold in a bamboo species (Arundinariapygmaea) as treated with silicon dioxide nanoparticles. Global Ecol. Conser. 24, e01306.

Falco, W.F., Botero, E.R., Falcão, E.A., Santiago, E.F., Bagnato, V.S., Caires, A.R.L., 2011. In vivo observation of chlorophyll fluorescence quenching induced by gold nanoparticles. J. Photochem. Photobiol. Chem. 225 (1), 65–71.

Feichtmeier, N.S., Walther, P., Leopold, K., 2015. Uptake, effects, and regeneration of barley plants exposed to gold nanoparticles. Environ. Sci. Pollut. Res. Int. 22, 8549–8558.

Fraceto, L.F., Grillo, R., de Medeiros, G.A., Scognamiglio, V., Rea, G., Bartolucci, C., 2016. Nanotechnology in agriculture: which innovation potential does it have? Front. Environ. Sci. 4, 20.

Fraceto, L.F., de Lima, R., Oliveira, H.C., Ávila, D.S., Chen, B., 2018. Future trends in nanotechnology aiming environmental applications. Energy, Ecol. Environ. 3 (2), 69–71.

Frazier, T.P., Burklew, C.E., Zhang, B., 2014. Titanium dioxide nanoparticles affect the growth and microRNA expression of tobacco (Nicotianatabacum). Funct. Integr. Genomics 14 (1), 75–83.

Gautam, R.K., Chattopadhyaya, M.C., 2016. Nanomaterials in the environment: sources, fate, transport, and ecotoxicology. Nanomaterials for Wastewater Remediation. Elsevier, Amsterdam, The Netherlands, pp. 311–326, ISBN 978-0-12-804609-8.

Goswami, A., Roy, I., Sengupta, S., Debnath, N., 2010. Novel applications of solid and liquid formulations of nanoparticles against insect pests and pathogens. Thin Solid Films 519, 1252–1257.

Gui, X., Rui, M., Song, Y., Ma, Y., Rui, Y., Zhang, P., et al., 2017. Phytotoxicity of CeO2 nanoparticles on radish plant (Raphanussativus). Environ. Sci. Pollut. Res. 24 (15), 13775–13781.

Hackenberg, S., Scherzed, A., Technau, A., Kessler, M., Froelich, K., Ginzkey, C., et al., 2011b. Cytotoxic, genotoxic and pro-inflammatory effects of zinc oxide nanoparticles in human nasal mucosa cells in vitro. Toxicol. In Vitro 25, 657–663.

Hobson, D.W., 2016. Nanotoxicology: the toxicology of nanomaterials and nanostructures. Int. J. Toxicol. 35, 3–4.

Hong, J., Rico, C.M., Zhao, L., Adeleye, A.S., Keller, A.A., Peralta-Videa, J.R., et al., 2015. Toxic effects of copper-based nanoparticles or compounds to lettuce (*Lactuca sativa*) and alfalfa (*Medicago sativa*). Environ. Sci. Process Impacts 17, 177–185.

Hossain, Z., Mustafa, G., Komatsu, S., 2015. Plant responses to nanoparticle stress. Int. J. Mol. Sci. 16 (11), 26644–26653.

Iqbal, M.S., Singh, A.K., Ansari, M.I., 2020. Effect of drought stress on crop production. New Frontiers in Stress Management for Durable Agriculture. Springer, Singapore, pp. 35–47.

Kah, M., 2015. Nanopesticides and nanofertilizers: emerging contaminants or opportunities for risk mitigation? Front. Chem. 3, 64.

Kah, M., Beulke, S., Tiede, K., Hofmann, T., 2013. Nanopesticides: state of knowledge, environmental fate and exposure modelling. Crit. Rev. Environ. Sci. Technol. 43, 1823–1867. Available from: https://doi.org/10.1080/10643389.2012.671750.

Kah, M., Kookana, R.S., Gogos, A., Bucheli, T.D., 2018. A critical evaluation of nanopesticides and nanofertilizers against their conventional analogues. Nat. Nanotechnol. 13 (8), 677–684.

Kairyte, K., Kadys, A., Luksiene, Z., 2013. Antibacterial and antifungal activity of photoactivated ZnO nanoparticles in suspension. J. Photochem. Photobiol. B: Biol. 128, 78–84.

Kaushal, M., Wani, S.P., 2017. Nanosensors: frontiers in precision agriculture. Nanotechnology. Springer, Singapore, pp. 279–291.

Knorre, D.A., Popadin, K.Y., Sokolov, S.S., Severin, F.F., 2013. Roles of mitochondrial dynamics under stressful and normal conditions in yeast cells. Oxid. Med. Cell Longev. 2013, 139491.

Kruszewski, M., Brzoska, K., Brunborg, G., Asare, N., Dobrzyńska, M., Dušinská, M., et al., 2011. Toxicity of silver nanomaterials in higher eukaryotes. Adv. Mol. Toxicol. 5, 179–218.

Kumari, M., Khan, S.S., Pakrashi, S., Mukherjee, A., Chandrasekaran, N., 2011. Cytogenetic and genotoxic effects of zinc oxide nanoparticles on root cells of *Allium cepa*. J. Hazard. Mater. 190 (1–3), 613–621.

Landa, P., Prerostova, S., Petrova, S., Knirsch, V., Vankova, R., Vanek, T., 2015. The transcriptomic response of *Arabidopsis thaliana* to zinc oxide: a comparison of the impact of nanoparticle, bulk, and ionic zinc. Environ. Sci. Technol. 49 (24), 14537–14545.

Lee, C.W., Mahendra, S., Zodrow, K., Li, D., Tsai, Y.C., Braam, J., et al., 2010. Developmental phytotoxicity of metal oxide nanoparticles to *Arabidopsis thaliana*. Environ. Toxicol. Chem. 29 (3), 669–675.

Lee, S., Kim, S., Kim, S., Lee, I., 2013. Assessment of phytotoxicity of ZnO NPs on a medicinal plant, Fagopyrumesculentum. Environ. Sci. Pollut. Res. 20 (2), 848–854.

Li, C., Wang, P., Lombi, E., Cheng, M., Tang, C., Howard, D.L., et al., 2018. Absorption of foliar-applied Zn fertilizers by trichomes in soybean and tomato. J. Exp. Bot. 69 (10), 2717–2729.

Li, S., Liu, J., Wang, Y., Gao, Y., Zhang, Z., Xu, J., et al., 2021. Comparative physiological and metabolomic analyses revealed that foliar spraying with zinc oxide and silica nanoparticles modulates metabolite profiles in cucumber (*Cucumissativus L.*). Food Energy Sec. 10 (1), e269.

Liman, R., Acikbas, Y., Ciğerci, İ.H., 2019. Cytotoxicity and genotoxicity of cerium oxide micro and nanoparticles by Allium and Comet tests. Ecotoxicol. Environ. Saf. 168, 408–414.

Liu, R., Lal, R., 2015. Potentials of engineered nanoparticles as fertilizers for increasing agronomic productions. Sci. Total Environ. 514, 131–139.

López-Moreno, M.L., De la Rosa, G., Hernández-Viezcas, Á., J., Castillo-Michel, H., Botez, C. E., Peralta-Videa, J.R., et al., 2010. Evidence of the differential biotransformation and genotoxicity of ZnO and CeO2 nanoparticles on soybean, *Glycine max* plants. Environ. Sci. Technol. 44, 7315–7320.

Ma, X., Geiser-Lee, J., Deng, Y., Kolmakov, A., 2010a. Interactions between engineered nanoparticles (ENPs) and plants: phytotoxicity, uptake and accumulation. Sci. Total Environ. 408 (16), 3053–3061.

Ma, Y., Kuang, L., He, X., Bai, W., Ding, Y., Zhang, Z., et al., 2010b. Effects of rare earth oxide nanoparticles on root elongation of plants. Chemosphere 78 (3), 273–279.

Ma, C., White, J.C., Dhankher, O.P., Xing, B., 2015. Metal-based nanotoxicity and detoxification pathways in higher plants. Environ. Sci. Technol. 49 (12), 7109–7122.

Magdolenova, Z., Collins, A., Kumar, A., Dhawan, A., Stone, V., Dusinska, M., 2014. Mechanisms of genotoxicity: a review of in vitro and in vivo studies with engineered nanoparticles. Nanotoxicology 8, 233–278.

Malaikozhundan, B., Vinodhini, J., 2018. Nanopesticidal effects of Pongamiapinnata leaf extract coated zinc oxide nanoparticle against the Pulse beetle, Callosobruchusmaculatus. Mater. Today Commun. 14, 106–115.

Mangalampalli, B., Dumala, N., Grover, P., 2018. *Allium cepa* root tip assay in assessment of toxicity of magnesium oxide nanoparticles and microparticles. J. Environ. Sci. 66, 125–137.

Martínez-Fernández, D., Barroso, D., Komárek, M., 2016. Root water transport of Helianthus annuus L. under iron oxide nanoparticle exposure. Environ. Sci. Pollut. Res. 23 (2), 1732–1741.

Mirzajani, F., Askari, H., Hamzelou, S., Farzaneh, M., Ghassempour, A., 2013. Effect of silver nanoparticles on *Oryza sativa L.* and its rhizosphere bacteria. Ecotoxicol. Environ. Saf. 88, 48–54.

Monreal, C.M., DeRosa, M., Mallubhotla, S.C., Bindraban, P.S., Dimkpa, C., 2016. Nanotechnologies for increasing the crop use efficiency of fertilizer-micronutrients. Biol. Fertil. Soils 52 (3), 423–437.

Morales, M.I., Rico, C.M., Hernandez-Viezcas, J.A., Nunez, J.E., Barrios, A.C., Tafoya, A., et al., 2013. Toxicity assessment of cerium oxide nanoparticles in cilantro (*Coriandrumsativum L.*) plants grown in organic soil. J. Agric. Food Chem. 61 (26), 6224–6230.

Nagaonkar, D., Shende, S., Rai, M., 2015. Biosynthesis of copper nanoparticles and its effect on actively dividing cells of mitosis in *Allium cepa*. Biotechnol. Prog. 31, 557–565.

Nair, P.M.G., Chung, I.M., 2014. A mechanistic study on the toxic effect of copper oxide nanoparticles in soybean (*Glycine max* L.) root development and lignification of root cells. Biol. Trace Elem. Res. 162 (1), 342–352.

Nair, P.M.G., Chung, I.M., 2015. Study on the correlation between copper oxide nanoparticles induced growth suppression and enhanced lignification in Indian mustard (*Brassica juncea L.*). Ecotoxicol. Environ. Saf. 113 (302), 313.

Nekrasova, G.F., Ushakova, O.S., Ermakov, A.E., Uimin, M.A., Byzov, I.V., 2011. Effects of copper (II) ions and copper oxide nanoparticles on Elodea densa Planch. Russ. J. Ecol. 42 (6), 458–463.

Oukarroum, A., Barhoumi, L., Pirastru, L., Dewez, D., 2013. Silver nanoparticle toxicity effect on growth and cellular viability of the aquatic plant Lemnagibba. Environ. Toxicol. Chem. 32 (4), 902–907.

Pacheco, I., Vanner, R., Buzea, C., 2010. Principles of nanotoxicology. In: Proceedings of the Association of Industrial Metallizers, Coaters and Laminators Fall Technical Conference and 24th International Vacuum Web Coating Conference 2010, Myrtle Beach, SC, 17–20 October 2010, pp. 1367–1393.

Pakrashi, S., Jain, N., Dalai, S., Jayakumar, J., Chandrasekaran, P.T., Raichur, A.M., et al., 2014. In vivo genotoxicity assessment of titanium dioxide nanoparticles by *Allium cepa* root tip assay at high exposure concentrations. PLoS One 9 (2), e87789.

Pandey, S., Giri, K., Kumar, R., Mishra, G., Raja Rishi, R., 2018. Nanopesticides: opportunities in crop protection and associated environmental risks. Proc. Natl. Acad. Sci., India Sect. B: Biol. Sci. 88 (4), 1287–1308.

Parisi, C., Vigani, M., Rodríguez-Cerezo, E., 2015. Agricultural nanotechnologies: what are the current possibilities? Nano Today 10 (2), 124–127.

Patlolla, A.K., Berry, A., May, L., Tchounwou, P.B., 2012. Genotoxicity of silver nanoparticles in Vicia faba: a pilot study on the environmental monitoring of nanoparticles. Int. J. Environ. Res. Public Health 9 (5), 1649–1662.

Pokhrel, L.R., Dubey, B., 2013. Evaluation of developmental responses of two crop plants exposed to silver and zinc oxide nanoparticles. Sci. Total Environ. 452, 321–332.

Prasad, T.N.V.K.V., Sudhakar, P., Sreenivasulu, Y., Latha, P., Munaswamy, V., Reddy, K.R., et al., 2012. Effect of nanoscale zinc oxide particles on the germination, growth and yield of peanut. J. Plant Nutr. 35 (6), 905–927.

Prasad, R., Bhattacharyya, A., Nguyen, Q.D., 2017. Nanotechnology in sustainable agriculture: recent developments, challenges, and perspectives. Front. Microbiol. 8, 1014.

Pulate, P.V., Ghurde, M.U., Deshmukh, V.R., 2011. Cytological effects of the biological and chemical silver-nanoparticles in *Allium cepa* L. Int. J. Innov. BioSci. 1, 32–35.

Raliya, R., Tarafdar, J.C., 2013. ZnO nanoparticle biosynthesis and its effect on phosphorous-mobilizing enzyme secretion and gum contents in Clusterbean (*Cyamopsistetragonoloba L.*). Agri. Res. 2 (1), 48–57.

Raliya, R., Nair, R., Chavalmane, S., Wang, W.N., Biswas, P., 2015. Mechanistic evaluation of translocation and physiological impact of titanium dioxide and zinc oxide nanoparticles on the tomato (*Solanumlycopersicum L.*) plant. Metallomics 7 (12), 1584–1594.

Raskar, S.V., Laware, S.L., 2014. Effect of zinc oxide nanoparticles on cytology and seed germination in onion. Int. J. Curr. Microbiol. Appl. Sci. 3, 467–473.

Rastogi, A., Zivcak, M., Sytar, O., Kalaji, H.M., He, X., Mbarki, S., et al., 2017. Impact of metal and metal oxide nanoparticles on plant: a critical review. Front. Chem. 5, 78.

Rico, C.M., Majumdar, S., Duarte-Gardea, M., Peralta-Videa, J.R., Gardea-Torresdey, J.L., 2011. Interaction of nanoparticles with edible plants and their possible implications in the food chain. J. Agric. Food Chem. 59 (8), 3485–3498.

Rico, C.M., Hong, J., Morales, M.I., Zhao, L., Barrios, A.C., Zhang, J.Y., et al., 2013. Effect of cerium oxide nanoparticles on rice: a study involving the antioxidant defense system and in vivo fluorescence imaging. Environ. Sci. Technol. 47 (11), 5635–5642.

Rico, C.M., Lee, S.C., Rubenecia, R., Mukherjee, A., Hong, J., Peralta-Videa, J.R., et al., 2014. Cerium oxide nanoparticles impact yield and modify nutritional parameters in wheat (*Triticumaestivum L.*). J. Agric. Food Chem. 62 (40), 9669–9675.

Rizwan, M., Ali, S., Qayyum, M.F., Ok, Y.S., Adrees, M., Ibrahim, M., et al., 2017. Effect of metal and metal oxide nanoparticles on growth and physiology of globally important food crops: a critical review. J. Hazard. Mater. 322, 2–16.

Rizwan, M., Ali, S., Ali, B., Adrees, M., Arshad, M., Hussain, A., et al., 2019. Zinc and iron oxide nanoparticles improved the plant growth and reduced the oxidative stress and cadmium concentration in wheat. Chemosphere 214, 269–277.

Rossi, L., Zhang, W., Ma, X., 2017. Cerium oxide nanoparticles alter the salt stress tolerance of *Brassica napus L.* by modifying the formation of root apoplastic barriers. Environ. Pollut. 229, 132–138.

Rui, M., Ma, C., Tang, X., Yang, J., Jiang, F., Pan, Y., et al., 2017. Phytotoxicity of silver nanoparticles to peanut (*Arachishypogaea L.*): physiological responses and food safety. ACS Sustain. Chem. Eng. 5, 6557–6567.

Sabo-Attwood, T., Unrine, J.M., Stone, J.W., Murphy, C.J., Ghoshroy, S., Blom, D., et al., 2012. Uptake, distribution and toxicity of gold nanoparticles in tobacco (Nicotianaxanthii) seedlings. Nanotoxicology 6 (4), 353–360.

Salehi, H., Chehregani, A., Lucini, L., Majd, A., Gholami, M., 2018. Morphological, proteomic and metabolomic insight into the effect of cerium dioxide nanoparticles to *Phaseolus vulgaris L.* under soil or foliar application. Sci. Total Environ. 616, 1540–1551.

Savi, G.D., Piacentini, K.C., de Souza, S.R., Costa, M.E.B., Santos, C.M.R., Scussel, V.M., 2015. Efcacy of zinc compounds in controlling Fusarium head blight anddeoxynivalenol formation in wheat (*Triticumaestivum L.*). Int. J. Food Microbiol. 205, 98–104.

Shaw, A.K., Hossain, Z., 2013. Impact of nano-CuO stress on rice (*Oryza sativa L.*) seedlings. Chemosphere 93 (6), 906–915.

Siddiqi, K.S., Husen, A., 2016. Biogenic fabrication of iron/iron oxide nanoparticles and their application. Nanoscale Res. Lett. 11 (1), 1–13.

Siddiqi, K.S., Husen, A., 2017. Plant response to engineered metal oxide nanoparticles. Nanoscale Res. Lett. 12 (1), 1–18.

Singh Brar, R., Kumar, A., Kaur, S., et al., 2021. Impact of metal oxide nanoparticles on cotton (*Gossypiumhirsutum* L.): a physiological perspective. J. Cotton Res. 4, 16.

Singh, N.B., Amist, N., Yadav, K., Singh, D., Pandey, J.K., Singh, S.C., 2013. Zinc oxide nanoparticles as fertilizer for the germination, growth and metabolism of vegetable crops. J. Nanoeng. Nanomanuf. 3 (4), 353–364.

Sohail, M.I., Ayub, M.A., urRehman, M.Z., Azhar, M., Farooqi, Z.U.R., Siddiqui, A., et al., 2021. Sufficiency and toxicity limits of metallic oxide nanoparticles in the biosphere. Nanomaterials: Synthesis, Characterization, Hazards and Safety. Elsevier, pp. 145–221.

Song, U., Jun, H., Waldman, B., Roh, J., Kim, Y., Yi, J., et al., 2013. Functional analyses of nanoparticle toxicity: a comparative study of the effects of TiO2 and Ag on tomatoes (*Lycopersiconesculentum*). Ecotoxicol. Environ. Saf. 93, 60–67.

Song, G., Hou, W., Gao, Y., Wang, Y., Lin, L., Zhang, Z., et al., 2016. Effects of CuO nanoparticles on *Lemna minor*. Bot. Stud. 57 (1), 1–8.

Subbaiah, L.V., Prasad, T.N.V.K.V., Krishna, T.G., Sudhakar, P., Reddy, B.R., Pradeep, T., 2016. Novel effects of nanoparticulate delivery of zinc on growth, productivity, and zinc biofortification in maize (*Zea mays L.*). J. Agric. Food Chem. 64 (19), 3778–3788.

Sundaria, N., Singh, M., Upreti, P., Chauhan, R.P., Jaiswal, J.P., Kumar, A., 2019. Seed priming with iron oxide nanoparticles triggers iron acquisition and biofortification in wheat (*Triticumaestivum L.*) grains. J. Plant Growth Regul. 38 (1), 122–131.

Tae Kim, S., Saha, K., Kim, C., Rotello, V.M., 2013. The role of surface functionality in determining nanoparticle cytotoxicity. Acc. Chem. Res. 46, 681–691.

Tanti, B., Kumar Das, A., Kakati, H., Chowdhury, D., 2012. Cytotoxic effect of silver-nanoparticles on root meristem of *Allium sativum L.* J. Res. Nanobiotechnol. 1, 1–8.

Tarafdar, J., Xiong, Y., Wang, W.-N., Quinl, D., Biswas, P., 2012. Standardization of size, shape and concentration of nanoparticle for plant application. Appl. Biol. Res. 14, 138–144.

Tiwari, E., Mondal, M., Singh, N., Khandelwal, N., Monikh, F.A., Darbha, G.K., 2020. Effect of the irrigation water type and other environmental parameters on CeO2 nanopesticide–clay colloid interactions. Environ. Sci.: Processes Impacts 22 (1), 84–94.

Vannini, C., Domingo, G., Onelli, E., Prinsi, B., Marsoni, M., Espen, L., et al., 2013. Morphological and proteomic responses of *Eruca sativa* exposed to silver nanoparticles or silver nitrate. PLoS One 8 (7), e68752.

Venkatachalam, P., Jayaraj, M., Manikandan, R., Geetha, N., Rene, E.R., Sharma, N.C., et al., 2017. Zinc oxide nanoparticles (ZnONPs) alleviate heavy metal-induced toxicity in Leucaenaleucocephala seedlings: a physiochemical analysis. Plant Physiol. Biochem. 110, 59–69.

Wang, H., Kou, X., Pei, Z., Xiao, J.Q., Shan, X., Xing, B., 2011. Physiological effects of magnetite (Fe3O4) nanoparticles on perennial ryegrass (*Loliumperenne L.*) and pumpkin (*Cucurbita mixta*) plants. Nanotoxicology 5 (1), 30–42.

Wang, H., Wu, F., Meng, W., White, J.C., Holden, P.A., Xing, B., 2013. Engineered nanoparticles may induce genotoxicity. Environ. Sci. Technol. 47, 13212–13214.

Wang, P., Lombi, E., Zhao, F.J., Kopittke, P.M., 2016a. Nanotechnology: a new opportunity in plant sciences. Trends Plant Sci. 21, 699–712. Available from: https://doi.org/10.1016/j.tplants.2016.04.005.

Wang, X., Yang, X., Chen, S., Li, Q., Wang, W., Hou, C., et al., 2016b. Zinc oxide nanoparticles affect biomass accumulation and photosynthesis in Arabidopsis. Front. Plant Sci. 6, 1243.

Wu, H., Tito, N., Giraldo, J.P., 2017. Anionic cerium oxide nanoparticles protect plant photosynthesis from abiotic stress by scavenging reactive oxygen species. ACS Nano 11 (11), 11283–11297.

Yang, F., Liu, C., Gao, F., Su, M., Wu, X., Zheng, L., et al., 2007. The improvement of spinach growth by nano-anatase TiO_2 treatment is related to nitrogen photoreduction. Biol. Trace Elem. Res. 119 (1), 77–88.

Yang, Z.Z., Chen, J., Dou, R.Z., Gao, X., Mao, C.B., Wang, L., 2015. Assessment of the phytotoxicity of metal oxide nanoparticles on two crop plants, maize (*Zea mays L.*) and rice (*Oryza sativa L.*). Int. J. Environ. Res. Public Health 12, 15100–15109.

Yanık, F., Vardar, F., 2018. Oxidative stress response to aluminum oxide (Al_2O_3) nanoparticles in *Triticumaestivum*. Biologia 73 (2), 129–135.

Yoon, S.J., Kwak, J.I., Lee, W.M., Holden, P.A., An, Y.J., 2014. Zinc oxide nanoparticles delay soybean development: a standard soil microcosm study. Ecotoxicol. Environ. Saf. 100, 131–137.

Yue, L., Ma, C., Zhan, X., White, J.C., Xing, B., 2017. Molecular mechanisms of maize seedling response to La_2O_3 NP exposure: water uptake, aquaporin gene expression and signal transduction. Environ. Sci.: Nano 4 (4), 843–855.

Zhang, H., Lu, L., Zhao, X., Zhao, S., Gu, X., Du, W., et al., 2019. Metabolomics reveals the "invisible" responses of spinach plants exposed to CeO2 nanoparticles. Environ. Sci. Technol. 53 (10), 6007–6017.

Zhao, L., Peng, B., Hernandez-Viezcas, J.A., Rico, C., Sun, Y., Peralta-Videa, J.R., et al., 2012. Stress response and tolerance of Zea mays to CeO2 nanoparticles: cross talk among H2O2, heat shock protein, and lipid peroxidation. ACS Nano 6 (11), 9615–9622.

Zhao, L., Hernandez-Viezcas, J.A., Peralta-Videa, J.R., Bandyopadhyay, S., Peng, B., Munoz, B., et al., 2013. ZnO nanoparticle fate in soil and zinc bioaccumulation in corn plants (*Zea mays*) influenced by alginate. Environ. Sci.: Process. Impacts 15 (1), 260–266.

Zhao, L., Zhang, H., White, J.C., Chen, X., Li, H., Qu, X., et al., 2019. Metabolomics reveals that engineered nanomaterial exposure in soil alters both soil rhizosphere metabolite profiles and maize metabolic pathways. Environ. Sci.: Nano 6 (6), 1716–1727.

Zulfiqar, F., Ashraf, M., 2021. Nanoparticles potentially mediate salt stress tolerance in plants. Plant Physiol. Biochem. 160, 257–268.

Zuverza-Mena, N., Armendariz, R., Peralta-Videa, J.R., Gardea-Torresdey, J.L., 2016. Effects of silver nanoparticles on radish sprouts: root growth reduction and modifications in the nutritional value. Front. Plant Sci. 7, 90.

Chapter 14

Role of metal oxide nanomaterials in the preservation of harvested crops

Venkatachalam Vasudevan[1,2] and Vidhya Arumugam[1]

[1]*Department of Biotechnology, A.V.C. College (Autonomous), Mayiladuthurai, Tamil Nadu, India,* [2]*Department of Biotechnology, Bharathidasan University, Tiruchirappalli, Tamil Nadu, India*

Chapter outline

14.1 Introduction

The term "nano" refers to anything microscopic, tiny, or atomic in nature, as well as anything on the nanoscale or one-millionth scale. This idea of nanomaterials incorporated with scientific technology led to the development of "nanoscience." Nanotechnology is like a magic spell that brings modification in every field of science and technology. Enormous applications are seen in the agri-food industry with recent advancements leading to rapid and steady progress in both the agriculture and food industries.

Agriculture is the backbone of developing countries as it generates the majority of income, with more than half of the people opting for it as a source of income. The scarcity of food and its spoilage is a primary concern for developing countries worldwide, whereas there is a surplus of food

Nanometal Oxides in Horticulture and Agronomy. DOI: https://doi.org/10.1016/B978-0-323-91809-1.00013-5

available in developed countries. It is the reason why the developing countries are aiming toward maximum pest-resistant crop yield with low inputs. In contrast, consumer demand for healthier, fresher, and nutrient-value-added foodstuffs drives the food industry in emerging countries. Scientists are constantly under pressure as crop yields are stagnating due to low nutrient utilization, multinutrient deficiencies, decreasing soil organic matter, unpredictably fluctuating climates, shrinking fertile lands, and water scarcity. Nanotechnology, a critical tool of the 21st century, is the frontier technology that will alter the agriculture industry in order to tackle these food security challenges. Nanotechnology makes the agriculture production systems value-added with a controlled and timely release of plant-promoting substances, herbicides, pesticides, and other nanoagrochemicals that support productivity and ensure environmental safety (Paul and Bhojane, 2021).

Nanotechnology improves the food industry with ultimate qualitative and quantitative production of high potential functional foods that are healthier, safer, and nutrient value-added. Because of the enhanced production of quality foods with increased shelf life and reduced contamination, the food market is moving toward nanotechnology as an alternative to traditional food technologies. This chapter focuses on the current state of biosynthesized nanomaterials in the food sector, including production, processing, packaging, security, storage, and application. Food bioavailability, taste, texture, appearance, shelf life, and nutritional value are a few applications of nanotechnology. The characteristics of nanoparticles (NPs) like particle size, shape, particle proportion, surface-to-volume ratio, rate of exposure, dosage, rate of accumulation, and surface charge play a significant role in their functional activity. Nanosensors and nanodevices have a synergistic action in food protection and safety with nanomaterials in innovative/intelligent food packaging and monitoring the stored foods for product authenticity (Paul and Bhojane, 2021). The methods used to assess nano foods in biological systems (human and environmental) are evaluated for their nontoxic impact on health and the environment, with the goal of developing regulatory laws to address safety issues and environmental considerations.

Research scientists and the food industry have recognized the potential applications of nanotechnology, although their viability is dependent on cost-effectiveness and public perceptions of potential advantages and hazards. Market usage of nanotechnology in the food and agriculture industry is marginal and uncertain. However, it has many advantages so its applications are high in other sectors like medicine, physical sciences, IT, and biotechnology. The significant initial investment without a proper regulatory framework and varied public views has some impact on the market coverage of nanoagrotech products. As large companies are patenting the applications of nanotechnology for their security in further operations, economic gain is also a bid deal. The general public seems unconcerned, preferring natural foods to nano-based foods and nanostructured additives. Industrialists are focusing on commercial applications of NPs in packaging rather than true nano meals as a result of public demand. While examining the safety issues, many regulatory organizations are recommending improvised nanotechnology-applied foods. According to studies on the adoption of nanotechnology in food, 30% of people prefer the benefits of nanotechnology to the hazards (Naseer et al., 2018). Individual choice of preference and the knowledge of science about the production of any novel food item governs consumer acceptance due to government regulations and trust issues. People mistake nanofood for genetically engineered food; thus, they avoid it out of fear of negative health consequences. Nanotechnology's applications in agriculture and the food industry are a boon to humankind's efforts to establish a sustainable future.

14.1.1 Nanotechnology in the food market

The majority of horticulture crops and their products are necessary for our daily nutrition as they provide us with micro and macronutrients for sustaining good health. Plant pathogenic microorganisms such as fungus and bacteria, which cause substantial losses during storage and transportation, are primarily responsible for diseases that arise after the harvest of crops. Various scientific approaches, including physical and chemical methods, preserve crops from infection and pest infestations, as well as from the ripening process after harvesting. For example, one of the most important strategies for lowering respiration is the edible coating and the application of metals to prevent disease and pest attack that generates numerous postharvest decaying of food products. In contrast to the chemical preservatives, metal oxide nanoparticles (MONPs) derived from plant extracts are a relatively new technique for inhibiting bacterial growth. MONPs are environmentally safe since they improve water uptake and vase life in cut flowers may be employed on a broad scale and are less toxic than their synthetic equivalents. Nanotechnology can help to retain the odor, color, and flavor of spices in the spice and aromatic industries. The gold NPs-antibody packaging can significantly reduce the contamination of highly dangerous aflatoxins that cause poor growth. Packaging using silver NPs and aluminum foils can be beneficial since they eventually create biofilms, allowing the quality of spices to be preserved and delivered to consumers in the same condition as at the time of harvest (Yan et al., 2020). Nanotechnology has become an emerging technology in all fields of science, including medicine and agriculture. Different nanomaterials, as well as NPs, are often used effectively at the scientific and industrial levels to transform traditional farming into sustainable agriculture, in particular, by avoiding postharvest problems with vegetables and fruits (Feregrino-Perez et al., 2018). Postharvest losses of fruits and vegetables range from 4.58% to 15.88%. Fresh fruits and vegetables lose between 10% and 15% of their market value and consumer acceptability due to shrink and decay. Animals and pests account for 18% of the losses. Pathogens account for 16%, and fungal infection specifically accounts for 72% to 85% of the losses (Oerke, 2006). Fungal diseases are responsible for most of the economic losses in the food industry. Postharvest production is variable, ranging from 30% to 50%, and is dependent mainly on good agricultural or cultural techniques as well as geographic location (Ladaniya, 2008; Youssef and Roberto, 2014a,b). According to the Food and Agriculture Organization, 33% of food delivered for human consumption is wasted during agricultural harvesting (Gastavsson et al., 2011). These losses spread throughout the production network as a result of microbial diseases, which account for the majority of food waste. Limiting these losses to a bare minimum will increase supply and lower overall expenses.

14.2 Controlling the growth of microorganisms

There are various pathogens that cause disease, but *Botrytis cinerea*, which causes gray mold disease, is a major postharvest pathogen worldwide. *B. cinerea* decomposes a wide range of commercially significant horticulture crops, particularly vegetables and fruits, and ornamental flowers such as marigolds. *B. cinerea* impacts the flower petals during the preharvest and postharvest stages. It is also a significant issue in long-distance transportation and storage. This disease infects leaves, stems, flowers, and fruits through primary or secondary infection (Sommer et al., 1992).

Edible coatings offer a variety of benefits, including the ability to reduce water loss, slow ripening, reduce chilling and mechanical damage, reduce decay, and add shine to coated items. Edible coatings can sometimes be antimicrobial composites, antioxidants, or antiripening compounds, which ultimately induce changes in flavor due to late ripening or the result of anaerobic changes. Proteins, lipids, polysaccharides, and other biomacromolecules can make coatings (Baldwin et al., 1996). Table 14.1 shows the various edible coatings suitable for fruits and vegetables. Chitosan is a chitin-based biopolymer (a polysaccharide constituent of crustacean shells). It also has antibacterial qualities, which keep horticultural crops from rotting after harvest (Benhamou, 1996; Gabler and Smilanick, 2001; Romanazzi et al., 2002; Liu et al., 2007). Chitosan prevents and manages preharvest and postharvest infections. It has negligible toxicity, would be safe for human consumption, and is eco-friendly, so it can also maintain the quality of fruits and vegetables (Rabea et al., 2003; Tripathi and Dubey, 2004). Chitosan also enhances tomato fruit resistance to gray mold disease caused by *B. cinerea*. It has excellent potential as a natural compound that can partially replace the usage of synthetic fungicides. Chitosan combined with 1% gum arabic on application to banana fruit slows down the ripening process and extends the shelf life for a few days when compared to uncoated bananas (Maqbool et al., 2011). A nanotechnological coating based on pullulan and polymeric nanocapsules containing an essential oil produced from thyme (*Thymus*

TABLE 14.1 Metal and metal oxide NPs as nanoemulsion, edible coatings for the preservation of fruits and vegetables.

Metal NPs	Vegetables/fruits/cut flower	Positive effect	References
Edible coating			
Titanium dioxide (TiO_2) NPs	*Cantaloupe*	Effective against the control of yeast and mold growth	Qiao et al. (2019)
Gelatin-chitosan/(Ag/ZnO)	Okra/lady's finger	Antimicrobial polyethylene packaging for the preservation of vegetable	Al-Naamani et al. (2018)
Chitosan-methyl cellulose/silica(SiO_2)	Loquat fruit	Effective in enhancing chilling tolerance	Song et al. (2016)
ZnO NPs	Cucumber	Carotene, zinc, and iron content protected	Seray et al. (2020)
TiO_2-chitosan	Grapes	Acts as a preservative and prevents the entry of microorganisms	Gu et al. (2020)
ZnO-chitosan	Apple and mango	Increases antioxidant activity and protects from microbial attack	Pérez-Moreno et al. (2021)
TiO-polyacrylonitrile	Tomato	Reduces ethylene synthesis	Yadav et al. (2021)
Chitosan-silver NPs composite	Mango	Antifungal activity against *Colletotrichum gloeosporioides*	Dean et al. (2012)
Silica NPs	*Grapes*	Antifungal activity	Hashim et al. (2019)
Nanoemulsion			
Tocopherol/nopal mucilage nanoemulsion	Red delicious-apple	Encapsulant	Gardesh et al. (2016)
Chitosan-nanoemulsion	Banana	Coating	Zahid et al. (2012)
Wax nanoemlusion	Papaya	Coating	Charych et al. (1996)
Enhance vase life of cut flowers			
Silver NPs (10 mg/L)	Gerbera	Maintained water content and hydraulic conductance	Motaghayer et al. (2019)

(Continued)

TABLE 14.1 (Continued)

Metal NPs	Vegetables/fruits/cut flower	Positive effect	References
Silver NPs (40 mg/L)	Lisianthus	Increasing quality and vase life	Kamiab et al. (2017)
Silver NPs (5 mg/L)	Mokara Red orchid flower	Increasing the vase life	Rahman et al. (2019)

vulgaris L.) extends the shelf life of table grapes. When compared to untreated grapes, grapes coated with the synthesized nanocapsules retained their color, firmness, and total soluble salts content for a more extended period. Gray mold was less common on coated grapes than on untreated grapes. The nanocapsule coating prevented volatile chemicals from evaporating too quickly and increased their persistence on grape surfaces, delaying the ripening process and extending shelf life (Pina-Barrera et al., 2019). *Colletotrichum gloeosporioides* is a significant disease that occurs postharvesting in several agricultural crops such as mango and guava. Fungal growth gets suppressed by just a silver NP composite size of 10–15 nm, which significantly inhibits conidial germination (Dean et al., 2012). *Penicillium digitatum*, which attacks the sweet orange "Valencia Late," was treated with a clay-chitosan nanocomposite. At a dosage of 20 g/mL, the clay-chitosan nanocomposite completely suppresses the pathogen and inhibits hyphal branching (Youssef and Hashim, 2020).

14.3 Metal nanoparticles for extending postharvest life of cut flowers

Cut flowers are commonly used to express gratitude, affection, and emotion on various occasions. Vase life and quality are the two most significant characteristics of a cut flower. Silver NPs, which extend the vase life of cut flowers, enter the vascular tissues of the stem, reducing ethylene synthesis and repressing the ethylene-producing genes ACO1 (Musa acuminata oxidase 1) and ACS1 (Musa acuminata synthase 1). Furthermore, they promote the uptake of the solution, hence extending the vase life of cut flowers. Because of the interrupted water intake caused by microbial development, cut flowers do not have a long shelf life after harvest. Various nanotechnological products, such as nano-silver, can be used to make antimicrobial agents and ethylene inhibitors. This preservative solution also increases the vase life of cut

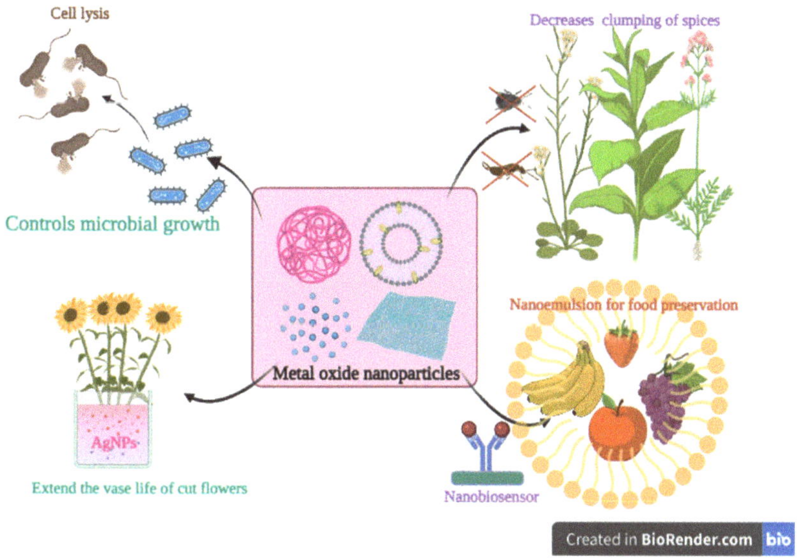

FIGURE 14.1 Application of silver nanoparticles in preservation of cut flowers.

flowers (Table 14.1). These silver NPs are safe to use, convenient to apply, and have a high level of stability. Silver NPs in variable concentrations are used to improve and maintain the quality of cut flowers in a variety of ways as shown in Fig. 14.1.

14.4 Metal nanoparticles for quality improvement of aromatic herbs and spices

Spices are the most significant substances for food flavoring since they improve flavor, enhance taste, and modify the color of food; they also inhibit germs and safeguard food from deterioration. Mycotoxins generated by the fungus *Aspergillus flavus* and *A. parasiticus* are known as aflatoxins. These poisons can hinder growth, lower immunity, and even cause liver cancer due to their harshness. The antibodies linked with NPs act as labels for recognizing aflatoxins B and G, as well as Aflatoxin M1 using a gold NPs-antibody label combination. With nanocoatings integrating metals with NPs or biopolymers, color enhancement, shelf-life extension, browning reduction, increased flavor, improved availability of antioxidants from spices, and other benefits are all feasible. Nanotechnology can increase the quality of spices and aromatic items. Spice clumping is a severe issue that shortens the shelf life and lowers the market quality. The NPs like nano SiO_2, nano Ca_2SO_4 Si, nano sodium aluminosilicate, and dicalcium phosphate have a wide range of applications in decreasing clumping and increasing the free-flow nature of spices (Fig. 14.1).

14.5 Nanotechnology in food packaging

The advancement in nanotechnology improves food packaging, especially the fruits. Blending polyethylene with nanopowder (nano-Ag, kaolin, anatase TiO_2, mineral TiO_2) yielded a novel technology of nanopacking materials with decreased relative humidity, oxygen transmission rate, and high strength, which improved the preservation quality of fruits during storage at 40°C. Compared to conventional packaging, nanopackaging can better preserve the sensory characteristics and chemical and physiological attributes of strawberry fruits (polyethylene bags). Furthermore, nanopackaging excels standard polyethylene packing in terms of the decay rate, anthocyanin, and malondialdehyde concentration.

The activity of polyphenol oxidase and pyrogallic acid oxidase is much lower in nanopacking than in traditional management. These findings suggested that nanopackaging could help to improve the fruit's preservation quality during long-term storage. Apples are put up for sale in the United States with a waxy covering that prevents the fruit from losing moisture and shrinking. For fruit, vegetables, and baked goods, edible coatings are currently 5 nm. The edible covering prevents gas exchange, protects color, flavor, and antioxidants, and keeps food fresh for a long time. Fuji apples coated with NanoSiOx/chitosan composite preservation agents exhibited a higher quality than those not coated (Fig. 14.1). When compared to standard packing, green tea packed with nanopacking retained more vitamin C, chlorophyll, polyphenols, and amino acids. Nanopackaging is also employed for food packaging for various purposes. Spice packaging is a complex subject since it not only extends shelf life but also preserves smell, color, and flavor and prevents microbiological contamination. Electrochemical anodization creates cylindrical nanopores with widths ranging from 15 to 100 nm on the surface of aluminum foil used for packing food. Then, the aluminum surfaces can be inoculated with cultures of *Escherichia coli* and *Listeria innocua*. Nanopores having a diameter of 15–25 nm effectively reduced both adhesion and biofilm growth, making them safe for spice packing.

14.5.1 Applications of nanoemulsions

An emulsion is a fine dispersion of minute droplets of one liquid in another in which it is not soluble or miscible. Nanoemulsions are nano-sized emulsions made to improve the delivery of active medicinal substances at the nanoscale. These are isotropic thermodynamic systems in which an emulsifying agent, such as surfactant and cosurfactant, mixes two immiscible liquids into a single phase. Nanoemulsion coating of Carnauba wax with a diameter of 42 nm (also known as Brazil wax or palm wax) is a feasible alternative. This wax was extracted at 2.4% from the leaves of the carnauba palm (*Copernicia prunifera*) and applied to papaya fruit. The fruits stored at 22°C for 9 days at a relative humidity of 60%–70% decayed less and lost less mass. Various nanoemulsions are employed on fruits to protect them from deterioration and microbial attacks. Nanolaminate is a fragile food-grade sheet with physically or chemically bonded dimensions (range 1–100 nm/layer). When it comes to making edible films, nanolaminate has a lot of advantages. Edible films appear in fruits and vegetables, as well as in chocolate, toffees, baked products, and French fries. Foods are protected by these layers from moisture, fats, and gas loss. These films help fruits in preserving their color, flavor, antioxidant, and nutrient content while improving their textural characteristics (Table 14.1).

The limited solubility, stability, and bioavailability of bioactive chemicals are technological barriers in generating food products. During food preparation, the majority of bioactive food components are susceptible to oxidative deterioration and destruction (Xianquan et al., 2005; Shahidi and Zhong, 2010). Some have a low solubility but a high metabolism, which reduces bioavailability, whereas others are volatile and sensitive to processing conditions (McClements and Li, 2010; Jin et al., 2016). The utilization of nanoemulsions to encapsulate bioactive chemicals in the food matrix can solve these problems. Bioactive chemicals are stabilized, bioavailable, and released at a controlled rate when encapsulated in an oil phase or emulsifier (McClements et al., 2007). The nanoemulsion-based delivery system must be compatible with the food matrix and have minimal impact on the food's organoleptic qualities, such as flavor, appearance, and texture. The encapsulation of a bioactive substance can protect it from processing conditions and keep it from degradation for a long time under storage conditions such as temperature, light, pH, and oxidative stress. The use of nanoemulsion-based food delivery systems necessitates that the process is economically feasible for large-scale production (Borthakur et al., 2016; Pathak, 2017).

Aldehyde, ketone, and esters are functional groups in food flavors and coloring chemicals, making them sensitive to oxidative and photolytic destruction. These substances can be encapsulated into nanoemulsions to prevent these adverse effects and extend their shelf life (Goindi et al., 2016). Citral is an aromatic chemical used as a flavoring agent in foods and cosmetics. It is an unsaturated aldehyde with one additional double bond. However, when it degrades, it creates chemicals that have an off-flavor. Citral nanoemulsions improve the stability of food. A nanoemulsion of citral combined with natural antioxidants such as β-carotene, tanshinone, and black tea extract shows excellent chemical stability during storage. The nanoemulsions are prepared with lecithin stabilized palm kernel fat at a 1:1 ratio of citral and antioxidants in a pH 3 buffer. Antioxidant encapsulation reduced the development of off-flavor molecules such as p-dimethyl styrene and p-methyl acetophenone (Yang et al., 2011). Similarly, the chemical stability of a nanoemulsion system containing citral and ubiquinol-10 (Q10H2) is also good. The ubiquinol-10 protects citral, which inhibits its chemical breakdown and oxidation. Off-flavor chemicals (p-cresol, p-dimethyl styrene, p-methyl acetophenone) and lipid degradation products are reduced. At a Q10H2/citral ratio of 1:1, the ubisemiquinone (Q10 •)/ubiquinone (Q10) redox transition occurs, resulting in pro-oxidant behavior of Q10H2 and thus it is able to maintain the stability of citral (Zhao et al., 2013). β-Carotene is a precursor to vitamin A and acts as a natural colorant and antioxidant in the food industry. β-Carotene on the other hand is easily degraded by heat, light, and oxygen. Attempts are being taken to increase the stability of β-carotene under a variety of processing conditions. The stability of β-carotene is further enhanced by using the emulsification-evaporation method. Emulsifying the organic solution of β-carotene in an aqueous phase containing an emulsifier yields a nanoemulsion. The weighted mean diameter of stable β-carotene nanodispersions ranged from 60 to 140 nm (Tan and Nakajima, 2005a). Polyglycerol esters of fatty acids (PGEs) with a mean diameter of 85–132 nm are used to improve the physicochemical properties and physical stability of O/W β-carotene nanodispersions. For highly stable β-carotene nanodispersions, PGE decaglycerol monolaurate (10 g /kg) with high glycerol polymerization can be employed (Tan and Nakajima, 2005b). The mean particle size of protein-stabilized β-carotene nanodispersions generated by emulsification-evaporation was 17 nm. At high concentrations, the emulsifier sodium caseinate reduces particle size and improves the polydispersity of the nanodispersions. At pH 7, the β-carotene nanodispersions possess a high potential of 27 mV, making them resistant to particle aggregation (Chu et al., 2007). The mean diameters of O/W-carotene nanoemulsions made with Tween 20 as an emulsifier range from 132 to 184 nm. Only 25% of the β-carotene in the nanoemulsions disintegrate after 4 weeks of storage at 4°C and 25°C, respectively (Yuan et al., 2008). A solvent displacement method can be used to prepare β-carotene nanodispersions of 30–206 nm using emulsifiers such as sodium caseinate, Tween 20, decaglycerol monolaurate, and sucrose fatty acid ester. Although starch casein stabilized nanodispersions possess big particles, they are the most oxidation resistant. This may be due to the physical barrier and antioxidant action of caseins(Yin et al., 2009). After the emulsification procedure, modified starch nanoemulsions stabilized by spray-dried powders enhance storage stability. β-carotene retention was high during the storage of modified starch with a low film oxygen permeability (Liang et al., 2013). The physicochemical properties of β-carotene nanoemulsions can be enhanced by treating them with starch caseinate and chitosan epigallocatechin-3-gallate conjugates (Wei and Gao, 2016). β-carotene nanoemulsions with high water dispersibility and chemical stability are prepared using a sonication-assisted technique in association with freeze-drying (Chen et al., 2017). The melon Cantaloupe is also used to extract carotenoids. The water solubility and color stability of nanoemulsions made from these carotenoids are improved. The average particle size of the o/w carotenoid nanoemulsions encapsulated in porcine gelatin and whey protein isolate was 70–160 nm. Gelatin increases the water solubility of carotenoids. For 60 days, the yogurt containing these nanoemulsions as a natural coloring ingredient remains stable (Medeiros et al., 2019).

Grape skins, blueberries, and raspberries contain a natural polyphenol that has antioxidant, anticancer, and antiobesity characteristics. Resveratrol is one such natural polyphenol widely used in nanotechnology. Nanoemulsion-based delivery techniques were used to encapsulate resveratrol. To encapsulate resveratrol as droplets, spontaneous emulsification will give a better result with a 10% oil phase (grapeseed oil and orange oil), 10% surfactant (Tween 80), and 80% of aqueous phase. The chemical stability of resveratrol against UV light degradation improves when encapsulated (Davidov-Pardo and McClements, 2015). O/W consumable nanoemulsions of vitamin D are often used to fortify dairy emulsions (cholecalciferol).

14.5.2 Application of nanobiosensors

Biodegradable nanotechnology-based sensors in the form of microchips, such as electronic tongue technology sensors, are now frequently used in food products. When the pH of a product changes due to deterioration, it alters its color. The preservative packing is controlled by a "Release on Command" nanobioswitch (Netherlands) (Yadollahi et al., 2009). Biochemical components such as sugars and proteins are used to create nanostructures for use as biosensors on food (Imafidon and Spanier, 1994). These biosensors have two functions: one is to detect food pathogens and other microorganisms, and the other is to track food products. Encapsulation systems that defend against environmental conditions can also benefit from nanotechnology. It can also be utilized to develop food elements like flavor and antioxidants (Haruyama, 2003). The primary goal is to increase the usefulness of these chemicals despite lowering their concentration. The use of these novel components in cuisines is gaining prominence (Lawrence and Rees., 2000). There will be more research on delivery controlled release methods for nutraceuticals (Delfino et al., 2020). Nanomaterials have a variety of optical and electrochemical properties in various hot and cold liquids, juices, and oils. Nanomaterials are used in the agriculture and food industries to create multiple biosensors. Virus recognition employing antibody sensor arrays on self-assembled nanoscale block copolymer and detecting food-borne toxins using multifunctional NPs are two examples of nanobiosensor applications. Nanobiosensors can also detect pore-forming toxins and molecularly imprinted polymers that can be used to identify plant and insect viruses.

14.6 Conclusion

Nanomaterials have a significant role in postharvest technology management. Simple postharvest measures such as early harvesting, clean washing, sorting, grading, packaging, pre-cooling, and proper transportation care can help to reduce postharvest losses in most horticultural commodities. Nanoedible coatings, which are less expensive, can help farmers maintain the quality of their vegetables and fruits by delaying the rate of respiration. By minimizing postharvest losses, nanocoatings consisted of zinc and silver protect crops from microbes. The farmer can also receive better market prices by using this edible coating and active packaging technology. Research on the precise nanopacking technology on preserved goods is also required. Nanotechnology will be able to be employed in a wider range of applications as a result of this. In the near future, the nanotechnology-based study will serve as a revolutionary tool for managing horticulture crop postharvest losses. Nanotechnology also has a direct positive impact on medicine and the environment. Nanotechnology has some unforeseen environmental implications, both within the human body and in the natural ecosystem. The scientific community must first explore the environmental and health risks before adopting this new technology widely for wellness and sustainability benefits. It can be challenging to determine the toxicity of NPs since not all designed NPs are more harmful than fine-size particles with the same chemical composition. Surface coatings on particles, UV exposure, and dispersion properties can all change particle behavior and even have negative consequences for the environment and human health. As a result, further research is necessary before nanomaterial-based technologies can be implemented.

References

Al-Naamani, L., Dutta, J., Dobretsov, S., 2018. Nanocomposite zinc oxide-chitosan coatings on polyethylene films for extending the storage life of okra (*Abelmoschus esculentus*). Nanomaterials 8 (7), 479.

Baldwin, E.A., Nisperos, M.O., Chen, X., Hagenmaier, R.D., 1996. Improving storage life of cut apples and potato with edible coating. Postharvest Biol. Technol. 9, 151–163.

Benhamou, N., 1996. Elicitor-induced plant defence pathways. Trends Plant. Sci. 1, 233–240.

Borthakur, P., Boruah, P.K., Sharma, B., Das, M.R., 2016. Nanoemulsion: preparation and its application in food industry. Emulsions 3, 153–191.

Charych, D., Cheng, Q., Reichert, A., Kuziemko, G., Stroh, M., Nagy, O.J., et al., 1996. A 'litmus test' for molecular recognition using artificial membranes. Chem. Biol. 3, 113–120.

Chen, J., Li, F., Li, Z., McClements, D.J., Xiao, H., 2017. Encapsulation of carotenoids in emulsion-based delivery systems: enhancement of β-carotene water-dispersibility and chemical stability. Food Hydrocoll. 69, 49–55.

Chu, B.S., Ichikawa, S., Kanafusa, S., Nakajima, M., 2007. Preparation of proteinstabilized β-carotene nanodispersions by emulsification–evaporation method. J. Am. Oil Chem. Soc. 84, 1053–1062.

Davidov-Pardo, G., McClements, D.J., 2015. Nutraceutical delivery systems: resveratrol encapsulation in grape seed oil nanoemulsions formed by spontaneous emulsification. Food Chem. 167, 205–212.

Dean, R., van Kan, J.A., Pretorius, Z.A., Hammond-Kosack, K.E., Di Pietro, A., Spanu, P.D., et al., 2012. The top 10 fungal pathogens in molecular plant pathology. Mol. Plant. Pathol. 13, 414–430.

Delfino, J.R., Da Silva, J.L., Marques, A.L.B., Stradiotto, N.R., 2020. Antioxidants detection in aviation biokerosene by high-performance liquid chromatography using gold nanoparticles anchored in reduced graphene oxide. Fuel 260, 116315.

Feregrino-Perez, A.A., Magaña-López, E., Guzmán, C., Esquivel, K.A., 2018. General overview of the benefits and possible negative effects of the nanotechnology in horticulture. Sci. Hortic. 238, 126–137.

Gabler, F.M., Smilanick, J.L., 2001. Post-harvest control of table grape gray mold on detached berries with carbonate and bicarbonate salts and disinfectants. Am. J. Enol. Vitic. 52, 12–20.

Gardesh, A.S.K., Badii, F., Hashemi, M., Ardakani, A.Y., Maftoonazad, N., Gorji, A.M., 2016. Effect of nanochitosan based coating on climacteric behavior and postharvest shelf-life extension of apple cv. GolabKohanz. LWT-Food Sci. Technol. 70, 33–40.

Gastavsson, J., Cederberg, C., Sonesson, U., 2011. Global Food Losses and Food Waste. Food and Agriculture Organization (FAO) of the United Nations, Rome, Italy.

Goindi, S., Kaur, A., Kaur, R., Kalra, A., Chauhan, P., 2016. Nanoemulsions: an emerging technology in the food industry. Emulsions 3, 651–688.

Gu, R., Yun, H., Chen, L., 2020. Regenerated cellulose films with amino-terminated hyper branchedpolyamic anchored nanosilver for active food packaging. ACS Appl. Bio Mater. 3, 602–610.

Haruyama, T., 2003. Micro- and nanobiotechnology for biosensing cellular responses. Adv. Drug. Deliv. Rev. 55, 393–401.

Hashim, A.F., Youssef, K., Abd-Elsalam, K.A., 2019. Ecofriendly nanomaterials for controlling graymold of table grapes and maintaining postharvest quality. Eur. J. Plant. Pathol. 154, 377–388.

Imafidon, G.I., Spanier, A.M., 1994. Unraveling the secret of meat flavor. Trends Food Sci. Technol. 5, 315–321.

Jin, W., Xu, W., Liang, H., Li, Y., Liu, S., Li, B., 2016. Nanoemulsions for food: properties, production, characterization, and applications. Emulsions 3, 1–36.

Kamiab, F., Shahmoradzadeh Fahreji, S., Zamani Bahramabadi, E., 2017. Antimicrobial and physiological effects of silver and silicon nanoparticles on vase life of lisianthus (Eustoma grandiflora cv. echo) flowers. Int. J. Hortic. Sci. Technol. 4 (1), 135–144.

Ladaniya, M.S., 2008. Citrus fruit. Academic Press, San Diego, CA.

Lawrence, M.J., Rees, G.D., 2000. Microemulsion-based media as novel drug delivery systems. Adv. Drug. Deliv. Rev. 45, 89–121.

Liang, R., Huang, Q., Ma, J., Shoemaker, C.F., Zhong, F., 2013. Effect of relative humidity on the store stability of spray-dried beta-carotene nanoemulsions. Food Hydrocoll. 33, 225–233.

Liu, J., Tian, S.P., Meng, X.H., Xu, Y., 2007. Effects of chitosan on control of postharvest diseases and physiological responses of tomato fruit. Postharvest Biol. Technol. 44, 300–306.

Maqbool, M., Ali, A., Alderson, P.G., Zahid, N., Siddiqui, Y., 2011. Effect of a novel edible composite coating based on Gum arabic and chitosan on biochemical and physiological responses of banana fruits during cold storage. J. Agric. Food Chem. 59, 5474–5482.

McClements, D.J., Li, Y., 2010. Review of in vitro digestion models for rapid screening of emulsion-based systems. Food Funct. 1, 32–59.

McClements, D.J., Decker, E.A., Weiss, J., 2007. Emulsion-based delivery systems for lipophilic bioactive components. J. Food Sci. 72, 109–124.

Medeiros, A.K.O.C., Gomes, C.C., Amaral, M.L.Q.A., Medeiros, L.D.G., Medeiros, I., Porto, D. L., et al., 2019. Nanoencapsulation improved water solubility and color stability of carotenoids extracted from Cantaloupe melon (*Cucumis melo* L.). Food Chem. 270, 562–572.

Motaghayer, M., Azizi, M., Teheranifar, A., 2019. Nanosilver, salicylic acid and essential oils effects on water relations of gerbera 'Rosalin' cut flowers. Adv. Hortic. Sci. 33, 271–281.

Naseer, B., Srivastava, G., Qadri, O.S., Faridi, S.A., Islam, R.U., Younis, K., 2018. Importance and health hazards of nanoparticles used in the food industry. Nanotechnol. Rev. 7 (6), 623–641.

Oerke, E.C., 2006. Crop losses to pests. J. Agric. Sci. 144, 31–43.

Pathak, M., 2017. Nanoemulsions and their stability for enhancing functional properties of food ingredients. Nanotechnol. Appl. Food. 87–106.

Pérez-Moreno, A., Fabián, F.L., Hermes, P.H., Edgar, V.N., Ileana, V.R., Ali, A., 2021. Nanoscience and nanotechnology regarding food packaging and nanomaterials to extending the postharvest life and the shelf life of foods. Food losses, Sustainable Postharvest and Food Technologies. Academic Press, pp. 313–384.

Pina-Barrera, A.M., Álvarez-Román, R., Báez-González, J.G., Amaya-Guerra, C.A., Rivas-Morales, C., Gallardo-Rivera, C.T., et al., 2019. Application of a multisystem coating based on polymeric nanocapsules containing essential oil of thymus vulgaris L. to increase the shelf life of table grapes (Vitis vinifera L.). IEEE Trans. NanoBioscience 18, 549–557.

Qiao, G., Xiao, Z., Ding, W., 2019. Effect of chitosan/Nano-titanium dioxide/Thymol and tween films on readyto-eat cantaloupe fruit quality. Coatings. 9 (12), 828.

Rabea, E.I., Badawy, M.E.T., Stevens, C.V., Smagghe, G., Steurbaut, W., 2003. Chitosan as antimicrobial agent: applications and mode of action. Biomacromolecules. 4, 1457–1465.

Rahman, M.M., Ahmad, S.H., Mohamed, M.T.M., Ab Rahman, M.Z., 2019. Improving the vase life of cut mokara red orchid flower using leaf extracts with silver nanoparticles. Proc. Natl Acad. Sci. India Sect. B: Biol. Sci. 89 (4), 1343–1350.

Romanazzi, G., Nigro, F., Ippolito, A., di Venere, D., Salerno, M., 2002. Effects of pre- and postharvest chitosan treatments to control storage greymould of table grapes. J. Food Sci. 67, 1862–1867.

Seray, M., Skender, A., Hadj-Hamou, A.S., 2020. Kinetics and mechanisms of Zn2 + release from antimicrobial food packaging based on poly (butylene adipate-co-terephthalate) and zinc oxide nanoparticles. Polym. Bull. 78 (2), 1021–1040.

Shahidi, F., Zhong, Y., 2010. Lipid oxidation and improving the oxidative stability. Chem. Soc. Rev. 39, 4067–4079.

Sommer, N.F., Fortlage, R.J., Edwards, D.C., 1992. Postharvest diseases of selected commodities. In: Kader, A.A. (Ed.), Postharvest Technology of Horticultural Crops, 3311. University of California Davis Division of Agriculture and Natural Resources, and Publication, USA.

Song, H., Yuan, W., Jin, P., Wang, W., Wang, X., Yang, L., et al., 2016. Effects of chitosan/nano-silica on postharvest quality and antioxidant capacity of loquat fruit during cold storage. Postharvest Biol. Technol. 119, 41–48.

Tan, C.P., Nakajima, M., 2005a. β-Carotene nanodispersions: preparation, characterization and stability evaluation. Food Chem. 92, 661–671.

Tan, C.P., Nakajima, M., 2005b. Effect of polyglycerol esters of fatty acids on physicochemical properties and stability of β-carotene nanodispersions prepared by emulsification/evaporation method. J. Sci. Food Agric. 85, 121–126.

Tripathi, P., Dubey, N.K., 2004. Exploitation of natural products as an alternative strategy to control postharvest fungal rotting of fruit and vegetables. Postharvest Biol. Technol. 32, 235–245.

Wei, Z., Gao, Y., 2016. Physicochemical properties of β-carotene bilayer emulsions coated by milk proteins and chitosan–EGCG conjugates. Food Hydrocoll. 52, 590–599.

Xianquan, S., Shi, J., Kakuda, Y., Yueming, J., 2005. Stability of lycopene during food processing and storage. J. Med. Food 8, 413–422.

Yadav, S., Mehrotra, G.K., Dutta, P.K., 2021. Chitosan based ZnO nanoparticles loaded gallic-acid films for active food packaging. Food Chem. 334, 127605.

Yadollahi, A., Arzani, K., Khoshghalb, H., 2009. The role of nanotechnology in horticultural crops postharvest management. In: Southeast Asia Symposium on Quality and Safety of Fresh and Fresh-Cut Produce. ISHS ActaHorticulturae, 875, 49–56.

Yan, C., Wang, Q., Yang, Q., Wu, W., 2020. Recent advances in aflatoxins detection based on nanomaterials. Nanomaterials 10 (9), 1626.

Yang, X., Tian, H., Ho, C.T., Huang, Q., 2011. Inhibition of citral degradation by oilin-water nanoemulsions combined with antioxidants. J. Agric. Food Chem. 59, 6113–6119.

Yin, L.J., Chu, B.S., Kobayashi, I., Nakajima, M., 2009. Performance of selected emulsifiers and their combinations in the preparation of β-carotene nanodispersions. Food Hydrocoll. 23, 1617–1622.

Youssef, K., Hashim, A.F., 2020. Inhibitory effect of clay/Chitosan nanocomposite against penicilliumdigitatum on citrus and its possible mode of action. Jordan. J. Biol. Sci. 13.

Youssef, K., Roberto, S.R., 2014a. Applications of salt solutions before and after harvest affect the quality and incidence of postharvest gray mold of 'Italia' table grapes. Postharvest Biol. Technol. 87, 95–102.

Youssef, K., Roberto, S.R., 2014b. Salt strategies to control botrytismold of 'benitaka' table grapes and to maintain fruit quality during storage. Postharvest Biol. Technol. 95, 95–102.

Yuan, Y., Gao, Y., Zhao, J., Mao, L., 2008. Characterization and stability evaluation of β-carotene nanoemulsions prepared by high pressure homogenization under various emulsifying conditions. Food Res. Int. 41, 61–68.

Zahid, N., Ali, A., Manickam, S., Siddiqui, Y., Maqbool, M., 2012. Potential of chitosan-loaded nanoemulsions to control different *Colletotrichum* spp. and maintain quality of tropical fruits during cold storage. J. Appl. Microbiology 113 (4), 925–939.

Zhao, Q., Ho, C.T., Huang, Q., 2013. Effect of ubiquinol-10 on citral stability and off-flavor formation in oil-in-water (o/w) nanoemulsions. J. Agric. Food Chem. 61, 7462–7469.

Chapter 15

Management of soil nutrient deficiency by nanometal oxides

Rajiv Periakaruppan[1,2], P. Vanathi[3] and D. Vidhya[4]
[1]*College of Horticulture, Nanjing Agricultural University, Nanjing, Jiangsu Province, P.R. China,* [2]*Department of Biotechnology, PSG College of Arts & Science, Coimbatore, Tamil Nadu, India,* [3]*Department of Biotechnology, Sri Ramakrishna College of Arts & Science, Coimbatore, Tamil Nadu, India,* [4]*TRM Biotech Private Limited, Namakkal, Tamil Nadu, India*

Chapter outline

Nanometal Oxides in Horticulture and Agronomy. DOI: https://doi.org/10.1016/B978-0-323-91809-1.00003-2

15.1 Introduction

Agronomy is the pillar of the domestic economy predominantly in emerging nations, with 60% of the population dependent on it for their source of revenue (McCauley et al., 2009; Brock et al., 2011). Currently, it is fronting abundant challenges, that is, decrease in agricultural land owing to lower crops yield with more significant effort cost, urbanization development that leads to insufficiency, and malnourishment of foods. Many factors such as a decrease in soil fertility and plant nutrient mismanagement are involved in lower crop production.

Over 90% of Indian soil samples showed lower nitrogen (N) and phosphorus (P) and 50% of potassium (K) content. These nutrients are highly required for the evolution and progress of plants. Similarly, among the micronutrients, copper (3%), zinc (49%), manganese (5%), boron (33%), molybdenum (13%), and iron (12%) are the most deficient nutrients in the plant (Singh, 2008a,b).

Soil nutrition is an essential factor for plant growth and crop yield. Soil contains many essential nutrients, minerals, water, and organic materials needed for crop improvement. If these nutrients are not available in sufficient quantity in soil or are deficient in the ground, it affects the crop development and yield and shows deficiency symptoms (McCauley et al., 2009). Healthy soil supplies the required amount of nutrients to the soil for their growth by biogeochemical cycle (Havlin et al., 1999). Overcoming nutrient deficiencies is one of the crucial obstacles in the modern agricultural system to produce increased quality attributes and growth of crops.

Plants need 18 prerequisite crucial nutrient supplements for growth and development. Fundamental nutrient elements are correspondingly essential nevertheless of their prerequisite or their existence in plants (van der Ploeg et al., 1999). Aron and Stout (1939) specified that the component should have the subsequent measures for the development of the plants (USDA NRCS, 2013).

1. The particular mineral element is necessary for the crop to accomplish its life cycle.
2. The mineral component should be explicit and not be substituted.
3. The specific mineral constituents should be unswervingly involved in the plant metabolism.

15.2 Cataloging of plant nutrients in a soil system

According to the requirement and prominence in the plant life cycle, nutrients are classified as macronutrients and micronutrients (Fig. 15.1).

15.2.1 Macronutrients

Macronutrients are essential for photosynthesis, production of cellular structure in plants, plant enzyme synthesis, functions, protein formation,

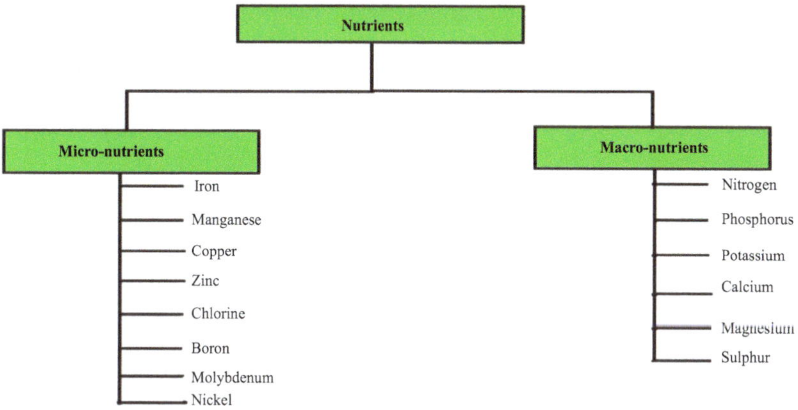

FIGURE 15.1 Classification of plant nutrients in the soil system.

and biochemical processes (Bernal et al., 2009). This is divided into the following categories.

1. *Primary nutrients*: This includes, nitrogen (N), phosphorus (P), and potassium (K). These minerals are provided through fertilizer to plants.
2. *Secondary nutrients*: It consists of calcium (Ca), magnesium (Mg), and sulfur (S).

15.2.2 Micronutrients

Micronutrients are utilized in smaller quantities by plants and is known as trace elements; nevertheless their absence can till now be a preventive factor. They are essential for enzyme production, protein synthesis, oxidation–reduction process, respiration process, plant pigment synthesis, RNA synthesis, germination, and development of plants. It includes zinc, iron, copper, manganese, boron, nickel, chlorine, and molybdenum.

Nutrient necessities differ from the harvest by nature and by plant development period. In many instances, there is a basis of nutrients like nitrogen and phosphorus already present in the soil. Organic nutrient sources contain soil biological matter and plant residues, whereas mineral nutrients include the chief minerals that emphasize ground and minerals like fertilizers (Sarkar et al., 2003; Karlen and Stott, 1994). The obtainability of nutrients can be exaggerated by abiotic soil constraints, such as pH, electrical conductivity, soil microbes, total nitrogen, and C/N ratio; soil microbes decompose the organic matter because plants cannot take it, so is converted to bioavailable forms. Traditional nutrients like nitrogen, phosphorus, and potassium (NPK) fertilizers are used as a nutrient conservancy, where both organic and inorganic nutrients are utilized to prevent micro—and macronutrient deficiencies (Haynes, 2005; Bronick and Lal, 2005; Burns et al., 2013).

15.3 Nutrient deficiency diagnosis and management

Nutrient deficiency is widespread in soil and plants, and it adversely affects soil health, reducing nutrient availability in agricultural products and causing financial losses to agriculturalists. Manures are applied in large quantities to upsurge crop growth and yield. However, the macronutrients are not readily available to plants: hence, plants take only half of the fertilizers applied for their development.

Moreover, maximum macronutrients are insoluble in water, and hence, it not directly utilized by the plants (Ulrich et al., 1993). Therefore, the remaining nutrients present in the soil leached out leading to soil and water pollution. Henceforth, the induction of chemical fertilizers in improving crop yield and productivity consequence in temporary economic profit, which stimulates prolonged effects on the agroecosystem.

The excessive usage of chemical fertilizers damages the soil nutrients, increases soil porosity, soil erosion, and soil infertility, decreases moisture content, damages the soil microorganisms, and disrupts the food web, which leads to mutation and causes changes in the ecosystem (Voss, 1993). The external addition of macronutrients (NPK) and micronutrients hinder plant growth and development in agricultural fields. Therefore, sustainable alteration to chemical fertilizers needs to develop, which supports the fundamental science and applies innovative technology to enhance the crop yield and their related biosphere for better nutrition and environmental remediation (Grundon, 1987) (Fig. 15.2).

The plant requires sufficient nutrients for normal functio

ning and growth. The nutrient supply to plants depends on the plant species and particular nutrients. Nutrient level exterior to the plant's adequacy assortment will cause whole crop development and deterioration owing to deficiency or toxicity. Nutrient deficit arises once the important nutrient is not available in the necessary amount to satisfy the requirement for the

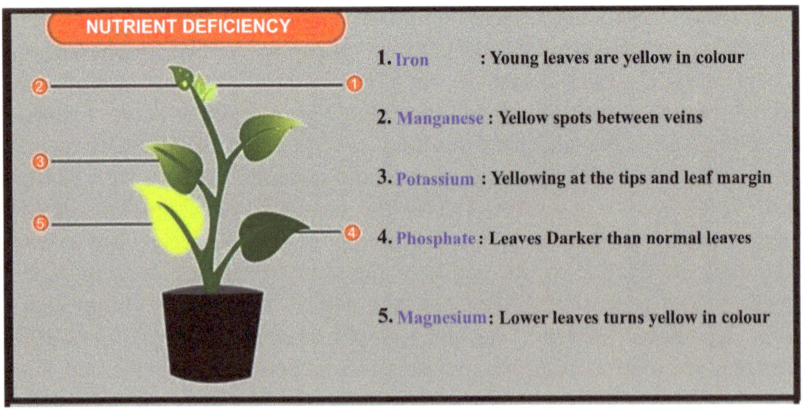

FIGURE 15.2 Symptoms of nutrient deficiencies in plants.

development of the plants (Rengel, 2001; Tisdale et al., 1985). Common nutrient deficiencies are N P K (nitrogen, phosphorus, and potassium) with some micronutrient deficiencies like zinc, copper, boron, sulfur, chloride, manganese, magnesium, and iron. There are three essential tools for analyzing nutrient deficiency.

1. Soil analysis
2. Crop analysis
3. Direct visual observation in fields

Quantitative tests such as soil testing and plant analysis are used for analyzing the availability of nutrients in a particular crop. Qualitative tests like visual observation are used to analyze the nutrient stress visually by observing the symptoms such as yellowing of leaves and short growth that occurs due to nutrient stress (Bennett, 1993; Jones, 1998).

15.4 Macronutrient deficiency in soil

Plant growth depends on the nutrients that are available in the soil through fertilizers. Nutrients in soil usually upsurge the development of florae and deficiency or excess nutrients may impair the soil structure and damage the plant cells, which inhibit the growth of the plant and yield production. Macronutrients are primarily used by plants for their growth and development. There are different types of macronutrients.

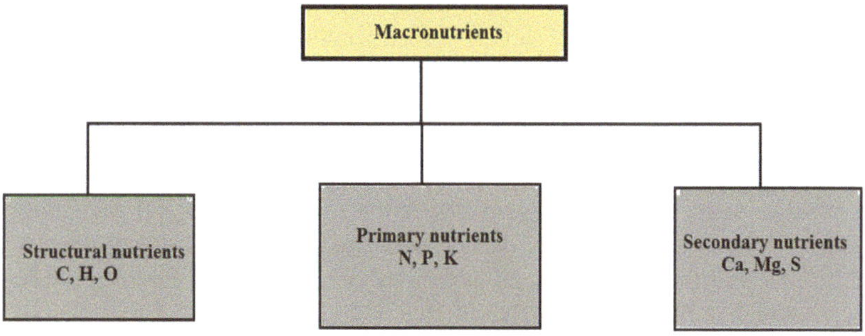

The structural nutrients consist of carbon (C), hydrogen (H), and oxygen (O). These components are attained from the air as CO_2 and H_2O, the source for simple carbohydrates namely, sugars and starch. These primary nutrients afford to strengthen the cell wall, leaves, and stem in addition to serving energy for plant growth and for the other living organisms that consume the plant for their energy development (Morgan and Connolly, 2013).

The primary nutrients comprise nitrogen (N), phosphorous (P), and potassium (K). These compounds are responsible for the assimilation of plant nutrient content, development of cell protein and chlorophyll, stimulation of root formation, necessary for enzyme action, carbohydrate, and protein synthesis, and the integrity of plant cells (Silva and Uchida, 2000).

The secondary nutrients consist of calcium (Ca), magnesium (Mg), and sulfur (S) required in small quantities that are responsible for protein formation, metabolic activities, improving plant vigor, improving cell wall constituents, and helping in the assimilation of fats and necessary for the plant survival. Deficiency of these macronutrients leads to shortening of the stem and leaves, slow growth, slow maturity, and the plant becoming dwarf (Silva and Uchida, 2000).

15.4.1 Nitrogen

The nitrogen is available to plants in the form of NO_3^- (nitrate) and NH_4^+ (ammonia). Nitrogen combines with carbon, hydrogen, oxygen, and sulfur for the synthesis of amino acids. Nitrogen is used in the formation of protoplasm, cell division, and enzymatic reactions in plants. It progresses the quality and amount of leafy grown vegetables and proteins in cereal crops. It is the central portion of the chlorophyll molecule and is essential for the photosynthesis process. Deficiency of nitrogen leads to the reduction in cell division, stunted growth of plants, chlorosis of leaves, the reduced protein content of seeds, a reduction in flowering, and prematurity in few crops, resulting in a decrease in harvest and quality of crops (Srinivasan et al., 2013).

15.4.2 Phosphorous

Phosphorous plays a significant part in energy storage and is transferred by means of ATP and ADP. It is responsible for photosynthesis and respiration. Phosphorous is the central part of genetic components (DNA and RNA) structures. A higher concentration of phosphorous was observed in seeds and younger plant cells, that is, shoot and root tips. It reduces the disease incidence in some plants and improves the quality of crops. Deficiency of phosphorous leads to slow, weak, and stunted growth of plants because phosphorous is needed in larger quantities at the early stage of cell division of plants. Phosphorous deficiency delays maturity, poor seed, and fruit development, and purpling of leaves and stem may occur in plants due to phosphorous deficiency (Nagula and Usha, 2016).

15.4.3 Potassium

Potassium is available to plants as K^+. Potassium plays an essential role in plant growth because it acts as an enzyme activator for plant growth. It contributes to inducing the uptake of water by plants by regulating the stomatal opening and closure process. Potassium promotes the translocation of photosynthates for plant growth. It helps in ATP production and protein synthesis. Potassium enhances the disease resistance in plants, improves the grain size, and improves the quality of seeds, vegetables, and fruits. The deficiency of potassium causes chlorosis, the stem becomes weak and thin, slows down the plant growth, size, and quality of fruits and grains are reduced (Nagula and Usha, 2016).

15.4.4 Calcium

Calcium is available as Ca^+. It plays a major role in cell wall formation and plasticity, maintaining cell integrity and permeability. It activates several enzymes that are responsible for carbohydrate and protein synthesis. It acts as a detoxifying agent by neutralizing the acid in plants. The deficiency of calcium leads to the acidity of the soil and weak stem, and leaves and root tip turns brown and die. Buds and blossoms fall prematurely in some plants (Srinivasan et al., 2013).

15.4.5 Magnesium

Magnesium is available to plants as the ion Mg^{2+}. It is a principal constituent of chlorophyll and is involved in photosynthesis. It acts as a cofactor in several enzymatic reactions that activate the phosphorylation processes. It helps in the stabilization of ribosome particles and nucleic acids. The deficiency of magnesium includes interveinal chlorosis, premature leaf drops, and the acidity of soil (Nagula and Usha, 2016).

15.4.6 Sulfur

Sulfur is available to plants as sulfate ion SO_4^{2-}. It is crucial for producing the plant protein for the constituent of various amino acids. Sulfur is essential for the metabolism of vitamin B and coenzyme A. It also helps in seed production, chlorophyll formation, nodal synthesis, and protein structure stabilization. The deficiency of sulfur leads to chlorosis of leaves, delayed maturity, retracted growth, and soil with low organic matter (Silva and Uchida, 2000) (Table 15.1).

TABLE 15.1 Macronutrient status, their uptake forms, and mobility in soil (Silva and Uchida, 2000).

Macronutrients	Uptake from	Soil flexibility
Carbon	CO_2, H_2CO_3	–
Hydrogen	H^+, OH^-, H_2O	–
Oxygen	O_2	–
Nitrogen	NO_3^-, NH_4^+	Moveable as NO_3^-, immovable as NH_4^+
Phosphorous	HPO_4^{2-}, $H_2PO_4^-$	Immovable
Potassium	K^+	Moderately movable
Calcium	Ca^{2+}	Moderately moveable
Magnesium	Mg^{2+}	Immovable
Sulfur	SO_4^-	Moveable

15.5 Micronutrient deficiency in soil

15.5.1 Boron

Soils in dry and semidry humid areas had the maximum contents of boron. Total boron content in Indian soil ranges from 7 to 630 boron mg/kg. Certain calcareous alluvial soils contain boron content equal to 83 mg/kg. Rough textured, percolated soils have subordinate boron content, and its accretion depends on the calcium carbonate ($CaCO_3$) content. On the other hand, Vertisol and Alfisol soils premeditated from granite, basalt, and limestone have 28–57 boron mg/kg (Alloway, 2008). The concentration of boron in Indian soil varies from 0.75 to 8.0 boron mg/kg. The boron deficiency in Indian soil ranged from 2% in alluvial soil to 68% in red soil. Averagely, 4%, 19.2%, and 21.5% of soils are critical deficient, deficient, and latent deficient of boron. Moreover, 12.1%, 11.4%, and 31.7% of soils were appropriate, adequate, and more available in boron.

Maximum boron deficiency is seen in the Alfisol soil of West Bengal and Assam as a result of diminution in boron content in the increase in rainfall. Intensification in the organic carbon content of 0.50%–0.75% induces the dosage of boron content in soil through 60% and it prevents leaching; hence, higher availability of nutrients will be available for the growth of the plant (Singh, 2007; Singh and Goswami, 2013). Over 60% of Indian states, namely Gujarat, Jammu and Kashmir, Odisha, Kerala, Maharashtra, Manipur, Meghalaya, Arunachal Pradesh, Bihar, Himachal Pradesh, Nagaland, Mizoram, West Bengal, and Jharkhand, were deficient in boron (Shukla et al., 2021).

15.5.2 Copper

The total copper content in Indian soil ranges from 1.8 and 285 Cu mg/kg. Cu content in soil is the result of molten rocks, shale, and granite, respectively. On average, around 2.10%, 2.10%, and 7.2% of Indian soils are critical, deficient, and latent deficient in Cu (copper) (Shukla et al., 2021). The total Cu content was improved with a rise in organic substance and clay composition in Indian soil. The copper content in Vertisols soil is wide ranging from 13 to 167 Cu mg/kg Alfisols and Entisols (Singh, 2007). The sufficient, adequate, and high quantity of Cu in soil ranges from 10.6%, 10.2%, and 67.6%. Among various chemical forms, 10.8% of total Cu in soil was poorly available to plants and the rest of the Cu contents remain in unavailing form (Table 15.2). Organic peat soil, alluvial soil, and hill soil showed Cu deficiency (Singh, 2012). In total, 25% of soil in states such as Goa, Haryana, Uttar Pradesh, Punjab, and Rajasthan are deficient in Cu.

15.5.3 Iron

The most common element in the Earth's crust and soil is iron. It is determined as the fourth important element on the Earth. The iron content in

TABLE 15.2 Micronutrient deficiency and availability in Indian soil (Shukla et al., 2021).

Properties	Boron (%)	Copper (%)	Iron (%)	Manganese (%)	Zinc (%)
Acute deficient	4.0	2.10	3.80,	1.10	7.90
Deficient	19.2	2.10	9.00	6.0	28.6
Latent deficient	21.5	7.2	6.40	10.3	14.7
Sufficient	12.1	10.6	9.60	13.2	12.7
Adequate	11.4	10.2	11.3	10.0	14.8
High level	31.7	67.6	59.9	60.4	21.3

Indian soil ranged from 0.46% to 27.3%. Only a small fraction of iron is utilized by plants for their progress and growth. The shortage of iron content in India ranges from 3.80% to 59%. When the iron content in soil is 3.80%, it shows acute deficiency. On average about 9% and 6.40% of soil were low and latent deficient in iron, whereas 9.60%, 11.3%, and 59.9% of iron content in soil were sufficient, adequate, and high. Iron deficiency is high in alkali soils than compared in calcareous soil and sandy soils (Table 15.2). More than 25% of soil in states such as Haryana, Andhra Pradesh, Goa, Gujarat, Karnataka, Rajasthan, and Telangana are deficient in iron (Fe). The plant shows chlorosis when it lacks iron micronutrients. Inorganic fertilizers like ferrous sulfate are supplied to treat deficiency of iron in the soil. The acidity of ferrous sulfate solution in the soil causes chlorosis in crops. The chelates like Fe-EDTA, farm yard manure, and FeS_2 (pyrite) are used to correct the iron deficiency (Shukla et al., 2021; Singh, 2007).

15.5.4 Manganese

The manganese (Mn) distribution in soil depends on the soil pH, organic matter, soil moisture, and $CaCO_3$ content in the soil. The processes namely leaching, soil erosion, and submergence alter the Mn content in the soil. The majority of our soils contain 200–2000 mg/kg Mn. Clay soil and alluvial soil contain high Mn deficiency than other soils. The Mn accretion transpires mainly in the superficial surface of the soil. Soil present in moist and moderately moist climates has 10%–15% of Mn content in soil related to arid and semiarid region soils (1%–5%). On average, 1.10%, 6%, and 10.3% of Indian soils remained severely deficient, deficient, and latent deficient of Mn present in soil (Table 15.2).

Comparatively, 20% of soil in states such as Jammu and Kashmir, Kerala, Bihar, Goa, Punjab, Nagaland, Rajasthan, Uttar Pradesh, and Tamil Nadu are deficient in Mn, and 13.2%, 0.10%, and 60.4% of Indian soils are marginally sufficient, adequate, and high available of Mn present in soil (Shukla et al., 2021). Interventional chlorosis occurs in plants due to the deficiency of Mn. Amongst various inorganic fertilizers, $MnSO_40.4H_2O$ is commonly used to treat Mn deficiency of soil because of its high solubility, easy availability, and low cost. Apart from this, other sources of fertilizers namely, manganese chelates, MnO_2, and $Mn_3(PO_4)$ are used for amending the Mn deficiency in soil (Singh, 2007).

15.5.5 Zinc

Among all micronutrients, zinc is the most predominant micronutrient required by the soil and plants. It is considered the fourth most deficient nutrient in India after nitrogen, phosphorus, and potassium. The lack of Zn in plants causes leaf discoloration, wilting, necrotic spots, and bronzing of leaves. More than 45% of Indian soil is deficient in Zn. The deficiency of Zn increased in the southern states of India attributes to the extreme usage of NPK inorganic manures.

Zinc deficiency in Indian soil will increase from 49% to 63% in 2025 (Singh, 2007). When the soil is more alkaline and sandier with more soil pH ($>$7.4), high $CaCO_3$ content, low organic matter, and soil erosion in topsoil cause Zn deficiency in the soil. In semiarid and arid climates, calcareous soil shows $>$15% of Zn deficiency. On average, 7.90%, 28.6%, and 14.7% of soils are acute deficient, deficient, and latent deficient of zinc, whereas 12.7%, 14.8%, and 21.3% of soils are adequate, moderately adequate, and more available of Zn in soil (Table 15.2).

Over 50% of Indian states, such as Goa, Karnataka, Rajasthan, Bihar, Odisha, Maharashtra, Tamil Nadu, Uttar Pradesh, Madhya Pradesh, Telangana, and Andhra Pradesh, have the soil deficient in Zn. Zinc availability for plants reduces from total zinc content due to properties such as organic matter, $CaCO_3$ content, soil pH, and soil moisture. Zinc availability increases by adding inorganic and synthetic fertilizers like zinc sulfate monohydrate and zinc sulfate heptahydrate, ZnO, zinc carbonate, Zn phosphate, and Zn chelates (Shukla et al., 2021; Arunachalam et al., 2013).

15.5.6 Impact of micronutrient deficiency in crops in India

The predominant soils present in India are red soil, alluvial soil, black soil, lateritic soil, desert soil, black-clay soil, and foothill soil. Most of the Indian soils have less nitrogen fertility. About 40%–55% soil is low in phosphorous, potassium, and sulfur.

Crops developed in half of the Indian soil grieve from micronutrient deficiency disorders (Alloway, 2008). India is the third major manure overriding country in the world, but the consumption of fertilizers is uneven. In 2050, India will increase its contemporary food manufacturing due to an increase in the high population (Singh, 2008a,b).

15.6 Nanotechnology in developing nanofertilizers

Farming is the support of the nationwide wealth, predominantly in emerging nations, with more than 60% of the inhabitants dependent on farming for their source of revenue. Currently, it is fronting several challenges like a decrease in farming land, climatic change owing to urbanization, and moderate efficiency of crops yield by means of more significant contribution that directs to destitution and malnourishment (Brock et al., 2011).

Among these provocations, poorer crop production and greater contribution to price is the foremost reason intended for agriculturalists' financial defeat in India. Almost 90% of Indian soil trials showed little phosphorus and nitrogen content that is important for improved plant development, advanced yield, and quality. Similarly, among the micronutrients, boron (33%), zinc (49%), iron (12%), copper (3%), manganese (5%), and molybdenum (13%) are the most underprovided nutrients related to further nutrients (Singh, 2008a,b; Adhikari et al., 2010). These multiple nutrient insufficiencies in soil cause inert levels in crop production with reduced soil organic matter, lesser fertilizer reaction ratio, and imbalanced fertilization.

Although external sources of nutrients have been developed as fertilizers, it improves only 50% of crop production. The efficiency of crops is low due to the nutrient loss over leaching, fixation, and microbial assimilation (Sharma et al., 2013). It assessed that about 40%–70% nitrogen, 80%–90% phosphorus, 50%–70% potassium, and above 95% of micronutrient availability in fertilizers pleaded to plants vanished in the habitat, and were not grasped by the plant in a required quantity that cause significant economic and resource losses with severe ecological pollution.

Hence, to upsurge the nutrient proficiency and to enhance the crop yield from torpidity point, protecting the plants from pathogens and diseases is needed using alternate advanced technology like nanotechnology (Chinnamuthu and Boopathi, 2009).

15.7 Progression of nanofertilizers

One of the present innovative technology used for developing new materials at the nanoscale level (1–100 nm) with a wide range of applications is nanotechnology. Nanotechnology has overcome these impediments to various agricultural fields like food safety, security, and productivity (Liu et al., 2006).

Micronutrients and macronutrients play an essential role in the protection of plants or crops against diseases. The use of nanoparticles can enhance the growth and yield of plants, improve stress tolerance in plants, and reduce the pathogen effect on plants.

This increases the need for nanotechnological mediations because of its specific properties like site-specific, highly targeted, and releases of nutrients in a controlled manner. This efficiently analyzes and prevents nutrient deficiency in farm soils (Kalpana Sastry et al., 2011).

Nanofertilizers comprise the nanoformulation of nutrients, applied to plants, to facilitate continued release and equal absorption. Nanofertilizers enhance plant productivity by increasing the nutrient uptake that facilitates the site-targeted controlled release of nutrients, decreases the soil toxicity, and mitigates adverse effects of excessive usage of chemical fertilizers as well as the frequency of application of fertilizers (Tarafdar et al., 2012c).

Sustainable agronomy involves minimal usage of agrochemicals that fortify the environment and protect different breeds of plants from eradication. Particularly, nanomaterials as nanofertilizers improve crop yields by improving the efficiency of agricultural inputs by facilitating the site-targeted controlled release of nutrients, endorsing the minimal usage of agro-inputs.

The main concern in agricultural production is to facilitate alteration and adaptation of plants to advanced climate change reasons like water shortage, salinity, alkalinity, extreme temperatures, and ecological pollution through toxic metals.

Therefore, nanotechnology not only reduces uncertainty but also coordinates agricultural production management as an alternative to conventional technologies. Agro-nanotech inventions deal with short-term technology to fix the problems faced in modern agriculture (Kalpana Sastry et al., 2010a,b; Liu et al., 2006; Tarafdar et al., 2012b; DeRosa et al., 2010).

The nanotechnological application is categorized into the following four parts:

1. Application in plant identification, health, growth, and development
2. Application associated with animal farming
3. Application related to soil management
4. Postharvest application and water, energy, and food management.

15.8 Plant nutrient management by nanotechnology

Best plant nutrient management practices include choosing the right source of fertilizers, the right rate, correct methods, and suitable time for improving the production and yield. Fertilizers play an essential role in exploiting agricultural fabrication, especially water-soluble and rapid-release fertilizers used by farmers. It comprises a greater quantity of nutrients but the fertilizers rapidly release nutrients and are going through several passageways such as leaching, volatilization, fixation, microbial assimilation, and run-off that create the lower nutrients efficacy. Nearly 40%–60% of nitrogen (N), 70%–90% phosphorus (P), and 60%–70% of potassium (K) of the standard fertilizers sprayed on plants are gone to the environment and not be fascinated by plants completely, which causes enormous economic depletion and various types of environmental pollution (Junejo et al., 2011).

The coating materials of controlled release nanomanure determine the quantity and period of release of the plant nutrient. The controlled release of fertilizer should have the following criteria. (i) More than 15% of the controlled release of nutrients should release within 24 h. (ii) However, 75% of nutrients are released within the specified release time of 40–360 days. (iii) 75% of them should release in 28 days. Currently, nanotechnology has become an important and advanced tool to the fertilizer manufacturing industry for making smart fertilizers over smart delivery systems, wherein slow or controlled fertilizers are established by techniques like encapsulation technique in which numerous surface coating materials like polymers, silica, and metal oxides are used (Trenkel, 1997; Zhao et al., 2014).

15.8.1 Slow and controlled release of nanomaterials

The slow and controlled release of nanoparticles is an alternative to soluble conventional fertilizers. Either natural or artificial polymers are used for the coating method. The acrylic acid, diatomite, and chitosan polymers are used as coating materials for the development of NPK compound fertilizers (Wu et al., 2008). This product has excellent slow and controlled release capacity and is nontoxic to soil and ecofriendly material. Encapsulation of urea, nitrogen, or phosphorus with sulfur nanocoating below 100 nm is useful for the controlled release of nutrients. It is mainly used for sulfur-deficient soil (Fernando et al., 2010).

Nowadays, clay minerals, for example, fertilizers, plant growth hormones, promoters, and pesticides are used as encapsulating materials because of their high surface area, modified surfaces, colloidal nature, and high absorption activity. The clay material like bentonite, kaolinite, urea, and montmorillonite is used due to their high water holding volume, distension rate, and slow and steady release properties (Basak et al., 2012).

In the entrapment method, zeolite particularly clinoptilolite is used as a surface material for the nanoparticle synthesis. Zeolites are crystalline with a three-dimensional lattice. It has a greater cationic exchange capacity in which the hydrated alumina silicates carry negative charges used as porous minerals. In contrast, ammonium and potassium are used as trap materials that are positively charged. This is responsible for the deliberate and controlled release of nutrients for the growth and development of plants (Preetha et al., 2014).

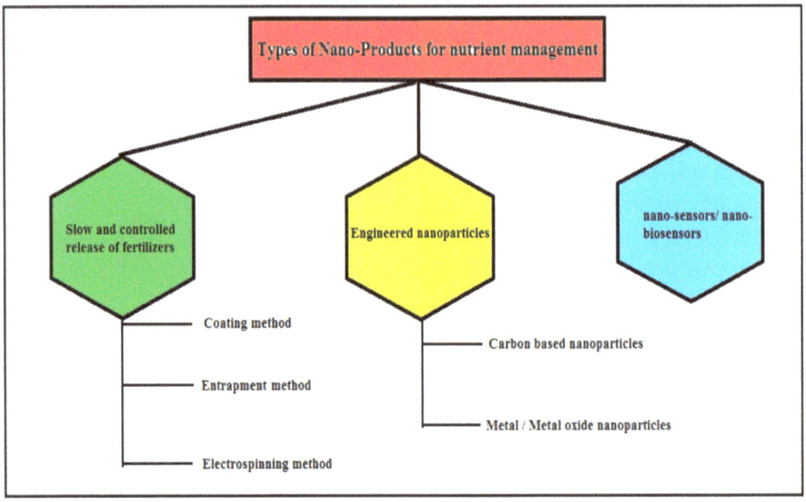

FIGURE 15.3 Types of nanoproducts for nutrient management.

Another critical research in developing nanoparticles by entrapment method is the research of deep core–shell nanoparticles used recently. This hollow core–shell nanomaterial is used in various applications like drug delivery, drug loading, pesticide development, and several biomedical and agricultural applications. $CaCO_3$ and magnesium carbonate are used for the preparation of hollow core–shell in different sizes, shapes, and shell thicknesses. These carbonates are utilized to trap the water-soluble fertilizers for the slow and steady release of nutrients by reducing the dissolution rate and increasing the efficacy of soil nutrients (Yuvaraj and Subramanian, 2015).

In the electrospinning method, electric forces are used to produce ultrafine fibers that are utilized as polymers for nanoparticle synthesis. It is produced from a single fluid (coaxial electrospinning) to two fluids (side–by–side electrospinning) that create nanofibers with new functions. Recently, the fibers synthesized from polyhydroxybutyrate and polylactic acid were used as shells and then mixed with fertilizers (NPK) through the electrospinning method (Kampeerapappun and Phanomkate, 2013) (Fig. 15.3).

15.8.2 Engineering of nanomaterials

In the present world, engineered nanomaterials develop into an essential and significant factor in the everyday life cycle and their usage increased rapidly in various industrial fields like cosmetics, medicine, machinery, painting, and agricultural sectors. These nanoparticles developed with specific properties like sizes, shapes, and surface properties. Engineered nanomaterials are categorized into four types conferring to USEPA namely metal-based materials, carbon-based materials, composites, and dendrimers. Amongst all types, metal nanoparticles seek more attention in various industrial applications and their toxicity assessment. One of the major advantages of the agricultural sector is the presence of a higher amount of nutrients in insoluble form and their conversion to a soluble form by size reduction and morphological changes through physical, chemical, and biological preparation methods. Therefore, it increases the efficacy of the nutrients.

15.8.2.1 Metal oxide nanoparticles

Several types of nanoparticles are used for the development of plants for better crop production. Metal oxide nanomaterials namely gold, silver, titanium, iron, zinc, copper, aluminum, and manganese influence seed germination and plant growth. Nanofertilizers are prepared by different methods: physical (top-down), chemical (bottom-up), and biosynthetic (biological) methods. These nanoparticles synthesized by various techniques include:

- Physical method
- Chemical method
- Biological method

15.8.2.1.1 Physical method

The physical method generally, termed a top-down approach, consists of thermal, diffusion, irradiation, and decomposition processes. The evaporation-condensation and laser ablation are the main physical practice used for the synthesis of nanomaterials. In the evaporation-condensation method, different inert gases are used for nanoparticle synthesis (Swihart, 2003) (Fig. 15.4).

15.8.2.1.2 Chemical method

The chemical reduction method is an example of the bottom-up approach used for nanoparticle synthesis. In this method, different metal elements like zinc, copper, molybdenum, manganese, and iron are reduced to form nanoparticles. This process interests significant attention between industries for the large-scale production of nanoparticles (Mohanpuria et al., 2008).

15.8.2.1.3 Biological method

The biosynthesis of nanoparticles fascinated several researchers due to their physical and chemical processes. This method has more advantages over the other method of nanoparticle synthesis. It is a nontoxic, cost-effective, and ecological approach. In this method, the nanoparticles were synthesized

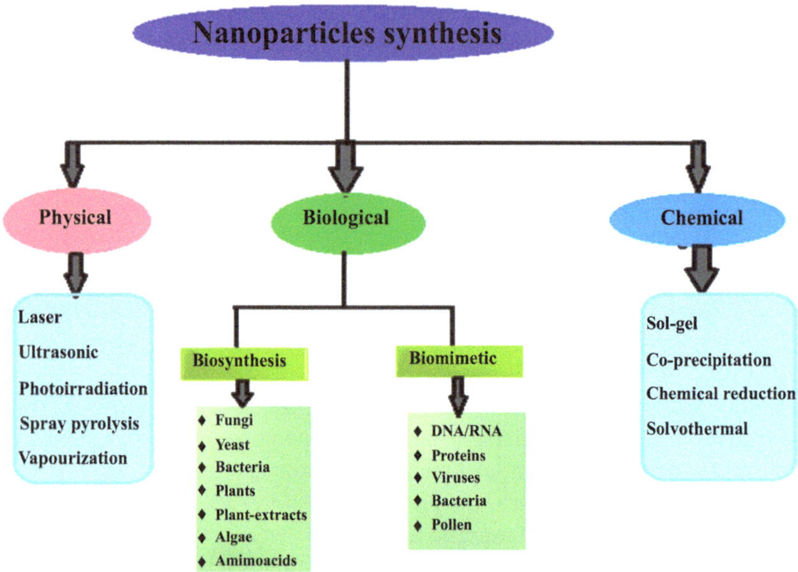

FIGURE 15.4 Schematic representation for nanoparticle synthesis (Thakkar et al., 2010).

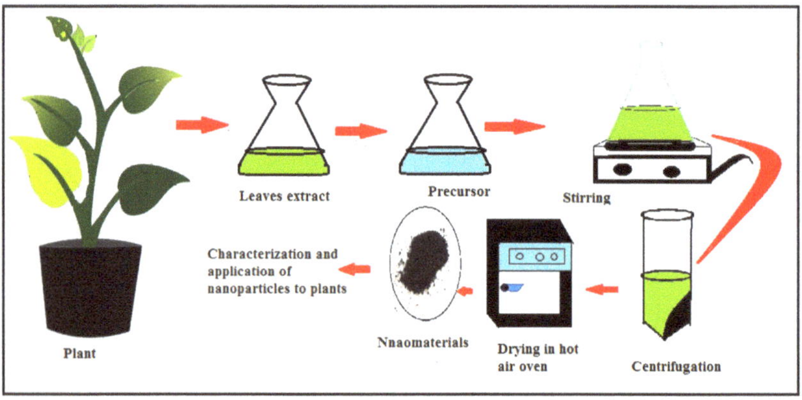

FIGURE 15.5 Schematic representation of methods for the production of nanoparticles.

using various sources like bacteria, fungi, seaweed, oil, bioconstituents, and plant extract (Thakkar et al., 2010) (Fig. 15.5).

15.8.2.2 *Carbon-based nanoparticles*

Carbon, a nonmetallic element that is used in various applications in the form of wood charcoal, activated charcoal for the water purification process and decolorization process. In recent years, these carbon-based nanomaterials are used in optical devices, quantum computers, tissue engineering, and the medicinal field as drug delivery systems, and also used in the fullerene cage developmental process. Currently, carbon nanomaterials are used in the agricultural sector to increase crop productivity, chlorophyll content, soil microorganisms' assimilation, and superoxide dismutase (SOD) activity as nanotubes (Kole et al., 2013).

15.8.3 Nanosensors and nanobiosensor for plant and soil nutrient management

Nanosensors and nanobiosensors received attention in several areas of science and industrial sectors. A nanosensor is an analytical device immobilized as a layer in sensors with DNA/RNA or proteins of microorganisms. It analyzes the biological signal and converts to an electrical signal. These sensors are used to diagnose cancer cells. It is also used in agricultural fields to sense the fertilizers, herbicides, pesticides, moisture content, soil pH, and pathogens. Nanosensors are linked with GPS to detect and monitor disease, soil conditions, soil moisture content, plant health, crop yield, water deficit, and the soil nutrient availability and depletion (Monreal et al., 2015).

Nanotechnology is an essential technology to soil science because several natural components in soil are nanosize in the structure. Nowadays, nanofertilizers have a significant impact on enhancing nutrient efficacy through brilliant slow and steady release. Moreover, soil pollution is controlled by preventing the wastage of fertilizers through leaching, volatilization, immobilization, and run-off. The controlled release of nutrients into rhizospheres is accomplished by the utilization of zeolite, polymer-based materials, and engineered nanomaterials. Soil is one of the main sources of our life cycle; without soil, there would be no microbes, no nutrients, no plants, and no human life.

In the agricultural sector, soil is responsible for plant development and nutrient absorption properties. Currently, nanoparticles are used for protecting soil nutrients and water management. Nanosand was established by layering sand with additives to inhibit water drainage under the plant root especially, in drylands, sustaining underground water tables and slow release of manures for plant progress and development. Along with constructive properties, adverse effects of nanomaterials in soil are also present. Nanoparticles like ZnO, CuO, TiO_2, and SiO_2 are toxic to gram-positive and gram-negative microbes. Likewise, silver (Ag) nanoparticles cause propagative damage to earthworms (*Eisenia foetida*). ZnO nanoparticles decrease the soil bacteriological community, biomass, and assortment. The enzyme activity, especially the dehydrogenase enzyme, reduces when soil is exposed to ZnO nanoparticles. Shoot and root length also decreased and showed phytotoxicity by immobilization of nanoparticles in soil (Tripathi et al., 2016).

Soil microorganisms and fauna play a vital role in soil nutrient mineralization. The porosity of soil is increased by soil fauna (earthworm) that promotes the transfer of nanoparticles to the rhizosphere (root) of plants, and then microorganism present in the rhizosphere plays a leading role in the chemical behavior of nanoparticles. Nanoparticles in soil undergo numerous processes such as the oxidation and reduction process and biogeochemical transformation process. The overall mechanism of surface interaction of nanoparticles to soil environment is given in Table 15.2, and the organic functional group present in nanoparticles and their associated reactions with mechanisms are displayed in Table 15.3 (Tables 15.4 and 15.5).

TABLE 15.3 Benefits of nanoenrichers (Cui et al., 2010a).

S. no	Properties	Conventional fertilizers	Nanofertilizers
1.	Nutrient acceptance efficacy	The chemical or conservative fertilizers cause macronutrients and micronutrients to become deficient by discharge, fixation, and bacteriological assimilation; hence, the nutrient ratio turns out to be deprived.	The nanofertilizer aids to rise the nutrient usage efficacy and uptake ratio by reducing the dimensions of the element that without difficulty enter into the root and leaf cuticular cells via soil and foliar spray application method.
2.	Controlled release	The loss of nourishment owing to the rapid eruption of nutrient discharge moreover through soil or by means of foliar application. It besides generating the difficulty in plant development and yielding because of deprived nutrient usage and noxiousness properties but as well as afford the great effects on losses of wealth and resources, conservational pollution, and the natural stability of the soil.	Nanoencapsulated fertilizers contribute the nutrients in deliberate and precise release through a long period and increase the nutrient efficacy. In this case, water-soluble fertilizers are coated with nanosized polymers and metal oxides. Superficial coverings of nanomaterials on fertilizer units embrace the elements sturdily wing to superior superficial pressure than the predictable shells and hence support in precise release.

(*Continued*)

TABLE 15.3 (Continued)

S. no	Properties	Conventional fertilizers	Nanofertilizers
3.	Effect during a controlled release	The essential nutrients for the completion of the plant life cycle are macro—and micronutrients. Plant nutrients are required in higher amount for their growth, and development but a lack of nutrients occur eventually in the growing season because of surplus proclamation of nourishments at the beginning of the plant growth stage, and it is lost at the end of the development through leaching, volatilization, and fixation.	Nanofertilizer preparations are superlative for the controlled release of nutrients for a sustained interval of time for plant growth, which not only enhances the crop productivity besides diminishing the lack of nutrients, which sequentially decreases the ecological pollution.
4.	Solvability and distribution of micronutrients	In plant manure conservation, micronutrient fertilizers in the composition of nitrate and sulfate are consumed for growing the efficiency and quality of the crops. Although these forms of fertilizers comprise a lesser quantity of nutrient content than other types of fertilizers such as copper, iron, manganese, and zinc oxides. Besides, the element size of the conservative manures is huge and it could not easily be uptake by the crops.	The absorption of the active constituent of micronutrients is improved by rising the solvability and narrowing the fixation capability in the soil by the oxide formation of fertilizers, which comprises developed manure content and can be rehabilitated to the resolvable formation by reducing their size and shape.

15.9 Types of nanofertilizers

Nanofertilizers are made from bulk material (macro—and micronutrient) by an inorganic or conventional method or from plant extracts with an encapsulation process or surface coating of fertilizers by nanoparticles for a slow and steady release of nutrients. Nanofertilizers have unique physiochemical properties and

TABLE 15.4 Overall mechanism of surface interaction on nanoparticles in soil environment (Valdes Solis et al., 2007).

Surface interaction	Overall mechanisms
Van der Waals force	A short arrangement of electromagnetism among molecules has a neutral charge only
Bridge force	Development of a bridge between the polymer folder and particles
Repulsive force	Overlay of hydrogen bond between water particle and hydrophilic superficial of particles
Electric double layer overlap	Electric interaction between the electric double layer overlay on particles
Depletion	Having less affinity between the surface of polymers or solute and solvent
Steric interaction	Short range interaction among adsorbed polymer layer and particles

TABLE 15.5 Nanoparticle's organic functional group and their reaction mechanisms in soil (Valdes Solis et al., 2007).

S. no	Organic functional group	Mechanisms	Examples
1.	Aromatic and alkyl C	Van der Waals force	Kaolinite, aryl clay with a neutral site
2.	Amines, carboxyl, phenyloxyl, heterocycle N	Hydrogen bonding	Kaolinite with organic surface
3.	Carboxylate, amines, heterocyclic N, ring-NH	Cation bridging	Aluminosilicates, allophane, imogolite
4.	Carboxylate, amines, carbonyl, and alcoholic −OH	Water bridging	Smectite, illite, vermiculite
5.	Amines, heterocyclic N, ring-NH	Anion exchange—pH dependent	Smectite, illite, vermiculite
6.	Carboxylate, phenolate	Cation exchange	Aluminosilicates, allophane,
7.	Amines, carboxyl,	Protonation	Aluminosilicates, allophane, Fe and Al oxides

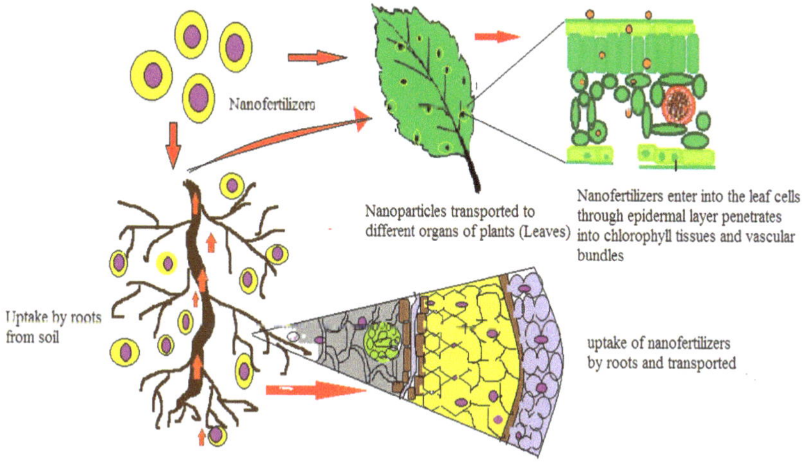

FIGURE 15.6 Nanofertilizers absorb uptake translocate to plant cells from soil.

release nanosized fertilizer in a slow and controlled manner to improve the efficacy of plants by increasing the absorption of nutrients by plants.

Nanofertilizers increase the bioavailability of nutrients owing to their large surface area and more reactivity with less size. Nanomaterials prevent leaching, decrease volatilization, and reduce environmental hazards. The efficiency of nanofertilizers is governed by essential factors such as intrinsic factors, extrinsic factors, and mode of administration. Intrinsic factors consist of nanoformulation preparation, particle size, and surface coating. Extrinsic factors include soil pH, soil moisture, temperature, soil depth, microbial activity, and organic matter availability, which affect the probable utilization of nanofertilizers (Fig. 15.6). Different kinds of nanoparticles such as carbon-based nanomaterials, Zn, Cu, Fe, polymeric nanomaterials, and silica nanomaterials are used as fertilizers.

15.9.1 Macronutrient based nanofertilizer for soil and plant nutrient management

Plant growth depends on the number of macronutrients and micronutrients present in the soil including oxygen, carbon, nitrogen, phosphorus, potassium, calcium, sulfur, magnesium, copper, iron, zinc, and manganese. However, all the nutrients are needed for the growth and development of plants, and primary macronutrients like nitrogen, phosphorus, and potassium are considered as the primary fertilizers.

15.9.1.1 Nitrogen fertilizers

It is the first nutrient essential for the growth of the plants as it is crucial for protein synthesis and energy metabolism. The vast usage of urea (chemical nitrogen fertilizers) has numerous environmental concerns due to the excessive loss of nitrogen through denitrification, leaching, and volatilization (Gao et al., 2015). Innumerable numbers of research studies stated that nitrogen-based fertilizers have more potential to increase production.

Nanomaterials and carriers like zeolites, chelates, and clay are used to synthesize nanofertilizers for the steady and controlled release of nutrients to plants and to rise the uptake of nourishments by plants. Manikandan and Subramanian (2014) reported that formulation of interchelated nanofertilizers increases the growth, quality, yield, and nutrient availability in maize crops, when compared to conventional urea. Rajonee (2016) stated that zeolite-based nitrogen nanofertilizers increase the accumulation of nitrogen in plants and show better pH and soil moisture than chemical fertilizers.

15.9.1.2 Phosphorous fertilizers

Phosphorous is responsible for the root growth, flowering stage, transport, and storage of energy and organic compound formation. The availability of phosphorous is higher in soil, but it is restricted to plants by some factors.

15.9.2 Micronutrient-based nanofertilizer for soil and plant nutrient management

Micronutrients namely Zn, Cu, B, Mn, Mb, Fe, and Ni are required in trace amounts for the nutrient management in soil and the growth and development of the plants.

15.9.2.1 Iron (Fe) nanofertilizers

Iron plays a vital role as a cofactor in enzymes for the biological process of plants. Soil is primarily composed of iron from the Earth's crust, which contains 5% of iron. Besides, more amount of iron is unavailable to plants because it is present in an insoluble form. The promising approach to make available of iron by nanoformulation is responsible for the steady and controlled release of manures to plants. Iron-based nanofertilizers and iron chelate nanofertilizers are highly stable and provide a slow and steady release of iron in a wide range of pH. Iron is absorbed by roots as Fe^{2+} and Fe^{3+} ions.

The sufficient iron quantity for plants ranges from 50 to 250 ppm. Iron is the essential structural component for substances such as cytochrome, heme, protein, and FeS protein, which are involved in oxidation–reduction reactions in photosynthesis and the respiration process. Additionally, Fe is used as the catalyst for chlorophyll synthesis and is included as an important constituent of nitrogenase enzyme, responsible for nitrogen fixation by N_2 fixing microorganisms. When the iron content is <50 ppm, the symptoms like interveinal chlorosis occur in young leaves, leaves become white, and necrosis occurs.

15.10 Effects of nanofertilizers in soil

The use of nanoenrichers in the soil increases the efficacy of elements, diminishes the harmfulness of soil, decreases the nutrient dosage, and reduces the application of chemical fertilizers to plants. Production and application of nanoparticles to plants absorb the nanofertilizers and store and fix the nutrient shortage. For instance, iron nanofertilizers are compatible with the environment and agricultural farms to improve the organic matter in the soil. Clays in soil are submicrometric elements that are covered by phyllosilicate constituents with base silicate polymer, which are nanodimensional at one level. Plants rapidly and completely absorb nanomaterials. When nanoparticles are applied to the soil, they penetrate the seeds, are uptake by plants through roots, and translocate to other parts of plants (Fig. 15.7).

Plants are in direct contact with nanomaterials that enter the plants from the soil in very high concentrations. After the nanoparticles move from the soil, as absorbed by roots, they are distributed to stems and leaves and affect plant physiological conditions.

The distribution of nanoparticles from soil to plants depends on the soil type, structure, and thickness of soil. Fang et al. (2009) stated that the TiO_2 nanoparticles transmit in soil. They showed that the mobility of silver nanowires was significantly reduced due to the large and easily absorbed by soil structures. Rahmatpour et al. (2018) reported that the mobility of silver nanoparticles in alkaline or calcareous soil is limited both in saturated and unsaturated conditions.

Nanoparticles from the soil are first diffused to seeds of plants and are transported to root and stem, then to leaves and localize in the cells. The nanoparticles penetrate the base, increase the absorption, and increase the seed

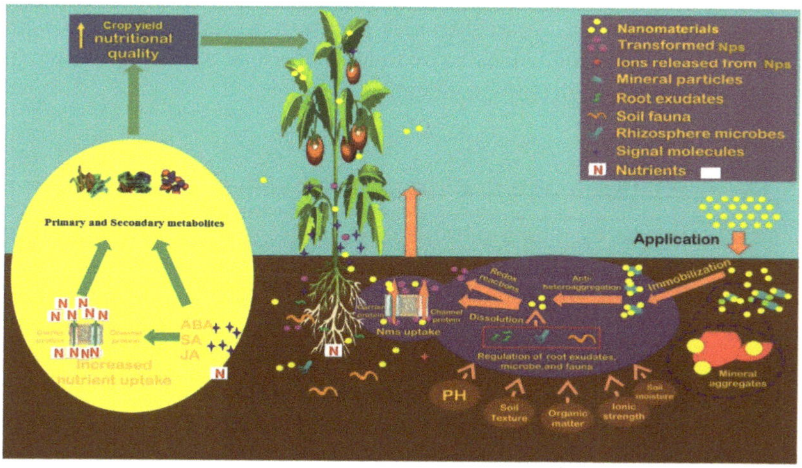

FIGURE 15.7 Migration of nanoparticles from soil to various organs of plants (Wang et al., 2013).

germination. The second way can be absorbed directly from soil to the roots of plants and spread to other organs. Another way is to transport nanomaterials in the soil or water used for irrigation to plant organs (Wang et al., 2013).

15.11 Phytotoxicity of nanoparticles and influencing factors

The intrinsic application of nanoparticles plays important role in the determination of interaction between plants and nanoparticles. The inherent properties of nanoparticles are namely biochemical composition, hydrophobicity, dimensions, form, hydrophilicity, structure, and angles. The toxicological properties of nanoparticles depend on the magnitude and the external surface area of the nanomaterials. As the size of the nanoparticle declines, the surface area of the nanoparticles increases, which upsurges the number of particles per-unit surface area. Numerous research studies described that the deadliness of nanoparticles in soil or plants is more toxic than compared to bulk nanoparticles (Judy et al., 2012; Larue et al., 2012).

The superficial properties, for example, constancy, valency in the external layer, zeta potential, particle aggregation, and coating, perform an important part in the phytotoxicity of nanoparticles. The phytotoxicity of different nanoparticles is due to the direct uptake by plants. Nanoparticles with positively charged surfaces are actively involved in the induction of cell membrane endocytosis while the negatively charged nanoparticles get transferred into vascular bundles (Schwab et al., 2016). The surface coatings help in controlling nanoparticle suspension and the amount of release into the soil.

Alternative perilous feature throughout nanoparticle application is their ideal dosage, which diverges from species to species. For instance, Taran et al. (2016) suggested that the biogenic nanoparticles, for instance, silver, copper, iron, zinc, and manganese nanomaterials at 120 mg/L, reduce oxidative stress significantly. Therefore, it is necessary to use nanoparticles for supportable crop making, which are extremely effectual and benign in bringing agrochemicals, particularly nutrients of plants.

15.12 Conclusion and imminent prospective

Nanotechnology is used as a substitute technique in the field of agriculture for soil fertility management, nutrient management, and food security purposes. Nanoparticles are proved to be a competent transporter structure for a slow and stable release of nanoparticles as nutrients to the target either soil or plants. Nanofertilizers are less toxic than compared to other chemical fertilizers. It concluded that the nanoparticles in soil nutrient management and plant science are used efficiently in the agriculture sector as nanotransporters for the slow and fixed release of nutrients. Nanoparticles have the capability to persuade stomas in the cell wall.

As nanoparticles move into the various organs of the plants, they interact with biologically active molecules in cells. As a result, a number of physiological effects may occur due to the interaction of nanoparticles with plants. The nanoparticle can rapidly move in plants and migrate from one organ to another. Based on the interaction of nanoparticles with plants, it is essential to evaluate the environmental risk that happens through the diffusion of nanoparticles into the environment. It is possible to obtain suitable approaches to regulate nanoparticle usage in the agricultural sector as nanofertilizers in the forthcoming years. Over these progressions in fertilizers, it is probable to modernize conservative agriculture to recent, competent, sustainable agriculture that eventually leads to food safety and nutrient management in soil.

References

Adhikari, T., Biswas, A.K., Kundu, S., 2010. Nano—fertilizer—a new dimension in agriculture. Indian J. Fert. 6 (8), 22–24.

Alloway, B.J., 2008. Zinc in Soils and Crop Nutrition., second ed. IZA and IFA, Brussels, Belgium and Paris, France.

Arnon, D.I., Stout, P.P., 1939. The essentiality of certain elements in minute quantity for plants with special reference to copper. Plant. Physiol. 14, 1460–1470.

Arunachalam, P., Kannan, P., Prabukumar, G., Govindaraj, M., 2013. Zinc deficiency in Indian soils with special focus to enrich zinc in peanut. Afr. J. Agric. Res. 8, 6681–668812.

Basak, B.B., Pal, S., Datta, S.C., 2012. Use of modified clays for retention and supply of water and nutrients. Curr. Sci. 102, 1272–1279.

Bennett, W.F. (Ed.), 1993. Nutrient Deficiencies and Toxicities in Crop Plants. APS Press, St. Paul, MN, p. 202.

Bernal, M.P., Clemente, R., Walker, D.J., 2009. Interaction of heavy metals with soil organic matter in relation to phytoremediation. In: Navarro-Aviño, J.P. (Ed.), Phytoremediarion: The Green Salvation of the World. Research Signpost, Trivandrum, India, pp. 109–129.

Brock, D.A., Douglas, T.E., Queller, D.C., Strassmann, J.E., 2011. Primitive agriculture in a social amoeba. Nature 469, 393–396. Available from: https://doi.org/10.1038/nature09668.

Bronick, C.J., Lal, R., 2005. Soil structure and management: a review. Geoderma 124, 3–221.

Burns, R.G., DeForest, J.L., Marxsen, J., Sinsabaugh, R.L., Stromberger, M.E., Wallenstein, M. D., et al., 2013. Soil enzymes in a changing environment: current knowledge and future directions. Soil. Biol. Biochem. 58, 216–234.

Chinnamuthu, C.R., Boopathi, P.M., 2009. Nanotechnology and agroecosystem. Madras Agric. J. 96 (6), 17–31.

Cui, H.X., Sun, C.J., Liu, Q., Jiang, J., Gu, W., 2010a. Applications of nanotechnology in agrochemical formulation, perspectives, challenges and strategies. In: International Conference on Nanoagri, Sao Pedro, Brazil, June, 20–25.

De Rosa, M.R., Monreal, C., Schnitzer, M., Walsh, R., Sultan, Y., 2010. Nanotechnology in fertilizers. Nat. Nanotechnol. J. 5, 91–96.

Fang, J., Shan, X.-Q., Wen, B., Lin, J.-M., Owens, G., 2009. Stability of titanium nanoparticles in soil suspensions and transport in saturated homogeneous soil columns. Environ. Pollut. 157, 1101–1109. Available from: https://doi.org/10.1016/j.envpol.2008.11.006.

Fernando, W.J.N., Ahmad, N.N.R., Ahmad, A.L., 2010 Release rates of phosphate from coated fertilizer through nano CaCO3-wax mixtures. In: Book of Abstracts International

Conference on Environmental Engineering and Applications (ICEEA 2010), Singapore, 2010, pp. 34–37.

Gao, W., Yang, H., Kou, L., Li, S., 2015. Effects of nitrogen deposition and fertilization on N transformations in forest soils: a review. J. Soils Sediment. 15, 863–879.

Grundon, N.J., 1987. Hungry Crops: A Guide to Nutrient Deficiencies in Field Crops. Queensland Government, Brisbane, Australia, p. 246.

Havlin, J.L., Beaton, J.D., Tisdale, S.L., Nelson, W.L., 1999. Soil Fertility and Fertilizers, sixth ed Prentice-Hall, Inc, Upper Saddle River, NJ, p. 499.

Haynes, R.J., 2005. Labile organic matter fractions as central components of the quality of agricultural soils: an overview. Adv. Agron. 85, 221–268.

Jones Jr., J.B., 1998. Plant Nutrition Manual. CRC Press, Boca Raton, FL, p. 149.

Judy, J., Unrine, J., Rao, W., Wilrick, S., Bertsch, P., 2012. Bioavailability of gold nanomaterials to plants: importance of particle size and surface coating. Environ. Sci. Technol. 46, 8467–8474. Available from: https://doi.org/10.1021/es3019397.

Junejo, N., Khanif, M.Y., Hanfi, M.M., Yunus, W.M.Z., Dharejo, K.A., 2011. Role of inhibitors and biodegradable material in mitigation of nitrogen losses from fertilized lands. Afr. J. Biotechnol. 10, 3504–3514.

Kalpana Sastry, R., Rashmi, H.B., Rao, N.H., Ilyas, S.M., 2010a. Integrating nanotechnology nto agri-food systems research in India: a conceptual framework. Technol. Forecast. Soc. Change 77, 639–648.

Kalpana Sastry, R., Rashmi, H.B., Rao, N.H., 2010b. Nanotechnology patents as R&D indicators for disease management strategies in agriculture. J. Intellect. Prop. Rights 1 (5), 197–205.

Kalpana Sastry, R., Rashmi, H.B., Rao, N.H., 2011. Nanotechnology for enhancing food security in India. Food Policy 36, 391–400.

Kampeerapappun, P., Phanomkate, N., 2013. Slow-release fertilizer from core-shell electrospun fibers. Chiang Mai J. Sci. 40 (4), 775–782.

Karlen, D.L., Stott, D.E., 1994. A framework for evaluating physical and chemical indicators of soil quality. In: Doran, J.W., Coleman, D.C., Bezdicek, D.F., Stewart, B.A. (Eds.), Defining Soil Quality for a Sustainable Environment. SSSA Special Publ 35 Soil Sci Soc Amer, Madison, WI, pp. 53–72.

Kole, C., Kole, P., Randunu, K.M., 2013. Nanobiotechnology can boost crop production and quality: first evidence from increased plant biomass, fruit yield and phytomedicine content in bitter melon (Momordicacharantia). BMC Biotechnol. 13, 37.

Larue, C., Veronesi, G., Flank, A.-M., Surblé, S., Herlin Boime, N., Carriere, M., 2012. Comparative uptake and impact of TiO2 nanoparticles in wheat and rapeseed. J. Toxicol. Environ. Health. Part. A 75, 722–734. Available from: https://doi.org/10.1080/15287394.2012.689800.

Liu, X., Feng, Z., Zhang, S., Zhang, J., Xiao, Q., Wang, Y., 2006. Preparation and testing of cementing nano-subnano composites of slow-or controlled release of fertilizers. Sci. Agric. Sin. 39, 1598–1604.

Manikandan, A., Subramanian, K.S., 2014. Fabrication and characterization of nanoporouszeolite-based N fertilizer. Afr. J. Agric. Res. 9, 276–284.

McCauley, A., Jones, C., Jacobsen, J., 2009. Soil pH and organic matter. Nutrient Management Modules, Vol. 8. Montana State University Extension Service, Bozeman, MT, pp. 1–12.

Mohanpuria, P., Rana, N.K., Yadav, S.K., 2008. Biosynthesis of nanoparticles: technological concepts and future applications. J. Nanopart. Res. 10, 507–517.

Monreal, C.M., DeRosa, M., Mallubhotla, S.C., Bindraban, P.S., Dimkpa, C., 2015. The application of nanotechnology for micronutrients in soil-plant systems. In: VFRC (Virtual Fertilizer Research Center) Report 2015/3, Washington, DC, p. 44.

Morgan, J.B., Connolly, E.L., 2013. Plant-soil interactions: nutrient uptake. Nat. Educ. Knowl. 4 (8), 2.

Nagula, S., Usha, P.B., 2016. Application of nanotechnology in soil and plant system with special reference to nanofertilizers. Adv. Life Sci. 5 (14), 5544–5548, Print: ISSN 2278–3849.

Preetha, P.S., Subramanian, K.S., Rahale, C.S., 2014. Characterization of slow release of Sulphur nutrient-a zeolite-based nano-fertilizer. Int. J. Dev. Res. 4, 229–233.

Rahmatpour, S., Mosaddeghi, M.R., Shirvani, M., Šimůnek, J., 2018. Transport of silver nanoparticles in intact columns of calcareous soils: the role of flow conditions and soil texture. Geoderma 322, 89–100. Available from: https://doi.org/10.1016/j.geoderma.2018.02.016.

Rajonee, A.A., 2016. Synthesis of nitrogen nano fertilizer and its efficacy. Environ. Sci. Pollut. R.

Rengel, Z., 2001. Genotypic differences in micronutrient use efficiency in crops. Commun. Soil. Sci. Plant. Anal. 32 (7–8), 1163–1186.

Sarkar, S., Singh, S.R., Singh, R.P., 2003. The effect of organic and inorganic fertilizer on soil physical condition and the productivity of rice-lentil cropping sequence in india. J. Agric. Sci. 140 (4), 419–425.

Schwab, F., Zhai, G., Kern, M., Turner, A., Schnoor, J.L., Wiesner, M.R., 2016. Barriers, pathways and processes for uptake, translocation and accumulation of nanomaterials in plants—critical review. Nanotoxicology 10, 257–278. Available from: https://doi.org/10.3109/17435390.2015.1048326.

Sharma, S.B., Sayyed, R.Z., Trivedi, M.H., Gobi, T.A., 2013. Phosphate solubilizing microbes: sustainable approach for managing phosphorus deficiency in agricultural soils. Springerplus 2, 587–600. Available from: https://doi.org/10.1186/2193-1801-2-587.

Shukla, A.K., Behera, S.K., Prakash, C., et al., 2021. Deficiency of phyto-available sulphur, zinc, boron, iron, copper and manganese in soils of India. Sci. Rep. 11, 19760. Available from: https://doi.org/10.1038/s41598-021-99040-2.

Silva, J.A., Uchida, R., (Eds.), 2000. Approaches for tropical and subtropical agriculture. In: Plant Nutrient Management in Hawaii's Soils. College of Tropical Agriculture and Human Resources, University of Hawaii at Manoa, 3, 31–55.

Singh, M.V., 2007. Micro and secondary and pollutant element re-search in India. AICRP Micro Secondary Pollut. Elem. Soils Plants, IISS, Bhopal 31, 1–112.

Singh, M.V., 2008a. Micronutrient's deficiencies in crops in India. In: Alloway, B.J. (Ed.), Micronutrient Deficiencies in Global Crop Production. Springer, Dordrecht, pp. 93–126.

Singh, M.V., 2008b. Micronutrient deficiencies in crops and soils in India. Micronutrient Deficiencies in Global Crop Production. Springer, Dordrecht, pp. 93–125.

Singh, M.V., 2012. Spread of micro-nutrient deficiencies specially boron in India and response of field crops. In: Brain Storming Workshop on Soil Test-Based Nutrients Including Boron and Other Micro Nutrients. Organized by ICRISAT—Agriculture Directorate, Karnataka—Rio Tinto India at Bangalore, India.

Singh, M.V., Goswami, V., 2013. Efficiency of boron fortified NPK fertilizer in correcting boron deficiency in some cereal and oilseeds crops in India. In: 17th International Plant Nutrient Colloquium 17–18 August, Istanbul, Turkey.

Srinivasan, K., et al., 2013. J. Chem. Chem. Sci. 3 (3), 107–116.

Swihart, M.T., 2003. Vapor-phase synthesis of nanoparticles. Curr. Opin. Colloid. Interface Sci. 8, 127–133.

Tarafdar, J.C., Raliya, R., Rathore, I., 2012b. Microbial synthesis of phosphorus nanoparticles from Tri-calcium phosphate using Aspergillus tubingensis TFR-5. J. Bionanoscience 6, 84–89.

Tarafdar, J.C., Xiang, Y., Wang, W.N., Dong, Q., Biswas, P., 2012c. Standardization of size, shape and concentration of nanoparticle for plant application. Appl. Biol. Res. 14, 138–144.

Taran, N., Batsmanova, L., Kovalenko, M., Okanenko, A., 2016. Impact of metal nanoform colloidal solution on the adaptive potential of plants. Nanoscale Res. Lett. 11. Available from: https://doi.org/10.1186/s11671-016-1294-z.

Thakkar, K.N., Mhatre, S.S., Parikh, R.Y., 2010. Biological synthesis of metallic nanoparticles. Nanomedicine 6, 257–262. Available from: https://doi.org/10.1016/j.nano.2009.07.002.

Tisdale, S.L., Nelson, W.L., Beaton, J.D., Havlin, J.L., 1985. Soil Fertility and Fertilizers, fifth ed. Macmillan Publishing Company, New York.

Trenkel, M.E., 1997. Controlled-Release and Stabilized Fertilizers in Agriculture. International Fertilizer Industry Association, Paris.

Tripathi, D.K., Shweta, Singh, S., Singh, S., Pandey, R., Singh, V.P., et al., 2016. An overview on manufactured nanoparticles in plants: uptake, translocation, accumulation and phytotoxicity. Plant. Physiol. Biochem (16), S0981–S9428. Available from: https://doi.org/10.1016/j.plaphy.2016.07.030. 30297–30292.

Ulrich, A., Moraghan, J.T., Whitney, E.D., 1993. Sugar beet. Nutrient Deficiencies and Toxicities in Crop Plants. APS Press, St. Paul, MN, p. 202.

USDA NRCS. 2013. Agronomy Technical Note No. 7 "Adaptive Nutrient Management Process."

van der Ploeg, R.R., Böhm, W., Kirkham, M.B., 1999. History of soil science: on the origin of the theory of mineral nutrition of plants and the law of the minimum. Soil. Sci. Soc. Am. J. 63, 1055–1062.

Voss, R.D., 1993. Corn. Nutrient Deficiencies and Toxicities in Crop Plants. APS Press, St. Paul, MN, p. 202.

Wang, P., Menzies, N.W., Lombi, E., McKenna, B.A., Johannessen, B., Glover, C.J., et al., 2013. Fate of ZnO nanoparticles in soils and cowpea (*Vigna unguiculata*). Environ. Sci. Technol. 47 (23), 13822–13830.

Wu, L., Liu, M., Liang, R., 2008. Preparation and properties of a double-coated slow-release NPK compound fertilizer with superabsorbent and water-retention. Bioresour. Technol. 99 (3), 547–554.

Yuvaraj, M., Subramanian, K.S., 2015. Controlled-release fertilizer of zinc encapsulated by a manganese hollow core shell. Soil. Sci. Plant. Nutr. 61, 319–326.

Zhao, L., Peralta-Videa, J.R., Rico, C.M., Hernandez-Viezcas, J.A., Sun, Y., Niu, G., et al., 2014. CeO(2) and ZnO nanoparticles change the nutritional qualities of cucumber (*Cucumis sativus*). J. Agric. Food Chem. 62, 2752–2759.

Chapter 16

Nano metal oxide as nanosensors in agriculture and environment

Valentin Romanovski[1,2], Iryna Matsukevich[3], Elena Romanovskaia[2] and Rajiv Periakaruppan[4,5]

[1]Center of Functional Nano-Ceramics, National University of Science and Technology "MISIS," Moscow, Russia, [2]Department of Materials Science and Engineering, University of Virginia, Charlottesville, VA, United States, [3]Institute of General and Inorganic Chemistry, National Academy of Sciences of Belarus, Minsk, Belarus, [4]College of Horticulture, Nanjing Agricultural University, Nanjing, Jiangsu Province, P.R. China, [5]Department of Biotechnology, PSG College of Arts & Science, Coimbatore, Tamil Nadu, India

Chapter outline

16.1 Introduction

Sustainable development affects all aspects of human life in the modern world. It is essential to highlight the environmental protection and the rational use of natural resources, including the development of waste-free technologies/clean technologies. The creation of cleaner production is a goal, and therefore, at the intermediate stage, it is vital to introduce technologies that minimize the impact on the environment. This also includes the recycling of waste. Research on the production of materials from waste that is not inferior in quality to those made from commercial reagents is widespread, for example, pigments (Martsul et al., 2012; Zalyhina et al., 2021a,b), building materials

Nanometal Oxides in Horticulture and Agronomy. DOI: https://doi.org/10.1016/B978-0-323-91809-1.00016-0

(Kamarou et al., 2020, 2021a,b,c), water treatment materials (Romanovskii and Khort, 2017; Propolsky et al., 2020; Romanovski, 2020, 2021a,b) and wastewater treatment (Ramanouski and Andreeva, 2012; Romanovskii and Martsul, 2009, 2012; Romanovski, 2021), secondary raw materials (Romanovskaia et al., 2021), etc. Also, sustainable agriculture is an equally important area. Here, at almost every stage of the product life cycle, more and more sensors are being introduced and used to optimize processes and quality control.

In modern publications, there are a lot of exciting review articles on the analysis of materials used as sensors in agriculture. According to the study of the information provided, the largest share of sensors in agriculture is occupied by organic compounds. Among metals, the most common in publications are silver and gold nanoparticles and materials based on them. The smallest number of articles is devoted to the use of metal oxides. This can be probably explained by the toxicity of most nano-metal oxides for plants. At the same time, nano metal oxides can be widely used in all auxiliary processes.

The objectives of this study are to (1) present a characterization of nano metal oxide as nanosensor, (2) describe methods of the synthesis of nano metal oxides used as nanosensors, (3) provide examples of the use of nano metal oxides as nanosensors in agriculture.

16.2 Nano metal oxide as nanosensor

Among chemical sensors, semiconductor sensors of the resistive type are the most widely used. The working principle of semiconductor chemical sensors is based on a change in the electrophysical properties of the sensitive layer of a semiconductor sample with a diference in the composition of the analyzed medium. The property of a resistive semiconductor sensor, which depends on the concentration of the analyte, is the electrical conductivity of the sensitive layer. This layer is a polycrystalline film of semiconductor metal oxides. The output analytical signal of a resistive sensor is either the resistance value of the sensitive layer or an analog voltage signal. Depending on the type of the detected impurity (electron donors or acceptors) and the conductivity type of the semiconductor (n- or p-type), the resistance of the sensitive layer of the sensor increases or decreases. During the adsorption of acceptor particles (O_2) on the surface of a semiconductor with n-type conductivity (e.g., ZnO), the resistance of the sensitive layer increases (acceptor signal), and upon adsorption of donor particles (H_2), it decreases (donor signal); for semiconductors with p-type conductivity (e.g., NiO), the relationship is inverse (Hansford et al., 2005).

Currently, there are specific requirements for metal oxide materials for sensors, including a high sensory response, selectivity, frequency and temperature stability, good adhesion to electrode materials, and substrates. At the same time, in addition to severe economic constraints, the sensitivity coefficient of a modern sensor, as a rule, should not be less than 1.5 (150%), and the power consumption should not exceed 200 mW (Haiduk et al., 2019).

The characteristics of the sensor are determined by the material of the sensitive layer, chemical composition and crystal structure, and depend on the properties of all constituent elements of the sensor: substrate material, material and geometry of the heater and measuring electrodes, and sensor design. In the sensitive layer of a resistive semiconductor sensor, when exposed to a substance, a set of interrelated processes occurs: electronic processes, surface and volume diffusion of adsorbed particles, and transfer of charge carriers between grains of polycrystalline samples. Accordingly, the description of the response of a semiconductor gas sensor is carried out based on theories establishing the relationship between molecular and electronic processes on the semiconductor surface, mainly according to the electronic theory of chemisorption (Guerin et al., 2006). Models describing the signal of a sensor operating in the air take into account the presence of adsorbed oxygen on the sensor surface (Williams, 1999). The charge transfer process between crystallite grains is calculated within the framework of the percolation theory, the model of surface traps, and barrier conductivity (Williams, 1999); to describe diffusion in a sensitive layer, approaches developed in heterogeneous catalysis are used. Diffusion of adsorbed particles and charge transfer between crystallites are considered in relation to the crystal structure and morphology of the sensitive layer (Bendahan et al., 2007; Boulmani et al., 2007). Today, only a few issues have been resolved regarding the relationship between the preparation conditions of sensors and their characteristics (Domènech-Gil et al., 2015). Creating sensors poses the problem of simulating the sensor signal depending on the parameters of the sensitive layer and obtaining, under controlled conditions, sensitive layers with the required parameters.

As sensitive layers for the detection of micro concentrations of O_3, NO_x, Cl_2, ClO_2, and HCl semiconductor sensors use mainly oxides: In_2O_3 (Zhang et al., 2015; Wang et al., 2018), WO_3 (Guerin et al., 2006; Bendahan et al., 2007), ZnO (Xing et al., 2017; Zhu et al., 2018; Jali et al., 2019; Chen et al., 2020; Jainulabdeen et al., 2020; Paliwal et al., 2017), SnO_2 (Zhu and Ma, 2019; Hu et al., 2018; Zhu et al., 2019), TiO_2 (Yu et al., 2019; Moon et al., 2011), and Co_3O_4 (Zhang et al., 2020a; Cao et al., 2020) both pure and with various alloying additions. An alloying addition is added to control the type of conductivity and the amount of electrical conductivity. The nature of the signal of metal oxide sensors with n-type conductivity, caused by the micro concentrations of O_3, Cl_2, ClO_2, NO_2, NO, is, as a rule, acceptor: their appearance in the gas phase leads to an increase in the sensor resistance.

The sensitive layer of the nanosensor can be made based on complex oxides. For example, nickel cobaltite $NiCo_2O_4$, which combines high electrical conductivity, good structural stability, easy electrolyte penetration, and high electrochemical activity, can be used to detect CO, Cl_2, and O_3 (Joshi et al., 2016).

The sensor based on nanostructured zinc ferrite $ZnFe_2O_4$ demonstrates a high sensitivity for detecting low concentrations of toluene, exhibiting a good linearity ranging from 1 to 100 ppm with high response and selectivity (Lai et al., 2018; Dong et al., 2017). Ternary oxide $SrTi_{0.85}Fe_{0.15}O_3$, obtained in the form of nanofilms by electron beam deposition (Da Silva et al., 2015) demonstrated high sensitivity to gaseous ozone, which makes it a promising material for gas ozone sensors.

To increase the specific surface area (the ratio of the surface to the volume of the sensitive material), all of the above materials can be obtained in the form of nanoparticles (0D), nanowires (1D), nanotubes (1D), nanoribbons (1D), nanoplates (2D), and nanocubes (3D) (Lee, 2009). It has been shown that the sensory response increases dramatically when the particle size becomes comparable to or less than the Debye radius (usually a few nanometers). Fig. 16.1 shows several examples of 1D nanostructures, as well as types of hierarchical structures that can be assembled from them (Comini et al., 2013).

Among 1D nanostructures, nanowires, nanopillars, and nanotubes are the most promising materials for gas sensors due to their unique electrical and optical properties, as well as their vast specific surface area. Hierarchical nanostructures are high-dimensional structures that are assembled from low-dimensional nano blocks such as nanoparticles (0D), nanowires, nanorods and nanotubes (1D), and nanoplates (2D). Hierarchical nanostructures show a well-oriented porous structure without loosing its highly developed surface. Hollow nanostructures with thin shells also have a high specific surface area. Thus, not only high sensitivity, but also fast response can be achieved when using well-designed, hierarchical, and hollow oxide structures as a material for sensitive layers of gas sensors, since the number of active gas adsorption sites on their surface significantly increases. In Lee (2009), various hierarchical structures are classified according to the sizes of nano blocks and

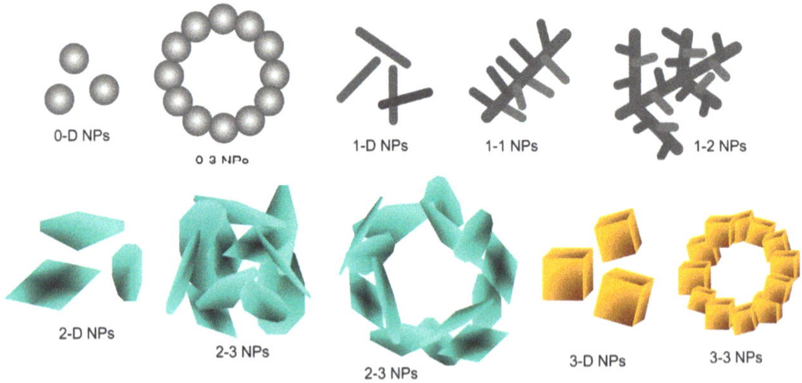

FIGURE 16.1 Morphology of some nanostructures.

emerging hierarchical structures (Fig. 16.1). For example, a 1–3 hedgehog structure means that 1D nanowires or nanorods are assembled into a 3D spherical hedgehog-shaped shape; "2–3 flower" denotes a flower-shaped 3D hierarchical structure built from many 2D nanosheets. With this approach, hollow spheres can be viewed as assemblies of 0D nanoparticles in 3D hollow spherical shapes. Thus, 0–3 hollow spheres should be considered as one of the types of hierarchical structures. There are many different technologies that can be used to create metal oxide 1D—3D nanostructures. These include growing nanostructures in the gas and liquid phases, chemical synthesis, electrochemical deposition, etc. (Comini et al., 2013).

Hierarchical structures include phase-inhomogeneous hybrid nanocomposite materials, which are increasingly becoming the subject of research in recent years as nanosensor-sensitive components. To improve the selectivity of wide gap semiconducting oxides, their surface is modified by creating complex inhomogeneous systems—nanocomposites $Me^{I}O$-$Me^{II}O$ by introducing catalytic additives into a highly dispersed oxide matrix: platinum group metals Pt, Pd, Ru, Rh or oxide catalysts such as Fe_2O_3, La_2O_3, Cr_2O_3, Co_3O_4, V_2O_5, NiO, CuO, MoO_3, CeO_2. Considering the high specific surface area of ultradispersed oxide semiconductors and the low solubility of catalytic additives in them, it is assumed that the surface distribution of modifiers is predominantly.

In general, surface modification involves not only changing the composition and size of semiconductor oxide nanocrystallites but also some additional possibilities, among which the most important are as follows:

- selection of a modifier catalytically active in the investigated solid-gas interaction;
- changing the reactivity of the material by changing the concentration of the modifier, due to the interaction "cluster—cluster";
- selection of the "cluster—matrix" pair; the nature of the semiconducting oxide can affect the configuration of transition metal d-electrons localized on the surface and change the surface activity.

Despite attempts to select dopants for sensor materials using the correlation parameters of electronegativity, electron affinity, thermodynamic characteristics of adsorption, work function, etc. (Takahata, 1988; Souteyrand, 1997), the most promising are the "collective" and "local" centers. In this case, the catalytic activity of the modifier is directly related to the valence state of dopants in the oxide matrix and their effect on the concentration of charge carriers in the semiconductor. On the other hand, the "local" center approach is based on the concept of an inhomogeneous surface. The interaction of semiconductor oxide with the gas phase, in this case, is represented through the formation of surface complexes. The chemical nature of the modifier and its reactivity in acid-base or redox reactions play a decisive role. A preliminary choice of the modifier can be made based on the analysis of the properties of the molecules of the detected gas and a large number of experimental data obtained in the field of heterogeneous catalysis on platinum group metals and metal oxides (Table 16.1) (Idriss and Barteau, 2000). The regularities established in the study of catalysts may not be observed when applied to sensor materials and should be verified experimentally.

Ultrathin 3D hierarchical heterostructures assembled based on nanolayers are promising materials for the manufacture of high-performance sensors. This conclusion was made by Rumyantseva et al. (2008), in which they prepared 3D hierarchical ZnO/In_2O_3 heterostructures assembled from ultrathin nanolayers synthesized using a simple solvothermal method. The results showed that the ZnO (20%)/In_2O_3 heterostructure consists of many ultrafine nanoparticles consisting of tightly bound ZnO and In_2O_3 nanoparticles, and has a specific surface area of 137.1 m^2/g. A sensor based on this material has shown excellent selectivity, good long-term stability, and a high recovery rate to ethanol (Rumyantseva et al., 2008). Examples of phase inhomogeneous hybrid nanocomposite materials are shown in Fig. 16.2.

Thus, the following groups of metal oxide nanosensor materials can be distinguished: individual oxides (e.g., In_2O_3, WO_3, ZnO, SnO_2, TiO_2, Co_3O_4), complex oxides (e.g., $NiCo_2O_4$, $ZnFe_2O_4$, $SrTi_{0.85}Fe_{0.15}O_3$), hybrid nanocomposite materials (Zhang et al., 2020b; Ma et al., 2020; Reddy et al., 2020; Vijayakumar and Vadivel, 2019; Pargoletti et al., 2020).

16.3 Fabrication of nano metal oxide nanosensor

The choice of the method for the synthesis of metal oxide nanoparticles is of great importance, since the method of preparation has a direct effect on their morphology, size, stability, and physicochemical properties (Fig. 16.3).

TABLE 16.1 Principles of choosing a modifying additive in the development of a selective sensor based on tin oxide (Idriss and Barteau, 2000).

Detected gases	Modifying component of the nanocomposite
Reducing gases: CO, H_2, CH_4	Precious metal additives: Pt, Pd, Ru, Rh, Au
Lewis base: NH_3	More acidic than SnO_2 metal oxides: V_2O_5, MoO_3, WO_3
Complex molecules with different functional groups	More basic than SnO_2 metal oxides: Fe_2O_3, In_2O_3, La_2O_3
Oxidizing gases: O_2, O_3, NO_2	Electron donor supplements—Sb_2O_5 Noble metal additives: Pt, Pd, Ru, Rh, Au

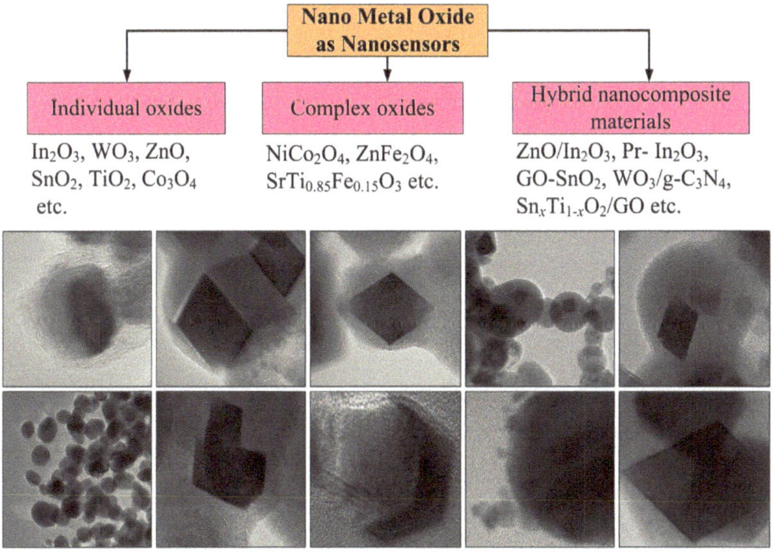

FIGURE 16.2 Classification of metal oxide nanosensor materials.

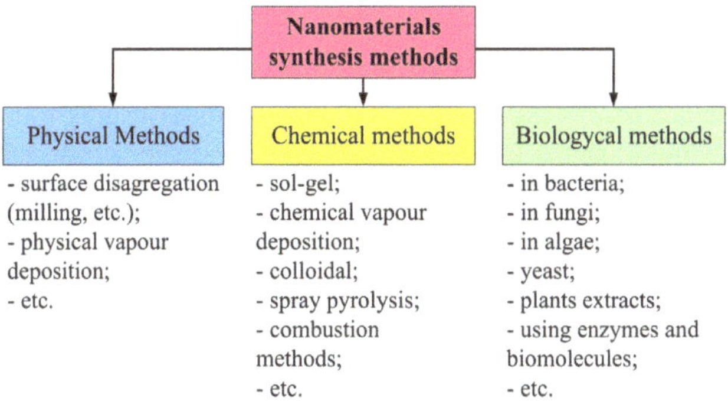

FIGURE 16.3 Methods for the synthesis of metal oxide nanosensor materials.

There are two approaches proposed today to create nanosensors: "top-down" and "bottom-up" (Fig. 16.4). The "top-down" approach is based on reducing the size of physical bodies by mechanical or ionic processing up to obtaining objects with nanometer parameters, for example, using lithography technologies. The bottom-up approach consists in assembling the created structure directly from atoms, molecules, structural fragments of biological cells, etc., arranged in the required order. These include a variety of gas-phase synthesis, chemical and physical deposition methods, and molecular self-assembly (Cao and Wang, 2004). The bottom-up approach is considered to be the most beneficial due to its simplicity and cost-effectiveness.

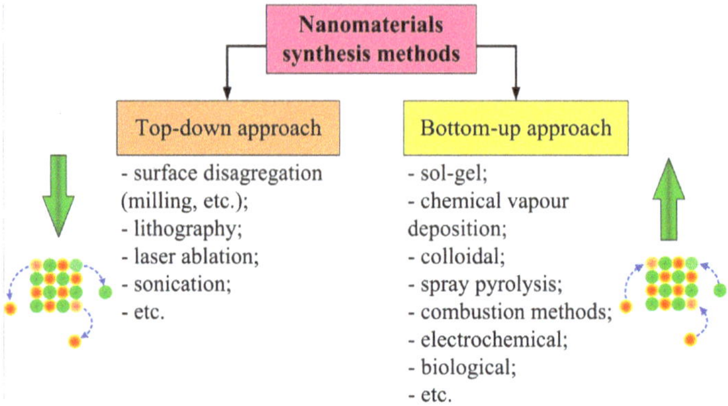

FIGURE 16.4 Top-down and bottom-up approaches.

In addition, two main groups of methods for synthesizing metal oxide nanosensors can be distinguished: chemical and physical. Methods for the chemical synthesis of nanoparticles include inorganic, organometallic, and organic synthesis with processes of heterogeneous phase formation in colloidal or similar systems.

16.3.1 Physical methods

Most of the procedures associated with forming nanostructures with arbitrary sizes, distribution, and orientation are not reliable for practical applications due to their low reproducibility. Recent advances in electron beam lithography, nanolithography, and patterned nanostructure growth technologies allow well-ordered nanostructures to be created, requiring expensive equipment and complex manufacturing operations. However, there are relatively new vacuum deposition techniques that enable the fabrication of well-ordered and complex 1D and 2D nanostructures. These include the methods of magnetron deposition of metal and metal oxide sensitive layers of gas sensors in sliding geometry (Robbie and Brett, 1997; Khatko et al., 2001), sequential deposition of ultrathin films of two different types of materials (Khatko et al., 2000; Khatko, 2013; Vallejos et al., 2008), and deposition of a sensitive layer both with interruption (Khatko et al., 2009) and with a change in the power of the magnetron discharge (Khatko, 2009).

The grazing geometry method, which allows the fabrication of sensitive layers based on an ordered region of WO_3 nanopillars, consists in placing the substrate on which the sensor structure is formed, to the incident ion beam at a grazing angle of less than 10 degrees (Robbie and Brett, 1997). This makes it possible to create homogeneous, with a high density, vertically aligned 1D nanopillars with an average size of their length and diameter of ≈ 400 and ≈ 50 nm, respectively. The formation of a partially isolated columnar structure can be explained by enhancing the effects of screening and limiting the diffusion of atoms, which occur when the flux of deposited ions and atoms falls onto the substrate at small grazing angles. At minimal grazing angles, atomic shielding increases sharply, especially under the condition of sputtering materials with low atomic mobility, which includes materials with a high melting point, or deposition of the material on a substrate at room temperature. The deposited stream of atoms and ions cannot fill the entire surface of the substrate, which contributes to the formation of pores in the deposited layer. This leads to the formation of a porous columnar microstructure of the sensitive layer of the gas sensor. The magnitude of the mobility of atoms and the deposition conditions determines the size of the isolated columns of a given material. The deposition of a material with high atomic mobility promotes the formation of nanopillars with a larger diameter than materials with low atomic mobility.

A way to create metal and metal oxide sensitive layers of sensors with nanoscale grains is to use sequential, step by step, deposition of ultrathin films of two different types of materials with an island structure. In this case, at a particular stage of growth of metal clusters of one material, clusters of another material (e.g., metal oxide) can form, limiting the coalescence of metal clusters and the formation of an extended (continuous) film. This method has been successfully applied in developing catalytic layers of silicon microcalorimetric gas sensors manufactured using microsystem technologies (Khatko et al., 2000, 2001). The layer structure consists mainly of metal (Pd) islands 1–2 nm in size and amorphous oxide (SiO_2) particles ≈ 1 nm in size, lying between palladium islands, and single palladium crystals 4–10 nm in size. When forming sensitive layers of a gas sensor with nanometer grain sizes, the main idea is to develop a special mode for the deposition of thin sensor layers of a metal oxide or metal by magnetron sputtering. Two varieties of this mode can be used. In the first case, at a constant magnetron discharge power, the process of condensation of sputtered particles on the substrate is interrupted several times by blocking the beam of incident particles with a remarkable shutter (Vallejos et al., 2008). In this case, an additional interphase ("extra") boundary can be created in the volume of the growing thin film. The reasons for forming an equilibrium surface at a given interface during interruption can be the saturation of free surface bonds with atoms from the residual atmosphere of structural relaxation of this surface. The subsequent continuation of the condensation process after the interruption causes a repetition of the initial stage of thin-film growth but is already at a new "extra" boundary. Consequently, the grain size in this film should be smaller than in the deposited film without interruption. In the second case, the formation of an "extra" boundary in the volume of the film is carried out by changing the power of the magnetron discharge during the sputtering process. The second mode provides not only obtaining a sensitive sensor layer with a minimum grain size in the nanometer range (≈ 11 nm), but it also allows creating an additional number of adsorption centers for molecules of toxic gases (Khatko et al., 2009; Khatko, 2009).

Chemical methods include a whole group of solutions: coprecipitation, microemulsion, solvothermal, sol-gel, and thermal decomposition. The deposition (coprecipitation) method is often used to obtain nanosized simple oxides In_2O_3, WO_3, ZnO, SnO_2, Co_3O_4, and composites based on them (Lanje et al., 2013; Singh et al., 2019; Sun et al., 2020; Fomekong et al., 2018; Bhattacharjee et al., 2011). The method is based on precipitation from salt solutions by both organic and inorganic compounds that play the role of precipitators. Most often, nitrate (Lanje et al., 2013) and acetate (Singh et al., 2019) are used as sources of indium, tungsten, zinc, tin, and cobalt, since in these cases, after calcination, no impurities remain in the product as a result of complete decomposition of ligands into gaseous products. Alkali metal hydroxides are used as inorganic precipitants: sodium (Moazzen et al., 2012), potassium (Jayakumar et al., 2006), less often lithium (Meulenkamp, 1998), and ammonium hydroxide, as well as sodium carbonates (Milenova et al., 2013), and ammonium (Yuan and Ryu, 2004). Primary, secondary, and tertiary amines (Cheng-Hsien, 2007) or urea (Bagheri et al., 2013) can also be used as precipitants. The role of a structure-forming agent in the synthesis of nanopowders can be played by various high-molecular-weight organic compounds, including water-soluble starch (Lanje et al., 2013), sodium dodecyl sulfate, polyacrylic acid of different molecular weights (Aimable et al., 2010a), and polyvinylpyrrolidone (Udayakumar et al., 2012). These organic reagents are used to control particle growth during deposition.

In addition to high molecular weight compounds, hydrothermal treatment is used to control the formation and growth of sol particles (Aimable et al., 2010b; Meng et al., 2018; Fu et al., 2019), which allows one to control particle size over a wide range. When obtaining ZnO nanoparticles by the deposition method, most authors use relatively low heat treatment temperatures of 90°C–120°C (Hu et al., 2003; Pompermayer et al., 2013), which is undoubtedly its positive side to its low energy consumption. However, this synthesis method requires the introduction of components that are difficult to separate from the product into the reaction mixture.

For the synthesis of nanoparticles by the microemulsion method, as a rule, reverse microemulsions are used, the micelles of which act as nanoreactors during the formation of ultrafine zinc oxide. The synthesis technique is as follows: at the first stage, two microemulsions are prepared, micelles of one contain a solution of a metal salt, the other microemulsion contains a precipitant; at the second stage, as a result of mixing these emulsions, the micelles open and merge (Sarkara et al., 2011).

As sources of metals, soluble salts acetates or nitrates are used, and organic metal derivatives can also be used. For example, in Inoguchi et al., 2008, zinc ethylate was used as a starting material, with absolute ethanol playing the role of a polar solvent. For the formation of micelles during the synthesis of oxide nanoparticles, surfactants are often used that are traditional for the classical microemulsion method, such as AOT (sodium salt of di- (2-ethyl) hexyl ester of sulfosuccinic acid) (Altıntas and Durucan, 2010), Triton X-100 (Sarkara et al., 2011), tregitol N-9 (Khatoon et al., 2012), and CTAB (cetyltrimethiammonium bromide) (Liu et al., 2011).

The main advantage of the microemulsion method is the ability to control the physicochemical properties of future powders at the stage of micelle formation. A striking example is a work (Altıntas and Durucan, 2010), where two microemulsions of different compositions were used: ME1 (AOT/glycerol/heptane = 5/5/90) and ME2 (AOT/glycerol/heptane = 30/5/65). As a result, ZnO nanoparticles of various shapes were obtained: in the first case, spherical particles were received, in the second, rods. Thus, it was shown that the use of surfactants with other systems leads to the formation of micelles of different structures, and, consequently, to the formation of ZnO particles of different shapes with different sizes and size distributions.

From the point of view of academic science, an exciting preparative method, the microemulsion method has many significant disadvantages that impede its practical implementation. This is the use of expensive chemical reagents, their increased fire hazard, and low product yield due to the thermodynamic features of the process. These disadvantages make it challenging to develop this method and its adaptation to industrial conditions.

The solvothermal method is a method of obtaining various chemical compounds and materials using physicochemical processes occurring in liquid media at high temperatures and pressures in closed systems. It is known that an increase in temperature and pressure can significantly increase the solubility of those substances that are insoluble under standard conditions and, by changing the synthesis conditions, it is possible to control the crystallization process from solution, and thus obtain particles of various morphologies. A particular case of the solvothermal method is the hydrothermal one. Using the solvothermal method, it is possible to get oxide nanoparticles of different shapes: spherical (Mishra et al., 2010), in the form of rods (Hu and Chen, 2008), and fibers on substrates (Fan et al., 2016). It was found that the nature of the surfactant has a significant influence on the morphology of the resulting powders (Hu and Chen, 2008; Geetha and Thilagavathi, 2010).

The sol-gel method, in combination with the subsequent heat treatment of reaction products, is widely used to obtain all kinds of oxide composite materials (both individual one-component oxides and two-, three- or more-component systems). In the first stage of the sol-gel process, using hydrolysis and polycondensation reactions, a colloidal solution of sol particles is formed, the size of which does not exceed several tens of nanometers. The second stage is the formation of a gel, accompanied by the formation of a spatial network of the gel. In an acidic medium, hydrolysis proceeds faster than polycondensation. Therefore, in the initial stages, chains are predominantly formed, and then there is a branching of chains and their cross-linking. In an alkaline medium, large spherical particles are formed (Brinker and Scherer, 1990). It is a composite material, the structure of which is interpenetrating solid and liquid phases. The third stage is the aging of the gel. In aging, the gel structure becomes denser, the network shrinks, and the solvent is released from the gel.

The processes of removing the solvent from the gel (drying) play a significant role in the synthesis. Depending on the method of their implementation, various synthesis products can be obtained (xerogels, ambigels, cryogels, aerogels). The typical features of these products are the retention of the nanosize of structural elements and relatively high values of the specific surface area. However, the bulk density can differ hundreds of times. When the solvent is removed under supercritical conditions, an aerogel is formed, and when heat treatment is carried out at an elevated temperature, a denser structure, a xerogel, is included. It was found that the calcination temperature has a significant influence on the morphology of nanosensor materials obtained by the sol-gel method, since the growth of particles during calcination is determined by surface interaction, and the growth kinetics depends on the conditions for the synthesis of nanoparticles. For example, in Hayata et al. (2011), Zak et al. (2011), and Zhang et al. (2009), different dependences of the crystallite size on the temperature and duration of calcination were established. In the publication by Zhang et al. (2009), gel pyrolysis was applied. A slight increase in the size of ZnO particles from 18 to 21 nm was observed with increasing temperature from 600°C to 750°C.

In recent years, self-sustaining exothermic methods for the synthesis of nano-oxide materials have been increasingly used. These methods make it possible to obtain nanoscale metals (Sdobnyakov et al., 2020; Romanovski et al., 2020; Khort et al., 2021; Romanovskiy et al., 2018) and oxides (Matsukevich et al., 2021a,b), as well as metal-graphene structures (Khort et al., 2020a,b) in one stage. These methods are characterized by simplicity, high reaction rate, and short preparatory operations and synthesis times (Romanovski, 2021).

After the synthesis of nanopowders of a sensor material for a sensitive element, it remains a difficult task to apply it to the substrate (membrane) of the sensor.

16.3.2 Electrochemical anodizing method

It allows the direct formation of both nanoporous metal oxide (TiO_2, WO_3, SnO_2, In_2O_3, etc. (Kasach et al., 2019, 2020; Osipenko et al., 2021)) sensitive layers and layers with 1D nanostructures. The issue of porous electrochemical anodization of many refractory metals was widely studied in 2005 in works of these authors (Tsuchiya et al., 2005; Tsuchiya and Schmuki. 2004; Macak et al., 2005; Tsuchiya and Schmuki, 2005; Sieber et al., 2005), where porous layers of WO_3, ZrO_2, TiO_2, HfO_2, and Nb_2O_5 were obtained by electrochemical anodizing of metals in various electrolytes. In 2008, gas sensors were investigated, the sensitive layers of which were ordered regions of 1D WO_3 nanopillars formed by electrochemical anodization of the Al-W thin-film system (Mozalev et al., 2008; Calavia et al., 2009). The idea of using 1D nanostructures as a sensitive layer of a gas sensor turned out to be highly fruitful, as evidenced by a large number of publications in this area carried out in subsequent years, some of which are presented in a review by a group of Italian researchers (Comini et al., 2013). It should be noted that the use of sensitive layers based on nanoporous anodic oxides of metals or ordered regions of nanopillars of metal oxides solves the problem of creating reliable metal contacts to them since the sensor heater and the contact system are completed before the formation of the sensitive layer (Mozalev et al., 2008; Calavia et al., 2009). At the same time, during the construction of a sensitive layer based on 1D nanostructures from a gas or liquid phase, the problem of contacts is quite acute due to the chaotic distribution of nanostructures, which causes the absence of reference of a significant part of them with the electrodes. This led to the need to develop new technology for manufacturing a gas sensor with a sensitive layer based on 1D nanostructures, including a relatively large number of photolithographic operations (Comini et al., 2013).

16.4 Nano metal oxides as nanosensor in agriculture and environment

Nanotechnology has found its place in agriculture, both in the main and in all auxiliary processes. The main areas of application of nanomaterials are various additives to improve soil properties and increase yields, as well as the use of nano metallic oxides as sensors (Fig. 16.5).

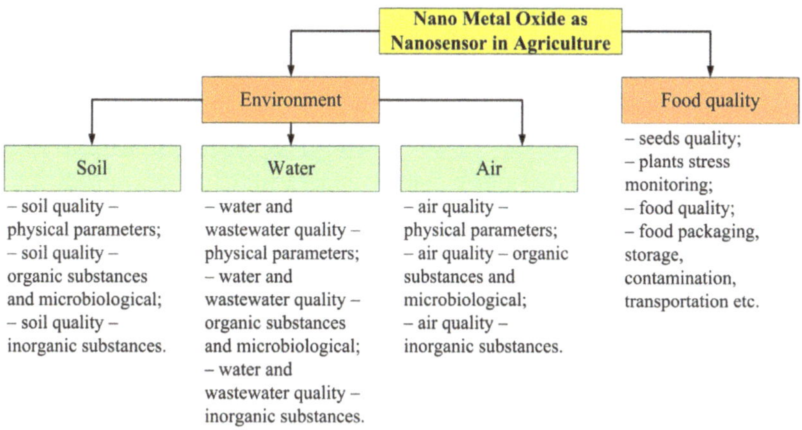

FIGURE 16.5 Application of nano metal oxide sensors in agriculture.

The use of nano metal oxide nanosensors contributes to the accurate measurement of nutrient content, soil properties, moisture content, pollutants, environmental conditions, exposure to stress, the ingress of agrochemicals into the soil system, evaporation of volatile components, and their residual content in the soil, onset of disease, appearance of pathogens, growth of plants, and much more. Timely detection of these factors provides more opportunities for their detection with higher specificity, sensitivity, and efficiency. For example, seasonal monitoring of cultivated fields is achieved using a global positioning system based on nanosensors. Similarly, nano carrier-based controlled release technology is tracked using wireless nanosensor devices (El Beyrouthya and El Azzi, 2014; Fraceto et al., 2016; Afsharinejad et al., 2016). The high sensitivity of some bioagents in detecting environmental conditions is combined with nanoparticles, which increases the efficiency of nanosensors up to several times.

The use of nanomaterials to improve soil properties and increase yields is associated with many safety issues and risks. When using metal oxides as nanosensors, things are easier if they are used to develop analytical instruments for quality control, bio/safety, and safety, not only in agriculture, but also in the food industry. As mentioned in Section 16.2, metal nano oxides can have sizes of no more than 100–200 nm and various shapes: nanoparticles, nanowires, nanotubes, and nanocrystals of different shapes and configurations. They are used to optimize signal transmission to sensitive elements in response to external influences of different nature. The main areas of their improvement are availability in production, improved thermal, electrical, and optical properties, increased sensitivity, reduced response time, etc.

One of the existing directions in this area is the use of nanosensors in plants, which make it possible to control the development of plants, the occurrence of stress conditions for a plant, etc. This will allow quickly responding to emerging harmful environmental conditions and correcting them to maximize the efficiency of the crop.

In agriculture, nanosensors can be used throughout the entire chain of growing and supplying food, from planting and growing to harvest, monitoring its quality, and transporting it to the consumer. Among them are nanosensors for dynamic measurement of soil parameters (pH, moisture, nutrient content, the content of toxic pollutants, etc.). This allows to maintain optimal conditions in the soil for plant growth. Apart from the soil, the same applies to the requirements of the air and water used for irrigation (Table 16.2). Combined with an automation system, this allows to monitor and maintain optimal conditions for plant growth quickly. Also, nanosensors can be used to assess the quality of food products, for example, the content of carbohydrates, amino acids, amino acids, vitamins, minerals, etc., as well as the range of pollutants.

An existing aspect of using nanosensors in agriculture is the early detection of stress conditions for plants. In crop production, this helps to minimize

TABLE 16.2 Nano metal oxides as nanosensors for different applications in agriculture system.

Used NPs	Target	References
ZnO based	*Trichoderma harzianum*	Siddiquee et al. (2014)
ZnO	Acetone	Li et al. (2017)
ZnO	Ethanol	Wan et al. (2004)
NiO-ZnO	Methanol and formaldehyde	Na et al. (2012)
ZnO	Ethanol	Tang et al. (2007)
RGO and three-dimensional ZnO nanoflower	2-Propanol	Acharyya et al. (2018)
ZnO nanorod	NO_2	Oh et al. (2009)
tube-like ZnO	NO_2	Chen et al. (2011)
ZnO-PMMA	H_2S	Kitture et al. (2017)
ZnO/CNTs/ionic liquid paste electrode	Bisphenol A	Najafi et al. (2014)
ZnO	Bisphenol A	Najafi et al. (2014), Zhang et al. (2014c)
Pt/ZnO based	Melamine and urea	Ai et al. (2009)
CuO-ZnO	CO	Wu et al. (2005)
Au-ZnO	CO	Chang et al. (2008)
Zn–In oxide nanowire		Singh et al. (2010)
Zn–Al layered double hydroxide	2,4-Dichlorophenoxyacetate (2,4-D)	bin Hussein et al. (2005)
Based on SnO_2	Humidity	Kuang et al. (2007), Gong et al. (2018), Zhang et al. (2016), Parthibavarman et al. (2011), Zhuo et al. (2013)
Fe-SnO_2	Humidity	Toloman et al. (2017)
NiO-SnO_2	Humidity	Pascariu et al. (2016)
Zn-SnO_2	Humidity	Md Sin et al. (2013), Sin et al. (2007)
SnO_2	Chloramphenicol	Muhammad et al. (2018)

(*Continued*)

TABLE 16.2 (Continued)

Used NPs	Target	References
In-WO_3/SnO_2	Acetone	Tomer et al. (2017)
Graphene-SnO_2	Cataluminescence gas sensors	Song et al. (2011)
SnO_2 and/or TiO_2	CO	Ram et al. (2005)
SnO_2	NO_2	Sahm et al. (2004)
SnO_2 based	CO	Zhao et al. (2007)
V-SnO_2	CO	Wang and Chen (2010)
SnO_2-Au	CO	Manjula et al. (2011)
SnO_2 based	CO	Habibzadeh et al. (2010)
SnO_2	Methanol, ethanol, and propanol	Chiu and Yeh (2007)
Based on SnO_2	Ethanol	Liang et al. (2004), Wisitsoraat et al. (2006), Liu et al. (2006), Chen et al. (2006)
SnO_2-In_2O_3	NO_2	Starke and Coles (2002)
SnO_2 nanowires	NO_2	Choi et al. (2008)
SWNT–SnO_2	Ammonia	Hoa et al. (2007)
SnO_2 based	NO_2	Zhang et al. (2014a), Zhang et al. (2014b)
Eu-SnO_2 nanobelt	Acetone	Chen et al. (2017b)
SnO_2 based	NO	Liang et al. (2004)
SnO_2-SrO, CaO, BaO, Bi_2O_3 and Sm_2O_3	N_2O	Kanazawa et al. (2001)
SnO_2-Sm_2O_3	Ethanol	Habibzadeh et al. (2010)
Based on Fe_xO_y		
Fe_2O_3	*Salmonella*	Zou et al. (2019)
Fe_2O_3	*Salmonella*	Hu et al. (2019)
Fe_3O_4@Au	Ochratoxin A	Song et al. (2018)
Eu- α-Fe_2O_3	Acetone	Cheng et al. (2017a)
Nanostructured α-Fe_2O_3-NiO	Acetone	Haq et al. (2018)
Perovskite PrFeO hollow nanofibers	Acetone	Ma et al. (2018)
Pt-decorated $CuFe_2O_4$ nanotubes	Acetone	Zhao et al. (2018)

(*Continued*)

TABLE 16.2 (Continued)

Used NPs	Target	References
Pseudo-peanut-shaped Fe_2O_3 nanoparticles	Aliphatic alcohols	Das et al. (2016)
$BaFe_{12}O_{19}$	Acetone and ethanol	Karmakar et al. (2014)
Fe_3O_4	*Campylobacter jejuni*	Huang et al. (2010)
Magnetic nanoparticles	*Escherichia coli*	Varshney and Li (2007)
Magnetic nanoparticles	*Escherichia coli*	Maalouf et al. (2008)
Fe_3O_4-graphene oxides	Aflatoxins M1 (AFM1) in milk	Gan et al. (2013), Gan et al. (2016)
α-Fe_3O_4/CS	As, Pb, and Ni ions from aqueous solution	Ahmed and Fekry (2013)
$LnFeO_3$ (Ln = La, Nd and Sm)	Methane, propane, and n-hexane	Giang et al. (2011)
Polyaniline/n-TiO_2	Liquefied petroleum gas	Dhawale et al. (2008), Dhawale et al. (2010)
TiO_2 based	Polyaniline	Gong et al. (2010)
TiO_2 based	Ammonia	Tai et al. (2008)
TiO_2 based	Ammonia	Tai et al. (2007)
TiO_2	*Listeria monocytogenes*	Wang et al. (2009a)
Anatase TiO_2	Ethanol	Das et al. (2017a)
Rutile TiO_2 nanoparticles	Isopropanol	Das et al. (2017b)
$BaTiO_3$ and polymer material	Humidity	Wang et al. (2002)
TiO_2	Atrazine	Yu et al. (2010)
Mn^{2+}-TiO_2	Urea	Pandey et al. (2010)
TiO_2/g-C_3N_4/rGO	NO_2	Dong et al. (2019)
TiO_2 based	Parathion	Li et al. (2006)
IrO_x or CrO_2/Ta_2O_5	pH	Uria et al. (2016)
RuO_2	pH	Liao and Chou (2008)
ZrO_2/Au	Parathion	Wang and Li (2008)
CdTe–SiO_2–MIPs	Deltamethrin	Ge et al. (2011)
$SmBO_3$	Phenol	Hu et al. (2014)
CuO	Hydrazine and bisphenol A	Teymoori et al. (2018)
CuO	Hydrazine and bisphenol A	Teymoori et al. (2018)

(Continued)

TABLE 16.2 (Continued)

Used NPs	Target	References
IrO_x or CrO_2/Ta_2O_5	*Escherichia coli*	Uria et al. (2016)
$Bi_{12x}La_xFe_{12x}Co_xO_3$ ($x = 0$, 0.05, 0.1)	Ethanol, butanol, acetone, petrol, diethyl ether	Glinskaya et al. (2021)
CeO_2	Ethanol	Khan et al. (2011)
CeO_2 based	Pesticides and pathogen detection	Jiang et al. (2012), Zhang et al. (2013)
HfO_2NP_2	AFM1	Pal et al. (2015)
WO_3 based	Ammonia	Bittencourt et al. (2006)
Polypyrrole-coated TiO_2-ZnO	Ammonia	Wang et al. (2009b)
WO_3 nanowires	Hydrocarbons	Rout et al. (2006b)
VO_2 nanorods	Methane	Li et al. (2016)
WO_3	CO	Bittencourt et al. (2006)
ZnO, In_2O_3, and WO_3 nanowire	N_xO_y	Rout et al. (2006a)
ZnO-SnO_2	NO_2	Lu et al. (2012)
V_2O_5-SnO_2	SO_2	Das et al. (2008)
SnO_2 with V_2O_5, MoO_3, Sb_2O_3, Al_2O_3, CeO_2, MgO	SO_2	Lee et al. (2011)
In_2O_3	Ethanol	Xiangfeng et al. (2004)
In_2O_3 micro-/nanostructures	Methanol, toluene, and formaldehyde	Liu et al. (2010)
Chestnut-like $CoFe_2O_4$ @ SiO_2 @ In_2O_3 nanocomposite	Acetone	Lin et al. (2018)
Rh-doped SnO_2 nanofibers	Acetone	Kou et al. (2018)
Two-dimensional SnO_2-ZnO ultrathin nanosheet and Ag-doped ZnO nanoneedles	Acetone	Al-Hadeethi et al. (2017a), Al-Hadeethi et al. (2017b)
WO_3-NiO core-shell nanorods	Acetone	Choi et al. (2017)

(Continued)

TABLE 16.2 (Continued)

Used NPs	Target	References
Pd-Sm-SnO_2	Acetone	Shaikh et al. (2017)
Carbon-WO_3 hollow spheres	Acetone	Shen et al. (2017)
Mn-ZnO and Mn-ZnO-CdO	Ethanol and acetone	Wang et al. (2017)
NiO and Zn-NiO	Formaldehyde, CO, H_2, NO_2, and NH_3	Fomekong et al. (2018)

agricultural losses. Unlike many existing remote control methods, for example, observation of the change in leaf color by chlorophyll, which gives a result after the accumulation of stress, the use of nanosensors makes it possible to quickly determine stress in plants at the earliest stages, by monitoring stress factors at the molecular level (tracking signal molecules, such as H_2O_2, NO, etc.), and the use of nanomaterials to expand the light spectrum and collect more electrons for plant photosynthesis (Antonaru et al., 2020; Loo et al., 2019; Wiesholler et al., 2019; So et al., 2015). downconversion nanoparticles (DCN) and upconversion nanoparticles (UCNP) nanoparticles can convert UV and IR light to visible light. Nanomaterials capable of capturing and donating electrons can be placed in chloroplasts to increase the delivery of electrons to the electron transport chain. Both approaches can improve plant photosynthesis, especially in low light conditions.

16.5 Conclusions and further perspectives

In addition to numerous auxiliary processes in agriculture, where nanosensors are widely used, the most important are: (1) determination of the quality of nutrients used; (2) determination of the quality of the environment (soil, water, air); (3) the presence of pollutants in the environment and agricultural products (residual content of antibiotics, pesticides, herbicides, heavy metals, pathogenic microorganisms, etc.). The use of nanosensors with improved optical, magnetic, and electrical/electrochemical properties is a significant advantage over traditional systems. The use of nanoscale materials promotes effective binding to target molecules even in complex systems or responses to physical parameters. The main directions of nanosensor development in agriculture are as follows:

- decrease in response time;
- increased sensitivity;
- increased selectivity;

- cost reduction;
- wireless nanosensors;
- multifunctionality;
- biosensors for early detection of stress;
- increasing plant resistance to stressful environmental conditions;
- minimization of the toxic effect.

The increased use of nanosensors in agriculture is directly related to the principles of sustainable development associated with growing populations and climate change. Also, the broader use of advanced nanosensors will increase the efficiency of the entire agricultural system as a whole, including many positive environmental, economic, and social effects.

References

Acharyya, D., Saini, A., Bhattacharyya, P., 2018. Influence of rGO cladding in improving the sensitivity and selectivity of ZnO nanoflowers-based alcohol sensor. IEEE Sens. J. 18, 1820–1827.

Afsharinejad, A., Davy, A., Jennings, B., 2016. Dynamic channel allocation in electromagnetic nanonetworks for high resolution monitoring of plants. Nano Commun. Networks. 1 (7), 2–16.

Ahmed, R.A., Fekry, A.M., 2013. Preparation and characterization of a nanoparticles modified chitosan sensor and its application for the determination of heavy metals from different aqueous media. Int. J. Electrochem. Sci. 8, 6692–6708.

Ai, K., Liu, Y., Lu, L., 2009. Hydrogen-bonding recognition-induced color change of gold nanoparticles for visual detection of melamine in raw milk and infant formula. J. Am. Chem. Soc. 131 (27), 9496–9497.

Aimable, A., Buscaglia, M.T., Buscaglia, V., Bowen, P., 2010a. Polymer-assisted precipitation of ZnO nanoparticles with narrow particle size distribution. J. Eur. Ceram. Soc. 30, 591–598.

Aimable, A., Strachowski, T., Wolska, E., Lojkowski, W., Bowen, P., 2010b. Comparison of two innovative precipitation systems for ZnO Al-doped ZnO nanoparticle synthesis. Process. Appl. Ceram. 4 (3), 107–114.

Al-Hadeethi, Y., Umar, A., Al-Heniti, S.H., Kumar, R., Kim, S.H., Zhang, X., et al., 2017a. 2D Sn-doped ZnO ultrathin nanosheet networks for enhanced acetone gas sensing application. Ceram. Int. 43, 2418–2423.

Al-Hadeethi, Y., Umar, A., Ibrahim, A.A., Al-Heniti, S.H., Kumar, R., Baskoutas, S., et al., 2017b. Synthesis, characterization and acetone gas sensing applications of Ag-doped ZnO nanoneedles. Ceram. Int. 43, 6765–6770.

Altıntas, O., Durucan, Y.C., 2010. Synthesis of zinc oxide nanoparticles elaborated by microemulsion method. J. Alloy. Compd. 506, 944–949.

Antonaru, L.A., Cardona, T., Larkum, A.W.D., Nürnberg, D.J., 2020. Global distribution of a chlorophyll f cyanobacterial marker. ISME J. 14, 2275–2287.

Bagheri, S., Chandrappa, K.G., Sharifah, B.A.H., 2013. Facile synthesis of nanosized ZnO by direct precipitation method. Der Pharma Chem 5 (3), 265–270.

Bendahan, M., Guerin, J., Boulmani, R., Aguir, K., 2007. WO_3 sensor response according to operating temperature: experiment and modeling. Sens. Actuators, B. 124, 24–29.

Bhattacharjee, C.R., Purkayastha, D.D., Bhattacharjee, S., Nath, A., 2011. Homogeneous chemical precipitation route to ZnO nanosphericals. Assam. Univ. J. Sci. Technol.: Phys. Sci. Technol. 7, 122–127.

bin Hussein, M.Z., Yahaya, A.H., Zainal, Z., Kian, L.H., 2005. Nanocomposite-based controlled release formulation of an herbicide, 2, 4-dichlorophenoxyacetate incapsulated in zinc–aluminium-layered double hydroxide. Sci. Technol. Adv. Mater. 6 (8), 956.

Bittencourt, C., Felten, A., Espinosa, E.H., Ionescu, R., Llobet, E., Correig, X., et al., 2006. WO3 films modified with functionalised multiwall carbon nanotubes: morphological, compositional and gas response studies. Sens. Actuators B Chem 115, 33–41.

Boulmani, R., Bendahan, M., Lambert-Mauriat, C., Gillet, M., Aguir, K., 2007. Correlation between rf-sputtering parameters and WO_3 sensor response towards ozone. Sens. Actuators B 125, 622–627.

Brinker, C.J., Scherer, G.W., 1990. Sol-Gel Science: The Physics and Chemistry of Sol-Gel Processing. Academic Press, Inc, San Diego, p. 908, 1990.

Calavia, R., Mozalev, A., Khatko, V., Gorokh, G., Vilanova, X., Correig, X., et al., 2009. A H_2 microsensor based on nanocolumnar tungsten oxide grown by template-assisted anodization. In: 15th International Conference on Solid-State Sensors, Actuators and Microsystems: "Tranducer-09," 2009, Denver, CO, T3P. 109, 1313–1316.

Cao, G., Wang, Y., 2004. Nanostructures and Nanomaterials Synthesis, Properties, and Applications. Imperial College Press, London.

Cao, J., Wang, S., Zhao, X., Xing, Y., Li, J., Li, D., 2020. Facile synthesis and enhanced toluene gas sensing performances of Co_3O_4 hollow nanosheets. Mater. Lett. 263, 127215.

Chang, S.J., Hsueh, T.J., Chen, I.C., Huang, B.R., 2008. Highly sensitive ZnO nanowire CO sensors with the adsorption of au nanoparticles. Nanotechnology 19, 175502–175506.

Chen, Y., Zhu, C., Wang, T., 2006. The enhanced ethanol sensing properties of multiwalled carbon nanotubes/SnO2. Nanotechnology 17, 3012–3017.

Chen, M., Wang, Z., Han, D., Gu, F., Guo, G., 2011. High-sensitivity NO2 gas sensors based on flowerlike and tube-like ZnO nanomaterials. Sens. Actuators B Chem. 157, 565–574.

Cheng, Y., Wang, Y., Zhang, J., Li, H., Liu, L., Li, Y., et al., 2017a. A comparison of Eu-doped α-Fe2O3 nanotubes and nanowires for acetone sensing. Nano 12, 1750138–1750148.

Chen, W., Qin, Z., Liu, Y., Zhang, Y., Li, Y., Shen, S., et al., 2017b. Promotion on acetone sensing of single SnO2 nanobelt by Eu doping. Nanoscale Res. Lett. 12, 405–412.

Chen, H., Li, C., Zhang, X., Yang, W., 2020. ZnO nanoplates with abundant porosity for significant formaldehyde-sensing. Mater. Lett. 260, 126982.

Cheng-Hsien, H., 2007. Spherical zinc oxide nano particles from zinc acetate in the precipitation method. J. Chin. Chem. Soc. 54, 31–34.

Chiu, H.C., Yeh, C.S., 2007. Hydrothermal synthesis of SnO2 nanoparticles and their gas-sensing of alcohol. J. Phys. Chem. C. 111, 7256–7259.

Choi, Y.J., Hwang, I.S., Park, J.G., Choi, K.J., Park, J.H., Lee, J.H., 2008. Novel fabrication of an SnO2 nanowire gas sensor with high sensitivity. Nanotechnology 19, 095508–095511.

Choi, S., Kyung, J., Woo, L., Lee, S., Lee, C., 2017. Acetone sensing of multinetworked WO3-NiO coreshell nanorod sensors. J. Korean. Phys. Soc. 71, 487–493.

Comini, E., Baratto, C., Concina, I., Faglia, G., Falasconi, M., Ferroni, M., et al., 2013. Metal oxide nanoscience and nanotechnology for chemical sensors. Sens. Actuators, B. 179, 3–20.

Da Silva, L.F., Mastelaro, V.R., Catto, A.C., Escanhoela, C.A., Bernardini, S., Zílio, S.C., et al., 2015. Ozone and nitrogen dioxide gas sensor based on a nanostructured $SrTi_{0.85}Fe_{0.15}O_3$ thin film. J. Alloy. Compd. 638, 374–379.

Das, S., Chakraborty, S., Parkash, O., Kumar, D., Bandyopadhyay, S., Samudrala, S.K., et al., 2008. Vanadium doped tin dioxide as a novel sulfur dioxide sensor. Talanta 75, 385–389.

Das, P., Mondal, B., Mukherjee, K., 2016. Hierarchical zinc oxide nano-tips and micro-rods: hydrothermal synthesis and improved chemi-resistive response towards ethanol. RSC Adv. 6, 1408–1414.

Das, P., Mondal, B., Mukherjee, K., 2017a. Simultaneous adsorption-desorption processes in the conductance transient of anatase titania for sensing ethanol: a distinctive feature with kinetic perception. J. Phys. Chem. C. 121, 1146–1152.

Das, P., Mondal, B., Mukherjee, K., 2017b. Chemi-resistive response of rutile titania nanoparticles towards isopropanol and formaldehyde: a correlation with the volatility and chemical reactivity of vapors. Mater. Res. Express. 4, 015503–015512.

Dhawale, D.S., Salunkhe, R.R., Patil, U.M., Gurav, K.V., More, A.M., Lokhande, C.D., 2008. Room temperature liquefied petroleum gas (LPG) sensor based on p-polyaniline/n-TiO_2 heterojunction. Sens. Actuators B Chem. 134, 988–992.

Dhawale, D.S., Dubal, D.P., More, A.M., Gujar, T.P., Lokhande, C.D., 2010. Room temperature liquefied petroleum gas (LPG) sensor. Sens. Actuators B Chem. 147, 488–494.

Domènech-Gil, G., Samà, J., Pellegrino, P., Barth, S., Gràcia, I., Cané, C., et al., 2015. Gas Nanosensors based on individual indium oxide nanostructures. Procedia Eng. 120, 795–798.

Dong, C., Liu, X., Xiao, X., Du, S., Wang, Y., 2017. Monodisperse ZnFe2O4 nanospheres synthesized by a nonaqueous route for a highly selective low-ppm-level toluene gas sensor. Sens. Actuators, B. 239, 1231–1236.

Dong, X., Wu, K., Zhu, W., Wu, P., Hou, J., Wang, Z., et al., 2019. TiO_2 nanotubes/g-C3N4 quantum dots/rGO Schottky heterojunction nanocomposites as sensors for ppb-level detection of NO2. J. Mater. Sci. 54, 7834–7849.

El Beyrouthya, M., El Azzi, D., 2014. Nanotechnologies: novel solutions for sustainable agriculture. Adv. Crop. Sci. Technol. 2 (03), 8863.

Fan, J., Li, T., Heng, H., 2016. Hydrothermal growth of ZnO nanoflowers and their photocatalyst application. Bull. Mater. Sci. 39, 19–26.

Fomekong, R.L., Kamta, H.M.T., Lambi, J.N., Lahem, D., Eloy, P., Debliquy, M., et al., 2018. A sub-ppm level formaldehyde gas sensor based on Zn-doped NiO prepared by a coprecipitation route. J. Alloy. Compd. 731, 1188–1196.

Fraceto, L.F., Grillo, R., de Medeiros, G.A., Scognamiglio, V., Rea, G., Bartolucci, C., 2016. Nanotechnology in agriculture: which innovation potential does it have? Front. Environ. Sci. 22 (4), 20.

Fu, X., Yang, P., Xiao, X., Zhou, D., Huang, R., Zhang, X., et al., 2019. Ultra-fast and highly selective roomtemperature formaldehyde gas sensing of Pt-decorated MoO_3 nanobelts. J. Alloy. Compd. 797, 666–675.

Gan, N., Zhou, J., Xiong, P., Hu, F., Cao, Y., Li, T., et al., 2013. An ultrasensitive electrochemiluminescent immunoassay for Aflatoxin M1 in milk, based on extraction by magnetic graphene and detection by antibody-labeled CdTe quantum dots-carbon nanotubes nanocomposite. Toxins 5 (5), 865–883.

Gan, W., Yang, H., Zhang, Y., Shi, S.Q., Lin, C., Pan, L., et al., 2016. Synthesis and characterization of sucrose-melamine-formaldehyde adhesives. Bioresources 11 (1), 2516–2525.

Ge, S.G., Lu, J.J., Ge, L., Yan, M., Yu, J.H., 2011. Development of a novel deltamethrin sensor based on molecularly imprinted silica nanospheres embedded CdTe quantum dots. Spectrochim. Acta A-Mol Biomol. Spectrosc. 79 (5), 1704–1709.

Geetha, D., Thilagavathi, T., 2010. Hydrothermal synthesis of nano ZnO structures from CTAB. Dig. J. Nanomater. Biostructures 5 (1), 297–301.

Giang, H.T., Duy, H.T., Ngan, P.Q., Thai, G.H., Thu, D.T.A., Thu, D.T., et al., 2011. Hydrocarbon gas sensing of nano-crystalline perovskite oxides LnFeO3 (Ln = La, Nd and Sm). Sens. Actuators B Chem. 158, 246–251.

Glinskaya, A., Petrov, G., Romanovski, V., 2021. Crystal structure, physicochemical and sensory properties of solid solutions $Bi_{1-x}La_xFe_{1-x}Co_xO_3$ (x = 0, 0.05, 0.1). J. Mater. Sci. Mater. Electron. 32, 22579–22587.

Gong, J., Li, Y., Hu, Z., Zhou, Z., Deng, Y., 2010. Ultrasensitive NH3 gas sensor from polyaniline nanograin enchased TiO_2 fibers. J. Phys. Chem. C. 114, 9970–9974.

Gong, M., Li, Y., Guo, Y., Lv, X., Dou, X., 2018. 2D TiO_2 nanosheets for ultrasensitive humidity sensing application benefited by abundant surface oxygen vacancy defects. Sens. Actuators B Chem. 262, 350–358.

Guerin, J., Aguir, K., Bendahan, M., 2006. Modeling of the conductionin WO_3 thin film as ozone sensor. Sens. Actuators, B. 119, 327–334.

Habibzadeh, S., Khodadadi, A.A., Mortazavi, Y., 2010. CO and ethanol dual selective sensor of Sm2O3-doped SnO2 nanoparticles synthesized by microwave-induced combustion. Sens. Actuators B Chem. 144, 131–138.

Haiduk, Yu.S., Khort, A.A., Lapchuk, N.M., Savitsky, A.A., 2019. Study of WO_3-In_2O_3 nanocomposites for highly sensitive CO and NO_2 gas sensors. J. Solid. State Chem. 273, 25–31.

Hansford, G.M., Freshwater, R.A., Bosch, R.A., Cox, R.A., Jones, R.L., Pratt, K.F.E., et al., 2005. A low cost instrument based on a solid state sensor for balloon-borne atmospheric O_3 profile sounding. J. Environ. Monit. 7, 158–162.

Haq, M., Wen, Z., Zhang, Z., Khan, S., Lou, Z., Ye, Z., et al., 2018. A two-step synthesis of nanosheetcovered fibers based on α-Fe2O3/NiO composites towards enhanced acetone sensing. Sci. Rep. 8, 1705–17017.

Hayata, K., Gondal, M.A., Khaleda, M.M., Ahmedc, S.S., Shemsid, A.M., 2011. Nano ZnO synthesis by modified sol gel method and its application in heterogeneous photocatalytic removal of phenol from water. Appl. Catalys. A 393, 122–129.

Hoa, N.D., Quy, N.V., Cho, Y.S., Kim, D., 2007. Nanocomposite of SWNTs and SnO2 fabricated by soldering process for ammonia gas sensor application. Phys. Status Solidi A 204, 1820–1824.

Hu, Z., Oskam, G., Searson, P.C., 2003. Influence of solvent on the growth of ZnO nanoparticles. J. Colloid Interface Sci. 263, 454–460.

Hu, Y., Chen, H.-J., 2008. Preparation and characterization of nanocrystalline ZnO particles from a hydrothermal process. J. Nanopart. Res. 10, 401–407.

Hu, P., Zhou, X., Wu, Q., 2014. A new nanosensor composed of laminated samarium borate and immobilized laccase for phenol determination. Nanoscale Res. Lett. 9, 1–7.

Hu, J., Wang, T., Wang, Y., Huang, D., He, G., Han, Y., et al., 2018. Enhanced formaldehyde detection based on Ni doping of SnO2 nanoparticles by one-step synthesis. Sens. Actuators, B. 263, 120–128.

Hu, J., Jiang, Y.Z., Tang, M., Wu, L.L., Xie, H.Y., Zhang, Z.L., et al., 2019. Colorimetric-fluorescent-magnetic nanosphere-based multimodal assay platform for Salmonella detection. Anal. Chem. 91, 1178–1184.

Huang, J.L., Yang, G.J., Meng, W.J., Wu, L.P., Zhu, A.P., Jiao, X.A., 2010. An electrochemical impedimetric immunosensor for label-free detection of C. jejuni in diarrhea patients' stool based on O-carboxymethylchitosan surface modified Fe3O4 nanoparticles. Biosens. Bioelectron. 25, 1204–1211.

Idriss, H., Barteau, M.A., 2000. Active sites on oxides: from single crystals to catalysts. Adv. Catal. 45, 261–331.

Inoguchi, M., Suzuki, K., Kageyama, K., Takagi, H., Sakabe, Y., 2008. Monodispersed and well-crystallized zinc oxide nanoparticles fabricated by microemulsion method. J. Am. Ceram. Soc. 91 (12), 3850–3855.

Jainulabdeen, S., Gopinathan, C., Mumtaz Parveen, A., Mahalakshmi, K., Karu, J., Saravanakumar, K., 2020. Investigations of ZnO nanorod films grown by hydrothermal method for ethanol gas sensor. Int. J. Nanosci. 19, 1–9.

Jali, M.H., Abdul Rahim, H.R., Johari, M.A.M., Ali, U., Johari, S., Mohamed, et al., 2019. Formaldehyde sensing using ZnO nanorods coated glass integrated with microfiber. Opt. Laser Technol. 120, 1–9.

Jayakumar, O.D., Salunke, H.G., Kadam, R.M., Mohapatra, M., Yaswant, G., Kulshreshtha, S. K., 2006. Magnetism in Mn-doped ZnO nanoparticles prepared by a coprecipitation method. Nanotechnology. 17, 1278–1285.

Jiang, L., Yao, M., Liu, B., Li, Q., Liu, R., Lv, H., et al., 2012. Controlled synthesis of Ceo2/graphene nanocomposites with highly enhanced optical and catalytic properties. J. Phys. Chem. C. 116, 11741–11745.

Joshi, N., da Silva, L.F., Jadhav, H., M'Peko, J.-C., Millan Torres, B.B., Aguir, K., et al., 2016. One-step approach for preparing ozone gas sensors based on hierarchical NiCo2O4 structures. RSC Adv. 6 (95), 92655–92662.

Kamarou, M., Kuzmenkov, M., Korob, N., Kwapinski, W., Romanovski, V., 2020. Structurally controlled synthesis of calcium sulphate dihydrate from industrial wastes of spent sulphuric acid and limestone. Environ. Technol. Innov. 17, 100582.

Kamarou, M., Korob, N., Hil, A., Moskovskikh, D., Romanovski, V., 2021a. Low-energy technology for producing anhydrite in the $CaCO_3$—H_2SO_4—H_2O system derived from industrial wastes. J. Chem. Technol. Biotechnol. 96 (7), 2065–2071.

Kamarou, M., Korob, N., Kwapinski, W., Romanovski, V., 2021b. High-quality gypsum binders based on synthetic calcium sulfate dihydrate produced from industrial wastes. J. Ind. Eng. Chem. 100, 324–332. Available from: https://doi.org/10.1016/j.jiec.2021.05.006.

Kamarou, M., Korob, N., Romanovski, V., 2021c. Structurally controlled synthesis of synthetic gypsum derived from industrial wastes: sustainable approach. J. Chem. Technol. Biotechnol. 96 (11), 3134–3141.

Kanazawa, E., Sakai, G., Shimanoe, K., Kanmura, Y., Teraoka, Y., Miura, N., et al., 2001. Metal oxide semiconductor N2O sensor for medical use. Sens. Actuators B Chem. 77, 72–77.

Karmakar, M., Mondal, B., Pal, M., Mukherjee, K., 2014. Acetone and ethanol sensing of barium hexaferrite particles: a case study considering the possibilities of nonconventional hexaferrite sensor. Sens. Actuators B Chem. 190, 627–633.

Kasach, A.A., Kharitonov, D.S., Romanovskii, V.I., Kuz'menok, N.M., Zharskii, I.M., Kurilo, I. I., 2019. Electrodeposition of Cu-Sn alloy from oxalic acid electrolyte in the presence of amine-containing surfactants. Russ. J. Appl. Chem. 92 (6), 835–841.

Kasach, A.A., Kharitonov, D.S., Wrzesińska, A., Bobowska, I., Predko, A.A., Romanovskii, V. I., et al., 2020. The effect of ultrasound treatment on physicochemical and tribological properties of electrolytic Cu–Sn–TiO_2 coatings. Prot. Met. Phys. Chem. 56, 385–391.

Khan, S.B., Faisal, M., Rahman, M.M., Jamal, A., 2011. Exploration of CeO_2 nanoparticles as a chemi-sensor and photo-catalyst for environmental applications. Sci. Total. Environ. 409, 2987–2992.

Khatko, V., 2009. Struktury metall—dioksid kremniya—poluprovodnik glya mikroskhem [Structure of metal-silicon dioxide-semiconductor for IC]. Minsk, BNTU, 234 p. (in Russian).

Khatko, V.V., 2013. [Pd/Al_2O_3 and Pt/Al_2O_3 multilayer stacks in thermocatalytic gas sensors on silicon]. Izvestsiya Natsyonal'noj Akademii Nauk Belarusi. Seriya fiz.-tech. Nauk. 4, 108–112.

Khatko, V.V., Logothetis, E.M., Soltis, R.E., Hangas, J.W., McBride, J.R., 2000. Development of highly active catalyst for Si- microcalorymetric gas sensor. In: Kruger, S., Gessner, W. (Eds.), Advanced Microsystems for Automotive Application. Springer, Berlin, pp. 27–37.

Khatko, V., Soltis, R., McBride, J., Nietering, K., 2001. Catalytic properties of Pd/SiO_2 and Pt/ SiO_2 multilayer stacks. Sens. Actuators, B. 77, 548–554.

Khatko, V., Vallejos, S., Calderer, J., Gràcia, I., Cané, C., Llobet, E., et al., 2009. Micro-machined WO_3-based sensors with improved characteristics. Sens. Actuators, B. 140, 356–362.

Khatoon, S., Ganguly, A., Ahmad, T., 2012. Fabrication of nano-sized solid solution of $Zn_{1-x}Mn_xO$ ($x = 0.05, 0.10, 0.15$) in reverse microemulsions: structural characterization and properties. Bull. Mater. Sci. 35 (3), 377–382.

Khort, A., Romanovski, V., Leybo, D., Moskovskikh, D., 2020a. CO oxidation and organic dyes degradation over graphene-Cu and graphene-CuNi catalysts obtained by solution combustion synthesis. Sci. Rep. 10, 16104.

Khort, A., Romanovski, V., Lapitskaya, V., Kuznetsova, T., Yusupov, K., Moskovskikh, D., et al., 2020b. Graphene@metal nanocomposites by solution combustion synthesis. Inorg. Chem. 59, 6550–6565.

Khort, A., Hedberg, J., Mei, N., Romanovski, V., Blomberg, E., Wallinder, I.O., 2021. Corrosion and transformation of solution combustion synthesized Co, Ni and CoNi nanoparticles in synthetic freshwater with and without natural organic matter. Sci. Rep. 11, 7860.

Kitture, R., Pawar, D., Rao, C.N., Choubey, R.K., Kale, S.N., 2017. Nanocomposite modified optical fiber: a room temperature, selective H2S gas sensor: studies using ZnO-PMMA. J. Alloy. Comp. 695, 2091–2096.

Kou, X., Xie, N., Chen, F., Wang, T., Guo, L., Wang, C., et al., 2018. Superior acetone gas sensor based on electrospun SnO2nanofibers by Rh doping. Sens. Actuators B Chem. 256, 861–869.

Kuang, Q., Lao, C., Wang, Z.L., Xie, Z., Zheng, L., 2007. High-sensitivity humidity sensor based on a single SnO2 nanowire. J. Am. Chem. Soc. 129, 6070–6071.

Lai, X., Cao, K., Shen, G., Xue, P., Wang, D., Hu, F., et al., 2018. Ordered mesoporous $NiFe_2O_4$ with ultrathin framework for low-ppb toluene sensing. Sci. Bull. 63, 187–193.

Lanje, A.S., Sharma, S.J., Ningthoujam, R.S., Ahn, J.-S., Pode, R.B., 2013. Low temperature dielectric studies of zinc oxide (ZnO) nanoparticles prepared by precipitation method. Adv. Powder Technol. 24, 331–335.

Lee, J.-H., 2009. Gas sensors using hierarchical and hollow oxide nanostructures. Sens. Actuators, B. 140, 319–336.

Lee, S.C., Hwang, B.W., Lee, S.J., Choi, H.Y., Kim, S.Y., Jung, S.Y., et al., 2011. A novel tin oxide-based recoverable thick film SO2 gas sensor promoted with magnesium and vanadium oxides. Sens. Actuators B Chem. 160, 1328–1334.

Li, C., Wang, C., Wang, C., Hu, S., 2006. Development of a parathion sensor based on molecularly imprinted nano-TiO_2 self-assembled film electrode. Sens. Actuators B Chem. 117 (1), 166–171.

Li, W., Liang, J., Liu, J., Zhou, L., Yang, R., Hu, M., 2016. Synthesis and room temperature CH4 gas sensing properties of vanadium dioxide nanorods. Mater. Lett. 173, 199–202.

Li, S.M., Zhang, L.X., Zhu, M.Y., Ji, G.J., Zhao, L.X., Yin, J., et al., 2017. Acetone sensing of ZnO nanosheets synthesized using room-temperature precipitation. Sens. Actuators B Chem. 249, 611–623.

Liang, Y.X., Chen, Y.J., Wang, T.H., 2004. Low-resistance gas sensors fabricated from multi-walled carbon nanotubes coated with a thin tin oxide layer. Appl. Phys. Lett. 85, 666–668.

Liao, Y.H., Chou, J.C., 2008. Preparation and characteristics of ruthenium dioxide for pH array sensors with real-time measurement system. Sens. Actuators B Chem. 128 (2), 603–612.

Lin, G., Wang, H., Li, X., Lai, X., Zou, Y., Zhou, X., et al., 2018. Chestnut-like CoFe2O4@SiO2@In2O3 nanocomposite microspheres with enhanced acetone sensing property. Sens. Actuators B Chem. 255, 3364–3373.

Liu, Y.L., Yang, H.F., Yang, Y., Liu, Z.M., Shen, G.L., Yu, R.Q., 2006. Gas sensing properties of tin dioxide coated onto multi-walled carbon nanotubes. Thin Solid. Films 497, 355–360.

Liu, J., Luo, T., Meng, F., Qian, K., Wan, Y., Liu, J., 2010. Porous hierarchical In2O3 micro-/nanostructures: preparation, formation mechanism, and their application in gas sensors for noxious volatile organic compound detection. J. Phys. Chem. C. 114, 4887–4894.

Liu, Y., Hua, L., Li, S., Xi, G., 2011. Preparation and characterization of mesoporous ZnO by polystyrene microemulsion. Desalin. Water Treat. 26, 297–300.

Loo, J.F.C., Chien, Y.H., Yin, F., Kong, S.K., Ho, H.P., Yong, K.T., 2019. Upconversion and down-conversion nanoparticles for biophotonics and nanomedicine. Coord. Chem. Rev. 400, 213042.

Lu, G., Xu, J., Sun, J., Yu, Y., Zhang, Y., Liu, F., 2012. UV-enhanced room temperature NO2 sensor using ZnO nanorods modified with SnO2 nanoparticles. Sens. Actuators B Chem. 162, 82–88.

Ma, L., Ma, S.Y., Shen, X.F., Wang, T.T., Jiang, X.H., Chen, Q., et al., 2018. PrFeO3 hollow nanofibers as a highly efficient gas sensor for acetone detection. Sens. Actuators B Chem. 255, 2546–2554.

Ma, Z.H., Yu, R.T., Song, J.M., 2020. Facile synthesis of Pr-doped In_2O_3 nanoparticles and their high gas sensing performance for ethanol. Sens. Actuators, B. 305, 127377.

Maalouf, R., Hassen, W.M., Fournier-Wirth, C., Coste, J., Jaffrezic-Renault, N., 2008. Comparison of two innovatives approaches for bacterial detection: paramagnetic nanoparticles and self assembled multilayer processes. Microchim. Acta 163, 157–161.

Macak, J.M., Sirotna, K., Schmuki, P., 2005. Self-organized porous titanium oxide prepared in Na_2SO_4/NaF electrolytes. Electrochim. Acta. 50, 3679–3684.

Manjula, P., Arunkumar, S., Manorama, S.V., 2011. Au/SnO2 an excellent material for room temperature carbon monoxide sensing. Sens. Actuators B Chem. 152, 168–175.

Martsul, V.N., Zalygina, O.S., Shibeka, L.A., Likhacheva, A.V., Romanovski, V.I., 2012. Some applications of galvanic manufacture waste. Proc. BSTU. Chem. Technol. Inorg. Substances 3, 66–70.

Matsukevich, I., Lipai, Y., Romanovski, V., 2021a. Cu/MgO and Ni/MgO composite nanoparticles for fast, high-efficiency adsorption of aqueous lead(II) and chromium(III) ions. J. Mater. Sci. 56 (8), 5031–5040.

Matsukevich, I., Kulak, A., Palkhouskaya, V., Romanovski, V., Jo, J.H., Aniskevich, Y., et al., 2021b. Comparison of different synthesis methods for $Li_2MTi_3O_8$ (M-Co, Cu, Zn) as electrode materials. J. Chem. Technol. Biotechnol. Available from: https://doi.org/10.1002/jctb.6992.

Md Sin, N.D., Samsudin, N., Ahmad, S., Mamat, M.H., Rusop, M., 2013. Zn-doped SnO2 with 3D cubic structure for humidity sensor. Procedia Eng. 56, 801–806.

Meng, D., Liu, D., Wang, G., Shen, Y., 2018. Low-temperature formaldehyde gas sensors based on NiO-SnO_2 heterojunction micro flowers assembled by thin porous nanosheets. Sens. Actuators, B. 273, 418–428.

Meulenkamp, E.A., 1998. Synthesis and growth of ZnO nanoparticles. J. Phys. Chem. B 102, 5566–5572.

Milenova, K., Stambolova, I., Blaskov, V., Eliyas, A., Vassilev, S., Shipochka, M., 2013. The effect of introducing copper dopant on the photocatalytic activity of ZnO nanoparticles. J. Chem. Technol. Metall. 48 (3), 259–264.

Mishra, S.K., Srivastava, S.K., Prakash, S.G., Yadav, R.S., Panday, A.C., 2010. Photoluminescence and photoconductive characteristics of hydrothermally synthesized ZnO nanoparticles. Opto-electron. Rev. 18 (4), 467–473.

Moazzen, M.A., Borghei, S.M., Taleshi, F., 2012. Synthesis and characterization of nano-sized hexagonal spherical nanoparticles of zinc oxide. J. Nanostruct. 2, 295–300.

Moon, H.G., Shim, Y.S., Su, D., Park, H.-H., Yoon, S.-J., Jang, H.W., 2011. Embossed TiO_2 thin films with tailored links between hollow hemispheres: synthesis and gas-sensing properties. J. Phys. Chem. C. 115, 9993–9999.

Mozalev, A., Khatko, V., Bittencourt, C., Hassel, A.W., Gorokh, G., Llobet, E., et al., 2008. Nanostructured columnlike tungsten oxide film by anodizing Al/W/Ti layers on Si. Chem. Mater. 20, 6482–6493.

Muhammad, N., Rahman, A., Younis, M.A., Subhani, Q., Shehzad, K., Cui, H., et al., 2018. Porous SnO2 nanoparticles based ion chromatographic determination of nonfluorescent antibiotic (chloramphenicol) in complex samples. Sci. Rep. 8, 1–11.

Na, C.W., Woo, H.S., Lee, J.H., 2012. Design of highly sensitive volatile organic compound sensors by controlling NiO loading on ZnO nanowire networks. RSC Adv. 2, 414–417.

Najafi, M., Khafilzadeh, M.A., Karimi-Maleh, H., 2014. A new strategy for determination of bisphenol A in the presence of Sudan I using a ZnO/CNTs/ionic liquid paste electrode in food samples. Food Chem. 158, 125–131.

Oh, E., Choi, H.Y., Jung, S.H., Cho, S., Kim, J.C., Lee, K.H., et al., 2009. High-performance NO2 gas sensor based on ZnO nanorod grown by ultrasonic irradiation. Sens. Actuators B Chem. 141, 239–243.

Osipenko, M.A., Kharitonov, D.S., Makarova, I.V., Romanovsky, V.I., Kurilo, I.I., 2021. Corrosion behavior of modified anodic oxide coatings on AD31 aluminium alloy. Prot. Met. Phys. Chem. 57, 550–558.

Pal, S., Sharma, M.K., Chatterjee, R., Bhand, S., 2015. Multi-platform nano-immunosensor for aflatoxin M1 in milk. Mater. Res. Exp. 2 (4), 045010.

Paliwal, A., Sharma, A., Tomar, M., Gupta, V., 2017. Carbon monoxide (C.O.) optical gas sensor based on ZnO thin films. Sens. Actuators, B. 250, 679–685.

Pandey, R.R., Saini, K.K., Dhayal, M., 2010. Using nano-arrayed structures in Sol-Gel delivered Mn_2+ doped TiO_2 for high sensitivity urea biosensor. J. Biosen. Bioelectr. 1, 1–4.

Pargoletti, E., Verga, S., Chiarello, G.L., Longhi, M., Cerrato, G., Giordana, A., et al., 2020. Exploring $Sn_xTi_{1-x}O_2$ solid solutions grown onto graphene oxide (GO) as selective toluene gas sensors. Nanomaterials. 10 (4), 761.

Parthibavarman, M., Hariharan, V., Sekar, C., 2011. High-sensitivity humidity sensorbased on SnO2 nanoparticles synthesized by microwave irradiation method. Mater. Sci. Eng. C. 31, 840–844.

Pascariu, P., Airinei, A., Olaru, N., Petrila, I., Nica, V., Sacarescu, L., et al., 2016. Microstructure, electrical and humidity sensor properties of electrospun NiO–SnO2 nanofibers. Sens. Actuators B Chem. 222, 1024–1031.

Pompermayer, N.B., Porto, M.B., Souza, E.F., 2013. Environmental analysis of the zinc oxide nanophotocatalyst synthesis, world academy of science. Eng. Technol. 7 (6), 380–385.

Propolsky, D., Romanovskaia, E., Kwapinski, W., Romanovski, V., 2020. Modified activated carbon for deironing of underground water. Environ. Res. 182, 108996.

Ram, M.K., Yavuz, O., Lahsangah, V., Aldissi, M., 2005. CO gas sensing from ultrathin nanocomposite conducting polymer film. Sens. Actuators B Chem. 106, 750–757.

Ramanouski, V.I., Andreeva, N.A., 2012. Purification of washing waters of iron removal stations. Proc. BSTU. Chem. Technol. Inorg. Subst 3, 62–65.

Reddy, C.S., Murali, G., Reddy, A.S., Park, S., In, I., 2020. GO incorporated SnO2 nanotubes as fast response sensors for ethanol vapor in different atmospheres. J. All. Compd. 813, 152251.

Robbie, K., Brett, M.J., 1997. Sculptured thin films and glancing angle deposition: growth mechanics and applications. J. Vac. Sci. Technol. A 15 (3), 1460–1465.

Romanovskaia, E., Romanovski, V., Kwapinski, W., Kurilo, I., 2021. Selective recovery of vanadium pentoxide from spent catalysts of sulfuric acid production: sustainable approach. Hydrometallurgy. 200, 105568.

Romanovski, V., 2020. New approach for inert filtering media modification by using precipitates of deironing filters for underground water treatment. Environ. Sci. Pollut. Res. 27, 31706–31714.

Romanovski, V., 2021. Agricultural waste based-nanomaterials: green technology for water purifications/aquananotechnology. In: Abd-Elsalam Muhammad Zahid, K. (Ed.), Applications of Nanomaterials for Water Purification, first ed. Elsevier, pp. 577–595.

Romanovskii, V.I., Martsul, V.N., 2009. Distribution of heteroatoms of synthetic ion exchangers in pyrolysis products. Russ. J. Appl. Chem. 82, 836–839.

Romanovskii, V.I., Martsul, V.N., 2012. Functional group distribution over the surface and in the bulk of particles of spent ion exchangers in the course of mechanochemical destruction. Russ. J. Appl. Chem. 85, 371–376.

Romanovskii, V.I., Khort, A.A., 2017. Modified anthracites for deironing of underground water. J. Water Chem. Technol. 39, 299–304.

Romanovski, V., Kolosov, A.Y., Khort, A.A., Myasnichenko, V.S., Podbolotov, K.B., Savina, K. G., et al., 2020. Features of Cu-Ni nanoparticle synthesis: experiment and computer simulation. Phys. Chem. Asp. Study Clust. Nanostruct. Nanomater. 12, 293–309.

Romanovski, V., Romanovskaia, E., Moskovskikh, D., Kuskov, K., Likhavitski, V., Mehmet, F. A., et al., 2021a. Recycling of iron-rich sediment for surface modification of filters for underground water deironing. J. Environ. Chem. Eng. 9 (4), 105712.

Romanovski, V., Klyndyuk, A., Kamarou, M., 2021b. Green approach for low-energy Direct synthesis of anhydrite from industrial wastes of lime mud and spent sulfuric acid. J. Environ. Chem. Eng. 9 (6), 106711.

Romanovskiy, V.I., Khort, A.A., Podbolotov, K.B., Sdobnyakov, N.Y., Myasnichenko, V.S., Sokolov, D.N., 2018. One-step synthesis of polymetallic nanoparticles in air ivironment. Izv. Vyssh. Uchebn. Zaved. Khim. Khim. Tekhnol. 61, 43–48.

Rout, C.S., Ganesh, K., Govindaraj, A., Rao, C.N.R., 2006a. Sensors for the nitrogen oxides, NO2, NO and N2O, based on In2O3 and WO3 nanowires. Appl. Phys. A. Mater. Sci. Process. 85, 241–246.

Rout, C.S., Govindaraj, A., Rao, C.N.R., 2006b. High-sensitivity hydrocarbon sensors based on tungsten oxide nanowires. J. Mater. Chem. 16, 3936–3941.

Rumyantseva, M.N., Kovalenko, V.V., Gas'kov, A.M., Pagnier, T., 2008. Metal-oxide based nanocomposites as materials for gas sensors. Russ. J. Gen. Chem. 78, 1081–1092.

Sahm, T., Mädler, L., Gurlo, A., Barsan, N., Pratsinis, S.E., Weimar, U., 2004. Flame spray synthesis of tin dioxide nanoparticles for gas sensing. Sens. Actuators B Chem. 98, 148–153.

Sarkara, D., Tikkub, S., Thaparb, V., Srinivasac, R.S., Khilarb, K.C., 2011. Formation of zinc oxide nanoparticles of different shapes in water-in-oil microemulsion. Colloids Surf., A. 381, 123–129.

Sdobnyakov, N., Khort, A., Myasnichenko, V., Podbolotov, K., Romanovskaia, E., Kolosov, A., et al., 2020. Solution combustion synthesis and Monte Carlo simulation of the formation of CuNi integrated nanoparticles. Comput. Mater. Sci. 184, 109936.

Shaikh, F.I., Chikhale, L.P., Mulla, I.S., Suryavanshi, S.S., 2017. Synthesis, characterization and enhanced acetone sensing performance of Pd loaded Sm doped SnO2 nanoparticles. Ceram. Int. 43, 10307–10315.

Shen, J.Y., Zhang, L., Ren, J., Wang, J.C., Yao, H.C., Li, Z.J., 2017. Highly enhanced acetone sensing performance of porous C-doped WO3 hollow spheres by carbon spheres as templates. Sens. Actuators B Chem. 239, 597–607.

Siddiquee, S., Rovina, K., Yusof, N.A., Rodrigues, K.F., Suryani, S., 2014. Nanoparticleenhanced electrochemical biosensor with DNA immobilization and hybridization of Trichoderma harzianum gene. Sens. Bio-Sens. Res. 2, 16–22.

Sieber, I., Hildebrand, H., Friedrich, A., Schmuki, P., 2005. Formation of self-organized niobium porous oxide on niobium. Electrochem. Commun. 7, 97–100.

Sin, M.L.Y., Chow, G.C.T., Wong, G.M.K., Li, W.J., Leong, P.H.W., Wong, K.W., 2007. Ultralow-power alcohol vapor sensors using chemically functionalized multiwalled carbon nanotubes. IEEE Trans. Nanotechnol. 6, 571–577.

Singh, N., Yan, C., Lee, P.S., 2010. Room temperature CO gas sensing using Zn-doped In2O3 single nanowire field effect transistors. Sens. Actuators B 150, 19–24.

Singh, G., Virpal, Singh, R.C., 2019. Highly sensitive gas sensor based on Er-doped SnO_2 nanostructures and its temperature dependent selectivity towards hydrogen and ethanol. Sens. Actuators, B. 282, 373–383.

So, M.C., Wiederrecht, G.P., Mondloch, J.E., Hupp, J.T., Farha, O.K., 2015. Metal-organic framework materials for light-harvesting and energy transfer. Chem. Commun. 51, 3501–3510.

Song, H., Zhang, L., He, C., Qu, Y., Tian, Y.F., Lv, Y., 2011. Graphene sheets decorated with SnO2 nanoparticles: in situ synthesis and highly efficient materials for cataluminescence gas sensors. J. Mater. Chem. 21, 5972–5977.

Song, D., Yang, R., Fang, S., Liu, Y., Long, F., Zhu, A., 2018. SERS based aptasensor for ochratoxin A by combining Fe3O4@Au magnetic nanoparticles and Au-DTNB@Ag nanoprobes with multiple signal enhancement. Microchim. Acta. 185, 491.

Souteyrand, E., 1997. In: Pijolat, C. (Ed.), Les Capteurs Chimiques. CMC2, Ecole Centrale, Lyon, pp. 52–62.

Starke, T.K.H., Coles, G.S.V., 2002. High sensitivity ozone sensors for environmental monitoring produced using laser ablated nanocrystalline metal oxides. IEEE Sens. J. 2, 14–19.

Sun, Y., Yang, H., Zhao, Z., Suematsu, K., Li, P., Yu, Z., et al., 2020. Fabrication of ZnO quantum dots@SnO_2 hollow nanospheres hybrid hierarchical structures for effectively detecting formaldehyde. Sens. Actuators, B. 318, 128222.

Tai, H., Jiang, Y., Xie, G., Yu, J., Chen, X., 2007. Fabrication and gas sensitivity of polyaniline–titanium dioxide nanocomposite thin film. Sens. Actuators B Chem. 125, 644–650.

Tai, H., Jiang, Y., Xie, G., Yu, J., Chen, X., Ying, Z., 2008. Influence of polymerization temperature on NH_3 response of PANI/TiO_2 thin film gas sensor. Sens. Actuators B Chem. 129, 319–326.

Takahata, K., 1988. In: Seiyama, T. (Ed.), Chemical Sensor Technology. Elsevier, Amsterdam, pp. 39–55.

Tang, H., Li, Y., Zheng, C., Ye, J., Hou, X., Lv, Y., 2007. An ethanol sensor based on cataluminescence on ZnO nanoparticles. Talanta 72, 1593–1597.

Teymoori, N., Raoof, J.B., Khalilzadeh, M.A., Ojani, R., 2018. An electrochemical sensor based on CuO nanoparticle for simultaneous determination of hydrazine and bisphenol a. J. Iran. Chem. Soc. 15, 2271–2279.

Toloman, D., Popa, A., Stan, M., Socaci, C., Biris, A.R., Katona, G., et al., 2017. Appl. Surf. Sci. 402, 410–417.

Tomer, V.K., Singh, K., Kaur, H., Shorie, M., Sabherwal, P., 2017. Rapid acetone detection using indium loaded WO3/SnO2 nanohybrid sensor. Sens. Actuators. B. Chem. 253, 703–713.

Tsuchiya, H., Schmuki, P., 2004. Thick self-organized porous zirconium oxide formed in H_2SO_4/NH_4F electrolytes. Electrochem. Commun. 6, 1131–1134.

Tsuchiya, H., Schmuki, P., 2005. Self-organized high aspect ratio porous hafnium oxide prepared by electrochemical anodization. Electrochem. Commun. 7, 49–52.

Tsuchiya, H., Macak, J.M., Sieber, I., Taveira, L., Ghicov, A., Sirotna, K., et al., 2005. Self-organized porous WO_3 formed in NaF electrolytes. Electrochem. Commun. 7, 295–298.

Udayakumar, S., Renuga, V., Kavitha, K., 2012. Synthesis and characterization of Ni-doped ZnO by chemical precipitation method. Int. J. Recent. Sci. Res. 3 (2), 118–122.

Uria, N., Abramova, N., Bratov, A., Muñoz-Pascual, F.X., Baldrich, E., 2016. Miniaturized metal oxide pH sensors for bacteria detection. Talanta. 15 (147), 364–369.

Vallejos, S., Khatko, V., Calderer, J., Gràcia, I., Cané, C., Llobet, E., et al., 2008. Micro-machined WO_3-based sensors selective to oxidizing gases. Sens. Actuators, B. 132, 209–215.

Varshney, M., Li, Y.B., 2007. Interdigitated array microelectrode based impedance biosensor coupled with magnetic nanoparticle-antibody conjugates for detection of Escherichia coli O157:H7 in food samples. Biosens. Bioelectron. 22, 2408–2414.

Vijayakumar, S., Vadivel, S., 2019. Fiber optic ethanol gas sensor based WO_3 and WO_3/gC_3N_4 nanocomposites by a novel microwave technique. Opt. Laser Technol. 118, 44–51.

Wan, Q., Li, Q., Yj, C., Wang, T.-H., Xl, H., Li, J., et al., 2004. Fabrication and ethanol sensing characteristics of ZnO nanowire gas sensors. Appl. Phys. Lett. 84, 3654–3656.

Wang, M., Li, Z.Y., 2008. Nano-composite ZrO2/Au film electrode for voltammetric detection of parathion. Sens. Actuators B-Chem 133 (2), 607–612.

Wang, C.T., Chen, M.T., 2010. Vanadium-promoted tin oxide semiconductor carbon monoxide gas sensors. Sens. Actuators B 150, 360–366.

Wang, J., Lin, Q., Zhou, R., Xu, B., 2002. Humidity sensors based on composite material of nano-$BaTiO_3$ and polymer RMX. Sens. Actuators B Chem. 81, 248–253.

Wang, R., Dong, W., Ruan, C., Kanayeva, D., Lassiter, K., Tian, R., et al., 2009a. TiO_2 nano-wire bundle microelectrode based impedance immunosensor for rapid and sensitive detection of Listeria monocytogenes. Nano Lett. 9, 4570–4573.

Wang, Y., Jia, W., Strout, T., Schempf, A., Zhang, H., Li, B., et al., 2009b. Ammonia gas sensor using polypyrrole-coated TiO_2/ZnO nanofibers. Electroanalysis 21, 1432–1438.

Wang, J., Yang, J., Han, N., Zhou, X., Gong, S., Yang, J., et al., 2017. Highly sensitive and selective ethanol and acetone gas sensors based on modified ZnO nanomaterials. Mater. Des. 121, 69–76.

Wang, S., Cao, J., Cui, W., Cui, W., Fan, L., Li, X., et al., 2018. Oxygen vacancies and grain boundaries potential barriers modulation facilitated formaldehyde gas sensing performances for In_2O_3 hierarchical architectures. Sens. Actuators, B. 255, 159–165.

Wiesholler, L.M., Frenzel, F., Grauel, B., Würth, C., Resch-Genger, U., Hirsch, T., 2019. Yb, Nd, Er-doped upconversion nanoparticles: 980 nm: vs 808 nm excitation. Nanoscale 11, 13440–13449.

Williams, D.E., 1999. Semiconducting oxides as gas-sensitive resistors. Sens. Actuators, B. 57, 1–16.

Wisitsoraat, A., Tuantranont, A., Thanachayanont, C., Patthanasettakul, V., Singjai, P., 2006. Electron beam evaporated carbon nanotube dispersed SnO2 thin film gas sensor. J. Electroceram. 17, 45–49.

Wu, N., Zhao, M., Zheng, J.G., Jiang, C., Myers, B., Li, S., et al., 2005. Porous CuO–ZnO nanocomposite for sensing electrode of high-temperature CO solid-state electrochemical sensor. Nanotechnology 16, 2878–2881.

Xiangfeng, C., Caihong, W., Dongli, J., Chenmou, Z., 2004. Ethanol sensor based on indium oxide nanowires prepared by carbothermal reduction reaction. Chem. Phys. Lett. 399, 461–464.

Xing, X., Xiao, X., Wang, L., Wang, Y., 2017. Highly sensitive formaldehyde gas sensor based on hierarchically porous Ag-loaded ZnO heterojunction nanocomposites. Sens. Actuators, B. 247, 797–806.

Yu, Z., Zhao, G., Liu, M., Lei, Y., Li, M., 2010. Fabrication of a novel atrazine biosensor and its subpart-per-trillion levels sensitive performance. Environ. Sci. Technol. 44 (20), 7878–7883.

Yu, A., Xun, H., Yi, J., 2019. Improving hydrogen sensing performance of TiO_2 nanotube arrays by ZnO modification. Front. Mater. 6, 70.

Yuan, F., Ryu, H., 2004. Microstructure of varistors prepared with zinc oxide nanoparticles coated with Bi_2O_3. J. Am. Ceram. Soc. 87 (4), 736–738.

Zak, K., Ebrahimizadeh Abrishami, M., Abd, et al., 2011. Effects of annealing temperature on some structural and optical properties of ZnO nanoparticles prepared by a modified sol-gel combustion method. Ceram. Int. 37, 393–398.

Zalyhina, V., Cheprasova, V., Belyaeva, V., Romanovski, V., 2021a. Pigments from spent Zn, Ni, Cu, and Cd electrolytes from electroplating industry. Environ. Sci. Pollut. Res. 28, 32660–32668.

Zalyhina, V., Cheprasova, V., Romanovski, V., 2021b. Pigments from spent chloride-ammonium zinc plating electrolytes. J. Chem. Technol. Biotechnol. 96 (10), 2767–2774.

Zhang, Y.L., Yang, Y., Zhao, J.H., Tan, R.Q., Cui, P., Song, W.J., 2009. Preparation of ZnO nanoparticles by a surfactant-assisted complex sol–gel method using zinc nitrate. J. Sol-Gel Sci. Technol. 51, 198–203.

Zhang, M., Yuan, R., Chai, Y., Wang, C., Xiaoping, W., 2013. Cerium oxide–graphene as the matrix for cholesterol sensor. Anal. Biochem. 436, 69–74.

Zhang, H., Feng, J., Fei, T., Liu, S., Zhang, T., 2014a. SnO2 nanoparticles-reduced graphene oxide nanocomposites for NO2 sensing at low operating temperature. Sens. Actuators B 190, 472–478.

Zhang, D., Tong, J., Xia, B., 2014b. Humidity-sensing properties of chemically reduced graphene oxide/polymer nanocomposite film sensor based on layer-by-layer nano self-assembly. Sens. Actuators B Chem. 197, 66–72.

Zhang, J., Zhao, S.Q., Zhang, K., Zhou, J.Q., 2014c. Cd-doped ZnO quantum dots-based immunoassay for the quantitative determination of bisphenol A. Chemosphere 95, 105–110.

Zhang, X., Gu, R., Zhao, J., Jin, G., Zhao, M., Xue, Y., 2015. Effects of particle size on the gas sensitivity and catalytic activity of In_2O_3. J. Mater. Eng. Perform. 24, 3815–3819.

Zhang, D., Chang, H., Li, P., Liu, R., Xue, Q., 2016. Fabrication and characterizationof an ultra-sensitive humidity sensor based on metal oxide/graphene hybridnanocomposite. Sens. Actuators B Chem. 225, 233–240.

Zhang, R., Gao, S., Zhou, T., Tu, J., Zhang, T., 2020a. Facile preparation of hierarchical structure based on p-type Co3O4 as toluene detecting sensor. Appl. Surf. Sci. 503, 144167.

Zhang, K., Qin, S., Tang, P., Feng, Y., Li, D., 2020b. Ultra-sensitive ethanol gas sensors based on nanosheet-assembled hierarchical ZnO-In_2O_3 heterostructures. J. Hazard. Mater. 391, 122191.

Zhao, L., Choi, M., Kim, H.S., Hong, S.H., 2007. The effect of multiwalled carbon nanotube doping on the CO gas sensitivity of SnO2-based nanomaterials. Nanotechnology 18, 445501–445505.

Zhao, C., Lan, W., Gong, H., Bai, J., Ramachandran, R., Liu, S., et al., 2018. Highly sensitive acetonesensing properties of Pt-decorated CuFe2O4 nanotubes prepared by electrospinning. Ceram. Int. 44, 2856–2863.

Zhu, K.M., Ma, S.Y., 2019. Preparations of Bi-doped SnO_2 hierarchical flowershaped nanostructures with highly sensitive HCHO sensing properties. Mater. Lett. 236, 491–494.

Zhu, L., Zeng, W., Yang, J., Li, Y., 2018. One-step hydrothermal fabrication of nanosheet-assembled NiO/ZnO micro flower and its ethanol sensing property. Ceram. Internat. 44, 19825–19830.

Zhu, K., Ma, S., Tie, Y., Zhang, Q., Wang, W., Pei, S., et al., 2019. Highly sensitive formaldehyde gas sensors based on Y-doped SnO_2 hierarchical flower-shaped nanostructures. J. All. Compd. 792, 938–944.

Zhuo, M., Chen, Y., Suna, J., Zhang, H., Guo, D., Zhang, H., et al., 2013. Humidity sensing properties of a single Sb doped SnO2 nanowire field effect transistor. Sens. Actuators B Chem. 186, 78–83.

Zou, D., Jin, L., Wu, B., Hu, L., Chen, X., Huang, G., et al., 2019. Rapid detection of Salmonella in milk by biofunctionalised magnetic nanoparticle cluster sensor based on nuclear magnetic resonance. Int. Dairy. J. 91, 82–88.

Chapter 17

Strengths and weaknesses of metal oxide nanoparticles in agriculture

Tawfiq Mohammad Al-Antary[1] and Alaa Yousef Ghidan[2]
[1]Department of Plant Protection, School of Agriculture, The University of Jordan, Amman, Jordan, [2]Department of Nanotechnologies and Microbiology, Research and Development Center, The Higher Council for Science and Technology, Amman, Jordan

Chapter outline

17.1 Introduction

Agriculture is one of the most important and economically stable sectors. Also, it has become a backbone of developing countries. Nanotechnology can be exploited in the value chain of the entire agriculture production system. Nanotechnology is the sixth revolution technology after the industrial revolution (the mid-1700s), nuclear energy revolution (1940s), green revolution (1960s), information technology (1980s), and biotechnology (1990s). Nanotechnology is now emerging, and it is a fast-growing field of science. Nanotechnology means like nanocapsules;

Nanometal Oxides in Horticulture and Agronomy. DOI: https://doi.org/10.1016/B978-0-323-91809-1.00004-4

nanoparticles (NPs) are examples of uses for the detection and treatment of diseases. Nanotechnology will revolutionize agriculture and the food industry such as (in the case of farming techniques) enhancing the ability of plants to absorb nutrients, disease detection, and control pests. The limited natural resources of production land, water, soil, forests, and minerals and the growing increase in the world population promoting agricultural development are economically important, viable, environmental, and efficient. This great change will be the major for achieving several factors in the present time (Johnston and Mellor, 1961; Pereire et al., 2007). Agricultural nutrient balances have differed noticeably with economic growth, especially from this prediction. The development of soil fertility is very much significant in developing countries (Campbell et al., 2014). The development of agriculture is a phenomenon to face poverty and hunger, which must be overcome from the recent status. As global food and nutritional security policies are fully embedded in the novel knowledge, clean sustainable agricultural practices increase and strengthen the real practical opportunities to overcome the dominant poverty and hunger of the people of developing countries (Yunlong and Smit, 1994). Agriculture is on the right track to be recovered; thus, environmental achievements are required and, at the same time, better participation of food chains in various ecosystems is required for the increasing growth of agricultural food production (Thornhill et al., 2016; Balakrishnan et al., 2020). It is predictable and no doubt that the sustainable growth of agriculture depends on the new and innovative techniques like nanotechnology, which has started to be used in several fields including agriculture. Afterward, nanotechnology developed more vivid way because more recent instruments were developed to consider or isolate nanomaterials in an accurate way (Feynman, 1996). Additionally, the number of publications related to the term "nano" has also grown exponentially. The developments of nanotechnology in materials and electronics have higher dynamics than other applications. Recently, food and agriculture also require an increased number of nanomaterials, especially in nanochemicals and packaging (Bonnell and Huey, 2001; Ion et al., 2010). Zinc, manganese, and iron nano oxides were synthesized via a green chemistry technique using a microwave-assisted hydrothermal method. The results from nano oxides showed a particle size of around 20–60 nm and small surface areas. These oxides were applied as a foliar nanofertilizer on plants (Ghidan et al., 2020a,b,c). The use of the prepared nano oxides as a foliar application improved the growth and the yield in comparison with untreated plants (Al-Antary et al., 2020; Ghidan et al., 2020a,b,c).

17.2 Nanotechnology and agricultural sustainable development

NPs are critical in developing future sustainable technologies for humanity and the environment. The green chemistry technique of plant NP synthesis links nanotechnology with plant biotechnology. Metal ions are bioreduced to create NPs using plant extracts (Ghidan et al., 2018a,b,c). Sugars, terpenoids, and other plant metabolites have been shown to have antimicrobial properties. Polyphenols, alkaloids, phenolic acids, and proteins are critical in reducing metal ions into NPs and the subsequent stability of these particles. The atoms of nanometric materials have properties that are significantly different from those of bulk materials. Traditionally, NPs have been synthesized using a chemical process, with chemicals being utilized due to their overall toxicity, instilling a strong desire to develop environmentally sustainable processes. As a result, biological alternatives are emerging to fill the hole, such as green synthesis, which uses them as the molecules of extracts produced from plant sources and shows superiority over chemical and/or biological processes. Several plants have been effectively employed for extracellular syntactic efficiency and speed of synthesis of gold and silver NPs. Geranium (*Pelargonium graveolens*), lemongrass, and *Emblica officinalis* fruit extract have demonstrated their ability to reduce Au(III) ions to generate gold NPs Au(0) and silver nitrate to form silver NPs Ag (0). Wheat (*Triticum aestivum*) and oat (*Avena sativa*) biomasses, alfalfa (*Medicago sativa*) biomasses, natural and chemically modified hop biomasses, and soaked Bengal gram bean (*Cicer arietinum)* residual water have also been used for synthesis of gold NPs. When discussing the benefits and drawbacks of nanotechnology, we must also consider what could be considered the technology's drawbacks. The potential loss of jobs in traditional farming and industrial industries is one of the many downsides of this science and its progress. Nanotechnology has also increased the risk of health; NPs, due to their small size, can cause breathing problems and many other fatal diseases by inhaling air containing NPs only for 60 s; nanotechnology is also known as nanopollution. This type of pollution is quite hazardous for the sake of living things. Some NPs have a higher proclivity for accumulating in body tissues. Unmodified copper or silver NPs, for example, can cause toxicity. Surface modification of NPs can result in agents that have considerably different chemical and/or biological properties than the "same" NPs with unaltered surfaces. It should also be noted that the risks connected with persistent insoluble NPs in the environment are likely to be far greater than those associated with human health evaluation. The protective goals and endpoints (i.e., person protection versus safeguarding the environment). Scientists found that the 48-h LC50 for uncoated water-soluble fullerenes NC60 in Daphnia magna is 800 ppb. After 48 h of exposure to 0.5 ppm fullerenes NC60, they found a significant increase in lipid peroxidation in the brain and glutathione depletion in the gill of juvenile largemouth bass (*Micropterus salmoides*), but not at 1 ppm. Scientists report the likely molecular explanation underlying these discoveries in their follow-up studies. Kai et al. discovered that fullerenes have antibacterial effects. However, a number of the abovementioned human toxicological studies have looked into the intake and effects of the drug. It is impossible to say if nanotechnologies are completely safe for human health or whether they are hazardous. Risks associated with chronic farmer exposure to nanomaterials, unknown life cycles, interactions with the biotic or abiotic environment, and possible amplified bioaccumulation effects have not been considered. These should be seriously considered before these applications move from the lab to the field. High processing costs, problems with the scalability of R&D for prototype and industrial production, and worries about public perception of environmental, health, and safety issues are all frequent challenges in commercializing nanotechnology. Before commercialization, governments all over the world should establish common and severe regulations and supervision.

Nanotechnology has an essential role in productivity through the control of nutrients (Mukhopadhyay, 2014). It can also be a part of the monitoring of pesticides, fertilizers, and water quality for clean sustainable development of agriculture (Prasad et al., 2014). Nanomaterials (NPs) have such diverse assets and activities that it is impossible to deliver a general assessment of their health and environmental risks (Prasad et al., 2014). Properties of NPs influencing toxicity include chemical composition, shape, surface structure, surface charge, behavior, and the extent of particle aggregation or disaggregation that may be associated with engineered NPs (Ion et al., 2010). For these reasons, even nanomaterials of the same chemical composition that have different sizes or shapes can exhibit their different toxicities. Nanomaterials not only directly catalyze the degradation of waste and toxic materials but also aids improve the efficiency of microorganisms in the degradation of waste and toxic materials. Bioremediation uses living organisms to break down or remove toxins and harmful substances from contaminated agricultural soil and water. In particular, some other terms are also generally used such as bioremediation of beneficial microbes, phytoremediation of plants, and microembolization of fungi and mushrooms. Thus, with bioremediation, the heavy metals can be removed from polluted soil and water environmentally and efficiently by microorganisms. Nanofertilizers may contain nano zinc, silica, iron and titanium dioxide, Zn, Cd, Se/ZnS core–shell quantum dots (QDs), InP/ZnS core–shell QDs, Mn/ZnSe QDs, gold nanorods, and core–shell QDs as well as should endorse control release and improve its quality. Studies of the uptake, biological fate, and toxicity of metal oxide NPs, namely Al_2O_3, TiO_2, CeO_2, FeO, and ZnONPs were carried out intensively in the present decade for agricultural production (Dimkpa, 2014; Zhang et al., 2016). The deficiency of zinc has been documented as one of the main problems in limiting agricultural productivity in the alkaline nature of soils (Sadeghzadeh, 2013).

Nanostructures materials have many advantages in logical sciences when used as transducers or as a part of the appreciation in a macrosized sensing device. In these facts, the gold NPs (AuNPs) have their intrinsic properties and may use as transducers for several improvements of agricultural products. Biological tests measuring the presence or activity of selected analytics become quicker, more sensitive, and flexible when nanoscale particles are put together (Kandasamy and Prema, 2015; Ramachandran et al., 2020).

Nontoxic and promising pesticide delivery systems were also developed for increasing global food production while reducing the negative environmental impacts on the ecosystem (Bhattacharyya et al., 2016; Grillo et al., 2016), and the toxicity of the NPs is normally exploited in the industry (Du et al., 2017). The toxicity of a metal depends upon several factors like solubility, binding specificity to a biological site, etc. Metal NPs exhibit antibacterial, anticandidal, and antifungal activities (Patra and Baek, 2017). Metal NPs exert cytotoxicity depending on the charge at the membrane surface. The efficiency of nanotoxicity effects of NPs depends on the structure of the targeted cell wall. Thus, the sensitive order should be mound > yeast > Gram-negative bacteria > Gram-positive bacteria. Nanotoxicity may be accredited to electrostatic interaction between NPs with membrane and their accumulation in cytoplasm (Aziz et al., 2016). Several NPs, particularly TiO_2, ZnO, SiO_2, and fullerenes, are photochemically active. When they are exposed to light, the excited electrons are generated, which then form superoxide radicals in the presence of oxygen by direct electron transfer (Hoffmann et al., 2007). Thus, this ecotoxicity is surprised in the act when organisms are simultaneously exposed to NPs and UV light since UV light has higher energy than visible light (Ghidan et al., 2020a,b,c).

In this case, the cells respond to oxidative stress by increasing the number of protective enzymatic or genetic constitutions that can easily be measured (Kovochich et al., 2005; Vanathi et al., 2014). Thus, the generation of reactive oxygen species (ROS) is an oxidative stress parameter that can be exploited in the determination of the context of toxicity and ecotoxicity. In vitro studies on the toxicity of NPs have confirmed the generation of ROS, for example, by TiO_2 and fullerenes (Sayes et al., 2004). On the other hand, some authors revealed that NPs of fullerenes and silicon NPs may protect against oxidative stress. Much more research studies related to interactions between cells and NPs as well as mechanistic facets of NPs metabolism in organisms and specific cells are needed to clarify this dichotomy.

The agricultural host plants and their fruits take the main part in the food chain. Recently, the plants do not only grow on agricultural lands but they are also developed on an aqueous medium. Naturally, several NPs of iron oxide (magnetite), a magnetic form of iron ore, can deposit in the plant host. It is interesting to propose that the iron (II, III) oxide NPs (Fe_3O_4 NPs) can accumulate in *Lepidium sativum* and *Pisum sativum* plants. Therefore, this type of observation proposes that the roles mentioned by NPs are present in the natural ecosystem (Bystrzejewska-Piotrowska et al., 2012; Abbas et al., 2016). Moreover, the uses of polymeric NPs in the agricultural field, especially loaded with insecticides of plant origin, are unique and increasingly permeated. No doubt that microorganisms play a crucial role in maintaining soil health, ecosystem, and crop productivity. Many advantages of physical–chemical properties of nanoscale materials are also exploitable in the field of biosensors development. Sagadevan and Periasamy (2014) stated that the sensitivity and performance of biosensors can be improved by using nanomaterials through new signal transduction technologies. The tremendous advancement in nanosensors is due to the great technological demand for rapid, sensitive, and cost-effective nanobiosensor systems in vital areas of human activity such as health care, agriculture, genome analysis, food and drink, the process industries, environmental monitoring, defense, and security. At present, nanotechnology-based biosensors are at the early stage of development. The improvement of tools and procedures used to fabricate, measure, and image nanoscale objects has led to the development of sensors. The nanomaterials such as metal (gold, silver, and cobalt) NPs, carbon nanotube (CNT), magnetic NPs, and QDs have been actively investigated for their applications in biosensors, which have become a new interdisciplinary frontier between biological detection and material science. Nanotechnology can supply bioactive ingredients in foodstuffs to hosts leading to improved knowledge of food materials at the nanoscale. Most of the time, nanomaterials are produced by bottom-up methods.

Recently, nanotechnology has been widely applied in food processing such as nanocarrier systems for delivery of nutrients and supplements and organic nanosized additives for food, supplements, and animal feed. Many food products already contain NP naturally. Milk contains casein, a form of milk protein present at the nanoscale, or meat is made up of protein filaments that are also classified into nanomaterial groups. The texture and properties of these products are determined by the organization and structures of proteins inside. Reduction in size to the nanoscale level results in an enormous increase of surface to volume ratio, so relatively more molecules of the chemical are present on the surface, thus enhancing the intrinsic toxicity. This may be one of the reasons why NPs are generally more toxic than larger particles of the same insoluble material when compared on a mass dose base. The expression of a dose–response relationship based on particle size resulted in a similar dose–response relationship between low solubility and low toxicity of particles of different sizes. In studies of low toxicity particles, TiO_2 induced more severe lung inflammation and particle lymph node burden compared to $BaSO_4$ when dosed at mass burden in milligrams. The surface area was, therefore, a driver for inflammation for these materials; the differences in severity of the response disappeared when the dose was expressed as surface area. Some foods and drinks were fortified with these NPs without affecting the taste or appearance. NP emulsions are being used in ice cream and spreads of this nanoemulsion can improve the texture and uniformity of the ice cream. KD Pharma BEXBACH GMBH (Bexbach, Germany) provides encapsulated omega-3 fatty acids in two different forms: suspension and powder. The capsulation technology used the resulted particles on nano- and microscale. In the food industry, maintaining some important factors such as quality, safe, freshness, and tasterequires the whole supply chain to package and label its products. The development of smart packages that can provide useful information is still a big challenge for researchers and producers. Recently, some packaging materials incorporated with "nanosensors" to detect the oxidation process in food have been produced and used in the food industry. The working scheme is quite simple: when the oxidation occurs in the food package, NP-based sensors indicate the color change and information about the nature of the packed foods can be observed. This technology has been successfully applied in a package of milk and meat.

17.3 Nanomaterials

Nanomaterials are defined in principle as materials in which a single unit is sized in one dimension between 1 and 100 nm ($1 \text{ nm} = 10^{-9}$ m). Materials with structure at the nanoscale often have unique physical and chemical properties.

NPs are of great scientific interest as they effectively act as a bridge between bulk materials and atomic or molecular structures. A bulk material should have constant physical properties regardless of its size, but at the nanoscale, this is often not the case. Nanomaterials such as copper oxide (CuONPs), zinc oxide (ZnONPs), magnesium hydroxide (MgOHNPs), and magnesium oxide (MgONPs) were synthesized by different physical and chemical methods (Kumar et al., 2013; Yousefi, 2014; Jiang et al., 2015). With the growing need to minimize or eliminate the use of environmental-risk substances, such as insecticides, the biosynthesis of NPs as an emerging highlight of the intersection of nanotechnology and biotechnology has received increasing attention in the last decade. The green synthesis techniques are generally synthetic routes that utilize relatively nontoxic solvents such as water, biological extracts, biological systems, and microwaves. The rate of reduction of metal ions using plants is much faster as compared to microorganisms and stable formation of NPs has been reported (Gunalan et al., 2012; Yallappa et al., 2013; Awwad and Ahmad, 2014; Sankar et al., 2014; Awwad and Albiss, 2015; Nasrollahzadeh et al., 2015; Sugirtha et al., 2015).

17.3.1 Copper oxide nanoparticles

Copper oxide (CuO) is an important metal oxide semiconductor with a narrow bandgap of 1.7 eV and stability in pesticide formulation and antibacterial agents. CuONPs are synthesized using different methods such as precipitation (Sahooli et al., 2012), sonochemical route (Safarifard and Morsali, 2012), sol−gel, hydrothermal approach, chemical bath deposition (Jiang et al., 2015), chemical reduction as shown in Fig. 17.1, nonvacuum and spin coating sol−gel techniques (Yahia et al., 2016), and reflux contestation (Bouazizi et al., 2015). Many plants' aqueous extracts have been reported such as *Gundelia tournefortii* leaves and stems (Nasrollahzadeh et al., 2015), *Tinospora cordifolia* leaves, *Calotropis gigan* tea leaf (Sharma et al., 2015), *Aloe barbadensis* leaves (Gunalan et al., 2012; Kumar et al., 2013), *Carica papaya* leaves (Sankar et al., 2014), *Glorio*

FIGURE 17.1 Graphical abstract of the green synthesis of copper oxide nanoparticles (CuONPs).

sasuperba L. (Naika et al., 2015), citrus limon juice (Mohan et al., 2015), *Tabernaemontana divaricate* leaves (Sivaraj et al., 2014), carob leaves (Awwad and Ibrahim, 2015), and *Malva sylvestris* leaves (Awwad et al., 2013).

17.3.2 Zinc oxide nanoparticles

Zinc oxide (ZnO) has potential applications in sensors, solar cells, UV lasers, semiconductors, hydrogen storage devices, and catalysts. These applications have led many researchers to develop different routes to the synthesis ZnONPs such as the chemical route (Fig. 17.2), hydrothermal route (Ipeksac et al., 2013), sol–gel template process (Kumari et al., 2010; El Ghoul et al., 2012), photoluminescence emission technique (Rocha et al., 2014), microwave-assisted hydrothermal and decomposition (Tseng et al., 2012), aerosol process (Ozcelik and Ergun, 2014), sonochemical synthesis (Zak et al., 2013), laser ablation (Thareja and Shukla, 2007), microemulsion method (Yildirım and Durucan, 2010), precipitation method (An et al., 2014), hydrolyzed in polar organic solvents, solid-state thermal decomposition, and microwave synthesis (Sutradhar et al., 2016). Different plants extracts have been reported in the open literature for green synthesis of ZnONPs such as *Olea europaea* (Awwad et al., 2014), *Eichhornia crassipes* leaves (Vanathi et al., 2014), *Aloe barbadensis* Miller leaves (Sangeetha et al., 2011), *Hibiscus subdariffa* leaves (Bala et al., 2015), *Solanum nigrum* leaves (Ramesh et al., 2015), *Camelia sinensis* leaves (Shah et al., 2015),

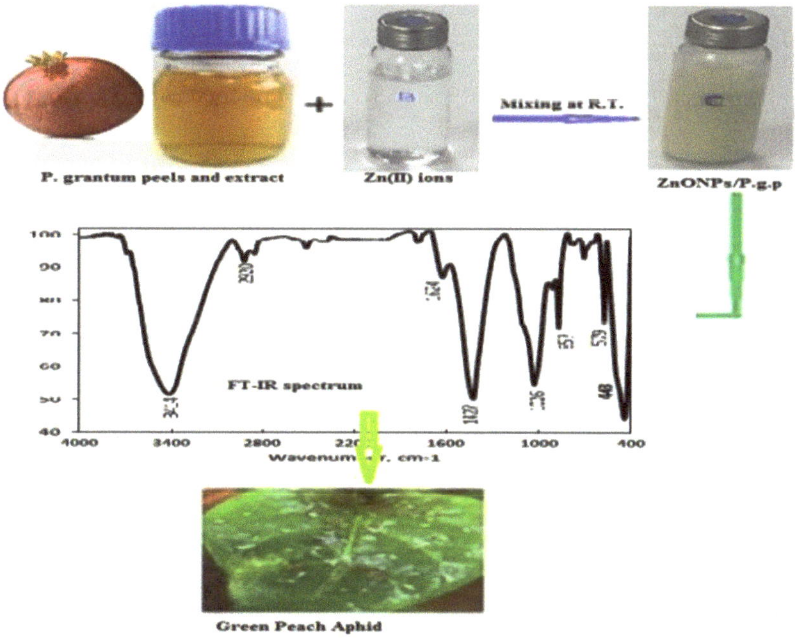

FIGURE 17.2 Graphical abstract of green synthesis of zinc oxide nanoparticles.

Hibiscus rosa-sinensis (Devi and Gayathri, 2014), *Cassia fistula* plant extract (Suresh et al., 2015), *Ocimum tenuiflorum* leaves (Raut et al., 2015), *Trifolium pratense* flower extract (Dobrucka and Dugaszewska, 2016), *Mimosa pudica* leaves and coffee powder (Fatimah et al., 2016), *Rambutan* peel extract (Karnan and Selvakumar, 2016) *Jacaranda mimosifolia* flowers (Sharma et al., 2016), six *Terminalia chebula* fruits (Rana et al., 2016), and *Azadirachta indica* (Madan et al., 2016).

17.3.3 Magnesium oxide (MgO) and hydroxide [$Mg(OH)_2$] nanoparticles

Magnesium oxide (MgO) (Fig. 17.4) and hydroxide $Mg(OH)_2$ (Fig. 17.3) are important inorganic materials with many applications such as adsorbent, fire retardant, catalyst support, advanced ceramics, toxic waste remediation, photoelectronic materials, antacid excipient in pharmaceuticals, and refractory materials. Therefore, various techniques and routes for synthesis of MgOHNPs and MgONPs have been reported in the literature such as the hydrothermal route (Yan et al., 2008; Dhaouadi et al., 2011; Hanlon et al., 2015); water-in-oil microemulsion, ultrasound-assisted method batch reaction crystallization (Song et al., 2011), precipitation and coprecipitation processes (Xu and Deng, 2006; Jiang et al., 2009; Wang et al., 2011), microwave

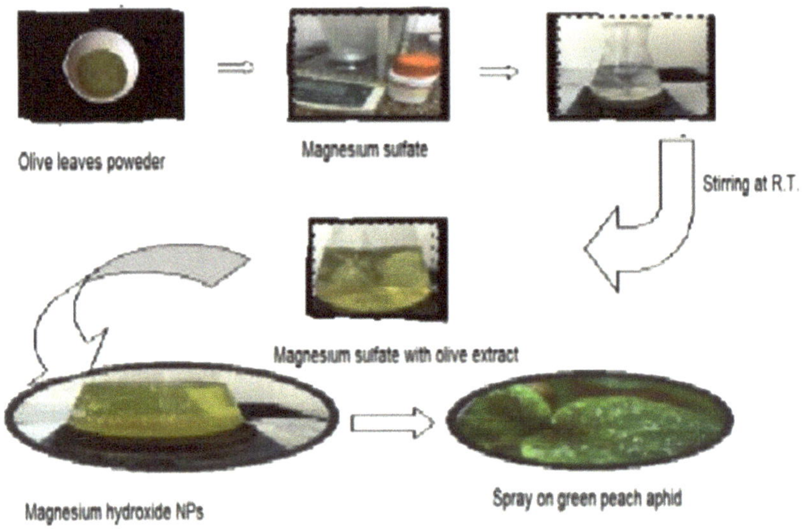

FIGURE 17.3 Graphical abstract of green synthesis of magnesium hydroxide nanoparticles (MgOHNPs).

reaction (Yousefi, 2014), via a facile template-free solvothermal route (Jinghui et al., 2009), stream– rotating packed bed reactor, and spinning desk reactor (Tai et al., 2007). MgOH was synthesized by green methods using nontoxic and ecofriendly such as neem leaves extract (Moorthy et al., 2015), *Citrus limon* leaves extract (Awwad and Ahmad, 2014), acacia gum, *Brassica oleracea*, *Punica granatum* peels (Sugirtha et al., 2015), and *Clitoria terna* tea (Sushma et al., 2016) (Fig. 17.4).

17.4 Comparison between effects of four nanoparticles on the green peach aphid at concentration of 8000 μg/mL

The highest mortality % in MgONPs of third and fourth late nymphal instars was at 8000 (μg/mL) concentration. It was 86% after 24 h, and then reached 83% and 98% after 48 and 72 h, respectively (Table 17.1).

17.5 Toxicity of nanoparticles

This may be one of the reasons why NPs are generally more toxic than larger particles of the same insoluble material when compared on a mass dose base. The expression of a dose–response relationship based on particle size resulted in a similar dose–response relationship between low solubility and low toxicity in particles of different sizes (Ghidan et al., 2020a,b,c).

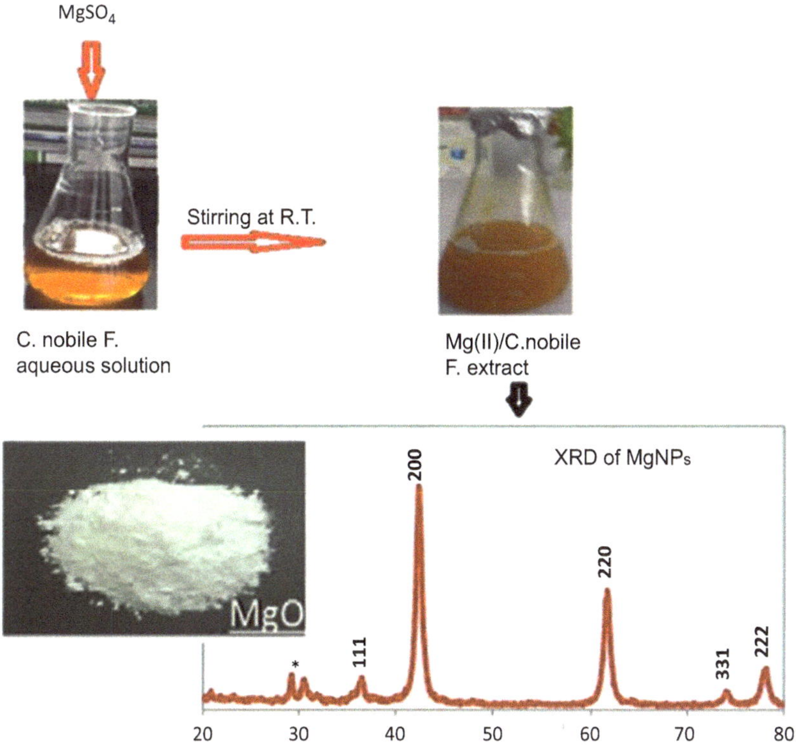

FIGURE 17.4 Graphical abstract of synthesis of magnesium oxide nanoparticles.

In studies of low toxicity particles, TiO_2 induced more severe lung inflammation and particle lymph node burden compared to $BaSO_4$ when dosed at mass burden in milligrams.

Due to nanostructure, NPs are good barriers to diffusion gases such as oxygen and carbon dioxide; hence, they can be exploited in food packaging. Some drinks such as beer and soda waters naturally have to keep appreciate the amount of carbon dioxide which can be packaged in the bottles made with nanocomposites because of minimization of CO_2 loss, decrease in weight of packaging materials, and increase in shelf life. Another exploitation way is the incorporation of NPs in packaging. This technology will slow down some biochemical processes such as oxidation and degradation. This helps to extend the shelf-life of food products. In the food packaging industry, the most used materials are plastic polymers that can be incorporated or coated with nanomaterials for improved mechanical or functional properties. Moreover, nanocoatings on food contact surfaces act as a barrier or antimicrobial properties. Silver NPs have been successfully embedded in the plastic for making food storage bins, and this acts like disinfection of bins, for minimizing harmful bacterial growth. Therefore, nanotechnology is a forward-looking process; it acts as agricultural biosecurity.

TABLE 17.1 Means of mortality percents (%) of early and late nymphal instars at a concentration of 8000 μg/mL of different nanomaterials after 24, 48, and 72 h at room conditions for three periods.

Treatment	Mortality %					
	24 h		48 h		72 h	
	Early nymphal instars	Late nymphal instars	Early nymphal instars	Late nymphal instars	Early nymphal instars	Late nymphal instars
CuONPs	$86^{a} \pm 1.0$	$86^{a} \pm 1.0$	$87^{b} \pm 3.0$	$83^{a} \pm 4.0$	$100^{a} \pm 0.0$	$96^{b} \pm 2.0$
ZnONPs	$76^{c} \pm 1.0$	$74^{b} \pm 1.0$	$90^{b} \pm 3.0$	$65^{b} \pm 4.0$	$100^{a} \pm 0.0$	$100^{a} \pm 0.0$
MgHNPs	$82^{b} \pm 1.0$	$50^{c} \pm 1.0$	$100^{a} \pm 0.0$	87a ± 4.0	$100^{a} \pm 0.0$	$94^{b} \pm 2.0$
MgONPs	$86^{a} \pm 1.0$	$86^{a} \pm 1.0$	$88^{b} \pm 3.0$	$83^{a} \pm 4.0$	$100^{a} \pm 0.0$	$98^{a} \pm 2.0$

Notes: Means within the same column sharing the same letter do not differ significantly at the 5% level using Fisher's Protected LSD test.

NPs in precise have revealed broad-spectrum antibacterial properties against both Gram-positive and Gram-negative bacteria. ZnONPs were found to inhibit *Staphylococcus aureus* and AgNPs exhibit concentration-dependent antimicrobial activity against *Escherichia coli*, *Aeromonas hydrophila*, and *Klebsiella pneumonia* (Aziz et al., 2016; Mariadoss et al., 2019). The antimicrobial mechanism of action of NPs is typically considered as of few prototypes such as oxidative stress and cell damage, metal ion release, or nonoxidative mechanisms (Wang et al., 2011). These mechanisms can happen concurrently. Firm studies have suggested that AgNPs prompt neutralization of the surface electric charge of the bacterial membrane and change its permeability, ultimately leading to apoptosis. Moreover, the generation of ROS prevents the antioxidant defense system and causes physiochemical damage to the intrinsic cell membrane. According to the current investigation, the major processes causal to the antibacterial effects of NPs are as follows: disruption of the bacterial cell membrane; generation of ROS (Sankaran et al., 2020), penetration of the bacterial cell membrane by passive or facilitated diffusion, and induction of intracellular antibacterial effects, including interfaces with DNA replication and inhibition of protein synthesis (Aziz et al., 2016; Wang et al., 2011). Nanosensors help in food labeling and combination with NP-based intelligent inks or reactive nanolayers may provide smart recognition of relevant food products. Printed labels in the food package that can indicate the following highlights: temperature, time, pathogens, freshness, and humidity. Nano barcode particles with different patterns of gold, which can form templates and also with silver stripes, are possible to synthesize themselves. Moreover, the contaminant or nutrient sorption on NPs surfaces has attracted the attention of researchers for more studies on soil chemistry, showing that NPs have high sorption capacities for metal and anionic contaminants. It was found that the contaminant sequestration was accomplished mainly by a surface complication. The sorbet surface species can likely be encapsulated within interior surfaces of NPs. A phenomenon with significant consequences for contaminant dispersion or remediation processes can exhibit. Moreover, metallic species such as Ni can be linked to natural short-ordered alumina silicates, TiO_2 surfaces, humic acids, and aromatic compounds and may be considered very potent bioremediation in nanoagricultural systems (Kahlel et al., 2020).

To measure the phytotoxicity of the nanomaterials on sweet paper, transplants (cv. Sonar) were planted in pots filled with peat moss in wooden cages covered with white muslin and placed under glasshouse conditions ($30°C \pm 5°C$) in May 2017 (Ghidan et al., 2016).

There were four treatments (nanomaterials). The nanomaterials were MgONPs, MgOHNPs, ZnONPs, and CuONPs. Concentrations of the different nanomaterials were separately tested on the pepper plants. The phytotoxicity scale for the nanomaterials on pepper from 0 to 10 was established. Control (only water) treatments were used with five replicates for each treatment.

17.5.1 Nanomaterial residues on the pepper plants

The experimental work that was mentioned previously was used. Different concentrations of the nanomaterials were sprayed (CuONPs, ZnONPs, MgOHNPs, and MgONPs) using Atomizer Sprayer at the maximum dose which killed about 100% of the GPA (late nymphal instar) on the pepper plants with five replicates (Ghidan et al., 2020a,b,c). Control (deionized distilled water) treatment was also used, with five replicates. Samples for the treated fruits (5 g) and leaves (5 g) were taken separately after 14 days of the treatment for residue analysis. The leaves and fruits of green pepper samples were dried in an oven at 60°C for 24 h. The concentrations of metal NPs were determined using scanning electron microscopy equipped with energy-dispersive X-ray spectroscopy (EDS). The experimental work was carried out at the University of Jordan and the Royal Scientific Society. The NP residue analysis was conducted at the Jordan University of Science and Technology at the nanotechnology center.

17.5.2 Effect of nanoparticles on green pepper leaves and fruits

The effect of synthesized NPs on green pepper leaves and fruits in the greenhouse was evaluated. 400–600 ppm of MgONPs and ZnONPs recorded significant increases in NPs absorption compared with control plants, as shown by the analysis of leaves and fruits by EDS spectroscopy (Fig. 17.5). Greener

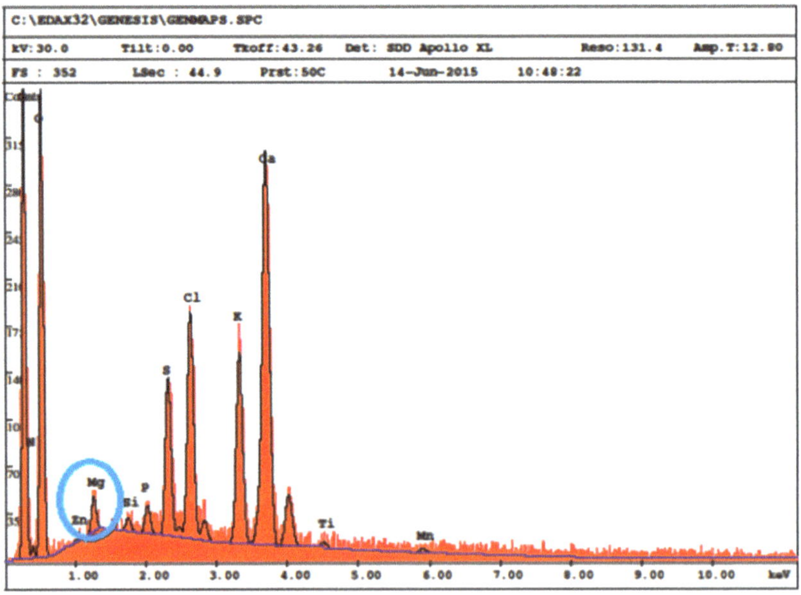

FIGURE 17.5 EDS analysis of leaves of green pepper.

leaves of green pepper were observed in the plants treated with 400–600 ppm of MgOHNPs and ZnONPs. These observations are a result of the interaction of MgONPs and ZnONPs absorbed by leaves and the organic compounds of green pepper tissues forming organic-Mg and organic-Zn compounds, which help in building the chlorophyll and nitrogen content of the leaves.

Fig. 17.6 shows that MgONPs, ZnONPs, and CuONPs as the concentration of NPs increased significantly in leaves. Traces of NPs studied were detected in fruits (>2 ppm) (Ghidan et al., 2017a,b). Moreover, Fig. 17.5 shows that residues of sprayed concentrations of the nanomaterials on pepper fruits were zero or about zero, except in the highest applied concentration (600 ppm). This concentration had residues 75 times more than the highest concentration, which killed 100% of the early nymphal instar of the GPA after 72 h of the treatment. Bearing in mind that all these nanomaterials are contacted in action, which can easily be removed by washing by water tap, in case of having few residues on the fruit by the misuse of the farmer application. However, finding residues of these nanomaterials on leaves could be useful due to their conversion to act as fertilizer. Moreover, using these nanomaterials is cheaper as well as safer for the environment than using chemicals. In addition, these used nanomaterials might penetrate or be absorbed by the plant epidermis, which facilitates its use of it as leaf (foliage) fertilizer due to very small particles.

17.6 Future perspectives

Clean sustainable agriculture must be taken as an ecosystem method, where abiotic–biotic living beings live in accord with the coordinated stability of food chains and their related energy balances. Use of agrochemicals, such as

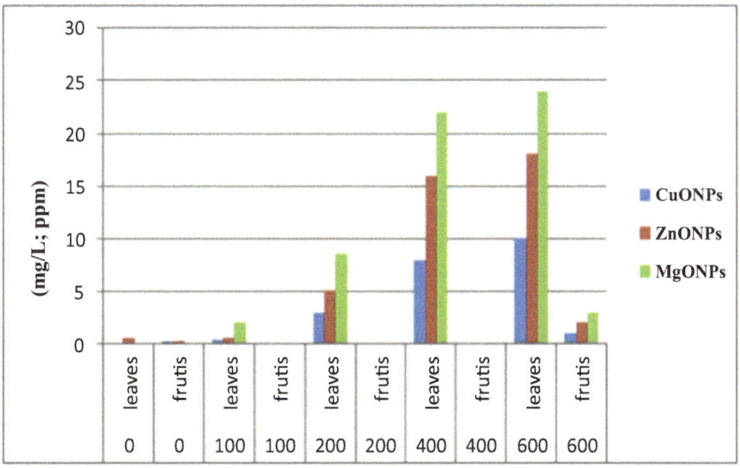

FIGURE 17.6 Means of concentrations (ppm) of nanoparticle content in leaves and fruits.

nanofertilizers, nanopesticides, nanoherbicides, precision farming techniques, intelligent feed, enhancement of food texture and quality, and bioavailability/nutrient values, packaging, and labeling, around the world of the agricultural sector are needed to be focused. Some focused areas may need more attention in the near future by researchers in the field of agricultural nanotechnology or nanofoods as follows.

- New environmental and safety delivery systems for carrying special food/feed compounds and plant nutrients. These systems can also have pharmaceutical application potential (Ghidan and Al Antary, 2019).
- The (bio)sensors related to nanotechnology have an effective role in insect pest control and food products of agriculture. Consumers can always get actual information about the state of certain food products via intelligent food packaging corporate with nanosensors.
- The properties of nanomaterials, such as size, dose, exposure time, surface chemistry, structures, immune response, accumulation, and retention time, and other effects should be accessed carefully. New analytical methods are needed to develop to detect, validate, and access the effects of each nanomaterial/nanofood in whole ecosystems. Life cycle analysis of nanomaterials/nanofood should be done. Improvement of wide-ranging databank as well as international collaboration for policy, idea, and regulation is needed for manipulation of this knowledge. Additionally, the authorities should provide clear guidelines and roadmaps for reducing the risks of the use of nanotechnological products.
- New communication channels and debates should be opened with the participation of different sides such as consumers, researchers, authorities, and industrial sectors to discuss the impacts of this technology on human life, economy, and science. This technology in the long term may provide innovative and economic development routes for human nutrition worldwide.

17.7 Conclusions

Despite having relative advantages in agriculture process, still developing countries are suffering from a lack of high importance of food products. Despite a lot of information about individual nanomaterials being available, toxicity level of many NPs is still indefinable. Thus, the application of these materials is limited due to the lack of knowledge of risk assessments and their effects on human health. The development of a comprehensive database and alarm system, as well as international cooperation for regulation and legislation, is necessary for the exploitation of this technology. The privilege of using such nutrients in the nanoscale is to increase its ability toward diffusion through the pores of the plant cell. In addition, mutagenic can occur in the plant by these nanomaterials improving its growth and yield. However, green methods are considered as saving techniques for synthesis of NPs environmentally.

NPs are very important in developing sustainable technologies for the future, humanity, and the environment. Synthesis of NPs by plants is a green chemistry approach that interconnects nanotechnology and plant biotechnology. Plant extracts are used for the metal ions bioreduction to form NPs. It has been demonstrated that plant metabolites like sugars, terpenoids, polyphenols, alkaloids, phenolic acids, and proteins play an important role in metal ions reduction into NPs and in supporting their subsequent stability. The properties of the atoms of the nanometric material are quite different from those of the bulk materials. Conventionally NPs are synthesized by the chemical method in which chemicals are used because of their general toxicity, engendering the serious concern to develop environment-friendly processes. Thus, to solve the objective, biological approaches are coming up to fill the void; for instance, green synthesis uses biological as the molecules of extracts derived from plant sources exhibiting superiority over chemical and/or biological methods, which several plants have successfully used for efficient and rapid extracellular synthesis of gold and silver NPs. Geranium (*P. graveolens*), leaf extracts of lemongrass (*Cymbopogon flexuosus*), *Cinnamomum camphora*, neem (*A. indica*), *Aloe vera*, tamarind (*Tamarindus indica*), and fruit extract of *E. officinalis* have proven potential in reducing Au (III) ions to form gold NPs Au(0) and silver nitrate to form silver NPs Ag(0). Biomasses of wheat (*T. aestivum*) and oat (*A. sativa*), alfalfa (*M. sativa*), native and chemically modified hop biomass, and remnant water collected from soaked Bengal gram bean (*C. arietinum*) have also been used for gold NPs synthesis. Alfalfa (*M. sativa*), *Chilopsis linearis*, and *Sesbania* seedlings showed the synthesis of gold NPs inside living plant parts. However, alfalfa (*M. sativa*) sprouts and *B. juncea* germinating seeds are used for silver and Ag–Au–Cu alloy NP synthesis. When tackling the advantages and disadvantages of nanotechnology, we will also need to point out what can be seen as the negative side of this technology. Including the list of disadvantages of this science and its development is the possible loss of jobs in the traditional farming and manufacturing industry. Nanotechnology has also increased the risk of health; NPs, due to their small size, can cause inhalation problems and many other fatal diseases by inhaling the air containing NPs for just 60 s and damage the lungs easily. Presently, nanotechnology is very expensive, and developing it can cost you a lot of money. It is also pretty difficult to manufacture, which is probably why products made with nanotechnology are more expensive. Nanotechnology has raised the standard of living, but at the same time, increased pollution, which includes water pollution and air pollution. The pollution caused by nanotechnology is known as nanopollution. This kind of pollution is very dangerous for living organisms. Some NPs are especially likely to accumulate in bodily tissues. For instance, unmodified copper or silver NPs can exhibit toxicity risks. Surface modifications of NPs can create agents with very distinct chemical and/or biological properties from the "same" NPs with unmodified surfaces. It cannot be claimed with certainty either those nanotechnologies are fully safe for health or that they are harmful. Risks associated with chronic exposure of farmers to nanomaterial, unknown life cycles, interactions with the biotic or abiotic environment, and their possible amplified bioaccumulation effects have not been accounted for and these should be seriously considered before these applications move from laboratories to the field. The common challenges related to commercializing nanotechnology are high processing costs, problems in the scalability of R&D for prototype and industrial production, and concerns about public perception of the environment, health, and safety issues. Governments across the world should form common and strict norms and monitoring before commercialization and bulk use of these nanomaterials due to bioaccumulation, toxicity, unpredictability, unknown effects, and difficulty in cleaning. Different limitations are recorded in the literature according to the method of preparation but it is very safe to use green methods, for example, metal oxides when using plant leaves, fruits, roots, flowers, and seeds for the synthesis of MgO, MgOH, and CuO. ZnO, nickel oxide, cadmium, and Cd using a low concentration such as 5 ppm will give activation for plants growth; also gold NPs may be used in a certain concentration (Ghidan et al., 2017a,b).

17.8 Recommendations

For safe, sustainable, responsible, and socially acceptable development and use of nanotechnologies, it is, therefore, required timely to undertake a general reflection on the adequacy of existing risk assessment methodologies as concern nanotechnologies and they are engineered and adventitious products and to identify any gaps in current knowledge which may be an obstruction to the demonstration of their safety.

There are different facets of pharmaceuticals and drug delivery systems, diagnostics, and medical technology, where the terms nanomedicine and bionanotechnology are already commonplace. Nanotechnology products may also offer novel challenges for the reduction of environmental pollution.

References

Abbas, S.S., Haneef, M., Lohani, M., Tabassum, H., Khan, A.F., 2016. Nanomaterials used as a plants growth enhancer: an update. Int. J. Pharm. Sci. Rev. Res. 5, 17–23.

Al-Antary, T.A., Kahlel, A., Ghidan, A., Asoufy, H., 2020. Effects of nanotechnology liquid fertilizers on fruit set and pods of broad bean (*Vicia faba* L.). Fresenius Environ. Bull. 29 (6), 4794–4798.

An, Y., Lian, X., Zou, Y., Deng, D., 2014. Synthesis of porous ZnO structure for gas sensor and photocatalytic applications. Colloids Surf. A: Physicochem. Eng. Asp. 447, 81–87.

Awwad, A., Ahmad, A., 2014. Biosynthesis, characterization, and optical properties of magnesium hydroxide and oxide nanoflakes using citrus limon leaf extract. Arab. J. Phys. Chem. 1, 65–70.

Awwad, A., Albiss, B., 2015. Biosynthesis of colloidal copper hydroxide nanowires using Pistachio leaf extract. Adv. Mater. Lett. 6, 51–54.

Awwad, A., Ibrahim, Q., 2015. Optical and X-ray diffraction characterization of biosynthesis copper oxide nanoparticles using carob leaf extract. Arab. J. Phys. Chem. 2, 20–24.

Awwad, A., Salem, N., Abdeen, A., 2013. Green synthesis of silver nanoparticles using carob leaf extract and its antibacterial activity. Int. J. Ind. Chem. 4, 29–35.

Awwad, A., Albiss, B., Ahmad, A., 2014. Green synthesis, characterization and optical properties of zinc oxide nanosheets using olea europea leaf extract. Advance Mater. Lett. 5, 520–524.

Aziz, N., Pandey, R., Barman, I., Prasad, R., 2016. Leveraging the attributes of Mucor hiemalis-derived silver nanoparticles for a synergistic broad-spectrum antimicrobial platform. Front. Microbiol. 7, 1984. Available from: https://doi.org/10.3389/fmicb.2016.01984.

Bala, N., Saha, S., Chakraborty, M., Maiti, M., Das, S., Basu, R., et al., 2015. Green synthesis of zinc oxide nanoparticles using *Hibiscus subdariffa* leaf extract: effect of temperature on synthesis, anti-bacterial activity and anti-diabetic activity. RSC Adv. 5, 4993–5003.

Balakrishnan, K., Casimeer, S., Ghidan, A., Ghethan, F., Venkatachalam, K., Singaravelu, A., 2020. Bioformulated hesperidin-loaded PLGA nanoparticles counteract the mitochondrial-mediated intrinsic apoptotic pathway in cancer cells. J. Inorg. Organomet. Polym. Mater. September, 16. Available from: https://doi.org/10.1007/s10904-020-01746-9.

Bhattacharyya, A., Duraisamy, P., Govindarajan, M., Buhroo, A.A., Prasad, R., 2016. Nano-bio-fungicides: emerging trend in insect pest control. In: Prasad, R. (Ed.), Advances and Applications through Fungal Nanobiotechnology. Springer International Publishing, Cham, pp. 307–319. Available from: 10.1007/978-3-319-42990-8_15.

Bonnell, D.A., Huey, B.D., 2001. Basic principles of scanning probe microscopy. In: Bonnell, D.A. (Ed.), *Scanning Probe Microscopy and Spectroscopy: Theory, Techniques, and Applications*. Wiley-VCH, New York.

Bouazizi, N., Bargougui, R., Oueslati, A., Benslama, R., 2015. Effect of synthesis time on structural, optical and electrical properties of CuO nanoparticles synthesized by reflux condensation method. Advance Mater. Lett. 6, 158–164.

Bystrzejewska-Piotrowska, G., Asztemborska, M., Steborowski, R., Polkowska-Motrenko, H., Danko, J., Ryniewicz, B., 2012. Application of neutron activation for investigation of Fe3O4 nanoparticles accumulation by plants. Nukleonika 57, 427–430.

Campbell, B.M., Thornton, P., Zougmoré, R., van Asten, P., Lipper, L., 2014. Sustainable intensification: what is its role in climate smart agriculture? Curr. Opin. Environ. Sustain. 8, 39–43. Available from: https://doi.org/10.1016/j.cosust.2014.07.002.

Devi, R., Gayathri, R., 2014. Green synthesis of zinc oxide nanoparticles by using Hibiscus rosa-sinensis. Int. J. Curr. Eng. Technol. 4, 2444–2446.

Dhaouadi, H., Chaabane, H., Touati, F., 2011. Mg(OH)2 nanorods synthesized by a facile hydrothermal method in the presence of CTAB. Nano Micro Lett. 3, 153–159.

Dimkpa, C.O., 2014. Can nanotechnology deliver the promised benefits without negatively impacting soil microbial life? J. Basic. Microbiol. 54, 889–904. Available from: https://doi.org/10.1002/jobm.201400298.

Dobrucka, R., Dugaszewska, J., 2016. Biosynthesis and antibacterial activity of ZnO nanoparticles using *Trifolium pratense* flower extract. Saudi J. Biolog. Sci. 23, 517–523.

Du, W., Tan, W., Peralta-Videa, J.R., Gardea-Torresdey, J.L., Ji, R., Yin, Y., et al., 2017. Interaction of metal oxide nanoparticles with higher terrestrial plants: physiological and biochemical aspects. Plant. Physiol. Biochem. 110, 210–225. Available from: https://doi.org/10.1016/j.plaphy.2016.04.024.

El Ghoul, J., Barthou, C., El Mir, L., 2012. Synthesis by sol−gel process, structural and optical properties of nanoparticles of zinc oxide doped vanadium. Superlattices Microstructures 51, 942−943.

Fatimah, I., Pradita, R., Nurfalinda, A., 2016. Plant extract mediated of ZnO nanoparticles by using ethanol extract of *Mimosa pudica* leaves and coffee powder. Procedia Eng. 148, 43−48.

Feynman, R.P., 1996. No Ordinary Genius: The Illustrated Richard Feynman. W.W. Norton & Company, New York.

Ghidan, A.Y., Al-Antary, T., Awwad, A., 2016. Green synthesis of copper oxide nanoparticles using Punica granatum peels extract: effect on green peach aphid. Environ. Nanotechnol., Monit. Manag. 6 (12), 95−98.

Ghidan, A.Y., Al-Antary, T., Awwad, A., Akash, M., 2017a. Aphicidal potential of green synthesized magnesium hydroxide nanoparticles using *Olea europaea* leaves extract. ARPN J. Agric. Biol. Sci. 7 (10), 292−301.

Ghidan, A.Y., Al-Antary, T., Salem, N., Awwad, A., 2017b. Facile green synthetic route to the Zine oxide (ZnONPS) nanoparticles: effect on green page 4 of 7 peach aphid and antibacterial activity. Netherland J. Agric. Sci. 9 (2), 131−138.

Ghidan, A.Y., Al Antary, A., Ghidan, H.Y., 2018a. Novel experimental work on non- beneficial insect: the role of Islamic law. Fresenius Environ. Bull. 27 (12A), 9414−9419.

Ghidan, A.Y., Al Antary, T., Awwad, A., Ayad, J., 2018b. Physiological effect of some nanomaterials on pepper (*Capsicum annuum* L.) plants. Fresenius Environ. Bull. 27 (11), 7872−7878.

Ghidan, A.Y., Al-Antary, T.M., Awwad, A.M., Ghidan, O.Y., Araj, S., Ateyyat, M.A., 2018c. Comparison of different green synthesized nanomaterials on green peach aphid as aphicidal potential. Fresenius Environ. Bull. 27 (10), 7009−7016.

Ghidan, A.Y., Al Antary, Tawfiq M., 2019. Applications of nanotechnology in agriculture. Applications of Nanobiotechnology. Intech Open. Available from: http://doi.org/10.5772/intechopen.88390.

Ghidan, A.Y., Kahlel, A., Al-Antary, T., 2020a. Effect of nanotechnology liquid fertilizers on yield and nitrogenous compounds of broad bean (*Vicia faba* L.). Fresenius Environ. Bull. 29 (6), 4124−4128.

Ghidan, A.Y., Kahlel, A., Al-Antary, T., Asoufi, H., 2020b. Efficacy of nanotechnology liquid fertilizers on weight and chlorophyll of broad bean (*Vicia faba*) L. Fresenius Environ. Bull. 29 (6), 4793. 4789.

Ghidan, A.Y., Ghethan, F.Y., Alghanmi, M., Sangeetha, C.C., David, E., Hadadin, M.S., et al., 2020c. Global impact and clinical management of severe respiratory syndrome coronovirus-2 (COVID-19). J. Appl. Adv. Res. 5, 11−18.

Grillo, R., Abhilash, P.C., Fraceto, L.F., 2016. Nanotechnology applied to bio-encapsulation of pesticides. J. Nanosci. Nanotechnol. 16, 1231−1234. Available from: https://doi.org/10.1016/j.tifs.2003.10.005.

Gunalan, S., Sivaraj, R., Venckatesh, R., 2012. Aloe barbadensis Miller mediated green synthesis of mono-disperse copper oxide nanoparticles: optical properties. Spectrochim. Acta A Mol. Biomol. Spectrosc. 97, 1140−1144.

Hanlon, J., Diaz, L., Balducci, G., Stobbs, B., Bielewski, M., Chung, P., et al., 2015. Rapid surfactant-free Hanlon synthesis of Mg(OH)2 nanoplates and pseudomorphic dehydration to MgO. Cryst. Eng. Comm. 17, 5672−5679.

Hoffmann, M., Holtze, E.M., Wiesner, M.R., 2007. Reactive oxygen species generation on nanoparticulate material. In: Wiesner, M.R., Bottero, J.Y. (Eds.), Environmental Nanotechnology. Applications and Impacts of Nanomaterials. McGraw Hill, New York, pp. 155−203.

Ion, A.C., Ion, I., Culetu, A., 2010. Carbon-based nanomaterials: environmental applications. Univ. Politehn. Buchar. 38, 129–132.

Ipeksac, T., Kaya, F., Kaya, C., 2013. Hydrothermal synthesis of zinc oxide (ZnO) nanotubes and its electrophoretic deposition on nickel filter. Adv. Mater. Lett. 100, 11–14.

Jiang, W., Hua, X., Han, Q., Yang, X., Lu, L., Wang, X., 2009. Preparation of lamellar magnesium hydroxide nanoparticles via precipitation method. Powder Technol. 191, 227–230.

Jiang, T., Wang, Y., Meng, D., Yu, M., 2015. Facile synthesis and photocatalytic performance of self-assembly CuO microspheres. Superlattices Microstruct. 85 (1–6), 81.

Jinghui, W., Xin, L., Junwei, Y., Yuan, L., Guiling, N., 2009. Facile controlled synthesis of Mg (OH)2 nanoparticles with novel shapes. J. Environ. Int. 58, 55–58.

Johnston, B.F., Mellor, J.W., 1961. The role of agriculture in economic development. Am. Econ. Rev. 51, 566–593.

Kahlel, A., Ghidan, A., Al-Antary, T., Alshomali, I., Asoufy, H., 2020.) Effects of nanotechnology liquid fertilizers on certain vegetative growth of broad bean (*Vicia faba* L.). Fresenius Environ. Bull. 29 (6), 4763–4768.

Kandasamy, S., Prema, R.S., 2015. Methods of synthesis of nano particles and its applications. J. Chem. Pharm. Res. 7, 278–285.

Karnan, T., Selvakumar, S., 2016. Biosynthesis of ZnO nanoparticles using rambutan (*Nephelium lappaceum* L.) peel extract and their photocatalytic activity on methyl orange dye. J. Mol. Struct. 1125, 358–365.

Kovochich, M., Xia, T., Xu, J., Yeh, J.I., Nel, A.E., 2005. Principles and procedures to assess nanoparticles. Environ. Sci. Technol. 39, 1250–1256.

Kumar, A., Saxena, A., Shankar, R., Mozumdar, S., 2013. Facile Synthesis Of Size-Tunable Copper and Copper Oxide Nanoparticles using Reverse Microemulsions, 3. The Royal Society of Chemistry Advance, pp. 5015–5021.

Kumari, L., Li, W.Z., Vannoy, C.H., Leblanc, R.M., Wang, D.Z., 2010. Zinc oxide micro- and nanoparticles: synthesis, structure and optical properties. Mater. Res. Bull. 45, 190–196.

Madan, H.R., Sharma, S.C., Udayabhanu, Suresh, D., Vidyad, Y.S., Nagabhushanae, H., et al., 2016. Facile green fabrication of nanostructure ZnO plates, bullets, flower, prismatic tip, closed pine cone: their antibacterial, antioxidant, photoluminescent and photocatalytic properties. Spectrochim. Acta A: Mol. Biomol. Spectrosc. 152, 404–416.

Mariadoss, A., Ramachandran, V., Vijayakumar, S., Herbert, F., Kumar, S., Ghidan, A., et al., 2019. Green Synthesis, characterization and antibacterial activity of silver nanoparticles by Malus domestica and its cytotoxic effect on (MCF-7) cell line. Microb. Pathog. 135, 103609.

Mohan, S., Singh, Y., Verna, D.K., Hassan, S.H., 2015. Synthesis of CuO nanoparticles through green route using citrus limon juice and its application as nanosorbent for Cr (VI) remediation: process optimization with RSM and ANN-GA based model. Process. Saf. Environ. Prot. 96, 156–166.

Moorthy, S.K., Ashok, C.H., Rao, K.V., Viswanathana, C., 2015. Synthesis and characterization of MgO nanoparticles by Neem leaves through green method. Mater. Today: Proc. 2, 4360–4368.

Mukhopadhyay, S.S., 2014. Nanotechnology in agriculture: prospects and constraints. Nanotechnol. Sci. Appl. 7, 63–71. Available from: https://doi.org/10.2147/NSA.S39409.

Naika, H.R., Lingarajua, K., Manjunath, K., Kumar, D., Nagaraju, G., Suresh, D., et al., 2015. Green synthesis of CuO nanoparticles using Gloriosa superba L. extract and their antibacterial activity. J. Taibah Univ. Sci. 9 (7–12), 82.

Nasrollahzadeh, M., Maham, M., Sajadi, S.M., 2015. Green synthesis of CuO nanoparticles by aqueous extract of *Gundelia tournefortii* and evaluation of their catalytic activity for the synthesis of N-monosubstituted ureas and reduction of 4- nitrophenol. J. Colloid Interface Sci. 455, 245–253.

Ozcelik, B.K., Ergun, C., 2014. Synthesis of ZnO nanoparticles by an aerosol process. Ceram. Inter. 40, 7107–7116.

Patra, J.K., Baek, K.-H., 2017. Antibacterial activity and synergistic antibacterial potential of biosynthesized silver nanoparticles against foodborne pathogenic bacteria along with its anticandidal and antioxidant effects. Front. Microbiol. 8, 167. Available from: https://doi.org/10.3389/fmicb.2017.00167.

Pereire, A., Ferreira, I., Marcelino, F., Valentao, P., Andrade, P., Seabra, R., et al., 2007. Phenolic compounds and antimicrobial activity of olive (*Olea europaea* L. cv *cobrancosa*) leaves. Molecules 12 (5), 1153–1162.

Prasad, R., Kumar, V., Prasad, K.S., 2014. Nanotechnology in sustainable agriculture: present concerns and future aspects. Afr. J. Biotechnol. 13, 705–713. Available from: https://doi.org/10.5897/AJBX2013.13554.

Ramachandran, V., Vijaya, A., Mariadoss, A., David, E., Venkatachalam, K., Vijayakumar, S., et al., 2020. antidiabetic activity of gold nanoparticles synthesized using wedelolactone in RIN-5F cell line. Antioxidants 9 (1).

Ramesh, M., Anbuvannan, M., Viruthagiri, G., 2015. Green synthesis of ZnO nanoparticles using *Solanum nigrum* leaf extract and their antibacterial activity. Spectrochim. Acta A Mol. Biomol. Spectrosc. 136, 864–870. Part B.

Rana, N., Chand, S., Gathania, A.K., 2016. Green synthesis of zinc oxide nanosized spherical particles using *Terminalia chebula* fruits extract for their photocatalytic applications. Inter. Nano Lett. 6, 91–98.

Raut, S., Thorat, P.V., Thakre, R., 2015. Green synthesis of zinc oxide (ZnO) nanoparticles using *Ocimum tenuiflorum* leaves. Inter. J. Sci. Res. 4, 1225–1228.

Rocha, L.S.R., Deus, R.C., Foschini, C.R., Moura, F., Gonzalez Garcia, F., Simões, A.Z., 2014. Photoluminescence emission at room temperature in zinc oxide nano-columns. Mater. Res. Bull. 50, 12–17.

Sadeghzadeh, B., 2013. A review of zinc nutrition and plant breeding. J. Soil. Sci. Plant. Nutr. 13, 905–927. Available from: https://doi.org/10.4067/S0718-95162013005000072.

Safarifard, V., Morsali, A., 2012. Sonochemical syntheses of a nano-sized copper (II) supramolecule as a precursor for the synthesis of copper (II) oxide nanoparticles. Ultrason. Sonochem. 19, 823–829.

Sagadevan, S., Periasamy, M., 2014. Recent trends in nanobiosensors and their applications—a review. Rev. Adv. Mater. Sci. 36, 62–69.

Sahooli, M., Sabbaghi, S., Saboori, R., 2012. Synthesis and characterization of mono sized CuO nanoparticles. Mater. Lett. 81, 169–172.

Sangeetha, G., Rajeshwari, S., Venckatesh, R., 2011. Green synthesis of zinc oxide nanoparticles by aloe barbadensis miller leaf extract: structure and optical properties. Mater. Res. Bull. 46, 2560–2566.

Sankar, R., Manikandan, P., Malarvizhi, V., Fathima, T., Shivashangari, K.S., Ravikumar, V., 2014. Green synthesis of colloidal copper oxide nanoparticles using Caricapapaya and its application in photocatalytic dye degradation. Spectrochim. Acta A Mol. Biomol. Spectrosc. 121, 746–750.

Sankaran, V., Murali, S.K., Ghidan, A., Al Antary, T., David, E., 2020. Oral cancer preventive potential of polydatin: a nanoencapsulation approach. J. Phytology 12, 109–116.

Sayes, C.M., Fortner, J.D., Guo, W., Lyon, D., Boyd, A.M., Ausman, K.D., et al., 2004. The differential cytotoxicity of water-soluble fullerenes. Nano Lett. 4, 1881–1887. Available from: https://doi.org/10.1002/btpr.707.

Shah, R.K., Boruah, F., Parween, N., 2015. Synthesis and characterization of ZnO nanoparticles using leaf extract of Camelia sinesis and evaluation of their antimicrobial efficacy. Int. J. Curr. Microbiol. Superlattices Microstruct. 85, 1–6.

Sharma, J., Akhtar, J.K., Ameen, M.S., Srivastava, S., Singh, P.G., 2015. Green synthesis of CuO nanoparticles with leaf extract of Calotropis gigantea and its dyesensitized solar cells applications. J. Alloy. Compd. 632, 321–325.

Sharma, D., Myalowenkosi, I., Sabela, M.I., Kanchi, S., Mdluli, P.S., Singh, G., et al., 2016. Biosynthesis of ZnO nanoparticles using Jacaranda mimosifolia flowers extract: synergistic antibacterial activity and molecular simulated facet specific adsorption studies. J. Photochem. Photobiol. B: Biol. 162, 199–207.

Sivaraj, R., Rahman, P.K., Rajiv, P., Salam, H., Venckatesh, R., 2014. Biogenic copper oxide nanoparticles synthesis using *Tabernaemontana* divaricate leaf extract and its antibacterial activity against urinary tract pathogen. Spectrochim. Acta A Mol. Biomol. Spectrosc. 133, 178–181.

Song, X., Sun, S., Zhang, D., Wang, J., Yu, J., 2011. Synthesis and characterization of magnesium hydroxide by batch reaction crystallization. Front. Chem. Sci. Eng. 5, 416–421.

Sugirtha, P., Divya, R., Yedhukrishnan, K., Suganthi, Anusha, N., Ponnusam, V., et al., 2015. Green synthesis of magnesium oxide nanoparticles using *Brassica oleracea* and, *Punica granatum* peels and their anticancer and photocatalytic activity. Asian J. Chem. 27, 2513–2517.

Suresh, D., Nethravathi, P., Udayabhan, U., Rajanaika, H., Sharma, S., 2015. Green synthesis of multifunctional zinc oxide (ZnO) nanoparticles using *Cassia* fistula plant extract and their photodegradative, antioxidant and antibacterial activities. Mater. Sci. Semiconductor Process. 31, 446–454.

Sushma, N.J., Prathyusha, D., Swathi, G., Madhavi, T., Deva, B., Raju, P., et al., 2016. Facile approach to synthesize magnesium oxide nanoparticles by using Clitoriaternatea-characterization and in vitro antioxidant studies. Appl. Nanosci. 6, 437–444.

Sutradhar, P., Debbarma, M., Saha, M., 2016. Microwave synthesis of zinc oxide nanoparticles using coffee powder extract and its application for solar cell. Synth. Reactivity Inorganic, Metal-Organic Nano-Metal Chem. 46, 1622–1627.

Tai, C., Tai, C., Chang, M.-H., Liu, H., 2007. Synthesis of magnesium hydroxide and oxide nanoparticles using a spinning disk reactor. Ind. Eng. Chem. Res. 46, 5536–5541.

Thareja, R., Shukla, S., 2007. Synthesis and characterization of zinc oxide nanoparticles by laser ablation of zinc in liquid. Appl. Surf. Sci. 253, 8889–8895.

Thornhill, S., Vargyas, E., Fitzgerald, T., Chisholm, N., 2016. Household food security and biofuel feedstock production in rural Mozambique and Tanzania. Food Sec 8, 953–971. Available from: https://doi.org/10.1007/s12571-016-0603-9.

Tseng, C.-C., Chou, Y.-H., Liu, C.-M., Liu, Y.-M., Ger, M.-D., 2012. Microwave assisted hydrothermal synthesis of zinc oxide particles starting from chloride precursor. Mater. Res. Bull. 47, 96–100.

Vanathi, P., Rajiva, P., Narendhran, S., Rajeshwari, S., Rahman, P.K., Venckatesh, R., 2014. Biosynthesis and characterization of phyto mediated zinc oxide nanoparticles: a green chemistry approach. Mater. Lett. 134, 13–15.

Wang, P., Li, C., Gong, H., Wang, H., Liu, J., 2011. Morphology control and growth mechanism of magnesium hydroxide nanoparticles via a simple wet precipitation method. Ceram. Int. 37, 3365–3370.

Xu, H., Deng, X.R., 2006. Preparation and properties of superfine Mg(OH)2 flame retardant. Trans. Nonferrous Met. Soc. China 16, 488–492.

Yahia, I.S., Farag, A., El-Faify, S., Yakuphanoglu, F., Al-Ghamdi, A., 2016. Synthesis, optical constants, optical dispersion parameters of CuO nanorods. Optik 127, 1429–1433.

Yallappa, S., Manjanna, J., Sindhe, M., Satyanarayan, N., Pramod, S., Nagaraja, K., 2013. Microwave assisted rapid synthesis and biological evaluation of stable copper nanoparticles using T. Arjuna bark extract. Spectrochim. Acta A: Mol. Bimol. Spectrosc. 110, 108–115.

Yan, H., Zhang, X., Wu, J., Wei, L., Liu, X., Xu, B., 2008. The use of CTAB to improve the crystallinity and dispersibility of ultrafine magnesium hydroxide by hydrothermal route. Powder Technol. 188, 128–132.

Yildirım, O.A., Durucan, C., 2010. Synthesis of zinc oxide nanoparticles elaborated by microemulsion method. J. Alloy. Compd. 506, 944–949.

Yousefi, M., 2014. Fast method for synthesis magnesium hydroxide nanoparticles, thermal stable and flame-retardant poly vinyl alcohol nanocomposite. J. Nanotechnol. Stud. 4, 383–388.

Yunlong, C., Smit, B., 1994. Sustainability in agriculture: a general review. Agric. Ecosyst. Environ. 49, 299–307. Available from: https://doi.org/10.1016/0167-8809(94)90059-0.

Zak, A.K., Majid, W.H., Wang, H.Z., Yousefi, R., Golsheikh, A.M., Ren, Z.F., 2013. Sonochemical synthesis of hierarchical ZnO nanostructures. Ultrason. Sonochem. 20, 395–400.

Zhang, Q., Han, L., Jing, H., Blom, D.A., Lin, Y., Xin, H.L., et al., 2016. Facet control of gold nanorods. ACS Nano 10, 2960–2974. Available from: https://doi.org/10.1021/acsnano.6b00258.

Index

Note: Page numbers followed by "*f*" and "*t*" refer to figures and tables, respectively.

B

C

D

E

F

G

H

I

J

K

L

M

T

FSC
www.fsc.org
MIX
Papier | Fördert
gute Waldnutzung
FSC® C083411

Kolibri 360 GmbH
Ferdinand-Jühlke-Straße 7
99095 Erfurt, Deutschland
produktsicherheit@kolibri360.de